ERRATUM

Because of a production error, the last page of Lesley B. Cormack's article, *"Good Fences Make Good Neighbors": Geography as Self-Definition in Early Modern England* (pages 64-85), was omitted from this volume. The missing page has been reproduced on the reverse side of this sheet so that it may be placed after page 85.

Thus chorography, in its various manifestations, aided the English in assuming power over their environment and, more important, in defining themselves as products of their place and time. Chorography, more than either mathematical or descriptive geography, provided those living in England with a mirror, a way of establishing their identities through description and classification.

CONCLUSION

The tripartite field of early modern geography was a complex and crowded place. I have only begun to describe the richness of detail and care with which Englishmen created their world by mapping and describing it. Mathematical geography allowed scholars to create an abstract, geometrical globe on which to place England and the rest of the known world. They reduced the world to a system they could understand and in so doing began to tame it. Descriptive geography helped the English develop a self-definition in contrast to exotic and foreign peoples and places. It supplied practical economic, political, and social information, while helping to create the "other" against which a typical English person could be assessed. Chorography gave these people a picture of themselves. It placed them in time and space, it named them, and it allowed them to begin the process of defining who they were.[63]

As well, these geographical subdisciplines demonstrate for us the beginning of a new ideology and methodology of science. They show this in three major ways. First, all three areas grew from a theoretical or classical foundation. Mathematical geography had Ptolemy and spherical geometry; descriptive geography hailed from Strabo and some of the classical historians; chorography developed from an unbeatable combination of English chronicles and humanist scholarship. Second, geography in all its guises was forced, by its very nature, to employ a methodology of incremental fact gathering, whether in the form of new latitude and longitude sightings, descriptions of far-off countries, or the observation of a new Roman inscription. Finally, early modern geography helped to promote an ideology of utility that encouraged the public use of scholarly knowledge for the good of the economic and political state. Within these three facets, not unique to geography, lies the key to the "new science" of the seventeenth century. Thus, I want to suggest that by concentrating on intermediate disciplines such as geography, historians of science can begin a new analysis of just how and why science changed in the late seventeenth century.

---

53), p. 167. This is part of a distancing of people from their world and their God seen in mathematical terms in chorography and in causal terms in history.

[63] Stephen Greenblatt, *Renaissance Self-Fashioning from More to Shakespeare* (Chicago: Univ. Chicago Press, 1980), discusses the self-fashioning of the sixteenth-century English person, largely through words and literature.

The Scientific Enterprise in
Early Modern Europe:
Readings from *Isis*

The Scientific Enterprise in Early Modern Europe

Readings from *Isis*

EDITED BY PETER DEAR

The University of Chicago Press

Chicago and London

The essays in this volume originally appeared in *Isis*. Acknowledgement of the original publication date can be found on the first page of each article.

The University of Chicago Press, Chicago 60637
The University of Chicago Press, Ltd., London
© 1952, 1973, 1974, 1975, 1982, 1984, 1985, 1986, 1987, 1988, 1991 by the History of Science Society
© 1997 by The University of Chicago
All rights reserved. Published in 1997
Printed in the United States of America
ISBN: (cl)   0-226-13946-8
ISBN: (pa)   0-226-13947-6

01 00 99 98 97   5 4 3 2 1

Library of Congress Cataloging in Publication Data

The scientific enterprise in early modern Europe : readings from Isis / edited by Peter Dear.
    p.   cm.
    Includes bibliographical references.
    ISBN 0-226-13946-8 (cloth). — ISBN 0-226-13947-6 (pbk.)
    1. Science—Europe—History—16th century. 2. Science—Europe—History—17th century. 3. Isis. I. Dear, Peter Robert. II. Isis.
Q125.S4357   1997
500.2′094′09031—dc20   97-42535
                                                                                  CIP

∞ The paper used in this publication meets the minimum requirements of American National Standard for Information Sciences—Permanence of Paper for Printed Library Materials, ANSI Z39.48.1984.

CONTENTS

PETER DEAR: *Introduction* — 1

ROBERT S. WESTMAN: *The Melanchthon Circle, Rheticus, and the Wittenberg Interpretation of the Copernican Theory* — 7

OWEN HANNAWAY: *Laboratory Design and the Aim of Science: Andreas Libavius versus Tycho Brahe* — 37

LESLEY B. CORMACK: *"Good Fences Make Good Neighbors": Geography as Self-Definition in Early Modern England* — 64

KEITH HUTCHISON: *What Happened to Occult Qualities in the Scientific Revolution?* — 86

MARGARET J. OSLER: *Galileo, Motion, and Essences* — 107

RICHARD S. WESTFALL: *Science and Patronage: Galileo and the Telescope* — 113

ALBERT VAN HELDEN: *The Telescope in the Seventeenth Century* — 133

BRUCE STANSFIELD EASTWOOD: *Descartes on Refraction: Scientific versus Rhetorical Method* — 154

CHRISTOPH MEINEL: *Early Seventeenth-Century Atomism: Theory, Epistemology, and the Insufficiency of Experiment* — 176

THOMAS S. KUHN: *Robert Boyle and Structural Chemistry in the Seventeenth Century* — 212

B. J. T. DOBBS: *Newton's Alchemy and His Theory of Matter* — 237

PETER DEAR: *Totius in verba: Rhetoric and Authority in the Early Royal Society* — 255

STEVEN SHAPIN: *The House of Experiment in Seventeenth-Century England* — 273

LONDA SCHIEBINGER: *Maria Winkelmann at the Berlin Academy: A Turning Point for Women in Science* — 305

APPENDIX: *Articles on Early Modern European Science in* Isis, *1970–1996* — 333

CONTRIBUTORS AND EDITORS — 337

Introduction

THE SUBJECT MATTER of this collection is what is often called the scientific revolution. That term refers to Europe in the sixteenth and seventeenth centuries; it begins with Copernicus, who put the earth in motion around the sun, and ends with Newton, who filled a boundless universe with gravity, thereby completing (so the story goes) the replacement of the previous finite, earth-centered universe of Aristotle. And yet it is unclear whether so neat a thing as a "scientific revolution" ever took place at all.

The heyday of the scientific revolution among historians of science was perhaps the 1950s and 1960s. The dominant view was that modern science could date its origins from the early modern period and that those origins were essentially intellectual: historians understood science either methodologically, as a particular way of learning about the natural world, or metaphysically, as a particular stance toward knowing nature.[1] In the 1970s, however, other approaches to understanding science came to be widely investigated. Science was held to be a broader historical enterprise—more than just ideas. The history of science, on this view, should encompass the social and cultural dimensions of scientific activity, paying heed to the who and the why and refusing to take for granted the motivations that led people to confront nature in the ways they did.

But whether intellectualist or contextualist in orientation,[2] treatments of the scientific revolution over the past several decades have dealt with the issue of the coherence (or lack thereof) of the period as a segment of the history of science. Was there an identifiable thing that we can call "the scientific revolution"? The answer used to be yes, but the shift away from exclusively intellectualist positions has encouraged the iconoclastic answer, no.

The intellectualist approach usually went along with a large-scale narrative of the period as a whole. In one form or another, this narrative concentrated on the physical sciences, and especially astronomy, taking Copernicus as the starting point and tracing the development of Isaac Newton's universe via the work of such great figures as Kepler

---

[1] The classic study focused on issues of methodology is Alistair C. Crombie, *Robert Grosseteste and the Origins of Experimental Science, 1100–1700* (Oxford: Clarendon, 1953); on metaphysical frameworks see, e.g., Alexandre Koyré, "Galileo and Plato," in his *Metaphysics and Measurement: Essays in Scientific Revolution* (1943; Cambridge, Mass.: Harvard Univ. Press, 1968), pp. 16–43; and Edwin A. Burtt, *The Metaphysical Foundations of Modern Physical Science*, rev. ed. (London: Routledge & Kegan Paul, 1932).

[2] On this terminology, see Steven Shapin, "History of Science and Its Sociological Reconstructions," *History of Science* 1982, 20:157–211.

and Galileo.[3] The assumption, or conviction, was that a controlling master narrative of this kind dealt adequately with the principal dynamics and content of early modern science. Even Marxist historians tacitly accepted the benchmarks on the road of progress that this story mapped out.[4] The newly fashionable departures of the 1970s (continuing up to the present), in starting to dissolve these older master narratives, have nonetheless continued to use the old framework of the scientific revolution to provide reference points.

The articles in this volume illustrate the central issue. Older studies were able to take for granted the broad legitimacy of the dominant intellectualist master narrative; article-length scholarship usually focused on topics whose relevance to that larger story was already established. But newer contextualist studies are no less in thrall to the old orthodoxy: these small-scale contextualist studies must themselves be understood in terms of a standard image of the period against which they define themselves. The historian's choice of subject matter and way of framing the questions to be asked of it are themselves always determined by reference to an existing historiographical picture. This means that there is, of necessity, a certain conservatism built into even the most revisionist historiography.

The earliest article in this book is Thomas Kuhn's on the chemistry of Robert Boyle. Kuhn is most famous as the author of *The Structure of Scientific Revolutions* (1962),[5] a book that in many ways can be seen as the crucial link between the intellectual and the sociocultural approaches to understanding science's past. Kuhn's starting point, as his article here shows, was the conceptual analysis of past scientific thought, an attempt to get inside an alien way of thought rather than assume that past science was just an undeveloped form of present-day science. Margaret Osler's article on Galileo uses a similar kind of approach. Later, in *Structure,* Kuhn argued that such conceptual frameworks—ways of thinking—should be understood in relation to the community that embodied and perpetuated them. Both methodological procedures and metaphysical commitments could be subsumed under the broad notion of a paradigm or disciplinary matrix, which was (in the view of many) to be understood sociologically.

More recent studies are still shaped by the older intellectualist grand narratives of the scientific revolution even when they criticize them. This remains a form of shaping, albeit now in a more negative sense, and is rather different from that kind of revisionist historiography which nonetheless relies heavily on certain aspects of a standard, accepted picture.[6] Steven Shapin's article, "The House of Experiment," illustrates the way in

---

[3] An exemplification of this approach is I. Bernard Cohen, *The Birth of a New Physics* (1960; rev. ed., New York: Norton, 1985).

[4] The classic instance is Boris Hessen, "The Social and Economic Roots of Newton's *Principia,*" in *Science at the Cross Roads,* ed. N. I. Bukharin et al. (1931; 2d ed., London: Cass, 1971), pp. 150–212; for another early example, see Edgar Zilsel, "The Origins of William Gilbert's Scientific Method," *Journal of the History of Ideas* 1941, *2*:1–32, and "The Sociological Roots of Science," *American Journal of Sociology* 1941–42, *47*:544–562. For an entrée into the historiographical situation engendered by such work, see Steven Shapin, "Discipline and Bounding: The History and Sociology of Science as Seen through the Externalism-Internalism Debate," *History of Science* 1992, *30*:333–369.

[5] Thomas S. Kuhn, *The Structure of Scientific Revolutions,* 3d ed. (Chicago: Univ. of Chicago Press, 1996).

[6] John Henry exemplifies this point in his criticisms of such revisionist accounts as those of James and Margaret Jacob and of Carolyn Merchant; see John Henry, "Occult Qualities and the Experimental Philosophy: Active Principles in Pre-Newtonian Matter Theory," *History of Science* 1986, *24*:335–381, esp. n. 80. Henry notes that these historians take for granted a conventional view of seventeenth-century matter theory as promoting an ontology of dead, inert matter. But Henry himself argues, in rather more fundamentally revisionist style, that such a view is simply untrue, at least for England, where Merchant and the Jacobs focus their

which this works: standard accounts of the scientific revolution give an important place to the early Royal Society of London, especially to Robert Boyle, and to their experimental program. Shapin uses this setting for his article and can be sure that readers will already grant the relevance and significance of his topical focus, simply because they are used to doing so. This is despite the fact that the point of his article is to represent Restoration experimental philosophy in a quite unintellectualist way: to investigate the social making of accredited experimental knowledge rather than treat experimentalism as a fairly unproblematic dimension of an empiricist ideology of science.

Any master narrative is intended by its very nature to subordinate and order the more localized understandings of particular events or themes that are typically the focus of the specialized article literature. But at the same time, that kind of scholarship always tends to evade the totalizing ambitions of a master narrative even while implicitly relying on one for its own claims to significance. Historical scholarship thus tends to proceed by a dialectical process wherein ambitious master narratives perennially appear, only to have their pretensions undermined by the objections of smaller-scale, more detailed research.

The existence of such tensions, always threatening the integrity of any large-scale historical account of the period, means that no cut-and-dried presentation of the findings of current scholarship is possible. What a collection such as this serves to do is to lay out the traces of active historical scholarship: the process of attempted understanding is what history as a scholarly endeavor is all about; it is not a prepackaged set of answers to prepackaged questions. Themes, rather than any particular master narrative, are what emerge as the lifeblood of the scholarship represented here, and if the relevance of those themes to some broader understanding may have been justified by such a master narrative, the themes nonetheless take on a life of their own.

Thus, the first two articles investigate issues surrounding post-Copernican astronomy up until the early seventeenth century. This is a theme directly related to the standard scientific revolution story of the period, in which Copernicus's astronomy and its fortunes were always central. The Westman piece is a classic and very important article that shows ways in which Copernicus's innovation was received in the second half of the sixteenth century, prior to Kepler, and in that sense it attempts to fill out a narrative that had previously seemed fairly straightforward when viewed from the intellectualist standpoint. No longer does one move from Copernicus straight to Tycho Brahe and thence to Kepler and his elliptical orbits; Westman investigates the realities of what happened to Copernicus's work in the specific contexts from which its influence was spread. Owen Hannaway's article elucidates Tycho's conception of astronomical knowledge and extends relevant considerations into other areas of philosophical endeavor. Once again, it is not really necessary for the author to justify the interest he shows in the topics he addresses, since they are already well accepted in the standard picture, even if their particular treatment here departs from the usual story line.

One theme that has come increasingly to the fore, especially in the wake of much of Westman's work on astronomy, is the early modern academic distinction between natural philosophy and mathematics. The latter term referred to a whole slew of subjects that would now be described as applied mathematics, including positional astronomy itself

---

accounts. Lorraine Daston adverts to a somewhat similar situation in "History of Science in an Elegiac Mode: E. A. Burtt's *The Metaphysical Foundations of Modern Physical Science*," *Isis* 1991, *82*:522–531. In Burtt's story, according to Daston, it is a matter of a romanticized image of the "organic" Middle Ages being presented as the antithesis of the new, soul-less seventeenth-century worldview.

along with such specialties as geometrical optics, theoretical mechanics, and cosmography and geography—broadly, mapmaking. Lesley Cormack discusses this last little-investigated historical area in an article whose importance lies in part in its illustration of the great breadth of mathematical sciences beyond astronomy and the existence of a community of mathematical practitioners whose work helped to redefine the character of natural knowledge at this time.

The following piece, by Keith Hutchison, is a recognized classic that provides a general frame for understanding some of the new departures in approaches to nature that replaced the older scholastic-Aristotelian views. Once again, it draws its rationale from a recognized issue in the master narratives, namely, the supposed rejection toward the end of the seventeenth century of so-called occult qualities in favor of mechanically intelligible properties of matter. Hutchison reassesses and revises that view to show how the central term was used during the century and how its connotations shifted. His analysis has applicability to a wide range of figures and issues in natural philosophy in the scientific revolution. Its conclusion is that the new philosophical departures of the seventeenth century were actually a matter of trying to provide explanations for properties previously pushed aside as occult—that is, as hidden and unknowable—rather than a rejection of them as unintelligible. He thus turns the received wisdom on its head, at the same time attesting to the dominance of that wider picture of which it has been a part. Margaret Osler's brief piece on the relation of Galileo's new science of motion to Aristotelian natural philosophy casts valuable light on the metaphysical implications of Galileo's new science, pointing up contrasts that supplement those of Hutchison's article.

Richard Westfall was one of the architects and refiners of the standard intellectualist master narrative of the scientific revolution. As such, he illustrates in his article the nonexclusive character of the intellectualist approach by integrating it with a sociocultural account of Galileo's use of patronage strategies surrounding the invention of telescopic astronomy. Patronage is an increasingly important theme in studies of the social and institutional structure of early modern science; this article also sheds interesting light on the background to Galileo's *Starry Messenger,* a standard (and historically important) text from the canon. Astronomy's importance in the history of science of this period, while it can easily be argued for, is nonetheless something that Westfall can take for granted in his article, thanks to the standard view. Thus, Albert Van Helden's article similarly comes with a ready-made relevance in providing a wider overview than Westfall's of the growth of telescopic astronomy during the seventeenth century and the difference it made to cosmological views. Bruce Eastwood's article also relates to optical concerns, introducing the theme of the rhetorical analysis of scientific texts (found also in Dear's "*Totius in verba*") as a means of learning about the target audience of a work and the role of literary form in shaping the technical presentation of innovation, here in the important case of Descartes. Eastwood's article shows the depth of investigation that is made possible by the existence of a master narrative that relieves the historian of the burden of justifying basic issues of topical selection.

The following three articles all relate to matter theory and epistemology, developing this central theme in various ways. Christoph Meinel investigates the foundations of the widespread belief in physical atomism in the seventeenth century, stressing the apparent flimsiness of the empirical evidence, while Kuhn looks at the structure of Boyle's matter theory as a conceptual system rather than as a theory in some strong sense entailed by the evidence (in this case, chemical). Somewhat orthogonally, Betty Jo Teeter Dobbs discusses the place of alchemy in Newton's ideas on matter—a classic example of a revisionist approach that necessarily piggybacks on an established master narrative. Newton's firmly established position as the hero and culmination of the scientific revolution is what renders Dobbs's story of interest: Newton was not just another alchemist, and she does not have to labor the point herself.[7]

The essays have now brought us into the later seventeenth century and to the England of the early Royal Society of London. The Royal Society, boasting such luminaries as Boyle and Newton, is another well-established site of scientific revolution historiography. Any article claiming to shed light on its workings or significance therefore has no particular need to justify itself. Dear's article is an attempt to reread the society's significance (without questioning it) in terms of apparent novelties in the textual form of the natural knowledge that it produced. Shapin, taking a different tack, considers the physical locations of experimental sites in that same time and place.[8] His article has the further feature, beyond its value for illuminating much material from the early Royal Society relating to experimental practices, of meshing with the themes of location and institutional context found in other items in the reader. These include especially Hannaway's regarding architectural space and Cormack's on the meaning of spatial specificities in cartography, as well as Westman's, Westfall's, and Londa Schiebinger's on various specific social and/or institutional contexts for scientific work and the meanings that they implied for the scientific work itself.

Schiebinger's article, bringing things into the early eighteenth century, rounds out the reader by in effect taking stock of the crystallization of social forums for scientific work at the end of the period—the "who" of early modern science. The scientific revolution remains an important historiographical referent framing her argument because Schiebinger can use it to identify the early eighteenth-century activities that she discusses with the enterprise of science in its modern acceptation—to which the label "scientific revolution" has always represented a direct appeal. Schiebinger's article gives a valuable picture of alternative institutional structurings of scientific activity in the period. In particular, it investigates the ways in which these structurings afforded, or in their newer forms denied, possibilities to women to participate in scientific endeavors.

There is a sense in which the themes touched on by these studies have retained some of the guiding assumptions of the older master narratives because these themes can now be treated as having no necessary connections among themselves. Not all of the articles included here (which span several decades) hold that position explicitly; some adhere unproblematically to the standard scientific revolution conceptualization. But the longevity of many of these assumptions, as I have argued, results from the extraction of topical themes from that master narrative and its usually intellectualist assumptions, so that they can have a life of their own. More recent historiography has tended to stress, more as a matter of empirical approach than out of some theoretical commitment, historical contingency in place of the unfolding of mutually implicated developments. The effect is to emphasize that all these occurrences have to be explained historically rather than as inevitable, preprogrammed elements of a restructured, totalizing scientific worldview.

---

[7] This point becomes quite clear in B. J. T. Dobbs, "Newton as Final Cause and First Mover," *Isis* 1994, *85*:633–643, in which the call to move away from the old image of Newton as the culmination of a physics-based scientific revolution nonetheless accompanies a continued assumption of Newton's enormous importance, even though now no longer clearly grounded.

[8] The article may usefully be read in conjunction with Steven Shapin and Simon Schaffer, *Leviathan and the Air-Pump: Hobbes, Boyle, and the Experimental Life* (Princeton, N.J.: Princeton Univ. Press, 1985), the most important and influential revisionist study on this period.

As a practical text for classroom use, therefore, this reader is centrally conceived as a way to cope with the continuing disparities between the still very influential received textbook accounts and the directions followed in more recent specialist scholarship. It should work especially well in concert with selected primary source material in translation and as a complement to a general textbook on the scientific revolution. The themes represented by these selections hold an important place in our current understanding of the period, while the articles have sufficient general import and breadth to lend themselves well to the deepening of textbook and lecture accounts; many will also serve well as focuses for classroom discussion of interpretive issues.

PETER DEAR

The Melanchthon Circle, Rheticus, and the Wittenberg Interpretation of the Copernican Theory

By Robert S. Westman

"ACCEPTANCE" OF A SCIENTIFIC THEORY AND THE WITTENBERG INTERPRETATION

CONTEMPORARY ASSESSMENT of a new scientific theory will always depend upon how that theory is perceived. This may seem an obvious point, yet there exists an understandable tendency among some historians and philosophers of science to treat a later, well-supported version of a theory as though it were the same account available to its earliest recipients. The language which historians often use to characterize the recipients of a new program of scientific research—categories such as Copernican, Newtonian, or Darwinian (not to mention their respective "isms")—all too often masks interesting differences in the meaning of "acceptance." Acceptance may connote provisional use of certain hypotheses (without commitment to truth content), or acceptance of certain parts of the theory as true while rejecting other propositions as false,[1] or the overzealous belief that all propositions of the theory are true even in the absence of sufficient confirmatory evidence, or acceptance of the theory as true without regarding it as a program for further research.

This article is a revised version of a paper, "The Wittenberg Interpretation of the Copernican Theory," which was read in April 1973 before a special collegium on "Science and Society in the 16th Century" as part of the fifth international symposium of the Smithsonian Institution and the National Academy of Sciences. This version appears, together with subsequent discussion, in *The Nature of Scientific Discovery*, ed. Owen Gingerich (Washington, D.C.: Smithsonian Institution, 1975). Other versions of the paper were read before the Los Angeles Psychohistory Study Group (July 1973) and the University of Pittsburgh Philosophy of Science Seminar (January 1974). I wish to acknowledge with appreciation the criticisms and suggestions that I received at those meetings as well as from my UCLA colleagues Peter Loewenberg and Amos Funkenstein, and also from Bruce Moran. I wish to express my gratitude for research support received from the American Philosophical Society (Penrose Fund, Grant No. 6450) and the Academic Senate of UCLA.

[1] On the question of when it is rational to work on a theory, see Noretta Koertge, "Theory Change in Science," *Conceptual Change*, ed. Glenn Pearce and Patrick Maynard (Dordrecht, Holland: Reidel, 1973), pp. 181–185; Israel Scheffler, *Science and Subjectivity* (New York: Bobbs-Merrill, 1967), pp. 86 ff.

NICOLAI COPERNICI

uel è conuerso. H igitur in lineam A B reclinabitur: alioqui accideret partem esse maiorē suo toto, quod facile puto intelligi. Recessit autem à priori loco secundum longitudinem A H retractam per infractam lineam D F H, æqualem ipsi A D, eo interuallo quo dimetiens D F G excedit subtensam D H. Et hoc modo perducetur H ad D centrum, q̃d erit in contingente D H G circulo, A B rectam lineam, dū uidelicet G D ad rectos angulos ipsi AB steterit, ac deinde in B alterum limitem peruenit, à quo rursus simili ratione reuertetur. Patet igitur è duobus motibus circularibus, & hoc modo sibi inuicem occurrentibus in rectam lineam motū

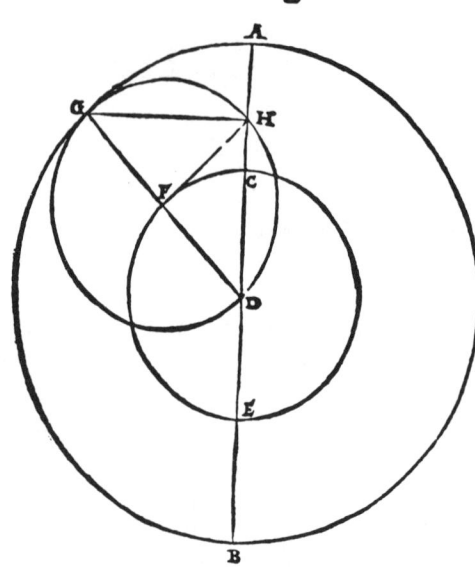

Motus in rectam lineam

• • •

Motus in diametrum.

A M ob causam uocare possumus motum hunc circuli in latitudinem, hoc est in diametrum, cuius tamen periodum & æqualitatem in circumcurrente: at dimensionem in subtensis lineis accipimus, ipsum propterea inæqualem apparere, & uelociorem circa centrum, ac tardiorem

Preservation of the axiom of uniform circular motion by means of libration mechanism: two rolling circles produce an oscillatory rectilinear motion of point H along the apsidal line ADB, used by Copernicus in his theory of trepidation to account for changes in the velocity of precession (De revolutionibus, Book III, Chapter 4, fol. 67ᵛ). At bottom and in margins are Erasmus Reinhold's annotations which explicate the components of this motion. (Courtesy of The Astronomer Royal for Scotland, The Crawford Library, Royal Observatory, Edinburgh.)

Beyond the important recognition that scientists generally perceive a new theory in a somewhat different way from its creator, a major problem for historians of science is to explain how particular interpretations of a theory are determined by considerations such as prior emotional expectations and procedural habits, the reactions of close colleagues and friends, as well as the context in which the new ideas are first encountered. Learning the essentials of a new scientific program directly from its discoverer, for example, will undoubtedly constitute an experience different from the once-removed occasion of reading a text. Nonetheless, if it is almost inevitable that different persons will perceive the same theory in somewhat diverse ways, commonality of professional training and contact will generate a counterforce which encourages the sharing of perceptions.

The idea of *sharing* is important. One of Thomas Kuhn's significant contributions to the discussion of scientific change was to stress shared elements in the scientific enterprise.[2] His early formulations, however, rather overemphasized the extent of agreement necessary for the pursuit of what he saw as mature, "paradigmatic" science: "law, theory, application and instrumentation together—provide models from which spring particular coherent traditions of scientific research."[3] Hence, revolutions in science came to be seen by Kuhn as radical, total displacements of one consensus by another,[4] and the Copernican Revolution provided one of the important historical cases upon which this account rested. It shall be one of our aims in the present study to show, however, that the existence of consensus in one area of a research program need not imply its presence in others; for methodological consensus among the astronomers who early considered the Copernican theory would not prevent fraternization across presumably hostile theoretical lines.

The "Wittenberg Interpretation" is a phrase which I shall use to designate a common methodological outlook or style, a consensus on how to "read" the newly published *De revolutionibus* which was shared by a group of young astronomers at the University of Wittenberg under the fatherly tutelage of the famous Protestant reformer Philipp Melanchthon (1497–1560). There were several features of this "reading" of Copernicus' innovation and, within this common orientation, certain shades of variation. By and large, the principal tenet of the Wittenberg viewpoint was that the new theory could only be trusted within the domain where it made predictions about the angular position of a planet[5] (although, in some cases, Copernicus' predictions were taken to be an improvement over those of Ptolemy). Beyond this basic attitude, some members of the Wittenberg circle believed that certain Copernican models—such

---

[2] *The Structure of Scientific Revolutions* (2nd ed., Chicago: Phoenix, 1970); but see also Michael Polanyi, *Personal Knowledge* (London: Routledge and Kegan Paul, 1958) and John Ziman, *Public Knowledge: The Social Dimension of Science* (Cambridge: Cambridge University Press, 1968).

[3] Kuhn, *Structure*, p. 10. For a textual analysis of Kuhn's diverse uses of the term *paradigm* see Margaret Masterman, "The Nature of a Paradigm," *Criticism and the Growth of Knowledge*, ed. I. Lakatos and A. Musgrave (Cambridge: Cambridge University Press, 1970), pp. 59–79.

[4] Kuhn, *Structure*, Chs. 9–10.

[5] This does not mean that, on occasion, Copernicus' values for the absolute distances of the sun and moon were not used in discussions of matters such as eclipses.

as those which replaced the Ptolemaic equant with epicyclic devices—were to be preferred. An important plank of the Wittenberg program, though one not to be realized until the succeeding generation, was the goal of translating these equantless devices into a geostatic reference frame. The least satisfactory Copernican claim was the assertion that the earth moved and that it moved with more than one motion. In some public summaries of the theory this claim is explicitly denied and taken to be refuted by the Aristotelian postulate that a body can have only one simple motion. Osiander's famous *Letter to the Reader* provided the slightly more moderate interpretation, however, that all celestial motions (including the earth's) were hypothetical.

But what the Wittenberg Interpretation *ignored* was as important as that which it either asserted or denied. In the writings both public and private of nearly every author of the generation which first received the work of Copernicus, the new analysis of the relative *linear* distances of the planets is simply passed over in silence. Ignoring the relative ordering of the planetary spheres does not mean, of course, that astronomers were uninterested in such matters. The familiar sixteenth-century woodcut of the geocentric cosmos with its four sublunary elements was graphic testimony to the unquestioned consensus on the true number and order of the planets.

The initial interpretation of the Copernican theory, therefore, was no mere repetition, as Pierre Duhem implied, of the ancient methodological formula ΣΩΖΕΙΝ ΤΑ ΦΑΙΝΟΜΕΝΑ;[6] it represented, that is to say, more than a position of epistemic resignation with regard to what one could know about actual celestial motions, while stopping short of a strong realist interpretation. Indeed, certain parts of the new theory were to be adopted and preferred as consistent with the foundations of astronomy *if* interpreted in a framework where the earth was at rest, while other aspects were rejected or ignored as irrelevant or as possessing low truth content.

The social context of this early response to the Copernican innovation was an informal circle of scholars drawn together under the leadership of Melanchthon, a generation of men who had been born in the period from about 1495 to 1525. The development of informal academies, especially in Italy, had already begun in the fifteenth century. Structurally they were composed of a patron with a surrounding circle of intellectuals, or a charismatic intellectual about whom gathered a group of scholars, or a group of intellectuals coming together for informal discussions.[7] Melanchthon's circle[8] most closely resembles the second type of organization, but unlike the Italian academies it evolved *within* the walls of the university. Lacking the symbols of autonomy and power, the bureaucratized organizational structure, tight membership criteria, and

---

[6] *To Save the Phenomena*, trans. E. Dolland and C. Maschler (Chicago/London: University of Chicago Press, 1969), pp. 87–91. (First published in 1908.)

[7] See Joseph Ben-David, *The Scientist's Role in Society* (Englewood Cliffs, N.J.: Prentice Hall, 1971), pp. 63 ff.

[8] I have borrowed this designation from Lynn Thorndike without always accepting what Thorndike has to say about it. *Cf. A History of Magic and Experimental Science*, Vol. V (New York: Columbia University Press, 1941), pp. 378–405.

control over publication which would characterize such a later, professionalized scientific society as the Paris Académie des Sciences, Melanchthon and his disciples yet exercised considerable influence on the discipline of astronomy by staffing many of the leading German universities with their pupils and by writing the textbooks that were used in those institutions. The effect of this informal scientific group on the early reception of the Copernican theory cannot be underestimated. Thanks to its efforts, the realist and cosmological claims of Copernicus' great discovery failed to be given full consideration.

There was but one notable exception to this methodological consensus: Georg Joachim Rheticus (1514–1574). For reasons that will become clear in the second half of this study, it was Rheticus alone who departed from the "split" interpretation of the innovation of Copernicus, an interpretation that generally characterized its earliest reception in the German universities and in many parts of Europe.

MELANCHTHON AND THE UNIVERSITY OF WITTENBERG

It was Kepler, writing some sixty-five years after the publication of *De revolutionibus*, heir to a half-century of critical commentary and discussion of the Copernican theory, and consciously committed to the construction of a new *kind* of physical astronomy, who first expressed the existence of two Copernicuses: the one, author of a truly sun-centered cosmology which was supported by a few physical arguments and a new, simpler argument for the ordering of the planets; the other, an inventor of planetary models calculated on the assumption that the planets revolve about a sun which is eccentric to the true center of the universe. Kepler wrote:

> Now, as to what body is to be found in the center [of the world]: whether there be none at all—as Copernicus wishes when he calculates, and Tycho as well, to some extent; or whether it be the Earth—as Ptolemy desires, and Tycho as well, to some extent; or finally, whether it be the Sun itself, as I myself believe, and Copernicus as well, that is, when he theorizes. I began to discuss this with physical reasons.[9]

It is hardly surprising that Kepler should have possessed such an accurate historical perspective, for he was not only the recipient of a bifurcated Copernican tradition but he was also self-conscious of his own role in bringing to it a new unity.[10] While the historical origins of this dissociated methodological

---

[9] Johannes Kepler, *Astronomia nova* (1609), *Gesammelte Werke*, ed. Max Caspar, Vol. III (Munich: C. H. Beck, 1937), p. 237.

[10] For the unity in Kepler's thought, see Robert S. Westman, "Kepler's Theory of Hypothesis and the 'Realist Dilemma,' " in *Internationales Kepler-Symposium Weil der Stadt 1971*, ed. F. Krafft et al. (Hildesheim: Gerstenberg, 1973), pp. 29–54, and also in *Studies in History and Philosophy of Science*, 1972, 3:233–264; Gerald Holton, "Johannes Kepler's Universe: Its Physics and Metaphysics," *Thematic Origins of Scientific Thought* (Cambridge, Mass.: Harvard University Press, 1973), pp. 69–90 (first published in 1956); Jürgen Mittelstrass, "Methodological Elements of Keplerian Astronomy," *Stud. Hist. Phil. Sci.*, 1972, 3:203–232.

viewpoint may be traced back to an earlier split between natural philosophy and mathematical astronomy in the Middle Ages and to Osiander's unique role in affixing his anonymous letter to *De revolutionibus,* the origins of its institutional entrenchment and promulgation must be sought in Melanchthon's Wittenberg circle.

Son of a sword cutler from Bretten in the Palatinate, a grand nephew of the great Hebraic and Cabalistic scholar Johannes Reuchlin (1455–1522), Melanchthon was early on disposed toward humanistic and Greek studies while also developing a deep interest in astrology and astronomy under the tutorship of Johann Stoeffler (1452–1531). He was an active leader in the humanistic movement at the University of Tübingen but left there in 1518, at the age of twenty-one, to assume a professorship of Greek at the newly formed University of Wittenberg.[11] There is evidence that he was the object of a vigorous recruiting campaign from two other universities, Ingolstadt and Leipzig, but he turned down these offers in spite of higher salaries and great faculty banquets in his honor. His biographer, Joachim Camerarius (1500–1574), adds that Philipp was "not so lusty a drinker as the Leipzig professors were."[12]

At Wittenberg the violent, popular energy of the Reformation was beginning to assert itself. Some of the more zealous followers of Luther demanded not only a return to the simplicity of the mass as in the early years of the Church, but the abolition of all education. Christ and his apostles had not been educated, it was argued, and the Gospel was intended for the simple not the wise.[13] Added to these conflicts was the very serious social disruption brought about by the Peasants' War (1524–1525). And in the years between about 1521 and 1536 attendance figures at the German universities generally tended to decline quite considerably and in some cases drastically.[14]

It was in this context that Melanchthon launched a vigorous and far-reaching campaign of educational reform which was to have profound effects on the structure and content of German learning. He involved himself in establishing and reforming the principles of organization for the Protestant universities which, beyond Wittenberg, included Tübingen, Leipzig, Frankfurt, Greifswald, Rostock, and Heidelberg. And the newly founded universities of Marburg (1527), Königsberg (1544), Jena (1548), and Helmstedt (1576) all reflected the Melanchthonian spirit of education. He wrote innumerable textbooks in

---

[11] See Clyde L. Manschreck, *Melanchthon, the Quiet Reformer* (New York/Nashville: Abingdon Press, 1958), pp. 19–26; Friedrich Paulsen, *German Education, Past and Present,* trans. T. Lorenz (New York: Scribner, 1908), pp. 60 f.; John Dillenberger, *Protestant Thought and Natural Science* (New York: Doubleday, 1960), pp. 28 ff.

[12] *Corpus Reformatorum, Philippi Melanthonis Opera quae supersunt omnia,* ed. C. G. Bretschneider, Vol. I (Halle: C. A. Schwetschke, 1834), p. 42 (hereafter cited as *CR*); Joachim Camerarius, *De vita Melanthonis narratio,* ed. G. T. Stroebel (Halle, 1777), p. 26.

[13] Manschreck, *Melanchthon,* p. 76.

[14] See Franz Eulenberg, *Die Frequenz der Deutschen Universitäten von Ihrer Grundung bis zur Gegenwart* (Abhandlung der Philologisch-Historischen Klasse der Königliche Sächsischen Gesellschaft der Wissenschaften, No. II) (Leipzig: B. Teubner, 1904), p. 288. Of the 14 universities surveyed by Eulenberg in this period Wittenberg consistently had the largest enrollment figures and began to recover its earlier losses by 1528.

a wide variety of areas ranging from the classical trivium to physics, astronomy, history, ethics, and, of course, theology. His prefaces appeared in editions of the *Sphere* of Sacrobosco, Erasmus Reinhold's Commentary on Peurbach's *New Theorics of the Planets* (1542), Euclid's *Elements,* and many other scientific treatises.[15] Most important of all, he gave special emphasis to the place of mathematics (and hence astronomy) in the university curriculum. In an oration in praise of astronomy, he links the study of nature with praise of the Creator:

> To recognize God the Creator from the order of the heavenly motions and of His entire work, *that* is true and useful divination, for which reason God wanted us also to behold His works. Let us therefore cherish the subject which demonstrates the order of the motions and the description of the year, and let us not be deterred by harmful opinions, since there are some who—rightly or wrongly—always hate the pursuit of knowledge . . . in the sky, God has represented the likeness of certain things in the Church. Just as the moon receives its light from the sun, so light and fire are transfused to the Church by the Son of God.[16]

The images remind us of similar passages in Kepler's writings where nature is even more explicitly conceived as a book which reflects God's will and where the study of astronomy is equated with the praise of God—and not surprisingly, since Kepler's theology instructor at Tübingen, Jacob Heerbrand (1521–1600), had once been a pupil of Melanchthon.[17] In short, Melanchthon's ideals were not merely contemplative. He was a dedicated teacher who was always close to his pupils. As he once remarked to his friend Camerarius: "I can assure you that I have a paternal affection for all my students and am deeply concerned about everything that affects their welfare."[18]

In practice Melanchthon's precepts and personal example inspired several generations of pupils and teachers to turn their energies to astronomy and the other natural sciences. Such a focusing of interests, however, ought not be mistaken for strict specialization: while many of Melanchthon's followers were competent to teach the elements of the sphere and sometimes wrote introductory treatises in astronomy, they not uncommonly held chairs in medicine, theology, Greek, or natural philosophy. Nonetheless, the overall stimulation of the mathematical disciplines was impressive. To Wittenberg in the 1530s came such important and later influential astronomers as Erasmus Reinhold (1511–1553) and Georg Joachim Rheticus (1514–1574), and, in 1540, their pupil, the future son-in-law of Melanchthon, Caspar Peucer (1525–1602).

---

[15] See William Hammer, "Melanchthon, Inspirer of the Study of Astronomy; With a Translation of His Oration in Praise of Astronomy (*De Orione*, 1553)," *Popular Astronomy*, 1951, *59*:308–319.
[16] *Ibid.,* p. 318.
[17] See, e.g., Kepler, *Mysterium cosmographicum, Gesammelte Werke,* Vol. I, p. 56; Westman, "Kepler's Theory of Hypothesis," pp. 243, 251; Jacob Heerbrand, *Compendium theologiae* (Tübingen, 1573), fol. /) (2ᵛ: "Hac ratione motus vir nostra aetate summus, & incomparabilis D. Phillipus Melanthon, Praeceptor olim meus fidelissimus. . . ." On Heerbrand and Kepler, see Jürgen Hübner, "Naturwissenschaft als Lobpreis des Schöpfers. Theologische Aspekte der naturwissenschaflichen Arbeit Keplers," in *Internationales Kepler-Symposium,* pp. 335–356.
[18] *CR* 3, p. 562; quoted and translated in Manschreck, *Melanchthon,* p. 152.

In addition to these relatively well known figures, there were several lesser lights who lectured at Wittenberg from the 1540s onward and either studied with one or all of the above or came under their influence. Between about 1540 and 1580, these *magistri* included at least the following: Sebastian Theodoricus of Winsheim, Hartmann Beyer, Matthias Lauterwalt, Bartholomaeus Schönborn, Johannes Praetorius (1537–1616), Andreas Schadt (1539–1602), and Caspar Straub.

More significant, however, was the large number of students and former professors who left Wittenberg to take up posts involving the teaching of astronomy at other universities.[19] At Leipzig were Melanchthon's close friend and biographer Camerarius, the theologian Victorinus Strigelius (d. 1569), and the astronomer Johannes Homelius (1518–1562), the latter a former pupil of Rheticus' (who later joined him at Leipzig) and one of Tycho Brahe's first teachers. At Tübingen one could find the former Wittenberger Samuel Eisenmenger (Siderocrates) and the theologian Jacob Heerbrand; the first was succeeded in 1569 by Philipp Apianus (1531–1589), whose pupil Michael Maestlin (1550–1631) became professor of mathematics at Heidelberg and then Tübingen (in 1584). Matthias Stoius and Friedrich Staphylus, both of whom had been present at Rheticus' lectures upon his return from Frombork in 1541, taught at the University of Königsberg. Hermann Witekind (1522–1603), one of Melanchthon's favorite students, took the chair in Greek at Heidelberg but later became professor of mathematics at Neustadt in 1581. At Jena, Michael Neander (1529–1581) became professor of Greek and mathematics in 1551 and produced a short work on the *Sphere* in 1561. Joachim Heller, a former classmate of Rheticus, succeeded Johann Schöner (1477–1547) at Nuremberg. At the newly founded Academy of Altdorf, Johannes Praetorius, a likely former student of Peucer and professor of mathematics at Wittenberg in the early 1570s, began a long and prolific career at the new university in 1576. And in Scandinavia, where Melanchthon's influence was particularly strong, a student of Peucer and Sebastian Theodoricus, Jørgen Dybvad (d. 1612), assumed the extraordinary chair of theology, natural philosophy, and mathematics at the University of Copenhagen in 1575.

Small wonder that Peter Ramus (1515–1572), the great French educational reformer, could remark with envy and admiration that Germany was "the

---

[19] A basic source of information on Wittenberg students and professors is *Album Academiae Vitebergensis ab A. Ch. MDII usque ad A. MDCII*, Vol. II (Halle, 1894). In addition, I have made use of the following references: Melchior Adam, *Vitae Germanorum philosophorum* (Heidelberg, 1615); Walter Friedensburg, ed., *Urkundenbuch der Universität Wittenberg*, Vol. I (Magdeburg: Selbstverlag der Historischen Kommission für die Provinz Sachsen, 1926), and *Geschichte der Universität Wittenberg* (Halle: Max Niemeyer, 1917); Karl Heinz Burmeister, *Georg Joachim Rhetikus, 1514–1574: Eine Bio-Bibliographie*, Vol. I (Wiesbaden: Guido Pressler, 1967); Thorndike, *History of Magic*, Vols. V–VI; Ernst Zinner, *Geschichte und Bibliographie der Astronomischen Literatur in Deutschland zur Zeit der Renaissance* (Leipzig: K. W. Hiersemann, 1941); Owen Gingerich, "From Copernicus to Kepler: Heliocentrism as Model and as Reality," *Proceedings of the American Philosophical Society*, 1973, *117*:516–520; Kristian P. Moesgaard, "How Copernicanism Took Root in Denmark and Norway," in *The Reception of Copernicus' Heliocentric Theory*, ed. J. Dobrzycki (Dordrecht, Holland: Reidel, 1973), pp. 117–119.

nursery of mathematics."[20] Only in Germany, said Ramus, were there learned princes, such as Wilhelm IV Landgrave of Hesse-Cassel (1532–1592), who were interested in fostering mathematical studies.[21] For Melanchthon, however, he reserved the greatest praise:

> Just as Plato revived the study of mathematics in Greece through the great power of his eloquence and erudition, so Melanchthon found [mathematical studies] already greatly encouraged in most academies of Germany,[22] with the exception of Wittenberg. Whereupon, through the force of much and varied instruction and through the example of a pious and upright life, which, at least in my opinion, no doctor or professor in that country had ever attained, he wondrously ignited [those studies]—with the result that Wittenberg became superior not only in theology and eloquence, in which fame it especially excels, but also in the studies of the mathematical discipline.[23]

It might be suggested, indeed, although space does not allow a full treatment here, that the relatively high state of astronomy in Germany in the sixteenth century, when compared with England, France, and perhaps even Italy, was stimulated in large measure by the reforms and charismatic leadership of Melanchthon, who in his own lifetime was rightly called *Praeceptor Germaniae*.

MELANCHTHON AND THE COPERNICAN THEORY

It is one thing to *encourage* expansion in a particular discipline, to recruit outstanding talent, and to solicit substantial financial support. These excellent qualities, however, do not necessarily guarantee a completely open and receptive attitude toward innovation. Great humanistic scholar, administrator, and pedagogue that he was, Melanchthon could not properly be called a working astronomer. His lectures on physics and astronomy, first published at Wittenberg in 1549 (although written several years earlier), reveal a firm grounding in the original texts of Aristotle and Ptolemy, a clear systematization of arguments in support of various claims, but a woeful lack of diagrams.[24] More significantly, Melanchthon was hardly a man just entering an age of youthful exuberance when he first learned of the Copernican theory through Rheticus' *Narratio prima*, which was sent to him on February 15, 1540.[25] At the age of forty-three

---

[20] *Scholae physicae praefatio; Collectaneae praefationes, epistolae, orationes* (Basel, 1577), p. 74; for an excellent summary of Ramus' views on the state of mathematical studies in Europe, see R. Hooykaas, *Humanisme, science et reforme: Pierre de la Ramée (1515–1572)* (Leyden: E. J. Brill, 1958), pp. 75–96.

[21] *Scholarum mathematicarum libri unus et triginta* (Frankfurt, 1627), Book II, p. 64; first published at Basel, 1569.

[22] With the possible exceptions of Tübingen, Nuremberg, and Ingolstadt in the 1520s, Ramus exaggerates the situation.

[23] *Scholarum mathematicarum*, p. 68.

[24] *Initia doctrinae physicae, dictata in Academia Witebergensi* (1549), in *CR* 13, pp. 216 ff. A manuscript in Melanchthon's hand, "De supputatione motus solis" (MS PR 5974), in the Huntington Library, San Marino, California, confirms this view.

[25] Georg Joachim Rheticus, *Narratio prima*, trans. Edward Rosen, in *Three Copernican Treatises* (3rd ed., New York: Octagon, 1971), p. 394. Rheticus' work is hereafter cited as *NP*.

he was certainly not old, but he had already been associated with the University of Wittenberg for more than two decades, he was the veteran of many academic battles, and he was at the height of his career.

His earliest reference to Copernicus, in a well-known letter to Mithobius, on October 16, 1541, is merely an incidental one and treats the new theory as a disturbance rather than a threat.[26] Later statements, however, take a harder line. In his published lectures of 1549 he perceives the new theory as an old and absurd "paradox" which Aristarchus had once defended and which young students ought to stay away from since it conflicted with Scripture.[27] Further passages clearly argue against the earth's motion from the Aristotelian doctrine of simple motion.[28] Contrasted with these views, however, we find evidence of a more positive and favorable interpretation. Thus he praises Copernicus' lunar theory because it is "so beautifully put together" (*admodum concinna*), although he quickly adds that students ought rather to learn the Ptolemaic viewpoint.[29] In several places he uses Copernican data for the solar apogee and for the apogees of the superior planets.[30] And, in the second edition of his *Initia doctrinae physicae*, as Emil Wohlwill first demonstrated, he tones down the negative allusions to Copernicus by deleting several phrases about those who argue, "either from love of novelty or from the desire to appear clever," that the earth moves.[31]

The customary dichotomy of "pro" and "anti" Copernican, then, becomes less than adequate as a description of Melanchthon's views and those of his disciples. To be sure, one can certainly discern a traditional component in Melanchthon's attitude toward the new theory. He recognizes that the earth's motion *could* be interpreted as a real one, and he explicitly rejects this possibility

---

[26] *CR* 4, p. 679. This is the interpretation given to the passage by A. Bruce Wrightsman in his unpublished doctoral dissertation "Andreas Osiander and Lutheran Contributions to the Copernican Revolution" (University of Wisconsin, 1970), pp. 345–346. *Cf.* Rosen's view: "when Rheticus returned from Fromb ork to Wittenberg on 16 October 1541 Philip Melanchthon, Luther's principal lieutenant, harshly condemned 'that Polish (Sarmatian) astronomer who set the earth in motion' " ("Biography of Copernicus," in *Three Copernican Treatises*, p. 400).

[27] *CR* 13, p. 216.

[28] *Ibid.*, p. 217.

[29] *Ibid.*, p. 244.

[30] *Ibid.*, pp. 225, 241, 262.

[31] The change in tone from the 1549 edition to the 1550 edition was first pointed out by Emil Wohlwill in "Melanchthon und Copernicus," *Mitteilungen zur Geschichte der Medizin und der Naturwissenschaft*, Vol. III (Hamburg/Leipzig, 1904), pp. 260–267, and it has passed down subsequently into the later German literature on the subject. *Cf.* Wilhelm Maurer, "Melanchthon und die Naturwissenschaft seiner Zeit," *Archiv für Kulturgeschichte*, 1962, 44:223; Konrad Müller, "Ph. Melanchthon und das kopernikanische Weltsystem," *Centaurus*, 1963, 9:16–28; Hans Blumenberg, *Die kopernikanische Wende* (Frankfurt am Main: Suhrkamp, 1965), pp. 100–121, 174 (the texts are reprinted in parallel columns). Although Thorndike made use of Wohlwill's discovery (*History of Magic*, Vol. V, p. 385), it was overlooked by Thomas Kuhn in *The Copernican Revolution* (New York: Vintage, 1957), pp. 191–192, 196, and by Marie Boas in *The Scientific Renaissance, 1450–1630* (New York: Harper, 1962), p. 126. The more moderate view of Melanchthon is found, however, in the recent work of younger scholars: J. R. Christianson, "Copernicus and the Lutherans," *Sixteenth Century Journal*, 1973, 4:1–10; Wrightsman, "Osiander," pp. 347–348; and Bruce Moran, "The Universe of Philip Melanchthon: Criticism and Use of the Copernican Theory," *Comitatus*, 1973, 4:1–23.

as contrary to the divine testimony of Scripture. By contrast with the position of Osiander, therefore, Melanchthon's interpretation is slightly less moderate, for Osiander's *praefatiuncula* nowhere rejects the motion of the earth with scriptural and physical arguments but simply awards to it an open hearing, namely, the same hypothetical status as the planetary theories of the ancients.[32] In the spirit of Osiander's *Letter,* however, there is also a *pragmatic* component in Melanchthon's evaluation of the Copernican theory. He seems to be saying that there may be something of value in this new teaching which will be useful for students and professors. On a personal level this viewpoint appears as a benign paternalism—tolerant, perhaps even amused by the revival of an old paradox, willing to encourage some experimentation, but ultimately wedded to established points of view. Not surprisingly, it was this same capacity for compromise and flexibility which had already gained for the *Praeceptor Germaniae* his reputation as an effective arbitrator of controversies, whether theological or personal.[33] While the florid and frequently dogmatic Luther had taken a characteristically strong (and, in this case, negative) position on a new idea about which he had only bare, hearsay knowledge,[34] Melanchthon adopted a typically moderate stance which absorbed the criticism of Luther while allowing considerable freedom to explore new pathways. Limited though it was, then, Melanchthon's attitude would permit and even encourage some articulation of the conceptual and empirical components of the Copernican theory: the improvement of observations, systematization of tables, discussion of the problem of precession, and use of certain Copernican planetary models in a geostatic framework.

THE PRINCIPAL DISCIPLES OF MELANCHTHON

Of the group of astronomers who pursued their discipline in the course of Melanchthon's tenure at Wittenberg, from about 1530 to 1560, none were more important than three figures: Erasmus Reinhold, Caspar Peucer, and Georg Joachim Rheticus. Over the years a strong father–son relationship had developed and under Melanchthon's aegis the science of astronomy flourished. Two of the three, Reinhold and Peucer, eventually held the high administrative

---

[32] In his copy of *De revolutionibus* Michael Maestlin wrote a critical marginal note by the anonymous preface, confirming Osiander's "neutral" position toward the Copernican theory: "the author of this letter, whoever he may be, while he wishes to entice the reader, neither boldly casts aside these hypotheses nor approves them but rather he imprudently squanders away something which might better have been kept silent." Maestlin's heavily annotated copy is located in the Stadtbibliothek, Schaffhausen, Switzerland.

[33] In 1525, e.g., Melanchthon intervened on behalf of a group of nuns who were threatened with expulsion from their convent. Mother Charitas Pirkheimer wrote in her memoirs: "He [Melanchthon] was more moderate and modest in his speech than any Lutheran I had yet heard. That the people were being subjected to force was very displeasing to him. . . . Thus he brought it about that the people desisted somewhat from hostility towards us and were no longer so violent concerning us." (*An Heroic Abbess of Reformation Days. The Memoirs of Mother Charitas Pirkheimer, Poor Clare of Nuremberg,* F. Mannhardt, intro. (St. Louis: Central Bureau, C.C.V. of A., 1930), p. 26.

[34] See Wilhelm Norlind, "Copernicus and Luther: A Critical Study," *Isis,* 1953, 44:273–276.

position of rector of the university, and each of the three in his own way helped to lay down a particular pattern for the reception of the Copernican theory.

Virtually nothing is known of the life of Reinhold.[35] Between 1541 and 1542 he had first become directly acquainted with the theory of Copernicus through his colleague Rheticus, who had only recently returned from Frombork. Already in his Commentary on Peurbach's *New Theorics of the Planets* (1542) a major source of his interest and curiosity about the new theory is discernible in his criticism of contemporary astronomy, where we note a dissatisfaction with the planetary theories that reminds us at once of a similar discontent in Copernicus.

> To make known the causes of the variegated appearances shown by the planetary motions, learned astronomers have, speaking generally, assumed or established either the eccentricity of the deferents or the multiplicity of the spheres. The numerousness of the spheres thus obtained must be attributed to the astronomer's art, or rather, to the weakness of our understanding.[36]

Somewhat later he refers to a new hope for the restoration of astronomy:

> I know of a recent author who is exceptionally skillful. He has raised a lively expectancy in everybody. One hopes that he will restore astronomy. He is just about to publish his work. *In the explanation of the phases of the moon he abandons the form that was adopted by Ptolemy. He assigns an epicyclic epicycle to the moon.* . . . [37]

Now we recall that Melanchthon had also mentioned Copernicus' lunar theory, but there was only one printed source from which Reinhold could have obtained this information by 1542: the recently published *Narratio prima* of Rheticus. Here Rheticus had written of Copernicus' handling of the moon's second inequality: "He assumes that the moon moves on an epicycle of an epicycle of a concentric."[38] And, on the following page, we grasp the larger context in which Rheticus writes:

> Furthermore, most learned Schöner, you see that here in the case of the moon we are liberated from an equant by the assumption of this theory, which, moreover, corresponds to experience and all observations. My teacher dispenses with equants for the other planets as well. . . . [39]

Liberation from the equant! How much pride Copernicus had taken in this achievement; and Reinhold leaves no doubt that it was this accomplishment which had initially impressed him so deeply. At the bottom of the title page

---

[35] See Owen Gingerich, "Erasmus Reinhold," in *Dictionary of Scientific Biography* (New York: Scribner's, forthcoming).
[36] Quoted in Duhem, *To Save the Phenomena*, p. 71.
[37] *Ibid.*, p. 72, my italics. I have modified Duhem's sometimes loose translations where necessary. In all cases I have reinterpreted the passages cited.
[38] *NP*, p. 134.
[39] *Ibid.*, p. 135.

of his own personal copy of *De revolutionibus* Reinhold wrote in beautiful and carefully formed red letters: *Axioma Astronomicum: Motus coelestis aequalis est et circularis vel ex aequalibus et circularibus compositus* (The Astronomical Axiom: Celestial motion is both uniform and circular or composed of uniform and circular motions).[40]

It was precisely this axiom, or boundary condition, which Copernicus had tried to satisfy, if not always successfully, in his construction of the planetary models and which Reinhold consistently singles out in his annotations. Thus, in Book V, Copernicus initiates his discussion of the planetary theories once more by attacking the ancients for conceding that "the regularity of the circular movement can occur with respect to a foreign and not the proper center; similarly and more so in the case of Mercury."[41] He then articulates this initial presupposition in terms of three particular models: a circle eccentric to an eccentric circle, an epicycle on an epicycle on a homocentric circle, and an eccentric circle carrying an epicycle.[42] Once again, Reinhold's pen boldly announces the point,[43] with a great flourish reserved for the orbit of Mercury: *Orbis Mercurij Eccentrj Eccentrus Eccentrepicyclus* (The orbit of Mercury is an eccentric on an eccentric on an eccentric circle carrying an epicycle).[44] And, in commenting upon Copernicus' model for the precession of the equinoxes—an oscillatory motion of the poles which Copernicus likened to "hanging bodies swinging over the same course between two limits"—Reinhold observes that "From two equal and regular circular motions, there is produced: I. motion in a straight line; II. reciprocal motion lest it be infinite; and, III. unequal or differing motion which is slower at the extremes than near the middle."[45] Such was the power of equantless astronomy which Ibn al-Shāṭir and others

---

[40] Reinhold's copy of the 1543 edition is presently located in the Crawford Library of the Royal Astronomical Observatory in Edinburgh. I should like to thank Owen Gingerich for generously allowing me to use his microfilm of the work which permitted me to check and supplement the notes which I took from the original in Nov. 1972. See also Owen Gingerich, "The Role of Erasmus Reinhold and the Prutenic Tables in the Dissemination of the Copernican Theory," *Studia Copernicana* VI (*Colloquia Copernicana* II) (Warsaw: Ossolineum, 1973), pp. 43–62, esp. pp. 56–58.

[41] Nicholas Copernicus, *On the Revolutions of the Heavenly Spheres*, trans. Charles Glenn Wallis, in *Great Books of the Western World*, Vol. XVI (Chicago: Encyclopaedia Britannica, 1952), Bk. V, Ch. 2, p. 740. I have checked all translations against the 1543 edition.

[42] *Ibid.*, V. 4, p. 742.

[43] Reinhold's copy of *De revolutionibus* (hereafter cited as *DR* Crawford), fol. 142r: "Primus Modus per Eccentrepicyclum;" fol. 142v: "Secundus Modus per Homocentrepicyclos."

[44] *Ibid.*, fol. 164v; Copernicus, *Great Books*, V.25, p. 785.

[45] *DR* Crawford, III.3, fol. 67v. Tycho Brahe, who owned at least three copies of *De revolutionibus*, extensively and carefully transcribed all the annotations from Reinhold's copy (including the passage cited above) on his visit to Saalfeld in 1575. Hence Reinhold's reading of *De revolutionibus* had a significant *private* impact on Tycho and, through him, on other sixteenth-century astronomers. I have recently discovered, at the University of Liège, what I believe to be the original exemplar used by Tycho when he copied out Reinhold's notes (1543 ed.). The other two copies located respectively at the Clementinum in Prague (1566 ed.) and in the Manuscript Division of the Biblioteca Apostolica Vaticana (1543), are discussed in Robert S. Westman, "Three Responses to the Copernican Theory: Johannes Praetorius, Tycho Brahe and Michael Maestlin," in *The Copernican Achievement*, ed. Robert S. Westman (Berkeley/Los Angeles: University of California Press, 1975); on the first identification of the Vatican copy see Owen Gingerich, "Copernicus and Tycho," *Scientific American*, 1973, *299*:86–101.

discovered in the fourteenth century and which now Reinhold, and later Tycho Brahe, cherished as a major development in the restoration of astronomy to its original foundations.

It is no wonder that literati later scorned the strange, seemingly pedantic activities of astronomers. For Reinhold, however, Copernicus had truly restored astronomy by attempting to clean up the morass of equants which had cluttered the art. This is evidently what he intends in his praise of Copernicus when he writes:

> All posterity will gratefully celebrate the name of Copernicus. The science of the celestial motions was almost in ruins; the studies and works of this author have restored it. God in His Goodness kindled a great light in him so that he discovered and explained a host of things which, until our day, had not been known or had been veiled in darkness.[46]

It was not the revolutionary cosmological arguments of Book I in *De revolutionibus*, therefore, that had brought such high esteem from Reinhold. For with the exception of a few underlinings and factual paraphrases, his own copy is practically devoid of interpretative comments.[47] And this seeming lack of interest is further confirmed by his unpublished commentary on Copernicus' work where he maintains what Aleksander Birkenmajer calls "the most perfect neutrality on the problem of geocentrism and heliocentrism."[48] There remained in Reinhold's view but one more major task: to systematize and recalculate the motions as set forth in *De revolutionibus* so that errors might be eliminated and convenient tables provided for the working astronomer. And this is precisely what he did. In an article which nicely complements our conclusions in this section it has been shown with the aid of a computer that the *Prutenic Tables* are based solely upon Copernican planetary mechanisms.[49]

Reinhold thus clearly became the leading figure in the Wittenberg circle in the years after Rheticus' departure for Leipzig in 1542. His great work on the new tables of motion received the full moral and financial support of Melanchthon, who wrote to Albrecht, Duke of Prussia, on his behalf.[50] And his *Tables* soon rapidly showed their competitive strength against other compilations of planetary data.[51] Through Reinhold the image of Copernicus

---

[46] Erasmus Reinhold, *Prutenicae tabulae coelestium motuum* (Tübingen, 1551), "Praeceptu calculi motuum coelestium," p. 21; quoted in Duhem, *To Save the Phenomena*, pp. 72–73.

[47] In Bk. I the longest note occurs in Ch. 6, fol. 4v, on the diurnal motion of the earth.

[48] Aleksander Birkenmajer, "Le commentaire inedit d'Erasmus Reinhold sur le *De revolutionibus* de Nicholas Copernic," *Études d'histoire de sciences en Pologne, Studia Copernicana* IV (Warsaw: Ossolineum, 1972), p. 765; first published in 1957. This neutrality had been foreseen almost exactly by Copernicus' friend Bishop Tiedemann Giese, who predicted that if Copernicus were to publish only his tables, without the hypotheses and principles underlying them, scarcely anyone would seek out those principles. As Rheticus reports Giese's views: "There was no place in science, he asserted, for the practice frequently adopted in kingdoms, conferences, and public affairs, where for a time plans are kept secret until the subjects see the fruitful results and remove from doubt the hope that they will come to approving the plans" (*NP*, p. 193).

[49] Gingerich, "Reinhold and the Prutenic Tables," pp. 48–50.

[50] Blumenberg, *Die kopernikanische Wende*, pp. 95–96.

[51] Gingerich, "Reinhold and the Prutenic Tables," pp. 51–55.

the Calculator and Copernicus the Reformer of Equant-Ridden Astronomy seeped inexorably into astronomical discussion both inside and outside the walls of the universities.

The man largely responsible for consolidating and institutionalizing the Wittenberg Interpretation within the universities of Germany was Caspar Peucer. Born in 1525, the son of an artisan, he showed such remarkable talents as a young man that he began to attend the University of Wittenberg at the age of fifteen, living at the home of Melanchthon.[52] In 1550 this informal relationship with the great Reformer was formalized when he married Magdalena, Melanchthon's daughter. A poem at the beginning of Peucer's *Elements of the Doctrine of the Celestial Circles* (1551) endearingly and not surprisingly dubs Melanchthon "Father," and a chronological list of "Astrologi," starting with the Creation and proceeding up to 1550, ends with Erasmus Reinhold, "Praeceptor mihi carissimus."[53] When Reinhold died of the plague in 1553 Melanchthon's son-in-law succeeded him. In 1559 he was named to the chair of medicine, and in 1560 he succeeded finally to the rectorate of the university upon the death of Melanchthon.

In this new post Peucer was in a powerful position from which to enforce and spread the teachings, especially the theological doctrines, of his father-in-law. The best way to do this, as he shrewdly recognized, was to give the principal chairs to the partisans of Melanchthon, or Philippists, as they were called, rather than to the more orthodox Lutherans. A lengthy power struggle ensued in which both sides attempted to curry the political support of the patron of the university, Augustus, the Elector of Saxony. The results were disastrous for Peucer. He was accused of being a crypto-Calvinist. Such charges were frequent during this period when there were increasing tensions among the different Protestant groups. As Clyde Manschreck has written:

> In the sixteenth century, arguments about the Lord's Supper were not academic tiffs; a man could be deprived of his job, or be exiled, or lose his life. At Rostock, participants in communion could not have beards; licking the blood of Christ from one's beard meant instant death! . . . Kongius of Hildesheim was expelled from his Office because he picked up a wafer. Melanchthon's son-in-law, Peucer, went to prison for 12 years for seeming to deny Christ's physical presence in the bread. If some wine accidentally spilled on a suit, it had to be cut out immediately and burned. If some fell on the ground, it had to be licked up by the priest.[54]

Amidst this obsessive and rigidly intolerant atmosphere, Peucer was incarcerated at Pleissenburg, near Leipzig, in 1576, in spite of efforts to have him released by Wilhelm IV, Landgrave of Hesse-Cassel with whom he had enjoyed

---

[52] The following biographical details are based mainly upon the articles on Peucer in *Nouvelle biographie generale*, pp. 767–770 and Thomas-Pope Blount, *Censura celebriorum authorum* (Geneva, 1710), pp. 735–737.

[53] Caspar Peucer, *Elementa doctrinae de circulis coelestibus, et primo motu, recognita et correcta* (Wittenberg, 1553), Preface.

[54] Manschreck, *Melanchthon*, pp. 229–230.

a stimulating scientific correspondence.[55] From an account of his long sufferings, written after his long confinement, we learn that he was allowed only a few books (his astronomical work effectively ceased after his imprisonment) and that in order to write he was forced to manufacture secretly a makeshift ink from the crumbs of roasted bread and from the dust dissolved in his beer. For paper he used the margins of his books.[56] The lesson of this episode, to which we shall return again, is that one was far more likely to face punishment in this period for holding theological views with *explicit* political consequences than for discussing the scriptural basis of the Copernican theory.[57]

The Wittenberg Interpretation, as reflected in Peucer's work and the writings of other Wittenberg professors, firmly echoes the views of Peucer's two mentors, Melanchthon and Reinhold. It is important to recognize, however, that there were different pedagogical levels at which the theory might be considered—especially since the teaching functions of medieval and early modern universities generally carried greater weight relative to their research functions.

In 1545 Melanchthon had reorganized the Faculty of Arts and therein defined the subject matter for the various curricula. Students in the natural sciences at the lower levels were to be instructed in arithmetic, physics, the second book of Pliny's *Natural History*, Aristotle's *Ethics*, and the *Sphere* of Sacrobosco, while those seeking the master's degree were to have lectures on Euclid, planetary theory (Peurbach), and Ptolemy's *Almagest*.[58] It is hardly surprising that the introduction of the Copernican theory into the scientific curriculum should follow the limits already prescribed by the existing system. Peucer, in his widely used introductory textbook, mentions the new theory briefly and cites quantitative material from it to illustrate various aspects of the celestial motions. He uses Copernican data, for example, on the absolute distances of the sun and moon from the earth in his discussion of eclipses and on the definition of the length of the day.[59] Just as in Melanchthon's lectures, Copernicus is portrayed as the reviver of the theory of Aristarchus,[60] from which, of course, students could learn nothing of the extensive arguments and demonstrations set forth by Copernicus. The only arguments to which beginning students were exposed (and these have often been taken by historians as representative of the general reaction to the new theory[61]) were those against the motion of the earth—that it is contrary to Holy Scripture and the Aristotelian laws of simple motion.[62] Just as today, when we do not include advanced and very recent research

---

[55] *Cf.* Caspar Peucer, *Historia carcerum C. Peuceri* (Zurich, 1604), p. 362.
[56] *Ibid.*, p. 363.
[57] At about this time Maestlin's teacher, Philipp Apianus, was dismissed from Ingolstadt on charges of being a secret Calvinist; Kepler himself later suffered for his own highly individualistic religious beliefs.
[58] Friedensburg, *Urkundenbuch*, Vol. I, pp. 256–57, and *Geschichte*, pp. 216–217.
[59] Peucer, *Elementa*, fol. 54; P2.
[60] *Ibid.*, fol. E4v; fol. G2v.
[61] See, e.g., Hugh Kearny, *Science and Change, 1500–1700* (New York: World University Library, 1971), pp. 101, 104; also n. 31.
[62] Peucer, *Elementa*, fol. E4v.

in introductory science texts, so Peucer and Melanchthon drew the line, perhaps a bit too strictly, at the treatment of planetary theory.

At the master's level the handling of planetary motions became more sophisticated. Here the basic text was not Sacrobosco but Peurbach's *Theoricae novae planetarum*. There seems little doubt that Wittenberg students used the version of Peurbach to which was affixed the commentary of Reinhold, at least until the middle of the 1550s.[63] Many later Wittenberg professors, such as Johannes Praetorius, Andreas Schadt, and Caspar Straub, who were all probably pupils of Peucer's, continued to use the Peurbachian framework for their lectures on planetary theory, and it was in this context that several aspects of the Copernican theory were accorded more extensive and serious consideration than in the introductory texts.

After about 1560 at least two generalizations may be made with regard to the subsequent articulation of the Wittenberg Interpretation. In the first place, the status of Copernicus' work was clearly ranked on the same level as Ptolemy's as a serious scientific theory. Peucer, Schadt, Straub, Praetorius, and others advised their students to consult *De revolutionibus* directly and often recommended the comparison of corresponding topics in the *Almagest*.[64] In a particularly influential and representative source of the Wittenberg viewpoint, Peucer's *Hypotyposes orbium coelestium*, it was suggested that Copernicus' precessional model could be "transferred," in principle, to a geostatic reference frame if ninth and tenth spheres were added. Peucer writes:

> . . . if these hypotheses were to be transferred (*hypotheses si transferantur*) to the eighth orb by means of two other spheres, the ninth and tenth, and if, in the same manner, there were to be established in the heavens a moving celestial equator with moving poles and axes as well as those same points where the equator intersects the ecliptic and those [points] which are most distant from it; and, assuming that the ecliptic with its poles always remains immobile with respect to the eighth sphere, then I believe that the same [effects] would be achieved without having to change the ancient hypotheses.[65]

Although Peucer does not take the trouble to work out the details of his proposed inversion of the Copernican precessional model, it is clear what tacit assumption he has made—namely, that the hypotheses of Ptolemy and Copernicus are geometrically equivalent. This presupposition likewise underlies a section entitled "Accommodation of these Hypotheses to the Copernican and Prutenic

---

[63] Editions appeared in 1542 and 1553. There were a few references to Copernicus in the edition of 1553.

[64] See, e.g., Caspar Straub, *Annotata in theorias planetarum Georgii Purbachi*, Nov. 2, 1575, Erlangen Universitätsbibliothek MS 840, fol. 2r; Andreas Schadt, *In theorias planetarum Purbachij annotationes Vitebergae privatim traditae*, 1577, ibid., fol. 72r; [Caspar Peucer], *Hypotyposes orbium coelestium, quas appellant theoricas planetarum: congruentes cum tabulis Alphonsinis & Copernici, seu etiam tabulis Prutenicis: in usum scholarum publicatae* (Argentorati, 1568), p. 301.

[65] Peucer, *Hypotyposes*, pp. 516–517. The edition cited above first appeared at Strassburg in 1568, both anonymously and with a preface by Conrad Dasypodius; however, it later appeared as *Hypotheses astronomicae* in 1571 at Wittenberg, where Peucer claimed the true authorship. For a discussion of the various publications of this work, see Zinner, *Geschichte und Bibliographie*, pp. 35, 244, 250, and Gingerich, "Reinhold and Prutenic Tables," pp. 59–60.

Canons" which follows his description of the Ptolemaic lunar theory. Here he summarizes Copernicus' bi-epicyclic hypothesis and then proceeds to compare it, angle by angle, with the Ptolemaic model. Peucer ends this section by advising students that if they wish to study the "demonstrations" of these hypotheses, then they ought to consult "Copernicus himself."[66]

A second basic feature of the evolution of the Wittenberg Interpretation was the conspicuous absence of urgency about the issue of cosmological choice. In the vast majority of advanced treatments of planetary theory we do not find that scriptural arguments were accorded high priority, certainly not the prominence that they received in the era of Galileo's conflict with the Church. Hence, it is quite unlikely that fear of censorship or other punishment from the religious and political authorities was a serious factor in preventing the emergence of a realist interpretation of the Copernican theory. As we have seen, Peucer, a man of great courage and astronomical skill, went to jail for his views on Holy Communion and not for his scientific beliefs. And crypto-Copernicanism, the holding of secret Copernican convictions, was a phenomenon of the seventeenth century rather than the sixteenth. The more profound reasons why debate over cosmological issues was delayed until the 1570s had less to do with fear of external reprisals than with the fact that questions about the Copernican ordering of the planets were not seen as important topics of investigation. In annotated copies of *De revolutionibus* which are datable to the period *circa* 1543–1570, passages in Book I extolling the newly discovered harmony of the planets and the eulogy to the sun, with its Hermetic implications, were usually passed over in silence. Copernicus was seen, in general, as the *reformer* of Ptolemaic astronomy, not in a revolutionary sense, however, as later thinkers such as Kepler would believe, but in an essentially conservative sense, as the admired inventor of new planetary hypotheses and an improved theory of precession.

And yet, while the Wittenberg reading of *De revolutionibus* persisted at least until the end of the sixteenth century, it was already being questioned and reformulated by a new generation of astronomers in the 1570s. For men such as Johannes Praetorius, Tycho Brahe, and Michael Maestlin, it seemed that their predecessors had overlooked the serious cosmological claims of Copernicus.[67] Of the men of the earlier generation, only one member had praised Copernicus for asserting an "absolute system" of the planets. Where Peucer and Reinhold had been dutiful "sons" of Melanchthon, the third "son," Rheticus, had not followed the cautious path laid down by the "father" of the Wittenberg circle. Rheticus had found a different father to idealize.

RHETICUS: PERSONALITY AND ACCEPTANCE OF THE COPERNICAN THEORY

There was something different about Rheticus. Karl Heinz Burmeister, his best and most recent biographer, suggests that he had an "abnormal personality,"

---

[66] *Hypotyposes*, pp. 299–301.
[67] For a more complete discussion of this generation, see Westman, "Three Responses," in *The Copernican Achievement*.

an insane genius characterized by "bionegative elements"—celibacy, childlessness, and homosexual tendencies.[68] Considerably more sympathetic than Burmeister's characterization, Arthur Koestler writes with the impressionistic flair of the novelist:

> Rheticus, like Giordano Bruno or Theophrastus Bombastus Paracelsus, was one of the knight errants of the Renaissance whose enthusiasm fanned borrowed sparks into flame; carrying their torches from one country to another, they acted as welcome incendiaries to the Republic of Letters. . . . [Rheticus was] an *enfant terrible* and inspired fool, a *condottiere* of science, an adoring disciple and fortunately, either homo- or bi-sexual, after the fashion of the time. I say 'fortunately' because the so afflicted have always proved to be the most devoted teachers and disciples, from Socrates to this day, and History owes them a debt.[69]

The comparisons with Paracelsus and Bruno are interesting although undeveloped. Rheticus certainly possessed a kind of Brunian-Paracelsian *Wanderlust*. Between 1538, when he temporarily left his post at Wittenberg, and 1554, when he settled in Cracow, his life was one of almost constant peregrinations. However, he was not fleeing the authorities because, like Paracelsus, he had thrown the works of Galen and Avicenna into a public bonfire,[70] nor like Bruno, because he was attempting to reconcile Protestants and Catholics in a Hermetic religion of the world.[71] The only journey which he undertook involuntarily, so far as we know, was his hurried departure from Leipzig in 1550, which was probably due to his having engaged in illegal sexual acts.[72] But unfortunately there is scarcely enough evidence for us to conclude anything certain about the general pattern of his sexual behavior.

Koestler is closest to the mark when he writes of the "enthusiasm" of Rheticus. If indeed one were to point to the single most prominent trait in Rheticus' personality, based upon the tone of his writings, the testimonies of his contemporaries, and his own life activities, one would have to seize upon his great energy and intensity—whether in the vitality of his work, in his widespread travels, or in his evident pursuit to lay to rest something inside himself. Now, in any individual who invests an unusual amount of psychic energy in a particular object or goal one can expect to find powerful, conflicting emotions. Rheticus in particular seems to have had an *inordinate* need to associate himself with famous men, to be pleasing (or displeasing) to them, and, in short, to be loved by them. Consider his decision to leave Wittenberg in 1538, at the age of twenty-four, after he had been teaching at the university for two years

---

[68] Burmeister, *Rhetikus*, Vol. I, p. 190.
[69] Arthur Koestler, *The Sleepwalkers* (New York: Grosset and Dunlap, 1963; first published in 1959), pp. 153–154.
[70] *Cf.* Walter Pagel, *Paracelsus: An Introduction to Philosophical Medicine in the Era of the Renaissance* (Basel/New York: Karger, 1958), pp. 20 f.
[71] See Frances A. Yates, *Giordano Bruno and the Hermetic Tradition* (London: Routledge and Kegan Paul, 1964).
[72] *Cf.* Ernst Zinner, *Enstehung und Ausbreitung der Coppernicanischen Lehre* (Sitzungsberichte der physikalisch-medizinischen Sozietät zu Erlangen, 74) (Erlangen, 1943), p. 259. Zinner is Koestler's source, *Sleepwalkers*, p. 188.

and was very well established. In a letter to Heinrich Widnauer in 1542 he recalls being "attracted by the fame of Johannes Schöner in Nuremberg, who had not only accomplished much in scientific subjects but had excelled in all the best things of life as well."[73] From Nuremberg Rheticus journeyed to Tübingen, where he visited the "famous pupils" of Johannes Stöffler (1452-1531),[74] and thence to Ingolstadt, where he met with Peter Apianus (1495-1552). "And finally," Rheticus concludes,

> I heard of the fame of Master Nicolaus Copernicus in the northern lands, and although the University of Wittenberg had made me a Public Professor in those arts, nonetheless, I did not think that I should be content until I had learned something more through the instruction of that man. And I also say that I regret neither the financial expenses nor the long journey nor the remaining hardships. Yet, it seems to me that there came a great reward for these troubles, namely, that I, a rather daring young man (*iuvenili quadam audacia*) compelled (*perpuli*) this venerable man to share his ideas sooner in this discipline with the whole world.[75]

Beyond the manifest purpose for visiting Copernicus—to learn about his theory—lay a deeper personal motivation. The "great reward" for Rheticus was to be found in his *relationship* with the older man, in causing him to share with the younger a treasure that no one else had been able to extract from him. This interpretation is reinforced by an interesting event which occurred near the end of Rheticus' life. A young mathematician named Valentine Otho (b. 1550), who had studied with the successors of Rheticus at Wittenberg, Peucer and Praetorius, came to visit him in 1574 in the town of Cassovia, located in the Tatra Mountains of Hungary.[76] In Otho the aging Rheticus saw himself as he had once been, almost as though he were looking into a mirror of the past. Otho reports:

> We had hardly exchanged a few words on this and that when, on learning the cause of my visit he burst forth with these words: "You come to see me at the same age as I was myself when I visited Copernicus. If I had not visited him, none of his works would have seen the light."[77]

Rheticus' personal relationship with Copernicus, therefore, was singular; and it had a closeness that no one at Wittenberg would ever fully comprehend. In his conversion to the Copernican theory—and in this case we can use the word with its religious and emotional overtones—the subjective meaning which the new ideas had for Rheticus, and his sense of sharing in their birth, would provide the emotional energy which fueled the arguments adduced in its favor.

---

[73] Burmeister, *Rhetikus*, Vol. III, p. 50.
[74] Stöffler had been succeeded by Philipp Imser.
[75] Burmeister, *Rhetikus*, Vol. III, p. 50.
[76] *Ibid.*, Vol. I, p. 175.
[77] Quoted in Leopold Friedrich Prowe, *Nicolaus Coppernicus*, Vol. I, Pt. 2 (Berlin: Weidmann, 1883-1884), p. 387; quoted and translated in Koestler, *Sleepwalkers*, pp. 189-190.

The *Narratio prima* itself, the little work written by Rheticus in 1540 as a kind of "trial balloon" for Copernicus, was not merely a "report" of Copernicus' ideas as the title seems to suggest. It contains ideas that were not found in *De revolutionibus* or in any of Copernicus' other writings,[78] it has at least one conceptual error born undoubtedly of excessive zeal,[79] and, most importantly, it gives selective emphasis to one Copernican argument in a way that goes well beyond the public position of its author. In short, as one reads the *Narratio prima* today, one gets a strikingly different perception of the Copernican theory from that reflected in the Wittenberg textbooks.

The work opens with a consideration of problems that were certainly of great interest to the Wittenbergers: the calendar, lunar motion, and the rejection of the equant. This is clearly the work of a man who has full command of the leading astronomical issues of his time. Yet there exists an unmistakable quality of enthusiasm in his writing that, perhaps not surprisingly, is distinctly absent from the university textbooks. Even more striking than its general tone, however, is the clear concentration upon one feature of the Copernican theory that had been consistently ignored by the Wittenberg astronomers: the demonstration of a system in the necessary interconnexity of the relative distances and periods of the planets. In the old hypotheses, Rheticus wrote, "there has not yet been established the common measure (*mensura communis*) whereby each sphere may be geometrically confined to its place" and where "they are all so arranged that no immense interval is left between one and the other."[80]

In these claims we find three assumptions about the planetary models: (1) that each planet is carried by a uniformly revolving sphere;[81] (2) that there are no gaps between the spheres (principle of plenitude); and (3) that the relative planetary positions are to be measured with respect to a common unit. Underlying these assumptions lay Copernicus' strong preference for unified explanations which he had invoked metaphorically when he criticized the Ptolemaic hypotheses on the grounds that they had not been able to demonstrate the "fixed proportionality" of the universe:

> With them, it is as though one were to gather the hands, feet, head and other members for one's images from diverse models, each excellently drawn, but not related to a single body, and since they in no way match each other, the result would be a monster rather than a man.[82]

---

[78] Particularly striking is his millennial prophecy based upon the motion of the sun's eccentric (*NP*, p. 122).

[79] As Curtis Wilson shows in a forthcoming article, Rheticus claimed erroneously that the Copernican account of precession could not be explained on the geostatic theory ("Rheticus, Ravetz and the 'Necessity' of Copernicus' Innovation," in *The Copernican Achievement*).

[80] *NP*, pp. 146, 147.

[81] Rheticus speaks *as though* the spheres were solid, although he nowhere calls them "material" or "solid." On the other hand, he nowhere says that they are merely an aid to the imagination, as does Melanchthon. His concern with the problem of plenitude and the force exerted by one sphere on another, however, suggests that he may have thought of them as solid (*cf.* esp. *NP*, p. 146).

[82] Nicholas Copernicus, *De revolutionibus orbium coelestium*, in *Nikolaus Kopernikus Gesamtausgabe*, Vol. II (Munich: R. Oldenbourg, 1949), p. 5.

The metaphor of the human body was only one of several with a long intellectual tradition in the West that expressed the architectonic unity of the parts in an integrated whole.[83] The human soul, the world, number, musical harmony, the well-proportioned dimensions of churches—each of these embodied the principle of unified, well-ordered wholeness. But now Copernicus had endowed this abstract value with powerful new concrete import. And Rheticus goes even further in his development of this theme. There are at least ten references to the harmony of the Copernican system in the *Narratio prima*, a number far greater than in *De revolutionibus* itself. Added to this, Rheticus ends his treatise with a conspicuously Platonic-Pythagorean passage on the harmony of the soul.[84] And it was probably his general belief in the necessity of the Copernican theory that led him to claim erroneously that the new account of precession could not be explained with the geostatic theory.[85]

None of Rheticus' Wittenberg colleagues had been struck by this deeper vision of the universe. None could write with the conviction of a man who personally had known the creator of a new intellectual system. Amplifying Copernicus' criticism of the Ptolemaic hypotheses and appealing to the criteria of unity and necessity, Rheticus wrote:

> . . . my teacher was especially influenced by the realization that the chief cause of all the uncertainty in astronomy was that the masters of this science (no offense is intended to divine Ptolemy, the father of astronomy) fashioned their theories and devices for correcting the motion of the heavenly bodies with too little regard for the rule which reminds us that the order and motions of the heavenly spheres agree in an absolute system. We fully grant these distinguished men their due honor, as we should. Nevertheless, we should have wished them, in establishing the harmony of the motions, to imitate the musicians who, when one string has either tightened or loosened, with great care and skill regulate and adjust the tones of all the other strings, until all together produce the desired harmony, and no dissonance is heard at all.[86]

The emotionally heightened Platonic-Pythagorean imagery of Rheticus' *Narratio prima*, fully infused with the rhetorical brilliance of the classical Renaissance style and with a sense that he had truly "seen the light" (a quality reminiscent of Paracelsus and Bruno), lies in stark contrast to the neutral tone of the Wittenberg circle. In a remarkable and revealing passage, where Rheticus explains his decision to adopt the Copernican theory, we see the link between the intellectual and the emotional basis for his conversion:

---

[83] For further discussion of this metaphor, see Westman, "Kepler's Theory of Hypothesis," pp. 249 ff.; also "Johannes Kepler's Adoption of the Copernican Hypothesis" (unpublished doctoral dissertation, University of Michigan, 1971), pp. 138-177; Leo Spitzer, *Classical and Christian Ideas of World Harmony*, ed. A. G. Hatcher (Baltimore: Johns Hopkins Press, 1963; first published in 1944-1945).

[84] With the exception of the gods, Rheticus says, only the human mind can understand harmony and number; diseased souls are cured by musical harmonies and happy republics are governed by rulers who possess harmonious souls. Earlier, he speaks of the connection between Copernicus' celestial harmony and the number six, the latter "honored beyond all others in the sacred prophecies of God and by the Pythagoreans and the other philosophers" (*NP*, pp. 196, 147).

[85] This is shown nicely by Wilson, "Rheticus, Ravetz," in *The Copernican Achievement*.

[86] *NP*, p. 138.

> I sincerely cherish Ptolemy and his followers equally with my teacher, since I have ever in mind and memory that sacred precept of Aristotle: "We must esteem both parties but follow the more accurate." And yet somehow I feel more inclined to the hypotheses of my teacher. This is so perhaps partly because I am persuaded that now at last I have a more accurate understanding of that delightful maxim which on account of its weightiness and truth is attributed to Plato: "God ever geometrizes";[87] but partly because in my teacher's revival of astronomy I see, as the saying is, with both eyes and as though a fog had lifted and the sky were now clear, the force of that wise statement of Socrates in the *Phaedrus:* "If I think any other man is able to see things that can naturally be collected into one and divided into many, him I follow after and 'walk in his footsteps as if he were a god.' "[88]

The statement attributed to Plato has clearly taken on a new and deeper significance for Rheticus within the context of the theory of Copernicus. For while the metaphor of an architectonic, geometrizing God had long been cited by philosophers, poets, and mathematicians, Copernicus had demonstrated it to possess new empirical content. By eliminating the earth's projected motion from the planetary models, the planets could be ordered continuously according to the criterion of their mean periodic motions; and once this had been done, the individual earth-sun-planet triangles could be compared with one another by the commonly shared earth-sun radius. If today we might defend the rationality of this argument on grounds of its empirical adequacy, its simplicity, and hence its considerable promise for future success, Rheticus went much further: he took it as evidence of the *absolute* truth of the entire theory of Copernicus. And the explanation for Rheticus' belief seems to lie in the *subjective* fact that Copernicus' act of unifying previously diverse fragments had a liberating, almost intoxicating effect upon him. In sharing in the birth of the theory of the great man, a veil that had covered an inner personal turmoil was lifted.

RHETICUS' CONFLICT: A POSSIBLE EXPLANATION

Let us pause for a moment and ask a new question: What was the nature and cause of Rheticus' inner conflict? Can we offer an adequate psychodynamic explanation for his behavior? Such questions of underlying motivation are seldom raised in the historiography of science, which has been dominated by a highly successful, but thoroughly intellectualist, methodology.[89] Yet, having broached the problem, one is faced immediately by an unfortunate paucity in the data with which historians of our period are all too familiar. A minimally satisfactory psychodynamic explanation, which assumes the importance of early childhood experiences and perceptions, of fantasies, unconscious wishes and

---

[87] As Rosen points out (*ibid.*, p. 168) this is not found in the Platonic corpus. The source is Plutarch's *Moralia: Questiones conviviales*, Bk. VIII, Ques. 2.

[88] *NP*, pp. 167–168.

[89] For a helpful survey of some issues in the psychohistory of science, see J. E. McGuire, "Newton and the Demonic Furies: Some Current Problems and Approaches in the History of Science," *History of Science*, 1973, 2:21–48.

dreams, must be able to draw, at the very least, upon sufficient information about significant *patterns* in an individual's early relationships with others so as to establish accurate *parallels* with later behavior. Conflicts which remain unresolved in early years will tend to reappear later on in different guises, and the historian's skillful identification of repetitive patterns can help to explain the cause of later actions which seem truly irrational, that is, inappropriate, exaggerated, unexpected, and even destructive in a particular individual. In addition to the general models of relating to mother, father, brothers, and sisters, *specific* events of traumatic severity can sometimes intensify a low-lying conflict or perhaps initiate a new one. The response to the devastating experiences of rape, warfare, or the observation of a violent death, for example, will depend upon factors such as age, basic ego strength, and external support, but the chances are that these events will be upsetting even for a normal individual. And yet, even with good evidence, the historian must be content with probable and intuitive conjectures, since the clinical procedure of testing against free associations in the present is not open to him.[90] Still, if the historian is at a disadvantage in his lack of immediate clinical evidence, he does possess one source of considerable value, as Frank Manuel has pointed out: "Historians . . . have a completed life before them, and the end always tells much about the beginnings."[91]

Now, in the case of Rheticus, we have virtually no information on the general patterns of his early familial relationships, but we do have evidence of one very important event in his life: the death of his father. And from this small piece of evidence we may form a possible account of what might have occurred in Rheticus' youth.

In 1528, when Rheticus was fourteen and had therefore just reached adolescence, his father, Georg Iserin, a doctor in the town of Feldkirch, Austria, was convicted on a charge of sorcery and beheaded.[92] Witch trials were not uncommon during this period.[93] Whatever had been the specific crime of Rheticus' father, however, the effects of his anti-social behavior (suspected or real) must have been difficult for the family. At least one painful result was that the alleged sorcerer's name could no longer be employed legally after his execution, and hence the unfortunate widow had to revert to her Italian maiden name, De Porris, and her son, Georg, later took on the additional name of Rheticus because he had been born within the old boundaries of the Roman province of Rhaetia.[94]

Georg Joachim Iserin, Jr. had lost not only his father but also part of his

---

[90] On the nature of psychoanalytic validation, see J. O. Wisdom, "Testing an Interpretation within a Session," and Clark Glymour, "Freud, Kepler, and the Clinical Evidence," in *Freud: A Collection of Critical Essays*, ed. Richard Wollheim (New York: Doubleday Anchor, 1974; first published in 1967), pp. 332–348, 285–304.

[91] *A Portrait of Isaac Newton* (Cambridge, Mass.: Harvard University Press, 1968), p. 5.

[92] Burmeister, *Rhetikus*, Vol. I, pp. 14–17.

[93] See A. D. J. MacFarlane, *Witchcraft in Tudor and Stuart England* (New York: Harper and Row, 1970) and H. C. Erik Midelfort, *Witch Hunting in Southwestern Germany, 1562–1684* (Stanford: Stanford University Press, 1972).

[94] See Edward Rosen, review of Burmeister, *Isis*, 1968, 59:231.

old identity. What must it have been like? We do not possess Rheticus' direct testimony, but we can imagine what a painful experience it must have been for the lad to have lost the man whom all young boys so greatly admire and fear. Indeed, we would predict that his later relationships with men, particularly with older men, would be marked by a deepened and intensified ambivalence. On the one hand, the manifest horror and grief over the power of aggression which had led to the violent death of his father would later appear in his need to atone for even the vaguest of hostile feelings. Together with this fear of his own aggression, we should expect to find determined efforts—in the search for wholeness, strength, and harmony—to unconsciously repair the damage earlier wrought on his father. On the other hand, there must have been an unconscious sense of liberation—at last he was freed from the tyranny of the old man—a feeling which was fully consonant with his later identification with intellectual rebellion in the persons of Copernicus and Paracelsus.[95]

Let us now see if we can find some support or refutation for these hypotheses in Rheticus' adult writings. We find evidence of ambivalence in his relationship with Johannes Schöner. The *Narratio prima* starts with the following dedication: "To the Illustrious John Schöner, as to his own revered father, G. Joachim Rheticus sends his greetings."[96] The motto on the title page is, however, of an entirely different tenor. It is really a sort of revolutionary manifesto embodied in a Greek quotation from Alcinous, which reads: "Free in mind must be he who desires to have understanding."[97] Reverence and rejection of authority: the ambivalence is announced at the beginning of the work and the themes reappear clearly in the text itself. Thus, he says to Schöner:

> Most illustrious and most learned Schöner, whom I shall always revere like a father, it now remains for you to receive this work of mine, such as it is, kindly and favorably. . . . If I have said anything with youthful enthusiasm (we young men are always endowed, as he says, with high rather than useful spirit) or inadvertently let fall any remark which may seem directed against venerable and sacred antiquity more boldly than perhaps the importance and dignity of the subject demanded, you surely, I have no doubt, will put a kindly construction upon the matter and will bear in mind my feeling toward you rather than my fault.

The tone is cautious, apologetic, and deferential, as if to implore his father unconsciously not to beat him for his aggressive sentiments. In the next sentence, we learn that Copernicus was also very respectful of ancient authority:

> . . . concerning my learned teacher I should like you to hold the opinion and be fully convinced that for him there is nothing better or more important than

---

[95] Rheticus mentions that he had met Paracelsus in 1532 and was impressed by him. It is interesting that he did not then decide on a career in medicine, a vocational choice that would have identified him directly with his father. But in 1554, at the age of forty and after his travels had ended in effect, Rheticus became a practicing physician in Cracow and his interest in Paracelsus revived. *Cf.* Burmeister, *Rhetikus,* Vol. I, pp. 35, 152–155.
[96] *NP*, p. 109.
[97] *Ibid.*, p. 108. A facsimile of the title page is printed in Burmeister, *Rhetikus,* Vol. II, pp. 58–59.

walking in the footsteps of Ptolemy and following, as Ptolemy did, the ancients and those who were much earlier than himself.[98]

Rheticus then describes the rebellious, innovative act of Copernicus almost as though it were beyond his control but ends with the quotation from Alcinous that had already appeared on the title page.

> However, when he [Copernicus] became aware that the phenomena, which control the astronomer, and mathematics compelled him to make certain assumptions even against his wishes, it was enough, he thought, if he aimed his arrows by the same method to the same target as Ptolemy, even though he employed a bow and arrow of a far different type of material from Ptolemy's. At this point, we should recall the saying: "Free in mind must he be who desires to have understanding."[99]

The imagery of the bow and arrow, symbolizing the new and different theoretical assumptions used by Copernicus in his attack on the phenomena, suggests at least part of the appeal which Copernicus' work held for the young Rheticus, for he himself had shared in the attack as well by influencing Copernicus to publish his new theory after so many years. In Copernicus, Rheticus had found a kind and strong father with a streak of youthful rebellion in him: a man who was different, as Rheticus' father had been; a father who could attack ancient authority with his intellectual weapons without himself being destroyed; a father who, *like the system he created*, had a head and a heart which were connected to the same body. With such a man Rheticus was not only eager to identify, but to idolize and deify.[100]

It is unfortunate that we do not have a detailed account of what transpired upon Rheticus' return to Wittenberg in September 1541. Burmeister conjectures that Rheticus simply did not stay in Wittenberg long enough for a "Copernicus school" to get off the ground.[101] While there is of course no way to know what might have happened had Rheticus stayed longer, it is relevant to note that such a group of followers did not form about him either in Leipzig or in Cracow; nor did either the *Narratio prima* or *De revolutionibus* itself have the effect of producing Rheticus-like conversions among those who remained at Wittenberg in the 1540s, 1550s, and 1560s. Although several students were present at Rheticus' lectures in 1541, after his return from Frombork and before his departure for Leipzig, none of these (Burmeister lists among them Peucer, Schreiber, Stoius, Homelius, Heller, Lauterwalt, Staphylus, and Acontius)[102] were to adopt a strong realist interpretation of the Copernican

---

[98] *NP*, p. 186.
[99] *Ibid.*, pp. 186–187.
[100] In 1557 Rheticus wrote of Copernicus, "whom I cultivated not only as a teacher, but as a father" (*quem non solum tanquam praeceptorem, sed ut patrem colui*). This letter (Rheticus to King Ferdinand I in Burmeister, *Rhetikus*, Vol. III, p. 139), which shows no animosity toward Copernicus, would seem to weigh against Koestler's interesting thesis that Rheticus felt betrayed by Copernicus when the latter failed to mention him in his great work.
[101] *Ibid.*, Vol. I, p. 72.
[102] *Ibid.*, pp. 71–72.

theory. Indeed, surviving annotated copies of *De revolutionibus* which I have been able to identify as belonging at one time to the first five mentioned above, as well as to Reinhold, show that in general Books III–VI were studied with some care while Book I apparently did not excite any profound interest.[103] Two exceptions were Rheticus' close friend the humanist physician and polymath Achilles Pirmin Gasser (1505–1577) and the Constanz physician Georg Vögelin, both of whom evidently accepted the Copernican theory as true but did not treat it as a program for further scientific research.[104] It is an interesting irony that, apart from Rheticus, those who did *not* accept the new theory as a true description of reality did more to articulate its mathematical structure than those who did so accept it.

From the fervor of Rheticus' writings and, by contrast, from the apparent pallidness of its reception, it is clear that others did not appreciate the nature of the personal experience which he had undergone. Still, this is perhaps understandable. We can empathize with a person who has undergone a terrible experience, yet few of us are conscious of the emotional scars that others have carried secretly with them since childhood. When Rheticus came back from Frombork he returned not only to report to Melanchthon, as Hans Blumenberg has written, but he returned full of fire, a believer and a convert.[105] As we might say, he came back "overheated." It is no wonder that Rheticus' great enthusiasm was not shared by the kindly and paternal Melanchthon, although his confidence in the young man's abilities continued unabated.[106] He could have had little way of knowing why the experience for Rheticus had been so relieving, nor was he predisposed to go against tradition and the status quo of which he was such an important part. As Melanchthon wrote to his friend Camerarius on July 25, 1542:

> I have been indulgent toward the age of our Rheticus in order that his disposition, which has been incited by a certain enthusiasm (*quodam Enthusiasmo*), as it were, might be moved toward that part of philosophy in which he is conversant. But at various times, I have said to myself that I desire in him a little more of the Socratic philosophy, which he is likely to acquire when he is the father of a family.[107]

But the Socrates whom Rheticus followed was the God-like Copernicus, the man who had lifted the fog for him and shot his arrows true to the target—not the quiet, moderate Socrates of Melanchthon. In the end, the vision of Rheticus was not censured; it was merely ignored and repressed from public consciousness.

---

[103] The locations of these copies are as follows: Peucer, Observatoire de Paris; Heller, Universitätsbibliothek Rostock; Schreiber (and Kepler), Universitätsbibliothek Leipzig; Stoius, Universitetsbiblioteket Copenhagen; Homelius (and Praetorius), Yale University Library; Reinhold, Crawford Library, Royal Observatory of Edinburgh. All editions are 1543.

[104] On Gasser and Vögelin, see Karl Heinz Burmeister, *Achilles Pirmin Gasser (1505–1577): Arzt und Naturforscher, Historiker und Humanist*, Vol. I (Wiesbaden: Guido Pressler, 1970), p. 72–80.

[105] Blumenberg, *Die kopernikanische Wende*, p. 109.

[106] He was elected to the deanship of the faculty of arts upon his return, and Melanchthon later recommended him highly for other positions. *Cf. CR* 4, p. 812; Burmeister, *Rhetikus*, Vol. I, pp. 67 ff.

[107] *CR* 4, p. 847.

CONCLUSIONS

Let us now attempt to formulate some conclusions which extend our specific claims to more general considerations. Broadly speaking, this paper has addressed itself to the problem of the choice between competing scientific theories from three basic perspectives: historical, philosophical, and psychosocial.

First, we have tried to demonstrate the historical fact that the University of Wittenberg, by far the most important mid-sixteenth-century German university, in no way supports the view of scholars such as Stillman Drake who say that "the doors of sixteenth-century universities were closed to new scientific ideas from the outside."[108] While one would not wish to describe the sixteenth-century universities as hotbeds of scientific revolutionary activity, the opposite view is equally misleading. Recognition of the great range of intermediate responses to innovation helps to make the historical landscape less rigidly compartmentalized.

A second important conclusion is closely related to the first. One needs to exercise great caution in defining just what it is that scientists are choosing between when they make what philosophers call a "theory choice." For it is clear from this study that a scientist's perception of a theory will affect the decisions that he makes about it. A beginning student at Wittenberg in the 1550s, for example, would have a rather misleading conception of the Copernican theory if he saw it only as the revival of Aristarchus' schematic picture of the universe; yet many nonastronomers probably perceived it in this manner. The master's candidate, like his professors, admired the theory of Copernicus for what it had to offer by way of new parameters and alternative models that might be translated into a geostatic framework. This helps to explain why commonly employed dichotomies, such as "pro-Copernican" and "anti-Copernican," "Copernican paradigm" and "Ptolemaic paradigm," fail to provide an accurate description of the choice situation in the earliest phase of the theory's reception. Theory or paradigm choice simply was not seen as a major problem by the majority of astronomers who constituted the circle of Melanchthon. They did not treat the Copernican theory as though it were a paradigm but merely made use of those parts of it with which they were in agreement.

If we were to look at the early reception of Copernican astronomy at Wittenberg through Kuhnian spectacles, therefore, we should have to make the paradoxical statement that it had been welcomed respectfully into the fold of Ptolemaic normal science—a situation which should never occur by Kuhn's reckoning. Kuhn himself actually gives different assessments of the historical situation depending upon whether he writes as an historian or as a philosopher. In the *Copernican Revolution*, when he is working with his historical spectacles, he is less apt to dichotomize, although the analysis is not always satisfactory. Hence, he can write of the reception of Reinhold's *Prutenic Tables:* "Every

---

[108] "Early Science and the Printed Book: The Spread of Science Beyond the Universities," *Renaissance and Reformation Journal*, 1970, 6:49.

man who used the *Prutenic Tables* was at least acquiescing in an implicit Copernicanism."[109] Here the notion of an "implicit Copernicanism" suggests a transitional process, even if the interpretation itself is misleading, since use of the *Prutenic Tables* in no way committed one to the paradigmatic features of the Copernican theory. On the other hand, Kuhn's philosophical lenses in *The Structure of Scientific Revolutions* tend to blind him somewhat to the intermediate steps between paradigms. Here he writes: "Copernicus' innovation was not simply to move the earth. Rather, it was a whole new way of regarding the problems of physics and astronomy, one that necessarily changed the meaning of both 'earth' and 'motion.' "[110] Kuhn's reading of Copernicus' innovation is actually a later, more fully rationalized interpretation, for even Copernicus himself barely glimpsed the new meaning of "earth" and "motion" implied by his theory; and "the whole new way of regarding the problems of physics and astronomy" was itself subject to the later interpretations of men such as Kepler, Galileo, and Descartes.

In short, we might venture to suggest that since the full implications of a new theory are never evident from the outset, attitudes toward old and new programs of research which develop early in the game will *always* be conditioned by some interpretation. And, as in the case of the Wittenberg Interpretation, the reading may be so conservative that while it may have the positive effect of increasing the exposure of parts of the theory to discussion and further use, it will do so at the price of ignoring its most fundamental presuppositions.

This brings us to a final general conclusion. It concerns the role of scientific institutions in affecting the choices between research programs. Social units control their members by a variety of sanctions.[111] As Warren O. Hagstrom has shown, the reward system in scientific groups may be viewed as analogous with gift giving in certain primitive societies, a system governed by reciprocity in which social recognition is exchanged for information.[112] Conformity with methodological standards in such groups can be enforced by the *withholding* of recognition. Now, when scientists disagree with one another, they do not usually lose their jobs (unlike theologians discussing the Lord's Supper in the sixteenth century), except perhaps under the extreme conditions of a Lysenko affair.[113] But they do run the risk of forfeiting full recognition for their contributions. By never challenging the paternalistic, familial framework established by Melanchthon, Reinhold and Peucer helped to institutionalize the moderate, pragmatic interpretation of the Copernican theory advocated by the *Praeceptor Germaniae*. The case of Rheticus suggests that recognition of his discovery of the cosmological power of the new theory was consciously or unconsciously withheld because he had deviated from the methodological

---

[109] *Copernican Revolution*, p. 188.
[110] *Structure*, pp. 149–150.
[111] *Cf.* Amitai Etzioni, *Modern Organizations* (Englewood Cliffs, N.J.: Prentice Hall, 1964), p. 58.
[112] Warren O. Hagstrom, *The Scientific Community* (New York: Basic Books, 1965), pp. 12–22.
[113] David Joravsky, *The Lysenko Affair* (Cambridge, Mass.: Harvard University Press, 1970).

consensus and because his identification with the revolutionary side of Copernicus may have seemed threatening or at least a childish enthusiasm which ought not be accorded serious consideration.

While it has not been our basic purpose here to evaluate the rationality of Rheticus' decision to become a Copernican in the context of recent theories of rationality, we are somewhat closer to the Kuhnian account of "conversion." Kuhn allows that the factors which actually motivate an individual's shift from one research program to another are not necessarily rational; and he is quite correct when he writes, "I do not expect that, merely because my arguments are logical, they will be compelling."[114] As historians we need to be able to explain why the same arguments will be evaluated in diverse ways by contemporaries no matter how we might judge their rationality from the advantage of retrospect. We have tried to argue that one important component in Rheticus' assessment of the Copernican innovation was unconscious: the overwhelming wish to put the pieces of his murdered father back together again, to restore wholeness and unity; and this wish was manifested on the intellectual level by his strong identification with a man who had no son and whose theory restored the fragmentation of the Ptolemaic monster into a whole and complete organism. If our psychohistorical reconstruction is correct, then it helps to explain in part why Rheticus was so strongly persuaded by the argument from harmony, and it also helps to explain why he was not drawn into the Wittenberg consensus. Indeed, one wonders how often correct choices—correct, that is, in light of a later theory of rationality—are made for reasons which are *felt* but not consciously understood, and for reasons which become much more comprehensible to us in light of a psychoanalytic theory of irrationality. As our study shows, there is a place for both conscious and unconscious factors in the history of science.

---

[114] Thomas S. Kuhn, "Reflections on My Critics," in *Criticism and the Growth of Knowledge*, p. 261. Although I agree with Kuhn that the decision to accept or not to accept a new theory may be motivated by nonrational considerations, it should be obvious that I do not agree with his description of scientific revolution as one which involves incommensurable conceptual worlds.

Laboratory Design and the Aim of Science

Andreas Libavius versus Tycho Brahe

By Owen Hannaway

THE HISTORY OF THE LABORATORY is an important but neglected aspect of early modern science. To ask where and when the laboratory first appeared as a distinctive place for the pursuit of the study of nature raises questions to which, as yet, there are no clear-cut answers. Even the etymology of the word *laboratory* is obscure: the Latin noun *laboratorium,* from which the vernacular cognates are derived, was certainly not of classical origin. There is evidence of medieval usage, but it appears that it was not until the late sixteenth century that the word took on something like its modern meaning. Indications are that the laboratory was at first linked exclusively with alchemy and chemistry; only gradually, it seems, was the term extended to describe all those distinctive places where the manipulative investigation of natural phenomena was carried out.[1] Certainly any broad investigation of the development of the laboratory in early modern science must include not only the chemistry laboratory but also the anatomy theater, the cabinet of curiosities, the botanical garden, and the astronomical observatory.

The etymological novelty of the word *laboratorium,* together with the laboratory's early associations with alchemy and chemistry, points to a deeper significance of the topic. The appearance of the laboratory is indicative of a new mode of scientific inquiry, one that involves the observation and manipulation of nature by means of specialized instruments, techniques, and apparatuses that require manual skills as well as conceptual knowledge for their construction and deployment. Not that the science of antiquity, or that of the Middle Ages, was

---

A version of this paper was presented at the conference entitled "Widening Perspectives in Renaissance Language," organized by Hugh Ormsby-Lennon and held at the Newberry Library, Chicago, in 1981. I thank Christoph Meinel, Bruce Moran, and Robert S. Westman for helpful comments on the present version.

[1] There is no entry in any of the standard dictionaries of classical Latin for *laboratorium*. There are two citations of medieval usage in Charles Du Cange, *Glossarium mediae et infimae Latinitatis,* new ed. enlarged by Leopold Favre (Paris: Librairie des Sciences et des Arts, 1938): the earlier (1347) means simply a task or labor; the second (1451) occurs in a monastic context and appears to mean a workshop. This latter usage is intriguing, since it suggests a parallel with *scriptorium* and *dormitorium* (a word that also occurs in the sentence cited in Du Cange, but I do not know of any supporting evidence, and the usual word for a monastic workshop is *officina.* A possible monastic link to the topic of this paper is suggested by the tantalizing but inconclusive thesis of J. R. Christianson, "Cloister and Observatory. Herrevad Abbey and Tycho Brahe's Uraniborg," University of Minnesota, Ph.D. 1964 (University Microfilms, Inc. Ann Arbor, Mich. 66–1685).

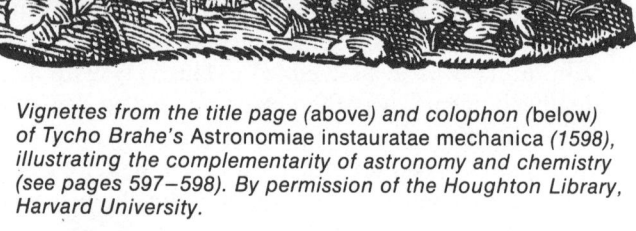

Vignettes from the title page (above) and colophon (below) of Tycho Brahe's Astronomiae instauratae mechanica *(1598), illustrating the complementarity of astronomy and chemistry (see pages 597–598). By permission of the Houghton Library, Harvard University.*

devoid of systematic observation or even instruments; but the emphasis placed upon this kind of activity was one of the hallmarks of the new science that emerged in the sixteenth and seventeenth centuries. With this emphasis there came a shift in the meaning of science itself: science no longer was simply a *kind of knowledge* (one possessed *scientia*); it increasingly became a *form of activity* (one did science). That there should have arisen in this period a place specially set aside for such activity and bearing a new name serves to measure the force of that shift.[2]

In order to illuminate some of the issues associated with the rise of laboratories, I propose to compare two of the most detailed sets of plans for laboratories that date from the early modern period. One is that of Uraniborg, Tycho Brahe's great castle-observatory that also housed a chemical laboratory, which the Danish astronomer described in a book about his astronomical instruments published in 1598[3]; the other is the Chemical House and laboratory that Andreas Libavius delineated in the first part of his commentaries to the second edition of his textbook of chemistry, which appeared in 1606.[4]

A number of facts concerning these plans should be made clear at the outset. In the first place, what are depicted are not simply laboratories, but whole dwelling places within which certain areas are designated for the practice of science; thus we are afforded a broader social context within which to assess the novel element of the laboratory. Second, whereas Uraniborg was an actual physical structure in which Brahe lived and worked between 1576 and 1597, Libavius's plans for the Chemical House were never realized and probably were never intended to be. While Libavius certainly carried out chemical experiments, possibly in a house rather like the one he describes, his plans are idealized. Third, the comparisons afforded by these descriptions and depictions are not fortuitous: just as Brahe had included an account of Uraniborg in that text where he described his astronomical instruments, so too did Libavius provide plans for an appropriate structure in which to house the chemist and his laboratory as a supplement to his description of chemical apparatus. More than this: I intend to show that Libavius tendered his plans as an explicit criticism of the meaning and

---

[2] Laboratories do not figure prominently in the historiography of early modern science, even in the literature on a number of well-established themes that might seem to be related, such as the history of scientific instruments, the history of experiment, the interaction of scholars and craftsmen in early modern science, and the history of scientific societies. In terms of the last category, an interesting description of an early laboratory (which may be idealized) is given in William Eamon and Françoise Paheau, "The Accademia Segreta of Girolamo Ruscelli: A Sixteenth-Century Italian Scientific Society," *Isis*, 1984, 75:327–342, esp. pp. 340–342.

[3] Tycho Brahe, *Astronomiae instauratae mechanica* (Wandsbek, 1598). A second, less elegant edition of this beautiful book appeared at Nuremberg in 1602, after the author's death. References in this paper are to the more readily available version in *Tychonis Brahe Dani Opera omnia*, ed J. L. E. Dreyer, H. Raeder, and E. Nystrom (hereafter **Brahe, *Opera***), 15 vols. (Copenhagen: Gyldendal, 1913–1929), Vol. V (1923), pp. 1–162. I have also consulted the English translation in *Tycho Brahe's Description of His Instruments and Scientific Work as given in* Astronomiae instauratae mechanica (*Wandsburgi, 1598*), trans. and ed. Hans Raeder, Eis Strömgren, and Bengt Strömgren (Copenhagen: Munksgaard, 1946).

[4] *Commentariorum alchymiae Andreae Libavii Med. D. pars prima sex libris declarata* (hereafter **Libavius, *Commentariorum pars prima***) (Frankfurt, 1606). The description of the chemical house and laboratory is in Bk. I, pp. 92–99. This work is the first of a two-part set of commentaries to the second edition of Libavius's *Alchymia* (Frankfurt, 1606), which first appeared under the title *Alchemia* in 1597. The best guide to the complicated bibliography of these works is J. R. Partington, *A History of Chemistry*, 4 vols. (London: Macmillan, 1961–1970), Vol. II, pp. 247–250.

style of science that Uraniborg exemplified. All of these facts aid my purpose, for my aim is not to compare the function and aesthetics of these buildings from an architectural point of view but to use the plans as a guide to explore the intellectual and ideological roots of a new mode of scientific life that arose for the first time in Western culture around the turn of the sixteenth century.[5]

Andreas Libavius, the instigator of this architectural polemic, was an unusual chemist. A Saxon physician and pedagogue, he was a product of late sixteenth-century Lutheran scholastic humanism. Much of his early career was spent in civic service at Rothenburg on the Tauber, where he was municipal physician and inspector of schools. He subsequently moved to Coburg, where he served as rector of the Academic Gymnasium from 1607 until his death in 1616. As befitted his intellectual training and station in life, Libavius was an indefatigable opponent of all enthusiasts, whether medical, pedagogical, or religious, and he defended what he regarded as rational orthodoxy in each of these domains in voluminous works that regularly fed the printing presses of Frankfurt.[6]

Many of Libavius's most forceful antienthusiastic polemics were directed against the followers of Paracelsus. The Paracelsians' claims to special illumination in the light of nature and of grace, together with their celebration of the intuitive wisdom of the peasant and the artisan, challenged the foundations of Libavius's professional world in education and medicine. He felt increasingly threatened when, toward the end of the sixteenth century, Paracelsianism, together with hermeticism, cabalism, and other forms of Renaissance magic and religious syncretism, penetrated some of the Protestant princely courts of Germany and—even more ominously, from Libavius's point of view—the imperial court at Prague during the reign of Rudolf II. He tended to see behind these developments a crypto-Calvinist conspiracy, posing as a benign irenicism, which threatened the political stability of the Holy Roman Empire, the survival of orthodox Lutheranism, and the humanistic tradition represented by Melanchthonian Aristotelianism. In Libavius's mind courts, cabals, and the chemical philosophy were the enemies.[7]

These strident polemics, however, represent just the negative force of the Libavian critique of Paracelsianism. The more constructive aspect is represented by his textbook on chemistry, the *Alchymia*. In this work Libavius sought to

---

[5] Whether Libavius actually had a laboratory in any of his houses is unclear. Włodzimierz Hubicki, who had an extensive knowledge of the manuscript sources relating to Libavius, has stated categorically that he had a laboratory in his houses at Rothenburg and Coburg; see Hubicki, "Andreas Libavius," *Dictionary of Scientific Biography*, ed. Charles C. Gillispie, 16 vols. (New York: Scribners, 1970–1980), Vol. VIII, pp. 309–312, on p. 311. But Bruce Moran, in a personal communication, has drawn my attention to a letter from Jacob Mosanus to Joseph Duchesne (Quercetanus), dated 22 Sept. 1606. It describes a recent two-month visit with Libavius in Coburg, undertaken at the behest of the Landgrave Moritz of Hessen-Kassel, to investigate Libavius's alchemical practice; Mosanus says Libavius did not have a furnace in his house. However, Libavius had just moved to Coburg at the time of the visit, and he may not yet have had the opportunity to set up his laboratory. The letter is referred to in Moran, "Privilege, Communication, and Chemistry: The Hermetic-Alchemical Circle of Moritz of Hessen-Kassel," *Ambix*, 1985, 32:110–126, on p. 113.

[6] On Libavius's life, see Hubicki, "Libavius." The cultural context of Libavius's life and work is discussed in Owen Hannaway, *The Chemists and the Word: The Didactic Origins of Chemistry* (Baltimore: Johns Hopkins Univ. Press, 1975), pp. 75–151.

[7] See Hannaway, *Chemists and the Word*, pp. 75–116. For court Paracelsianism on the continent of Europe, see R. J. W. Evans, *Rudolf II and His World: A Study in Intellectual History, 1576–1612* (Oxford: Clarendon Press, 1973), esp. pp. 196–242; and Moran, "Privilege, Communication, and Chemistry" (cit. n. 5).

wrest the initiative from the Paracelsians by exhaustively articulating chemical technique and practice: he undercut their claims to unique illumination by ranging encyclopedically over the literature on the chemical arts; and he rescued prescription from the secretiveness of the adept by bringing chemistry into the light of day and placing it within a historical tradition of established discourse. From the accumulated texts on the chemical arts, which, like rhetoric, were intended to move men to action, Libavius "invented" a dialectic of chemical technique and recipe that was meant to convince them intellectually.[8]

The plans for the Chemical House, which Libavius sets out in the first part of the commentary on the *Alchymia,* belong to his exposition of chemistry as art and science, and to his polemics. As a preface, so to speak, to the detailed treatment of furnaces and apparatus, Libavius describes an appropriate place in which to house this equipment and carry out the operations of chemistry—namely, a laboratory. This laboratory, in turn, is located in a house where the chemist lives as well as works. That matters are not quite so simple as this, however, is quickly made clear in Libavius's text. The far-from-subtle references to the luxury and splendor of Tycho Brahe's Uraniborg and the ironic references to mystical symbols that could govern the structural design of both buildings announce that the reader is about to be swept up into one of Libavius's many polemical exercises.[9]

In contrast to Libavius, Tycho Brahe (1546–1601) scarcely needs introduction. Flanked by Copernicus and Kepler, he is the middle figure in the great trinity of early modern mathematical astronomy. He is best remembered for his observations of the new star of 1572, which raised doubts about the immutability of the Aristotelian heavens; for his observations of the comet of 1577, which called in question the existence of the crystalline spheres as the moving agents of the planets; and as the author of the Tychonic system, that arrangement of planetary orbits that preserved geocentricity while permitting the other planets to circle the sun as it orbited the earth. But to present Brahe's accomplishments in these purely conceptual terms is to mask the true character of his life's work. This was devoted to the systematic renewal of the observational basis of mathematical astronomy. He had begun in 1563 when, as a sixteen-year-old student, he made his first observations of a conjunction of Jupiter and Saturn with a pair of compasses; and he continued to chart the heavens by increasingly complex and resourceful means until his death in 1601. In that time he transformed the instrumentation and accuracy of observational astronomy. Within a decade of his death the dramatic novelty of Galileo's telescope would overshadow Brahe's improvements in instruments for the naked eye; but Kepler's triumphs in mathematical astronomy owed little to the Pisan's telescope and almost everything to measurements made by means of the Dane's sextants, quadrants, parallactic rulers, and armillary spheres. These instruments, together with the castle-observatory, Uraniborg, were described in the *Astronomiae instauratae mechanica.*[10]

---

[8] Hannaway, *Chemists and the Word,* pp. 117–151.

[9] Libavius, *Commentariorum pars prima* (cit. n. 4), Bk. I, pp. 92–93. Libavius states that it is not necessary to give a mystical interpretation of the geometrical design of Uraniborg, in terms either of the shape of the great and the little world or of the hieroglyphic monad; it is sufficient to say that it is the best shape to admit the rays of the sun and the stars. He adds, in ironic vein, that he has not disdained Platonic "roundness" or ignored the triangles of the Homeric chain in the design of his own building (p. 93).

[10] The principal accounts of Brahe's life and work are J. L. E. Dreyer, *Tycho Brahe: A Picture of*

Just as Libavius was an unlikely chemist, so too was Brahe a most improbable astronomer. Brahe—or the Lord of Knudstrup, to give him his ancestral title—was a member of the feudal nobility of Denmark. In 1576 King Frederick II of Denmark granted Tycho the island of Hven for the specific purpose of building an observatory, Uraniborg, in which to pursue his astronomical studies.[11] Here Brahe lived and worked until 1597, when, following the accession of the young King Christian IV into his majority rule, relations with the crown so deteriorated that Brahe abandoned Hven and Denmark in search of a new location and new patronage for his instruments and work.[12] The first edition (1598) of the *Astronomiae instauratae mechanica* was an integral part of this relocation project. It was printed on Brahe's own press, brought from Uraniborg, and issued from the castle of the Holstein nobleman Heinrich von Rantzau at Wandsbek, near Hamburg, where Brahe was temporarily installed. This work contains detailed descriptions of the castle-observatory on Hven, together with the instruments it contained, and includes an account of Brahe's accomplishments in astronomy up to that time. Most of the extant copies are known to have been presentation copies to monarchs and important court figures: many have hand-colored plates, and a few are sumptuously bound in silk or velvet. The work was dedicated to the Emperor Rudolf II, who was a primary target of Brahe's overtures. These proved successful, in the end, when Brahe was invited to Prague by the emperor and granted the castle at Benatky as a site at which to reestablish his observatory and laboratory. Clearly this was no ordinary book—indeed, it may be unique in the annals of printing: it was the work of a nobleman of considerable means who enjoyed full control over its physical production; and it was tendered to an emperor, not in expectation of some lowly court position, but for nothing less than a feudal demesne.[13]

Scientific Life and Work in the Sixteenth Century (Edinburgh: Adam & Charles Black, 1890; New York: Dover, 1963); and Wilhelm Norlind, *Tycho Brahe: En levnadsteckning med nya bidrag belysande hans liv och verk* (Lund: Gleerup, 1970). See also Victor E. Thoren, "Tycho Brahe: Past and Future Research," *History of Science*, 1973, *11*:270–282.

[11] Brahe was a nobleman whose relationship to the crown was feudal; he was not a court figure and resisted becoming one. Although Brahe was given an outright sum of money for the construction of his residence on Hven, the grant of the island itself was in the form of a feudal deed, and the crown subsequently supported his astronomy by means of the award of various fiefdoms throughout Scandinavia. On the crown's support of Brahe, see J. R. Christianson, "Tycho Brahe and Patronage of Science, 1576–1597," *American Philosophical Society Year Book, 1972* (Philadelphia: American Philosophical Society, 1973), pp. 572–573 (a preliminary report); and Christianson, "Tycho Brahe's German Treatise on the Comet of 1577: A Study in Science and Politics," *Isis*, 1979, *70*:110–140, esp. pp. 117–118. See also n. 17 below.

[12] The circumstances surrounding Brahe's departure from Denmark are discussed in Dreyer, *Brahe*, pp. 214–276, and Norlind, *Brahe*, pp. 240–266 (both cit. n. 10). It seems important to locate this event in the growing power struggle between the crown and the Danish nobility, which came to a head in the reign of Christian IV and led to the abolition of the feudal nobility in Denmark in 1660. For an introduction to the issues involved, see Knud J. V. Jespersen, "Social Change and Military Revolution in Early Modern Europe: Some Danish Evidence," *Historical Journal*, 1983, *26*:1–13.

[13] A listing and description of the approximately forty extant copies, including dedications, is given in Norlind, *Brahe* (cit. n. 10), pp. 266–293. This builds on the earlier, but still useful, catalogue of B. Hasselberg, "Einige Bemerkungen über Tycho Brahes *Astronomiae instauratae Mechanica Wandesburgi 1598*," *Vierteljahrsschrift der Astronomischen Gesellschaft*, 1903, *38*:180–187, supplemented by various reports of additional copies cited by Hasselberg and others in "Weitere Exemplare von Tycho Brahes *Mechanica*," *Zentralblatt für Bibliothekswesen*, 1904, *21*:396–403. Among the recipients of presentation copies were Ferdinand de Medici, Grand Duke of Tuscany, Duke Frederick William of Saxony, Prince Maurice of Orange, the Archbishop of Salzburg, and Duke Heinrich Julius of Braunschweig. Rudolf II's copy was personally presented to him by Brahe's eldest son in January 1598; see Dreyer, *Brahe* (cit. n. 10), p. 265. In the negotiations with the emperor it is significant that,

I do not know whether Libavius ever handled one of the fine exemplars of the first edition of Brahe's *Astronomiae instauratae mechanica* or whether he saw it only in the more fustian garb in which it was put out in a second edition at Nuremberg in 1602, one year after Brahe's death. The format of this later trade edition, slightly smaller and less carefully printed, was certainly closer to the kind of book production that Libavius customarily dealt in. The commentary to the *Alchymia,* which contains Libavius's description of the Chemical House and its apparatus, is a large enough folio; but the cramped double columns and the thin paper betray its humble origins in a commercial printing house in Frankfurt. The contrast in the outward appearance of the two books carries over into their contents, and Libavius quickly reveals himself to be acutely conscious of the social distance that separates him and his book from the Danish nobleman and his works. In introducing his plans for the chemist's house and laboratory he explicitly contrasts his circumstances with those of Brahe: acknowledging that he lacks both the wealth and the patronage that would allow him to hire architects and realize their designs in stone and wood, he nevertheless proposes a plan, or an "idea," of an appropriate dwelling and workplace for a chemist.[14] The almost comic incongruity of the two buildings as depicted in their elevations and ground plans (compare Figure 2 with Figures 5, 6, and 7, below) graphically reinforces Libavius's words. In contrast to the extraordinary castle that Brahe's architects designed and constructed for him on Hven, Libavius tenders something more in the plain style—"not a royal residence," as he puts it, "but one for a private citizen of the middling class."[15] The crude drawings of the chemist's house— which lack scale, perspective, and a means of getting to the third floor—betray an absence of architectural skill and knowledge, but they do convey an idea. This is of a sixteenth-century German townhouse that has been prised apart from its neighbors and set in its own grounds. Is this a *bürgerlich* answer to *Burgwissenschaft?* Possibly. But in order to take full measure of the ideological thrust of this architectural polemic, we must first discuss Uraniborg on its island, Hven.

When King Frederick II deeded Hven to Tycho Brahe in 1576, it was for the purpose of keeping the astronomer in Denmark. This was at a point in Brahe's life when he was contemplating leaving his homeland for good and settling in Basel in order to devote himself to astronomy. Brahe recalled his sentiments at that time in the *Astronomiae instauratae mechanica:*

> I was of the feeling then, that it would not be convenient enough or prudent of me to pursue these studies in my native country [*patria*], especially if I remained fixed in Scania at my seat of Knudstrup, or elsewhere in one of the great provinces of Den-

---

although Brahe would receive a salary as imperial mathematician, he insisted on having his own demesne away from the court (Benatky was approximately 20 miles from Prague). As it turned out, he was unable to establish another Uraniborg there, in part owing to the lack of funds for construction; *ibid.,* pp. 278–287, 298.

[14] "Sed facultate utriusque auxilii [i.e., of patrons and architects] destituti simus, neque tamen putemus nos hanc operis partem posse salvo iure et dignitate chymica praeterire, reiectis ad artifices demonstrationibus, ad opulentos vero locupletesque exaedificatione materiali, quantum possumus in Idea delineanda elaborabimus, et quid nobis in hac re consultum videatur, proponemus"; Libavius, *Commentariorum pars prima* (cit. n. 4), Bk. I, p. 92.

[15] "Neque enim regiam molimur domum, sed privati hominis et civis mediocris"; *ibid.,* p. 96. Libavius adds, with characteristic irony, that this kind of house will do for the present—until some king or emperor comes along to build a more splendid and luxurious chemical house. All translations in this article are my own.

mark where there would be a continual coming-and-going of noblemen and friends who would disturb the peace necessary for philosophical study and so impede it.[16]

Hven, which provided Brahe with the seclusion he desired, is an island of about two thousand acres of chalk tableland, approximately a hundred and sixty feet above sea level. It is located in the Danish sound nine miles south of the straits between Elsinore and Helsingborg. The island was crown land that Frederick II granted to Brahe quit and free for the period of his lifetime, together with the rents of the crown's tenants on the island. The king also gave Brahe a grant of money to help defray the expenses of building the house and observatory.[17]

When Brahe took possession of the island, there was one church, a windmill, and a village of about forty dwellings, whose tenants shared a significant portion of the island as common grazing. These same tenants provided the labor for the improvements Brahe made. In addition to the main residence and observatory, a workshop was built for the artisans who constructed some of the instruments of the observatory. There were stocked fishponds, one of which was dammed up in order to drive a waterwheel constructed near the coast. This same wheel was used to make paper, which was consumed in the printing shop housed in one of the main buildings. The cornerstone of the castle and observatory was laid on 8 August 1576 by Charles Dançay, the French ambassador to the Danish court and Brahe's close friend. An inscription later placed on the stone to mark the event recorded that the house was built by Brahe, on the order of the king, "for the contemplation of philosophy, especially of the stars [*philosophiae imprimisque astrorum contemplationi*]."[18]

Figure 1 shows the main building of Uraniborg, enclosed in formal grounds of notable precision. The castle itself was set foursquare on the points of the compass, so that the corners of the enclosing square were oriented in the direction of the cardinal points and the paths on the diagonals ran precisely east-west and north-south. The front exposure of the house was to the east. The enclosing wall was of earth covered with stone. The portals at the east and west corners were, according to Brahe, in the "rustic Tuscan style." Atop each of them were kennels that housed a pair of large English dogs who signaled the approach of strangers. The building in the north corner of the wall (in style, something of a replica of the main building) was a residence for the servants; the corresponding building in the south corner was the printing house, which produced the publications for Uraniborg. The formal herb and flower gardens surrounding the house were separated by a fence from the arboretum, which Brahe says contained three hundred species of trees.[19]

---

[16] "Praesentiebam enim, me non satis commode et tuto haec studia in Patria excolere posse, praesertim si in Scania, atque sede mea Knudtstorpiana, aut alibi in ampla quadam Daniae parte haererem, ubi Nobilium et Amicorum creber esset concursus, qui talibus, otium Philosophicum interrumpendo, impedimento essent": Brahe, *Mechanica*, in *Opera* (cit. n. 3), Vol. V, p. 109. In another place in the *Mechanica* Brahe describes an equatorial armillary that can be dismantled for moving. This, he suggests, should be a feature of all instruments, since the astronomer must be a "cosmopolite," i.e., a citizen of the world. Following this happy expression, Brahe goes on to praise the virtue of the astronomer who, from his elevated position, can disdain the ignorance and interference of statesmen (*politici*) and seek immortality through his studies; *ibid.*, pp. 62–63.

[17] The deed granting Hven to Brahe is translated in Dreyer, *Brahe* (cit. n. 10), pp. 86–87. The description of the island that follows is based on Brahe, *Mechanica*, in *Opera*, Vol. V, pp. 150–152.

[18] The full inscription is given in Brahe, *Mechanica*, in *Opera*, Vol. V, p. 143.

[19] See "Arcis Uraniburgi, quoad totam capacitatem explicatio," *ibid.*, pp. 139–141.

Figure 1. Uraniborg within its walls and grounds. From Astronomiae instauratae mechanica. By permission of the Houghton Library, Harvard University.

Figure 2 shows the front (east) elevation of the main building, together with the ground plan. A precise symmetry is also manifest in the plan, which consists essentially of a central square flanked on the north and south sides by circular bays. The domestic quarters are contained largely within the central square, which consists of two floors and a basement. On the ground floor the southeast room (D on the ground plan) is described as the winter dining room. The other three rooms (E, F, and G) are designated as spare bedrooms, but that they contain tables for the students as well as beds (V and Y) indicates that the work of the observatory spilled over into this area. (Modifications were also made in the winter dining room to convert it into a small chemical laboratory.) A feature of the ground floor was the central water fountain (B), supplied from a pumped well in the basement of the northern bay under the kitchen (H), which sent water through pipes in the wall to the whole building. Most of the rooms on the second floor are specified by color rather than function: the northeast room is referred to as the red room; the southeast room is simply the blue room. To the west, over-

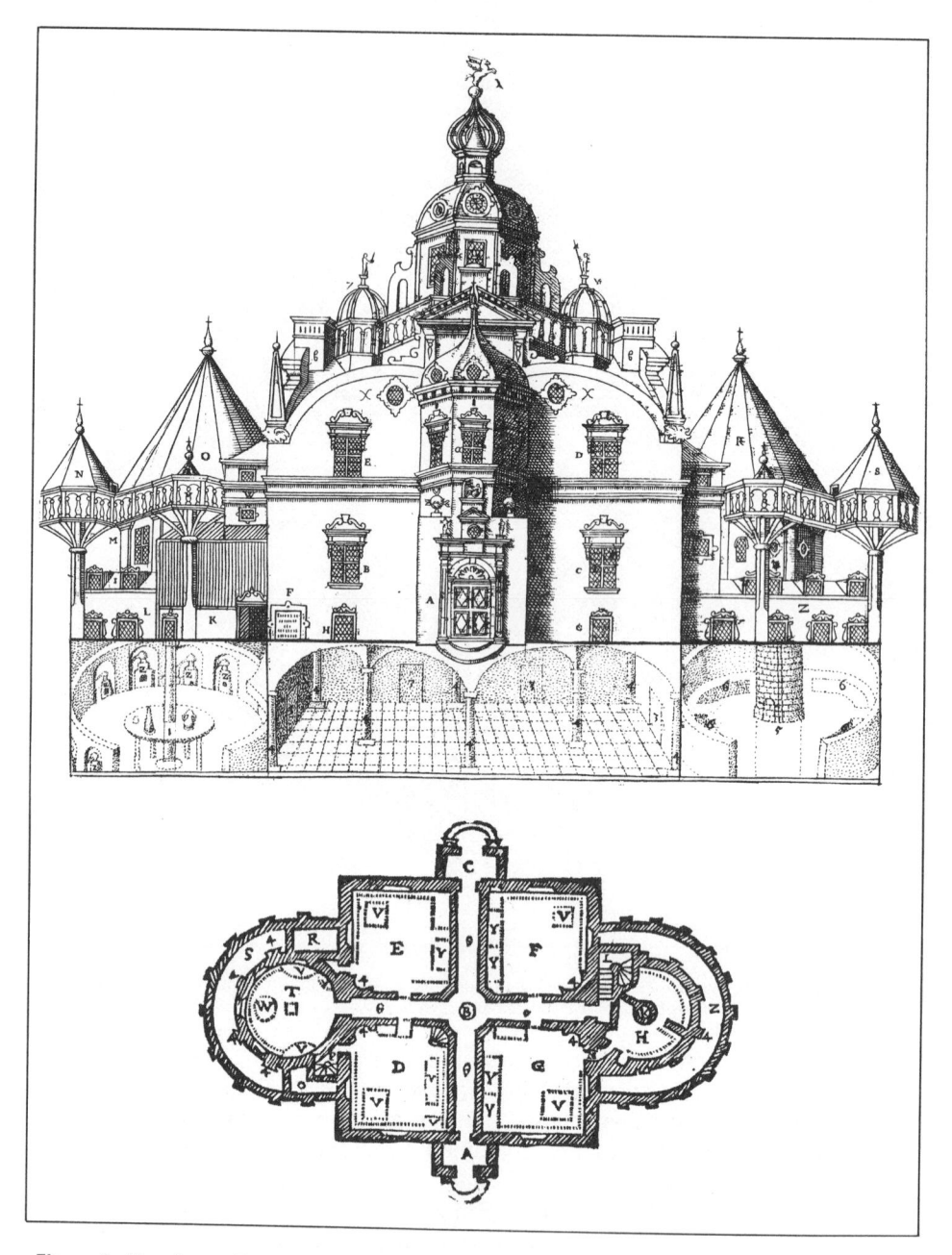

Figure 2. Uraniborg. Front (east) elevation of main building with ground plan. By permission of the Houghton Library, Harvard University.

looking the sound, a large room painted green, with a ceiling decorated with flowers and plants, is the only one specified as to function—the summer dining room. The small circular windows at the top of the house (*X* in the elevation) illuminate eight garret bedrooms for the students; the servants of the house, on the other hand, lived in the separate building set in the north angle of the wall.

The main structure was topped with an octagonal tower with a gallery, in which was placed a clock with a bell; a Pegasus weather vane pranced on high.[20]

The circular bays to the north and south were the chief work areas of Uraniborg. These areas are instantly recognizable in the elevation by the pyramidal roofs of the observation decks on which the instruments were placed. Each bay had one large deck connected by galleries to two smaller decks mounted on single pillars. The roofs of the decks were constructed so that triangular sections could be removed to expose various parts of the sky. Below the northern deck stood the kitchen, at ground level, and under that was a basement storage area with a well and a pump for the water supply. Under the observation decks of the south bay stood Brahe's own circular study, at ground level. Below that, at basement level, was the main chemical laboratory of Uraniborg. The chemical laboratory, like the study above it, was circular and had a central pillar around which a worktable was constructed. Brahe states that there were sixteen furnaces "placed round about in any way whatever." (The plan in elevation suggests that they were located around the peripheral wall.) Various classes of furnace are mentioned in the accompanying text, but none are specified as to their precise location. It is stated that most of them are built from Bergen stone brought from Norway, which withstands heat very well.[21]

Some years after the completion of Uraniborg, around 1584, Brahe had another structure built a short distance beyond the wall of the main compound. This was called the Stellaeburg ("star-castle") (see Figure 3) and was, in effect, an underground observatory. The semicircular domes of this structure pulled back to reveal some of the larger sighting instruments at the bottom of stepped circular terraces linked to a square central area that served as both a work station and a central heating installation for the whole unit. The reasons for constructing this subterranean observatory were twofold: it kept the larger instruments from vibrating in the wind, a problem when they were perched on the decks of Uraniborg itself; and it gave Brahe the opportunity to split up teams of his student-observers so that they could make independent sightings of the same phenomenon without comparing notes. The Stellaeburg was to be connected with the chemical laboratory of Uraniborg by an underground passage, but this was never completed.[22]

A vivid impression of the interior layout of the main work area of Uraniborg is provided by the depiction of one of the most remarkable instruments there, the great mural quadrant (see Figure 4). This famous picture from the *Astronomiae instauratae mechanica* shows sightings being made with a brass quadrant affixed to a wall. In the area inside the quadrant is a mural depicting Brahe and the interior of Uraniborg. Although Brahe's text does not give the location of this instrument, it does tell us that the quadrant was fastened to a wall whose plane lay exactly in a north-south direction; perpendicular to this wall was another one,

---

[20] See "Explicatio partium majoris et praecipuae domus," *ibid.*, pp. 143–145.
[21] Brahe refers to the chemical laboratory variously as the *laboratorium Chymicum* and the *laboratorium Pyronomicum; ibid.*, pp. 31, 142, 144, 145. Of the furnaces, he writes: "Numerus 2 [in Fig. 2] furnos varii generis circumcirca dispositos utcunque repraesentat. Erantque numero 16." He then goes on to classify the general categories of furnaces in the laboratory and to note what they are made of. He ends with the statement that if everything that was in the laboratory were to be described, it could not be done in a few words; *ibid.*, p. 145.
[22] See "Stellaeburgi Explicatio," *ibid.*, pp. 146–149.

Figure 3. Stellaeburg. General view and ground plan. By permission of the Houghton Library, Harvard University.

running east-west, that contained an aperture at the apex of the quadrant through which the star or planet could be sighted. J. L. E. Dreyer, Tycho's biographer, asserts that the quadrant was mounted on the west wall of the southwest room of the main building (*E* on the ground plan; see Figure 2). The instrument was used for two kinds of observation: to measure the altitude of a star and to measure the time of a transit across the meridian. In the first case, an observer such as the one shown at *F* in Figure 4 sighted the star through the aperture and read off the measurement from the quadrant; this was recorded by the collaborator at the

Figure 4. Uraniborg. The mural quadrant. By permission of the Houghton Library, Harvard University.

table. When the timing of a transit was involved, a third collaborator read out the time from the clocks at the instant of transit.[23]

In the painting within the area of the quadrant, Tycho Brahe is shown seated at a table with his right hand raised as if pointing to the aperture through which the sighting was made. This portrait is the work of the artist Tobias Gemperlin, whom Brahe had brought from Augsburg to Denmark. According to Brahe, it was a remarkable likeness, which gave a good impression of his height and stature.[24] At Brahe's feet rests one of his hounds, symbolizing both fidelity and sagacity. In the niche behind his head is a clockwork globe, which simulated the movement of the sun and moon. This is flanked by portraits of Brahe's benefactor and liegelord, King Frederick II, and his queen, Sophia. Above the globe is part of Brahe's library.

The backdrop to the portrait reveals the interior of Uraniborg: this was in fact the work of another artist, Johannes Stenwinckel of Emden, one of the architects

---

[23] The instrument and its uses are described in Brahe, *Mechanica, ibid.*, pp. 29–31; the description from Dreyer is in *Brahe* (cit. n. 10), pp. 99, 100. The use to which this instrument was put explains why the main building of Uraniborg was aligned with the points of the compass.

[24] Brahe, *Mechanica,* in *Opera,* Vol. V, p. 30.

of Uraniborg. What is shown may be considered a cross-section of the south bay of the main building. On the top level are depicted the galleries, with instruments and observers taking readings. Below this, on the middle level, is Brahe's study, with collaborators doing calculations at tables; behind the pillar is the great celestial globe that he had had made in Augsburg, on which was recorded his star map. In the vaulted basement is the chemical laboratory.[25]

With reference to this depiction of the chemistry laboratory, Brahe writes: "From my youth I have been interested in this subject, no less than in astronomy, and I have prosecuted it with great diligence and at no little expense." Elsewhere Brahe tells us that his interest in chemistry began in 1569, while he was in Augsburg, only to be suspended for a time when the appearance of the new star in 1572 diverted his attention more fully to astronomy.[26] But he never abandoned chemistry, and several allusions are made in his writings to his work at the furnaces. Despite these, the precise nature and goal of Brahe's chemistry remain somewhat elusive. There can be no doubt that he was influenced by the writings of Paracelsus, whose views surface in a number of contexts in Brahe's works. The first context is cosmological: in discussing the new star of 1572 and the comet of 1577, Brahe cites with seeming approval (but without wholehearted endorsement) Paracelsus's notion of fire as a heavenly element that could account for change in the celestial regions. In another place he refers to the aurora borealis as a sulfureous vapor presaging infectious diseases that can be cured by chemically prepared terrestrial sulfur; this efficacy is based on a Paracelsian principle of the like curing the like.[27] That Brahe indeed prepared chemical remedies is well attested by three related prescriptions he wrote out—for specific kinds of diseases, including plague, epilepsy, and serious skin infections—which taken collectively constitute a universal medicament.[28] A clear leitmotiv runs through Brahe's chemistry, both cosmological and medical—namely, belief in a correspondence between the powers and effects of phenomena in the heavens and those of things that grow on and under the earth. This relationship is expressed in a pair of vignettes and mottoes that are prominently displayed on the title page and colophon of the *Astronomiae instauratae mechanica*. One vignette shows a godlike figure, a putto at his side, who gazes skyward while leaning on a star globe with a pair of compasses in his hand; it carries the motto *suspiciendo despicio* ("in looking up, I look down"). The corresponding vignette of the colophon shows the godlike figure reclining on a drape-covered mound, holding some

---

[25] *Ibid.*, p. 31. The landscape at the top of the picture was the work of yet a third artist, John of Antwerp, the royal painter of Kronborg. Brahe's assistants, depicted in the mural, are discussed in Victor E. Thoren, "Tycho Brahe as the Dean of a Renaissance Research Institute," in *Religion, Science, and Worldview: Essays in Honor of Richard S. Westfall*, ed. Margaret J. Osler and Paul Lawrence Farber (Cambridge: Cambridge Univ. Press, 1985), pp. 275–295.

[26] Brahe, *Mechanica*, in *Opera*, Vol. V, pp. 31, 108.

[27] See Christianson, "Brahe on the Comet of 1577" (cit. n. 11), pp. 128, 133; and, on like curing like, Brahe as cited by Dreyer, *Brahe* (cit. n. 10), pp. 130–131.

[28] These medicaments have been analyzed with reference to existing manuscripts and printed versions in Karin Figala, "Tycho Brahes Elixier," *Annals of Science*, 1972, 28:139–176. Brahe's formulation of these recipes is closely linked to his move from Hven. The manuscript versions reproduced by Figala carry a dedication to Heinrich von Rantzau, in gratitude for the hospitality of his castle at Wandsbek, where Brahe stayed in 1597–1598 and where the *Astronomiae instauratae mechanica* was produced; the dedication stipulates that the recipe be kept a secret among von Rantzau and his intimates (pp. 170–173). One of the recipes was given to the Emperor Rudolf II in 1599, during an outbreak of the plague.

herbs, while the snake of Aesculapius coils round his arm; the putto, meantime, pulls aside the cover and peers underground at some furnaces and chemical apparatus. The motto now reads *despiciendo suspicio* ("in looking down, I look up").[29] Thus for Brahe the observation and contemplation of the heavens was complemented by the chemical scrutiny of the fruits of the terrestrial world—so much so that he referred to chemistry as "terrestrial astronomy."[30] The placement of the chemical laboratory at Uraniborg underground and the plan to link it with the subterranean observatory at the Stellaeburg, then, may have had symbolic meaning as well as a functional purpose.

If the details of Brahe's chemistry are somewhat vague, the cause of our ignorance about them is no mystery. At the end of the account of his accomplishments in the *Astronomiae instauratae mechanica* Brahe reveals his attitude toward chemistry:

> I shall not shrink from discussing these matters openly with noblemen and princes, as well as other distinguished and learned men who are interested in such questions and have some knowledge of them; and, at the appropriate time, I shall share some things with them, provided that I am convinced of their goodwill and of the fact they will keep these things secret. For it is not expedient or fitting that such things become common knowledge.[31]

This association of secretiveness with aristocratic aloofness was totally at odds with Libavius's ideal of chemistry as a truly liberal art based upon open access to shared information. Libavius first articulated this ideal in his published chemical letters of 1595. At that time Libavius had higher hopes of Brahe: he states that the castle of Uraniborg and the household of von Rantzau promise to be "a refuge of great hope" in the matter of chemical practice.[32] In fact, both von Rantzau and Brahe figure in a list of correspondents from whom Libavius claims to have derived information about chemistry, which appeared in the first edition (1597) of the *Alchemia*. This list was omitted from the second edition (1606). By then, of course, Libavius was fully aware of Brahe's attitudes about chemistry as

---

[29] Brahe explained the meaning of these mottoes in a published letter to Christopher Rothman, mathematician to Landgrave Wilhelm of Hessen-Kassel, dated 17 Aug. 1588; see Brahe, *Opera,* Vol. VI, pp. 144–146. Here he discusses in general terms the analogy between planets, parts of the human body, and metals, as well as correspondences between planets and stars on the one hand and gemstones, minerals, and plants on the other. He goes on to remark that ordinary physics does not teach these things; one has to go to "the Pyronomic School" to learn "what can only be felt with the hands, seen with the eyes and perceived by the exterior and interior senses."

[30] See *ibid.*, p. 145, and the following passage from the *Mechanica:* "Quin et in Spagyricis praeparationibus, seu Pyronomicis exercitiis non minimam impendi curam; ut hoc obiter hic indicem cum eae quas tractat materiae Coelestibus corporibus et influentiis analogae sint. Ideo terrestrem Astronomiam appellare soleo"; Brahe, *Opera,* Vol. V, pp. 117–118. Some of Brahe's other synonyms for chemistry are *ars spagyrica, labor pyronomicus,* and *exercitia pyronomica.*

[31] "De quibus cum Illustribus et Principibus viris, aliisque praestantibus et Eruditis, qui talibus afficiuntur, atque eorum cognitionem aliquam habent, ingenue conferre atque nonnulla iis communicare per occasionem tergiversabor; modo mihi de eorum voluntate constiterit, quodque ea secreta habituri sint. Talia enim vulgaria fieri nec expedit, nec aequum est"; *ibid.*, p. 118. Just how literally these remarks were intended is shown by Brahe's tendering his chemical recipes to the Holstein nobleman Heinrich von Rantzau and to the Emperor Rudolf II around the time they were written (see n. 28). Indeed, it is difficult in the circumstances not to see them as being directed at Rudolf, who was notorious as a seeker after alchemical secrets.

[32] See Libavius, *Rerum chymicarum epistolica forma liber primus* (Frankfurt, 1595), "Praefatio ad lectorem," n.p. For other interesting evidence of Libavius's early views on Brahe, von Rantzau, and Danish Paracelsianism in general, see Figala, "Tycho Brahe's Elixier" (cit. n. 28), p. 142, n. 15.

expressed above and of his association with that center of alchemical occultism that Libavius so detested, Rudolfine Prague.

Libavius clearly had two books to hand when he drew up his plans for a Chemical House: Brahe's *Astronomiae instauratae mechanica,* with its description of Uraniborg; and Vitruvius's *De architectura,* which supplied his want of architects and builders.[33] Libavius comes quickly to the point in enumerating the considerations to be taken into account in designing a suitable dwelling and workplace for the chemist. The passage is worth quoting in full:

> We do not want the chemist to neglect the exercises of piety or exempt himself from the other duties of an upright life, simply pining away amidst his dark furnaces. Rather we want him to cultivate *humanitas* in a civil society and to bring luster to his profession by an upright household, so that he may strive for every virtue and be able to assist with his friends as an aid and counsel to his country. Thus we are not going to devise for him just a *chymeion* or laboratory to use as a private study and hideaway in order that his practice will be more distinguished than anyone else's; but rather, what we shall provide for him is a dwelling suitable for decorous participation in society and living the life of a free man, together with all the appurtenances necessary for such an existence. Thus in addition to his country estates, to be looked after by trustworthy servants and the mother of the household, let him have a house in town and live in a body politic of strictest piety which cherishes the laws.[34]

This passage is a pointed and intended criticism of Brahe. In the *Astronomiae instauratae mechanica* Brahe related how he had sought seclusion from his fellow nobles and had even contemplated leaving the fatherland before isolating himself on the island of Hven. Retreat in order to devote oneself exclusively to science, Libavius implies, is a failure of civic responsibility. Instead of isolating his chemist in a fortress on an island, Libavius sets him down in a townhouse. Not that Libavius neglects privacy: in his plans the boundary between private and public certainly exists, but it is always possible to cross it; if this were not the case there could be no civil society. This is one of the great themes in Libavius's description of the Chemical House, repeated with many variations. While the house stands freely on its own grounds, it is not isolated. Libavius permits a neighboring property as close as possible on the north side, so long as the chemist "is not oppressed by it, and it does not interfere with the enjoyment of his own amenities or block his light."[35] Light is a second major theme in Libavius's

---

[33] Libavius's dependence on Vitruvius's *De architectura* is clearly seen in the vocabulary he employs and in his frequent references to Vitruvius: see Libavius, *Commentariorum pars prima,* Bk. I, pp. 92–99 and *passim.* I have used the Loeb Classical Library edition of Vitruvius, *On Architecture,* ed. and trans. Frank Granger (London: Heinemann; New York: Putnam, 1934).

[34] "Volumus chymicum nec pietatis exercitia intermittere, nec a caeteris vitae honesto [*sic*] officiis immunem furnulis duntaxat obscuris immori; sed in societate civili humanitatem colere, honestaque familia professionem suam illustrare, ut in omnem virtutem intentum habeat animum, et adminiculis consilioque patriam et sociis queat adesse. Itaque non χυμεῖον duntaxat seu laboratorium ei affingimus, quanquam hoc sit futurum, uti studium proprium et interius, ita etiam exercitium prae caeteris conspicuum, sed una honestae conversationis liberalisque vitae domicilium, quod et ea comprehendat, unde ad vivendum satis commode peti possint adminicula, eidem attribuimus. Praeter praedia itaque rustica per administros fidos matremque familias procuranda, habeat in urbe domum, vivatque in civitate orthodoxae pietatis et legum correctissimarum amantissima"; Libavius, *Commentariorum pars prima,* Bk. I, p. 92. The last sentence of this quotation may harbor a very personal attack on Brahe if Libavius was aware, as seems probable, that Brahe lived on Hven with his common-law wife of nonnoble birth and that many considered his children illegitimate: see Dreyer, *Brahe* (cit. n. 10), pp. 70–72.

[35] "Ab hac item parte habitator admittere quam proxime vicinam potest, modo ab ea non opprima-

design, and it is frequently intertwined with the first. Libavius sought to take chemistry out of Brahe's basement and bring it into the light of day. Public exposure was what science needed, not hidden isolation.

To take a tour of the Chemical House requires imagination. Libavius had obvious difficulties in relating his reading of Vitruvius's text to his experience of houses. What the elevations and ground plan of the chemical house (Figures 5, 6, 7) represent, I believe, is an attempt to fit certain elements of a Roman villa, as described by Vitruvius, into a German townhouse. The latter is characterized by its high, pointed gable and its steeply sloping tiled roof; but it has been detached from its immediate neighbors and set on its own grounds, as befits a true villa.[36] As the front (southeast) elevation shows (Figure 5), there are three main floors, corresponding to the three levels of Uraniborg; but, significantly, the first level of Libavius's house is above ground, and it is this level that contains the chemical laboratory. The rear (northwest) (Figure 6) elevation shows three towers at the back of the house. These are part of the laboratory complex, and their shape and number were no doubt inspired by the circular observation decks of Uraniborg. Libavius even includes in the grounds of the house a saltpeter works with trenches in the ground (Figure 8) that corresponds to Brahe's Stellaeburg.[37] As we enter through the front door we find ourselves in a circular turret (*B* of Figure 5, *C* of Figure 7) with a spiral staircase leading to the second floor. A second door takes us into the interior of the building. Once here, we must try to imagine an atrium (*F* in Figure 7), which rises to the rafters of the building. On the north and south aisles of the atrium stand separate rooms—some, but not all, with direct access from the atrium (*R, S, T, N,* and *Q, P, L* of Figure 7). We can imagine galleries running round the corresponding rooms on the second and third floors.[38] The atrium itself reaches deep into the house. It comes to a sudden end, however, at a wall: in that wall is a door, and behind the door is the laboratory area proper (*G* of Figure 7). The wall and the door are both important: the wall separates the atrium, the public space where friends, citizens, and clients may enter, from the *laboratorium,* the private space where our chemist practices his art; the door is a reminder that there should be no absolute barrier between

---

tur, suisque commoditatibus et luce libera perfrui negetur": Libavius, *Commentariorum pars prima*, p. 93.

[36] The ground plan (Fig. 7) attempts to delineate more of the grounds of the house than is depicted in the elevations (Figs. 5 and 6). Libavius had great difficulty giving his rooms the orientation advocated by Vitruvius. He justifies his departures from classical practice by appeals to convenience and to the fact that the Germans live a more indoor life than the Italians, protecting themselves from the harsher climate by means of fires (*ibid.*, p. 95). This is, of course, a partly ironic riposte to Vitruvius's introductory chapter on domestic architecture, where he elaborates on the doctrine of climatic influence on national temperaments; the advantage is to the Italians as opposed to all other tribes, especially the northern ones; see Vitruvius, *On Architecture* (cit. n. 33), Vol. II, pp. 10–20.

[37] Libavius, *Commentariorum pars prima*, Bk. I, p. 99. Libavius acknowledges that he has taken the illustration from Lazarus Ercker: see Ercker, *Treatise on Ores and Assaying,* trans. A. G. Sisco and C. S. Smith (Chicago: Univ. Chicago Press, 1951), p. 311.

[38] Although Libavius shows the spiral staircase at the front of the house (Fig. 5) and the rear of the laboratory projecting out of the back of the house (Fig. 6) as rising only to the height of the second floor, it would seem that in the interior of the building he envisaged the atrium and adjacent laboratory to rise through all three floors to the roof of the building, since he alludes to being able to open out a room in the southeast corner of the third floor (*K* of Fig. 5) onto a gallery, thus linking it with the atrium so that public performances of music, drama, and oratory could be given; see Libavius, *Commentariorum pars prima,* Bk. I, p. 96. This leaves unsolved the problem of access to the third floor.

the two. Indeed, the chemist is expected to pass through that door frequently, since devotion to the art at the expense of public duty is the sign of a poor friend and a bad citizen.[39]

Figure 5. *The Chemical House: front (southeast) elevation. Key. A: Doorway to entrance turret. B: Windows of spiral staircase. C: Window of vegetable storeroom. D: East window of master bedroom. E: East and south windows of storeroom. F: Winter dining room, which can also serve as a living room. G: Round windows to illuminate middle level of atrium. H: Guest room. I: Large window to illuminate the upper atrium. K: Window of hall with gallery, which can be used for public performances when opened out to the atrium. L: Gable windows. M: Window of wood storeroom. N: Window of coagulatorium. O: Bay window of preparation room. P: Window of bedroom of laboratory assistants. Q: Living room on middle level. R: Kitchen window with drain spout. S: Window of picture gallery. T: Vents for lower atrium. V: Vents of cellar. X: East cellar door. Y: Wine cellar door. Z: Barred windows of lower atrium. Courtesy of the National Library of Medicine.*

[39] The importance of doors in Libavius's house is signified by their being clearly marked in the ground plan (Fig. 7). It is also significant that Libavius includes an atrium in the house. Vitruvius states that atria are not necessary in the houses of people of common fortune, who fulfill their duties

All rooms in the ground plan and elevation are specified as to purpose. The rooms on the ground floor off the atrium are connected with the general running of the house or with the laboratory. The rooms on the second and third floors are the private rooms of the chemist and his household (see keys to Figures 5, 6, and 7).[40] In describing these upper rooms in terms of their domestic function (master bedroom, children's bedrooms, sitting room, dining room, kitchen, hall, family room, guest suite), Libavius creates a vivid picture of the chemist's family life—this in contrast to Brahe, who describes his domestic quarters principally in terms of the colors of the rooms and makes no mention at all of the members of his family, referring only to those who assist him in his scientific work. Uraniborg, by implication, was not only the hideaway of a poor friend and a bad citizen; it was also the home of a neglectful *paterfamilias*.

To return to the ground floor and the laboratory. On the north aisle of the atrium stands first a storage room for provisions, the Vitruvian *oporotheca* or fruit and vegetable storeroom (R of Figure 7).[41] Next to this is the bathing room, adjacent to the stripping room and privies (S and T). On the south aisle, the first area (Q) is designated as another general storage area, and next to it stands the wood storage room (P). The other rooms in the north and south aisles are connected with the activity of the laboratory. On the north side of the ground plan, N and K are storerooms that contain, respectively, the portable vessels and glassware, and the raw materials and chemicals used in the laboratory. Access to these two rooms can be gained only from the laboratory itself.

On the south aisle stand three separate work areas. On the ground plan (Figure 7), O designates a room called the *coagulatorium* or *filtrarium;* it is given over to procedures of purification and separation, such as filtration, crystallization, and precipitation, that require special apparatus but not the use of furnaces. Access to this area is from the adjacent room (L of Figure 7), which is called the *praeparatorium,* a general work area for the laboratory assistants. As can be seen in the front elevation, this room (O of Figure 5) has a bay window in order, as Libavius explains, to allow as much light as possible to shine down into it. This has a practical and a symbolic significance: functionally, the rays of the sun hasten certain chemical operations such as precipitation and crystallization; and symbolically, all chemical procedures should be exposed to the light of day.[42] It should be noted also that there is a doorway from this room into the atrium, so that even the laboratory assistants are not isolated from the public part of the house. The last room in this sequence (M of Figure 7) is the bedroom of the laboratory assistants, which is positioned so as to give them direct access to the laboratory and also has windows overlooking the laboratory area so that they could keep a vigilant watch over the processes going on there.[43]

---

by visiting others; only men who hold office require atria to receive guests; see Vitruvius, *On Architecture* (cit. n. 33), Vol. II, p. 36. Thus the presence of an atrium in the chemist's house underlines the fact that Libavius expected its owner to hold public office.

[40] For descriptions of these rooms see Libavius, *Commentariorum pars prima*, Bk. I, pp. 96–97.

[41] The description of the ground-floor rooms that follows is based upon *ibid.*, pp. 93–94.

[42] "Volumus autem, ut tam praeparatorium, quam coagulatorium ab anteriore tabulato reductum sit in hemicyclum moenianum seu podium, ut radii solares toto die excipi ad opera quaedam in Sole peragenda queant, cuius rei plenior et perfectior sit administratio in aulaeo hypostego"; *ibid.*, p. 93; see also p. 96 and n. 45 below.

[43] "Itaque ex praeparatorio et cubiculo tam ianuae, quam fenestrae in laboratorium tendant, ut

56   OWEN HANNAWAY

The laboratory itself (*G* of Figure 7) projects out of the rear of the building, somewhat like an apse. It is flanked on either side by two smaller turrets that, from the rear, look like apsidal chapels (see Figure 6). Indeed, the vocabulary Libavius uses when describing this part of the building evokes a temple or a church as much as a house (e.g., *aedes, adytum, cella*).[44] In introducing his

Figure 6. *The Chemical House: rear (northwest) elevation. Key.* A: *Study windows.* B: *The adytum or inner retreat of the chemist.* Γ: *Laboratory window.* Δ: *Probatorium or assay room.* E: *West window of laboratory assistants' bedroom.* Z: *Window of middle-level living room.* H: *Window of picture gallery.* Θ: *Unspecified room.* I: *Windows of chemical storage room.* K: *Window of apparatus storage room.* Λ: *Privies and sewer drain.* M: *Washroom or wood storage room.* N: *North window of master bedroom.* Ξ,Ο,Π: *Family bedrooms.* P: *Guest room, which can be divided into bedroom and sitting room.* Σ: *Upper-level pantry.* T: *Laboratory drain.* Y: *Air vents to control degree of heat in laboratory furnace.* Φ: *Laboratory chimney.* X: *Assay room chamber.* Ψ: *Adytum chamber.* Ω: *Gable windows.* 5: *Rear chimney. Courtesy of the National Library of Medicine.*

---

prospici operae queant, et vigilanti oculo observari, si quando assidere perpetuo non est necesse"; *ibid.*, p. 93.

[44] *Ibid.*, pp. 94–95. We meet here a favorite device of Libavius, the play on the double meaning of words. *Aedes* can mean any dwelling or habitation, but it is more specifically a dwelling of the gods, i.e., a temple; and it is the word Vitruvius uses in that context. Similarly, *adytum* can mean any hidden retreat or secret place, but it has a technical meaning as that part of the temple into which only priests can enter. Finally, *cella* can describe any chamber or closet for storing things, but it is also

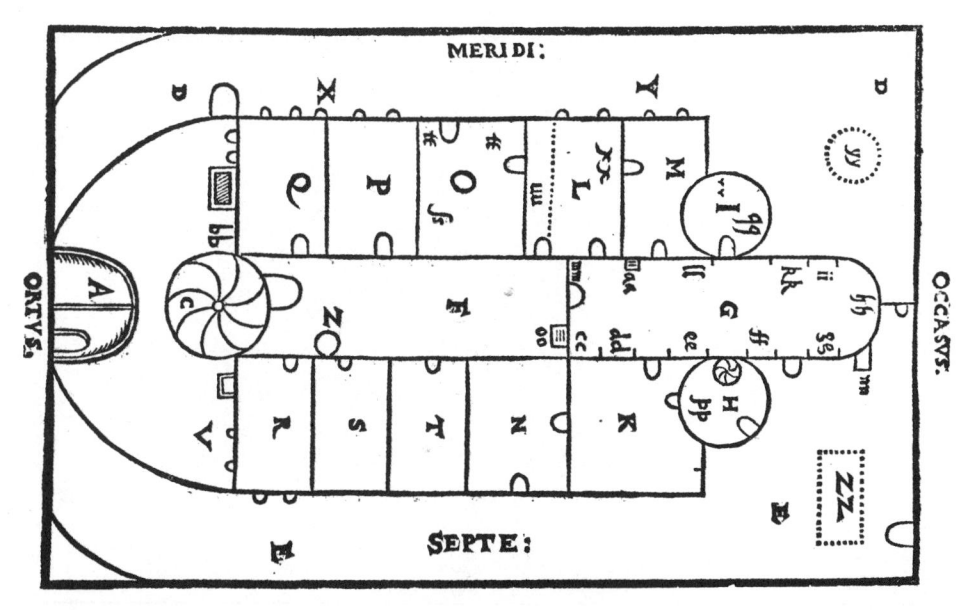

Figure 7. The Chemical House: ground plan. Some features in the ground plan, mainly exterior to the house, are not shown in the elevations. The symbol ⌒ indicates a door. Key. A: *East gate.* [B]: *Porticoed terrace (not marked, but presumably between gate and entrance to house).* C: *Spiral staircase to lower- and middle-level* atria. D: *Garden.* E: *Northern walkway.* F: *Lower atrium or vestibule of the laboratory.* G: *Laboratory.* H: *Adytum with spiral stair to study.* I: *Assay room tower.* K: *Storage room for chemicals.* L: *Preparation room.* M: *Laboratory assistants' bedroom.* N: *Apparatus storage room.* O: *Coagulatorium.* P: *Wood storage room.* Q: *South storeroom.* R: *Vegetable storage room.* S: *Wash room or wood storage room.* T: *Room for undressing.* V: *Cellar for provisions.* X: *Wine cellar.* Y: *Laboratory cellar.* [Z]: *Aqueduct (not marked).* aa: *Entrance to laboratory cellar.* bb: *Entrance to wine cellar.* cc: *Steam bath.* dd: *Ash bath.* ee: *Simple water bath.* ff: *Downward distillation apparatus.* gg: *Sublimation apparatus.* hh: *Central hearth* (focus communis). ii: *Reverberatory furnace.* kk: *Stepped-down distillation apparatus.* ll: *Serpentine distillation apparatus.* mm: *Dung bath.* nn: *Bellows.* oo: *Coal cellar.* pp: *Philosophical furnace.* qq: *Assay furnace.* rr: *Assay balance.* ss: *Vessels for coagulation.* tt: *Distillation using cloth fibers.* uu: *Press stand.* xx: *Desks, preparation tables, and mortars for grinding.* yy: *Fishpond.* zz: *Site for saltpeter, alum, and vitriol works. Courtesy of the National Library of Medicine.*

reader to this part of the building, he immediately distinguishes his laboratory from that of Brahe: Tycho had made his circular and placed it underground as if it were a shrine to Vulcan, condemned into the hands of the god of the underworld; his own laboratory is exposed to the light of the sun, whose rays shine down from the large windows on the west to illuminate every aspect of the work.[45] In contrast to Brahe, who had "crammed" sixteen furnaces indistinguishably round his circular laboratory, Libavius places each of his ten specialized furnaces in specifically designated places around the perimeter, like so many statues in their temple niches: at the head of the laboratory stands the common hearth (*focus communis*), which Libavius stipulates is to be used for domestic

---

that part of the temple in which the statue of the god stood and is so used by Vitruvius. Facilitating Libavius's use of double entendre here is the fact that the two main types of building Vitruvius describes are houses and temples. See also n. 47 below.

[45] "Tycho rotundum id fecit et subterraneum, veluti Vulcano sacrum, damnatumque orci manibus. Nos id Solis dignamur luce"; *ibid.*, p. 94. For a description of the rooms discussed in this paragraph, see *ibid.*, pp. 94–95.

Figure 8. The Chemical House. Saltpeter works. Courtesy of the National Library of Medicine.

purposes as well as for the services of the art (see key to Figure 7). The turret to the south (*I* of Figure 7) is the *probatorium*, the assaying laboratory where the quantitative proportions of metals in ores are determined: it contains a portable assay furnace and a balance within its own case. The corresponding turret to the north (*H* of Figure 7) is the *adytum,* which serves as the inner retreat for the chemist; it is equipped with its own furnace, and from it a spiral staircase leads to a study above.

In four short sentences Libavius sought to capture the contrast between Uraniborg and the Chemical House: "He [Brahe], through royal support, was able to sit in splendor amidst the dwellings of the heavenly beings. We lesser mortals (*homunciones*) are content with our little spark (*scintillula nostra*). In this art [i.e., chemistry] different things are [always] coming to the boil. These require more implements, and this in turn, means much more coming and going."[46] An image and a text, I believe, inform this passage, with its revealing use of repeated diminutives to draw attention to the social and cosmic contrast. The image is that on the mural quadrant at Uraniborg (see Figure 4). Here Brahe sits, reclining in his chair under the portraits of his royal patrons, pointing to the eternally shining stars in the heavens. In contrast, Libavius evokes the chemist running hither and thither to keep alive the ephemeral sparks in his furnaces, which only too obviously belong here below in the world of coming-to-be and passing-away. By clear implication, the contrast is between the life of contemplation (*vita contemplativa*) and the life of action (*vita activa*): the life of the astronomer philosopher, seated in his study, contemplating the eternal verities of divine things; and the life of the physically and socially active chemist, mindful of both his science and his civic responsibilities.

---

[46] "Illi licuit regio sumtu splendore aedium superiorum luxiari. Nos homunciones scintillula nostra sumus contenti. Fervent in hac operae variae. Itaque etiam adminicula desiderat plura, quae plures discursus et transitus requirunt"; *ibid.*, p. 94.

The text that I believe has influenced Libavius in this passage is by another figure who sought to argue that science need not be restricted to the study of the eternal forms of the heavens but could also embrace the transient forms of the world here below. It is Aristotle, justifying the study of living things in his work *On the Parts of Animals*. Aristotle's argument runs as follows: the ungenerated, imperishable, and eternal forms of the celestial bodies by their very nature attract us to their study; but their very inaccessibility makes them remote from our senses and hence from our knowledge. By contrast, the study of the transient forms of plants and animals, although inherently less attractive—and sometimes even repulsive—affords us far greater opportunity for detailed knowledge, living, as we do, amidst them. Aristotle concludes: "We therefore must not recoil with childish aversion from the examination of the humbler animals. Every realm of nature is marvellous: and as Heraclitus, when the strangers who came to visit him found him warming himself at the furnace in the kitchen and hesitated to go in, is reported to have bidden them not to be afraid to enter, as even in that kitchen divinities were present, so we should venture on the study of every kind of animal without distaste."[47] What divinities could possibly be present amid the sparks that flew from the furnaces that stand in their individual niches around the common hearth (*focus communis*) of Libavius's laboratory? To pose the question in this way invites an answer that our pious Lutheran schoolmaster does not give; they strongly suggest the *Lares et Penates*—the Roman gods of the hearth and household.[48]

We are now in a position to appreciate fully the ideological significance of the Chemical House as a response to Uraniborg. Libavius, the bulk of whose career was spent in municipal service as both physician and schoolmaster—and many of whose works were dedicated, not to princes, but to the free cities of the Holy Roman Empire—has drawn upon the full resources of civic humanism to mount his architectural challenge to Brahe's castle-observatory.[49] Hans Baron, in his classic work on the early Italian Renaissance, has delineated the major themes of that politico-cultural ideology. His characterization points to three fundamental elements. First, there is the conviction that a life of political and social engagement is superior to a life of withdrawal into scholarship and contemplation. Libavius's townhouse is quite explicitly designed and located to permit the chemist to engage in social intercourse: this in contrast to Uraniborg, where, in order that he might contemplate the stars, Brahe explicitly isolated himself in self-sufficient splendor in his island fortress, surrounded by thick walls and guarded by dogs who barked at the very approach of strangers. Second, civic humanism cherishes the family as the foundation of a sound society. Although within Libavius's house the work areas and the living quarters are distinct, there are connections between them. As Libavius guides us through the domestic quarters, his description of them readily evokes the chemist as *paterfamilias* living amongst his household and meeting his domestic responsibilities. In the description of Urani-

---

[47] Aristotle, *De partibus animalium*, trans. William Ogle, (Oxford: Clarendon Press, 1911), 1.5.

[48] The images of these deities stood in their little shrines around the hearth of the Roman dwelling. Thus elements of the temple and the house are fused together in this expression of domestic religion.

[49] Examples of Libavius's dedications are: *Rerum chymicarum epistolica forma liber primus* (Frankfurt, 1595), to the senators and consuls of Ratisbon; *Alchemia* (Frankfurt, 1597), to the senators and consuls of Augsburg; and *Dialectica Philippo-Ramaea* (Frankfurt, 1608), to the senators and consuls of Gotha.

borg we never meet Brahe's family or learn anything of his domestic arrangements: the only people we encounter are his students and assistants. Third, civic humanism rejects the notion that the perfect life is that of the sage and instead celebrates the citizen who, to quote Baron, "in addition to his studies, consummates his *humanitas* by shouldering man's social duties and by serving his fellow-citizens in public office." Baron's words echo perfectly those of Libavius that express his aspirations for the chemist: "We want him to cultivate *humanitas* in a civil society and to bring luster to his profession by an upright household, so that he may strive for every virtue and be able to assist with his friends as an aid and counsel to his country."[50] These sentiments contrast sharply with those of Brahe, who, as he tells us himself, fled the company of friends and noblemen to carry out his scientific work undisturbed by responsibilities to the *patria*.

It may be felt that it is stretching matters too far to extend Baron's characterization of civic humanism to cover a chemist's criticism of an astronomer. But Libavius, as a humanist, could call on an impeccable source in warning against the dangers posed by astronomy and mathematics to the pursuit of virtue in the active life. This was none other than Cicero; in *De officiis* he praises Gaius Sulpicius and Sextus Pompey, who, in spite of their respective enthusiasm for astronomy and geometry, did not allow their pursuit of scientific truth to draw them away from the active life, the only arena in which to pursue moral virtue. This depiction of the chemist as civic humanist is entirely consistent with the intellectual wellspring of Libavius's chemistry—the Ciceronianism that was at the heart of the Melanchthonian tradition of Lutheran humanism for which Libavius was such a vigorous spokesman. The celebration of the orator as philosopher underlay his dialectical and rhetorical approach to the organization of chemistry as knowledge and practice. It should not surprise us, therefore, that in portraying the chemist in his house and his laboratory, Libavius depicts him as resembling a Roman orator.[51]

To speak of oratory is to suggest another level of analysis of Libavius's laboratory design that will help us fit the ideological garb of his science more neatly on its intellectual form. Libavius belonged to that strain of northern pedagogical humanism that began with Rudolph Agricola and led through Philipp Melanchthon and Johann Sturm to Petrus Ramus. He was himself the author of a textbook of dialectic and rhetoric that explicitly described itself as Philippo-Ramist. As accounts of this episode in the history of logic and dialectic have

---

[50] Hans Baron, *The Crisis of the Early Italian Renaissance: Civic Humanism and Republican Liberty in an Age of Classicism and Tyranny*, 2nd ed. (Princeton, N.J.: Princeton Univ. Press, 1966), p. 7; Libavius, *Commentariorum pars prima*, Bk. I., p. 92; Latin quoted in n. 34 above.

[51] See Hannaway, *Chemists and the Word* (cit. n. 6), pp. 117–119; and Cicero, *De officiis* 1.6, trans. W. Miller (Loeb Classical Library) (1913; Cambridge, Mass.: Harvard Univ. Press, 1961), pp. 20–21. The context is the place of wisdom in the pursuit of virtue. Cicero acknowledges that the desire for wisdom is natural and morally right but notes that two errors must be avoided: first, treating the unknown as known, and second, devoting too much time and energy to matters that are difficult and obscure. If these errors are avoided, then time spent on problems that are morally right and worth solving will be rewarded. Cicero cites Gaius Sulpicius in astronomy and Sextus Pompey in mathematics as examples of figures who have achieved this. Gaius Sulpicius Gallus, praetor in 169 B.C. and consul in 166 B.C., predicted an eclipse of the moon on the eve of the battle of Pydna, thus preventing panic in the ranks. Sextus Pompey, the uncle of Pompey the Great, was personally known to Cicero, who admired his deep knowledge of mathematics and Stoic thought. See *Cicero On Moral Obligation*, trans. John Higginbotham (Berkeley/Los Angeles: Univ. California Press, 1967), p. 187.

frequently pointed out, the movement was characterized by an emphasis, not on the *Categories* or the *Analytics* of Aristotle's Organon, but on the *Topics*—an emphasis that was reinforced in no small way through Cicero's topical approach to rhetoric. The topics—or, as they were more frequently called, the *loci* or places—were those common headings or key notions to which resort was made in order to find what was available in the repository of knowledge on any given subject. These places housed the themes for speeches, the headings for lectures, and the arguments for cases. In this tradition, the art of all discourse—or, more properly, that part of it called "invention"—consisted of dislodging the arguments from their places so that they might properly be disposed in the second part of dialectic, known as judgment or disposition. The number and nature of the places underwent many variations in this episode of dialectical theory, but one theme runs through the whole development: an emphasis on the discreteness and the organizing function of the places. This process will culminate in the seemingly ever-branching expansions of Ramus's bracketed tables, the kind that Libavius employed to organize the art of chemistry.[52]

In moving through the Chemical House with Libavius as guide, one cannot but be struck by the frequent emphasis on the importance and distinctiveness of the places within it. At the end of that ironic passage in which Libavius mocks symbolic designs in architecture, he reveals the controlling consideration of his own ground plan: it is simply that the vessels (instruments) of chemistry stand out and reveal themselves most prominently (*potissimum prae se ferunt*).[53] The same consideration lies at the heart of his criticism of Brahe's underground laboratory: Libavius notes that the latter had crowded together (*frequentaverit stipaveritque*) his furnaces in his circular room—the implication being that they were not set apart from one another and were thereby indistinguishable. "We," continues Libavius, "have separated certain of ours from the crowd. For not everything can be carried out in one place."[54] We have seen how Libavius disposed his furnaces around his laboratory, allotting each specific type to its own specific chamber or niche (*cella*). Indeed, the whole Chemical House is a series of places within places, so arranged as to effect the most methodical practice of the art. Rudolph Agricola, the fountainhead of Renaissance topical logics, said of the dialectical places that they were so called "because all the instruments for estab-

---

[52] Among the many works dealing with this episode are the following: Wilbur S. Howell, *Logic and Rhetoric in England, 1500–1700* (Princeton, N.J.: Princeton Univ. Press, 1956); Walter J. Ong, *Ramus, Method, and the Decay of Dialogue* (Cambridge, Mass.: Harvard Univ. Press, 1958); Neal W. Gilbert, *Renaissance Concepts of Method* (New York: Columbia Univ. Press, 1960); W. Risse, *Die Logik der Neuzeit*, Bk. I (Stuttgart-Bad Cannstatt: Friedrich Fromman Verlag, 1964); C. Vasoli, *La dialettica e la retorica dell' umanesimo* (Milan: Feltrinelli, 1968); and Lisa Jardine, *Francis Bacon: Discovery and the Art of Discourse* (Cambridge: Cambridge Univ. Press, 1974), esp. Ch. I, pp. 17–58. For Libavius's relationship to this tradition and a discussion of his textbook of dialectic and rhetoric, see Hannaway, *Chemists and the Word* (cit. n. 6), pp. 134–151.

[53] "Hoc in nostra ichnographia observatum cupimus, quod etiam Chymica vasa potissimum prae se ferunt"; Libavius, *Commentariorum pars prima*, Bk. I, p. 93.

[54] "Quanquam Tycho suum laboratorium sedecim fornacibus frequentaverit stipaveritque, id, quod ei facere in rotunditate capaci licuit. Nos quasdam separavimus a turba"; *ibid.*, p. 94. In his reference here to the crowd (*turba*), I suspect Libavius is indulging in a triple play on words: he means he has brought order to his furnaces by putting each of them in its proper place; in arranging them around the laboratory he has separated them from the crowd, which can frequent the adjacent atrium; and finally, in so doing, he has distinguished his chemistry from that of the *turba philosophorum*, the host of alchemical philosophers whose allegorical sayings are collected in the well-known alchemical work of that name.

lishing conviction are located within them as in a receptacle or a treasure chest."[55] In the places of the Chemical House there were stored the instruments of the chemical art that would dislodge, not arguments, but the essences of material things so that they might be disposed in the light of day to display their virtues as the products of a truly liberal art. The laboratory itself, we might say, was the common place that housed the particular places of chemistry. In this light, the opening sentences of Libavius's description of his building take on their full meaning: "It is now the proper time," he writes, "to tackle the subject of furnaces and vessels. But we perceive that a place (*locum*) has not yet been invented (*inventum*) for the assembling and storing of them, whose ingenious arrangement (*constitutio*) can bring greatest benefit to the art."[56] The play on the double meaning of these technical terms in dialectical theory is characteristic of Libavius and is typically Ciceronian in its inspiration.

Finally, this architectural setting of the dialectical places recalls the art of memory, which was an integral part of classical rhetoric. In Quintilian's description of it in the *Institutio oratoria,* the orator was advised, as an aid to memorizing his speech, to imagine himself in a spacious house with which he was very familiar. Then, in various distinctive rooms around the house he was instructed to locate striking images related to the subject matter of his discourse. Subsequently, in the course of delivery, the speaker could envisage himself moving from room to room and use the images to recall parts of his speech.[57] Libavius came from an intellectual tradition that was opposed to such artificial aids to memory, and he was extremely hostile to the symbolic and emblematic mode of thought that was embodied in them; but in a curious way his construction of the Chemical House resembles the art of memory: we can term it a memory system inverted. Libavius's task was to dispose the instruments of his well-ordered art in the places of an *unfamiliar* building that he had to conjure up. We, if not he, can appreciate the symbolic significance for Western culture of his displacement of the ancient images from the familiar places of memory and his construction of a new building around the instruments of laboratory science.

What, then, does our comparison of laboratory designs tell us? In the first place, it suggests that we should not underestimate the intellectual and ideological resources that classical antiquity bequeathed to early modern science. While we continue to recognize the supreme importance of the mathematical way in the evolution of modern science, we must also recognize that there emerged other kinds of scientific practice that were not linked to mathematics. In attempting to comprehend the intellectual forms that shaped these new sciences, it is important

---

[55] Rudolph Agricola, *De inventione dialectica* (Paris, 1529) Bk. I, Ch. ii, p. 8, quoted and translated in Ong, *Ramus* (cit. n. 51), p. 118.

[56] "Ad fornacum vasorumque accedere disciplinam iam maturum erat. Sed intelligimus, extruendis reponendisque nondum inventum esse locum, cuius ingeniosa constitutio maximum potest ad exercitium huius artis afferre momentum"; Libavius, *Commentariorum pars prima*, Bk. I, p. 92. *Constitutio,* besides having the meaning of "arrangement" or "disposition," as here, has the technical meaning in dialectical theory of "definition" or "point in question."

[57] See Quintilian, *Institutio oratoria,* trans. H. E. Butler (Loeb Classical Library), 4 vols. (London: Heinemann; New York: Putnam, 1920–1922), Vol. IV, Bk. XI, Ch. ii, pp. 17–26. Quintilian was in fact quite skeptical about the usefulness of artificial aids to memory in speech making. On the art of memory in general, see Paolo Rossi, *Clavis universalis: Arti mnemoniche e logica combinatoria da Lullo a Leibniz* (Milan: R. Ricciardi, 1960); and Frances A. Yates, *The Art of Memory* (1966; Harmondsworth: Penguin Books, 1969), esp. pp. 37–41.

to look beyond the mathematical and analytical tradition to the dialectical and rhetorical tradition of ancient philosophy. While the former sought to convince with certainty about the divine, unchanging, and eternal beings in the cosmos, the latter sought to persuade intellectually about more transient things and to move men to action in the social and political realm. These two intellectual modes defined distinctive social contexts for the pursuit of knowledge. The one evoked a life of withdrawal devoted to the contemplative study and articulation of eternal verities; the other called forth a life of social intercourse and active engagement pursued for the betterment of mankind.

We have seen each of these modes of life espoused by our two protagonists: the contemplative mode by a figure we usually associate with astronomy, Tycho Brahe; the active mode by the chemist, Andreas Libavius. But it would be a mistake to identify each of the sciences involved with these respective categories —the *observational* science of astronomy with the contemplative life and the active life with the *laboratory* science of chemistry. The evidence does not support such a simple distinction. Libavius would surely have recognized, with Cicero, that astronomy, pursued in the proper spirit, could result in many useful social and political practices, such as the fixing of calendars, the casting of horoscopes, and the prediction of irregular cosmic events that would otherwise threaten social and political stability. Brahe, who embraced the contemplative life, on the other hand, worked at chemistry in his laboratory. The answer to this apparent paradox lies not in the subject matter of the sciences themselves, but in the goals for which they are pursued. Brahe practiced chemistry—his terrestrial astronomy—so that he might reveal in the darkness of his underground laboratory the divinely created cosmic harmony that linked the products of his furnaces with the shining bodies of the heavens that he gazed at from the decks of his observatory in the darkness of night. In looking down, he looked up; and in looking up, he looked down. His practice served contemplation.[58] By contrast, Libavius constructed his laboratory so that the products of his furnaces, exposed to the fullness of the light of day, might reveal, not their cosmic significance, but their benefits to mankind. His practice served social utility. Laboratories, it seems, can no more be identified exclusively with the active or the contemplative life than can specific sciences. But their designs may tell us much.

[58] I am aware that this interpretation runs counter to a popular one stemming from Frances A. Yates and others that identifies the magical, the Hermetic, and the Paracelsian with a newfound activist spirit that expresses man's domination over nature. But on this issue I side with R. J. W. Evans, who says of the reign of Rudolf II: "The conflict which played itself out in the Habsburg lands during those years was a political reflection of the whole intellectual confrontation between acceptance of nature and domination over it. And while the old cosmology with its magic and symbolism possessed elements which were favourable to a more active view of man's role in the world (we may see them at work in various departments of occultism), it nevertheless held basically to the traditional idea that the highest human purpose was one of contemplation"; *Rudolf II and His World* (cit. n. 7), p. 4. Brahe's clinging to feudal forms and the geocentricity of the old cosmology, as well as the design of Uraniborg, seem to me to embody precisely the tensions described here by Evans.

"Good Fences Make Good Neighbors"

Geography as Self-Definition in Early Modern England

By Lesley B. Cormack

OVER FIFTY YEARS AGO, Robert Merton claimed that early modern science could not be separated from the society in which it developed. This claim was to become the focus of one of the great historiographical debates about the Scientific Revolution.[1] Until recently most historians of science, following the lead of A. R. Hall and such essentialists as Alexandre Koyré, have maintained that theory and practice must be seen as two mutually exclusive aspects of knowing, with theory as the superior category. Even Thomas Kuhn's important article "Mathematical versus Experimental Traditions in the Development of Physical Science" attempted to divide knowledge of the physical world into two completely distinct categories, based in part on the observation and manipulation of nature.[2] In recent years, however, the artificiality of this separation has become apparent. Through the work of Robert Westman, J. A. Bennett, Simon Schaffer, and Steven Shapin, among others, we have begun to recognize the centrality of practice in defining science.[3] Early modern science cannot be separated from its social manifestations; rather, theory and practice must be seen to have existed in

---

I would like to thank Susan Lawrence, Roy Laird, and Andrew Ede for reading earlier versions of this paper.

[1] Robert K. Merton, *Science, Technology and Society in Seventeenth-Century England* (orig. publ. in *Osiris*, 1938, 4:360–632) (New York: Howard Fertig, 1988). Charles Webster, *The Great Instauration: Science, Medicine, and Reform, 1626–1660* (London: Duckworth, 1975), and Christopher Hill, *Intellectual Origins of the English Revolution* (Oxford: Clarendon, 1965), address both science and Puritanism. For some of the Mertonian debate see Gary A. Abraham, "Misunderstanding the Merton Thesis: A Boundary Dispute between History and Sociology," *Isis*, 1983, 74:368–387; John Morgan, "Puritanism and Science: A Reinterpretation," *Historical Journal*, 1979, 22:535–560; and esp. Thomas Gieryn, "Distancing Science from Religion in Seventeenth-Century England," *Isis*, 1988, 79:582–593, and Steven Shapin, "Understanding the Merton Thesis," *ibid.*, pp. 594–605.

[2] See, e.g., A. R. Hall, "The Scholar and the Craftsman in the Scientific Revolution," in *Critical Problems in the History of Science*, ed. Marshall Clagett (Madison: Univ. Wisconsin Press, 1959), pp. 1–22; Hall, *The Scientific Revolution* (London: Longman, 1954); Alexandre Koyré, *From the Closed World to the Infinite Universe* (Baltimore: Johns Hopkins Press, 1957); and I. B. Cohen, *The Birth of a New Physics* (New York: Anchor Books, 1960). Thomas Kuhn, "Mathematical versus Experimental Traditions in the Development of Physical Science," in *The Essential Tension: Selected Studies in Scientific Tradition and Change* (Chicago: Univ. Chicago Press, 1977), pp. 31–65.

[3] E. G. R. Taylor, *Mathematical Practitioners of Tudor and Stuart England* (Cambridge: Cambridge Univ. Press, 1954), discovered practical men; Robert Westman, "The Astronomer's Role in the Sixteenth Century: A Preliminary Study," *History of Science*, 1980, 18:105–147, set the path for

a symbiotic relationship. The importance of this relationship for the development of the sciences in the sixteenth and seventeenth centuries can best be demonstrated using the model of the interactive sciences—those studies that demanded an interaction between a practical experience of nature and a mathematical or theoretical framework. Geography, a study often overlooked by historians of science, provides a good example of the interrelation of politics, economics, and society with theories of the earth and its inhabitants.[4] Although the story of early modern geography could be told from the point of view of many European countries, the English tale is especially informative, leading as it did to ideologies of separateness and empire that would be so successfully reified in the late seventeenth and eighteenth centuries.

EARLY MODERN GEOGRAPHY

The Elizabethan world view encompassed an expanding globe and an enclosing nation. While more and more of the world lay within the grasp of those brave or foolhardy enough to venture forth, the English were increasingly defining themselves and their country as separate from the Continent and the rest of the terraqueous globe. This seemingly contradictory view of their world—at once expansive and exclusive—was developed by Englishmen in the sixteenth and early seventeenth centuries through the study of geography.

Geography was a complex and wide-ranging discipline in early modern England, providing a focus for both exploration and nation building. It was a lively study, involving the work of many men and closely followed by scores of students and scholars. It gave English investigators a yardstick of the "other" by which they could measure themselves, allowing them to fashion themselves into patriots and privateers. As well, an investigation of early modern geography is vital for historians of science, suggesting a new explanation for the growth of experimental science and perhaps of mechanical philosophy in the seventeenth century. Geography had both a theoretical underpinning and a practical purpose. Because of this, it corresponds neither to a model of disinterested natural philosophy nor to one of craft-oriented technology. Rather, geography was an intermediate discipline that combined aspects of theory and practice, adopting its own ideology and methodology. That new ideology had three components: an immense value placed on mathematics; an emphasis on the importance of gathering information in an incremental and inductive way; and a desire to make the knowledge so obtained into a public and useful science. Geography thus provides an example of the type of investigation that encouraged its seventeenth-century

---

investigating the interconnection of theory and practice, followed by work by J. A. Bennett, "The Mechanic's Philosophy and the Mechanical Philosophy," *Hist. Sci.*, 1986, 24:1–28; Steven Shapin and Simon Schaffer, *Leviathan and the Air-Pump: Hobbes, Boyle, and the Experimental Life* (Princeton, N.J.: Princeton Univ. Press, 1985); Schaffer, "Natural Philosophy and Public Spectacle in the Eighteenth Century," *Hist. Sci.*, 1983, 21:1–43; Bruce T. Moran, "German Prince-Practitioners: Aspects in the Development of Courtly Science, Technology, and Procedures in the Renaissance," *Technology and Culture*, 1981, 22:253–274; and Mario Biagioli, "Galileo's System of Patronage," *Hist. Sci.*, 1990, 28:1–62, *inter alia*.

[4] See David N. Livingstone, "The History of Science and the History of Geography: Interactions and Implications," *Hist. Sci.*, 1984, 22:271–302; and Thomas Glick, "In Search of Geography," *Isis*, 1983, 74:92–97, for a declaration of the importance of history of geography.

practitioners to develop a new, engaged approach to natural inquiry; such an approach is emblematic of the "new science."

In this article I will argue that geography developed in early modern England in three related directions and that in each of the resulting subdisciplines we can see the unfolding of this new ideology and methodology of science. Especially, the utility of geography shines through; part of this utility, I maintain, involved the development of a definition of things and people English, one that would aid the nation in its imperial and national strivings.[5]

GEOGRAPHY AND COSMOGRAPHY: DEFINITIONS

In sixteenth-century England geography was developing into a discipline distinct from the older study of cosmography. Cosmography, as John Dee proclaimed, "matcheth Heaven, and the Earth, in one frame," requiring "*Astronomie, Geographie, Hydrographie,* and *Musike*" to be complete. Geography, on the other hand, "teacheth wayes, by which, in sundry formes, (as *Sphaerike, Plaine,* or other), the Situation of Cities, Townes, Villages, Fortes, Castells, Mountaines, Woods, Havens, Rivers, Crekes, and such other things, upon the outface of the earthly Globe . . . may be described and designed."[6] In other words, while the subject of cosmography was the globe and its relationship with the heavens as a whole, picturing the earth as an integral part of the cosmos, geography had a narrower focus, planting people's feet firmly on an earth they were beginning to feel they might control. Geographers abstracted the globe from its surrounding cosmos and began to classify its parts by separation rather than by union. This study of geography thus reflects a growing self-absorption on the part of people in early modern England and in turn aided them in their attempts to master their mortal world.[7]

During the late sixteenth and early seventeenth centuries the study of geography gained popularity in England. There developed the three related branches of mathematical, descriptive, and chorographical geography, each with distinct practitioners and different topics of investigation. These distinctions had classical roots, especially in the work of Ptolemy, which were maintained and reinforced by contemporary taxonomists of knowledge like John Dee. Each branch helped develop the ideology of geography and the image of English men and women as unique and separate from other peoples.

From antiquity, the study of geography had followed two different traditions, one leading to mathematical geography and the other to descriptive geography,

---

[5] Although the story I am telling is an English one, parallels could and should be drawn to similar developments in Continental geography. Especially in the Netherlands, an area similar to England in many ways, geography and cartography developed in tandem with imperial desires and helped fashion the Dutch identity. The elaboration of this comparison would, of course, be the work of another article; an interesting starting point would be the provocative study of maps and images in seventeenth-century Dutch painting in Svetlana Alpers, *The Art of Describing: Dutch Art in the Seventeenth Century* (Chicago: Univ. Chicago Press, 1983).

[6] John Dee, *The Mathematicall Praeface to the Elements of Euclid* (London, 1570), fols. b3a, a4a. This distinction is repeated by Thomas Blundeville, *His Exercises, Containing Six Treatises* (London, 1594), P. 2; and later by Nathanael Carpenter, *Geography Delineated Forth in Two Books Containing the Sphaericall and Topicall Parts Thereof* (Oxford, 1625), p. 1.

[7] See Roy Porter, "The Terraqueous Globe," in *The Ferment of Knowledge: Studies in the Historiography of Eighteenth-Century Science,* ed. G. S. Rousseau and R. Porter (Cambridge: Cambridge Univ. Press, 1980), pp. 285–324, for a discussion of this shift into the eighteenth century.

with chorography as an offshoot of both. Ptolemy, perhaps the most famous ancient geographer, was interested in the mathematical mapping of the globe using the tools of astronomy. He separated the more elevated study of geography proper from chorography, the purview of which he defined as the description of ports, villages, peoples, rivers, and so forth.[8] Descriptive geography, on the other hand, owed its origins to Strabo's *De situ orbis,* where Strabo set out to describe every detail of the known world on the basis of his own extensive travels, fellow travelers' accounts, and authoritative sources. Thus the three branches of sixteenth-century geography owed their divisions and some of their similarities to these two ancient authorities.

Mathematical geography was most closely akin to the modern study of geodesy, the branch of applied mathematics that determines the exact positions of points and the figures and areas of large portions of the earth's surface, the shape and size of the earth, and the variations of terrestrial gravity and magnetism. Closely akin to mathematical geography was the practical art of cartography, the study of maps and mapmaking, although cartography depended far more on guild methods of transfer of knowledge and less on any systematic development of theories or models.[9] The development of mapping techniques in the early modern period focused on translating coordinates and measurements of actual coastlines onto charts suitable for use by navigators, while mathematical geography proper manipulated the globe as a more theoretical construct, consisting of an exact grid of coordinates and properties, which necessitated the use of exact mathematical formulas. Studied by a small group of men who were also interested in other mathematical topics, this was the most rigorously theoretical form of geography.

The second branch, descriptive geography, portrayed the physical and political structures of other lands, usually in an inductive and relatively unsophisticated manner. Because of this relative lack of rigorous analysis, and because its primary goal, as defined by Strabo, was to promote useful knowledge, descriptive geography was the most accessible of the three geographical subdisciplines. It encompassed everything from practical descriptions of European road conditions to outlandish yarns of exotic locales, providing intriguing reading and practical information alike.

The final type of geography, chorography, developed in the course of the late sixteenth century, combining a medieval chronicle tradition with the Italian Renaissance study of local description. Chorography was the most wide ranging of the geographical subdisciplines, since it included an interest in genealogy, chronology, and antiquities, as well as local history and topography. Chorography thus united an anecdotal interest in local families and wonders with the mathematically arduous task of genealogical and chronological research.[10]

---

[8] Claudius Ptolemy, *Geographia,* ed. Sebastian Munster (Basel, 1540); rpt. with an introduction by R. A. Skelton (Amsterdam: Theatrum Orbis Terrarum, 1966), fol. a1a.

[9] Thomas Smith, "Manuscript and Printed Sea Charts in Seventeenth-Century London: The Case of the Thames School," in *The Compleat Plattmaker,* ed. Norman Thrower (Berkeley: Univ. California Press, 1978), pp. 45–100.

[10] See Stanley J. G. Mendyk, "Painting the Landscape: Regional Study in Britain during the Seventeenth Century" (Ph.D. diss., McMaster Univ., 1983); now published as *"Speculum Britanniae": Regional Study, Antiquarianism, and Science in Britain to 1700* (Toronto: Univ. Toronto Press, 1989). See also F. J. Levy, "The Making of Camden's *Britannia,*" *Bibliothèque d'Humanisme et Renaissance,* 1964, *26*:76–92.

These three subdisciplines, emerging from two distinct classical traditions and combined with a third Renaissance development, were defined in part through practice—different networks of scholars studied and wrote about each of the areas.[11] As well, their development reflects a growing distinction made within the scholarly community, demonstrated by categorizations by taxonomists of knowledge such as John Dee.

Dee included geography and chorography as two of the mathematical arts derived from geometry in his *Mathematicall Praeface,* a complex categorization of the mathematical sciences. His definition of geography stressed both its mathematical precision and the practice of descriptive collection, while chorography was said to develop observational skills and local pride: "Chorographie seemeth to be an underling and a twig of *Geographie:* and yet nevertheless, is in practise manifolde, and in use very ample. This teacheth Analogically to describe a small portion or circuite of ground, with the contentes . . . in the territory or parcell of ground which it taketh in hand to make description of, it leaveth out . . . no notable or odde thing, above ground visible." Dee thus defined the function of chorography as the description of local detail, rather than the "commensuration it hath to the whole," which was the job of geography proper.[12]

Nathanael Carpenter reiterated this distinction between geography and chorography. He listed four differences between the topics: geography described the whole sphere, while chorography handled the lesser parts without regard for the whole; geography dealt in quantities where chorography used accidental qualities; chorography used the art of painting where geography had no need of it; and, finally, geography used the mathematical sciences, while chorography did not. Still, claimed Carpenter, "I see no great reason why *Chorographie* should not be referred to *Geographie;* as a part to the whole," thus placing chorography as one of the geographical subdisciplines.[13]

Further, Carpenter created a distinction within geography between the "Sphericall" or "mathematicall" and the "Topicall" or "historicall": "The former receiveth greatest light from *Astronomie,* whence some have called it the Astronomicall part: The later from *Philosophie* and *Historicall* observation, being (as we have said) a mixt Science, taking part of divers faculties." Carpenter, then, recognized the three geographical areas of mathematical geography, descriptive geography, and chorography.[14]

---

[11] For a full discussion of these networks of geographical interest see Lesley B. Cormack, "*Non sufficit orbem:* Geography as an Interactive Science at Oxford and Cambridge, 1580–1620" (Ph.D. diss., Univ. Toronto, 1988), esp. pp. 145–150.

[12] Dee, *Mathematicall Praeface* (cit. n. 6), fol. a4a. For a detailed analysis of the *Mathematicall Praeface* see Nicholas Clulee, *John Dee's Natural Philosophy: Between Science and Religion* (London: Routledge, 1988).

[13] Carpenter, *Geography Delineated* (cit. n. 6), pp. 1–4, on p. 4.

[14] *Ibid.,* p. 5. This, of course, was part of a Continent-wide discussion of the classification of the different subdisciplines of geography and was in no way unique to the English. Continental and English geographers followed Ptolemy's lead, elaborated by Peter Apian, *Cosmographia* (1529). Bartholomew Keckermann discussed the tripartite division in *Systema geographicum* (Hanover, 1612); and, of course, by 1650 Bernardus Varenius, *Geographia generalis* (Amsterdam, 1650), had defined the two areas of general and special geography, corresponding to mathematical and descriptive geography, with chorography deemed a separate discipline. By 1630, in fact, the question of classification had ceased to be of the first importance for English writers; William Premble, in *A Briefe Introduction to Geography Containing a Description of the Grounds, and Generall Part Thereof* (Oxford, 1630), could state these distinctions succinctly within the first page.

Geography was thus established as a mixed science by its sixteenth- and seventeenth-century taxonomists. These taxonomists considered geography a united science, held together by its common subject, the earth, although they all stressed its subdivision into three separate areas of emphasis.

MATHEMATICAL GEOGRAPHY: THEORY AT PRACTICE

Mathematical geography, essentially a branch of applied mathematics, developed as a study separate from both pure mathematics and descriptive geography. The study of mathematical geography involved more theoretical manipulation and mathematical skill than did the other subdisciplines of geography, but equally it required an application of that mathematical understanding to the world of practical affairs. This rigorous subdiscipline attracted a relatively homogeneous, dedicated, and comparatively small group of scholars who developed a new methodology of experimentation, the systematic tabulation of facts, and mathematical manipulation. This methodology was vitally important in the development of a new mathematical science that used practical knowledge to advance its aspirations to new and fruitful theories of the earth, allowing the application of scientific ideas to the practical world of human affairs.

Mathematical geography had its roots in Ptolemy's *Geographia*. He treated the celestial and terrestrial globes as equivalent, applying the same grid system to each and using the same spherical geometry to plot particular points thereon. He divided the globe into a series of parallel belts or "climates," separated by the equator, the tropics, and the arctic and antarctic circles. He developed a grid of longitude and latitude coordinates, creating a mapping system that has never been entirely superseded.[15] Ptolemy, then, was more concerned in *Geographia* with the organization of the globe than with the people inhabiting it, since such description could only be inexact and subjective.[16]

Sixteenth-century geography took up this distinction. With the rediscovery of Ptolemy's *Geographia* in the fifteenth century,[17] mathematical geography emerged as a subdiscipline differentiated by classical precedent from its descriptive cousin. Humanists and natural philosophers alike saw the advantages of a

---

[15] Ptolemy, *Geographia*, ed. Munster (cit. n. 8). Ptolemy discusses both the five climatic regions and the longitude-latitude grid, fols. C2–5. See D. W. Waters, *The Art of Navigation in England in Elizabethan and Early Stuart Times* (London: Hollis & Carter, 1958), p. 43; and the important recent work of J. B. Harley and David Woodward, eds., *The History of Cartography*, Vol. I: *Cartography in Prehistoric, Ancient, and Medieval Europe and the Mediterranean* (Chicago: Univ. Chicago Press, 1987).

[16] Ptolemy, *Geographia*, ed. Munster, fol. a1a. Margarita Bowen, *Empiricism and Geographical Thought: From Francis Bacon to Alexander von Humbolt* (Cambridge: Cambridge Univ. Press, 1981), p. 31, discusses the benefit of subjective analysis in historical and modern geography.

[17] Ptolemy's *Geographia* was not included in the Ptolemaic *opera* introduced into the West in the twelfth century. It was only rediscovered in the West ca. 1406, when it was translated into Latin by Jacobus Angelus in Florence. In addition to numerous manuscript copies, it appeared in six printed editions in the fifteenth century: Bologna, 1462 (1482?); Vicenza, 1475; Rome, 1478; Ulm, 1482; Ulm, 1486; and Rome, 1490. It appeared in numerous editions in the sixteenth century in both folio and quarto: twenty in Latin, six in Italian, and two in Greek. For a discussion of the rediscovery of Ptolemy see Leo Bagrow, *History of Cartography*, rev. R. A. Skelton (London: C. A. Watts, 1964), p. 77. For a thought-provoking discussion of the importance of Ptolemy's rediscovery for our perceptions of art and the world around us see Samuel Edgerton, Jr., "From Mental Matrix to *Mappamundi* to Christian Empire: The Heritage of Ptolemaic Cartography in the Renaissance," in *Art and Cartography: Six Historical Essays*, ed. David Woodward (Chicago: Univ. Chicago Press, 1987), pp. 10–50.

world made navigable and mathematically exact, and mathematical geography developed within the tradition of the quadrivium as a form of applied mathematics, much akin to astronomy. This branch of geography, which combined mathematical rigor and practical curiosity, encouraged a reevaluation of the natural world as based on both number and action. Thus mathematical geographers helped to develop a new methodology that combined mathematics, inductive experimentation, and practical application; this suggests that we should begin to reevaluate their role in the development of seventeenth-century science.[18]

Mathematicians in late Elizabethan and early Stuart England found this quantitative and exact study of the earth compelling. At a time when the English hoped to gain a foothold in the European race for new trade routes and colonies in the New World and the East, they needed accurate methods of navigation. Tales of hitherto unknown lands, moreover, cast doubt on the accuracy of Ptolemy's maps and called for new and improved charts and globes. While Ptolemy's *Geographia* might have inspired the study of mathematical geography, the very existence of this ancient source, especially with its later Byzantine maps, compelled geographers to compare it with current knowledge and to create a new picture of the globe that would supersede the one it offered. This discrepancy between Ptolemy and contemporary knowledge soon led mathematical geographers to abandon any slavish devotion to Ptolemy in favor of the investigation of problems of navigation and map projection.

Sixteenth-century English mathematical geographers developed three main areas of investigation, all centering on practical problems. The first and most pressing was the need to determine longitude at sea. This became essential as more transatlantic voyages were made; it was desirable to be able to plot the shortest path to the New World rather than the longer course along a fixed latitude. It was even more desirable to avoid foundering off the Scilly Islands on the return trip, as was the fate of many ships. Second, English geographers faced the difficult question of how to navigate in northern waters. In northern climes, most common maps distorted directions and distances, and extreme magnetic variation caused compass needles to point to many different directions as north. The solution lay in the creation of a polar projection map and the charting of the isogonic lines of compass variation. Third, geographers needed to develop a mathematical map projection that would allow sailing and rhumb lines to be drawn in a straight line. This promised to make navigation an easier and more exact art.

The problem of determining longitude at sea plagued geographers from the first Renaissance voyages until the eighteenth century.[19] One popular idea in England in the late sixteenth century was the possibility of using the variation of the compass to determine longitude. Compass needles, it was realized very early, did not point to true north. Furthermore, sixteenth-century mathematical geographers and navigators soon discovered that compass variation was not uniform throughout the world, but varied from place to place. It was commonly suggested

---

[18] See Bennett, "Mechanic's Philosophy" (cit. n. 3), for a discussion of geography and navigation as the foundation for seventeenth-century mechanical philosophy.

[19] This problem was solved only with the invention of an accurate chronometer, first tested by Captain James Cook on his famous voyages. See P. J. Marshall and Glyndwr Williams, *The Great Map of Mankind: British Perceptions of the World in the Age of Enlightenment* (London: J. M. Dent, 1982).

that this variation of compass bearing, if charted, could allow navigators to determine longitude, since compass variation was thought to vary consistently with geographical location. Some geographers believed that the compass needle was affected by large land masses, since, for example, it seemed to point to Africa throughout the circumnavigation of that continent.[20] Thus, it was believed that the determination of the degree of variation could be used to establish the degree of longitude, using mathematical calculations or following previously compiled charts.

Unfortunately, this solution to the problem of longitude did not yield easily to investigation. Robert Norman, a self-taught instrument maker, felt that compass variation was too erratic to be useful in finding longitude.

> This Variation is iudged by divers travailers to bee by equall proportion, but herein they are muche deceived, and therefore it appeareth, that notwithstandyng their travaile, they have more followed their bookes then experience in that matter. True it is, that Martin Curtes doeth allowe it to be by proportion, but it is a moste false and erroneous rule. For there is neither proportion nor uniformitie in it, but in some places swift and sudden, and in some places slowe.

He sought the answer in a related phenomenon, the dip of the compass needle (i.e., its pull toward the center of the earth). Norman claimed that this carefully observed dip might be used to determine at least latitudinal position—he used it to calculate the latitude of London as 71°50'—and perhaps longitudinal position as well.[21]

In 1581 William Borough, another self-taught mathematician, published *A Discourse on the Variation of the Cumpas,* usually bound with Norman's *New Attractive*. In it Borough explained the phenomenon of compass variation, "knowing the variation of the Cumpasse to bee the cause of many errours and imperfections in Navigation." Recognizing that the only way to understand variation was by collecting a mass of inductive data, Borough charged mariners and navigators to make continual, accurate observations. He blamed variation for wreaking havoc with chart construction, "for, either the partes in them contained, are framed to agree in their latitudes by the skale thereof, and so wrested from the true courses that one place beareth from an other by the Cumpas, or els in setting the parts to agree in their due courses, thei have placed them in false latitudes; or abridged, or overstretched the true distances betweene them."[22] Borough tentatively proposed variation as a longitude-finding tool, although he

---

[20] E.g., William Gilbert, *De magnete* (London, 1600). He used this idea to predict the existence of a Northeast passage and of a large *terra Australis*. Dutch pilots claimed that the line of 0° variation ran through Java, while Gilbert favored the Peloponnese; so claimed an anonymous reporter of Guillaume de Nautonier's work *Metrocomie,* in British Library (BL), Burney MS 368, fol. 32a–b. See E. G. R. Taylor, *Late Tudor and Early Stuart Geography, 1583–1650* (London: Methuen, 1934), p. 69.

[21] Robert Norman, *The New Attractive* (London, 1581; rpt. Amsterdam: Theatrum Orbis Terrarum, 1974), pp. 21 (quotation), 10 (latitude of London). The latitude of London is usually given as 51° 30'.

[22] William Borough, *A Discourse on the Variation of the Cumpas* (London, 1581; rpt. Amsterdam: Theatrum Orbis Terrarum, 1974), preface, fols. *2a, *3a, F2a. On Borough see *Dictionary of National Biography: From the Earliest Times to 1900,* 22 vols. (Oxford: Oxford Univ. Press, 1921), Vol. II, p. 867; and Taylor, *Mathematical Practitioners* (cit. n. 3), p. 68.

While he continued to struggle with the problem of longitude variation, Borough acknowledged that Mercator's "universal map," containing his new projection, had greatly simplified the navigator's task.

temporized that "if there might be had a portable Clocke that would continue true of space of forty or fifty houres together . . . then might the difference of longitude of any two places of knowen Latitudes . . . be also most exactly given."[23] He acknowledged that this was, for the time at least, an impossible dream.

In 1599 Edward Wright, perhaps England's foremost mathematical geographer, translated Simon Stevin's *The Haven-Finding Arte* from the Dutch. In this work Stevin claimed that magnetic variation could be used as an aid to navigation in lieu of the calculation of longitude. He set down tables of variation, means of finding harbors with known variations, and methods of determining variation. In his translation Wright called for systematic observations of compass variation to be conducted on a worldwide scale, "that at length we may come to the certaintie that they which take charge of ships may know in their navigations to what latitude and to what variation (which shal serve in stead of the longitude not yet found) they ought to bring themselves."[24]

Thus both Wright and Borough demonstrate the close connection between navigation and the promotion of a "proto-Baconian" tabulation of facts meant for both practical application and scientific advancement. Here appears the foundation of an experimental science, grounded in both practical application and theoretical mathematics, quite separate from any more traditional Aristotelian natural philosophy or Neoplatonic mathematics. Unfortunately, Wright's scheme was not entirely successful. By 1610, in the second edition of *Certaine Errors in Navigation,* Wright had constructed a detailed chart of compass variation—but he had also become more hesitant in his claims concerning the use of variation to determine longitude.[25]

Belief in the equivalency of variation and longitude or the use of variation in determining longitude died hard. In about 1620, in a manuscript called "A Magnetical Problem," the university-trained divine and mathematician Thomas Lydiat, following William Gilbert's lead, claimed that a compass touched by a lodestone resembled the earth, itself a great lodestone, and thus that each part of the compass represented a different part of the earth. Compass variation therefore mirrored the natural magnetic variation of the globe. "Nowe if this prove true," claimed Lydiat, "that thereby is given a most certain and redy means of measuring the Longitude, or East and West distances, and withal a most easie way of sayling by a great Circle."[26] It was not until 1634 that John Pell, John Marr, and Henry Gellibrand discovered that magnetic variation itself varied over time. Their discovery that the variation had significantly diminished at two observational sites since Edmund Gunter's measurements in 1622 and 1624 effectively laid to rest the chimera of variation as a longitude-finding tool.[27] While a great

---

[23] Borough, *Discourse on Variation,* fols. D1b–2b. Taylor claims that this was "a technique for finding the longitude by carrying a number of spring driven watches," not quite what Borough had in mind: *Mathematical Practitioners,* biography no. 58.

[24] Edward Wright, *The Haven-Finding Arte by the Latitude and Variation,* from the Dutch of Simon Stevin (London, 1599; rpt. Amsterdam: Theatrum Orbis Terrarum, 1968), p. 3, and preface, fol. B3a (quotation). On Wright see Taylor, *Mathematical Practitioners,* biography no. 100.

[25] Edward Wright, *Certaine Errors in Navigation: Detected and Corrected by Edward Wright* (2nd ed., London, 1610), fols. 2P1a–8a. See also Waters, *Art of Navigation* (cit. n. 15), p. 316.

[26] Oxford, Bodleian MS Bodl 313, fol. 63b.

[27] Taylor, *Mathematical Practitioners* (cit. n. 3), biography no. 158. This finding was announced in Henry Gellibrand, *A Discourse Mathematicall on the Variation of the Magneticall Needle, Together with Its Admirable Diminution Lately Discovered* (London, 1635). For an important interpretation of

deal of time had thus been spent on observations and calculations related to the problem of compass variation, to little avail, this investigation brought together theoretical and practical methods and ideas in a way that blended observational and mathematical concerns. Compass variation was interesting from the theoretical viewpoint of understanding the magnetic world, but both the data themselves and the motivation to collect them came from the practical world of navigation.

The difficulty of navigating in northern waters clearly fueled this English attempt to link longitude and variation. Because the English entered the Age of Discovery later than other European nations, their share of the New World was a northern one, and their only route to Cathay was through a hypothetical Northwest or Northeast passage. Much ink was spilled, in fact, convincing the various English monarchs that such a passage existed.[28] It soon became apparent to navigators, however, that the plane charts that were adequate for Mediterranean travel were useless for polar sailing. By extending the point of the North Pole into the top edge of the map, these charts made nonsense of compass navigation and sailing distance. Some northern charts at least acknowledged this problem by depicting two compass needles, one pointing north along Labrador and a second pointing along a convergent path following the coast of Greenland. The answer to this problem, which lay in stretching a point into a line consisting of an infinite number of points, and to the problem of conflicting sailing directions was a stereographic polar projection showing the pole at the center and the latitudinal lines as concentric circles. John Dee apparently invented such a projection, which he called the "Paradoxal Compass," in 1556.[29] Although this technique did not facilitate compass use, it did have the advantage of projecting correct spatial relations and allowed a navigator to use the radiating rhumb lines to plot an approximate great or lesser circle course.

Since Dee was much given to secrecy, it is not clear how many people knew of his innovative map projection. By 1594, at least, Captain John Davis included this projection and instructions for navigating with it in *The Seaman's Secrets;* and George Waymouth in *The Jewelle of Artes* (1604) offers "the demonstration of an Instrument to finde out the degree and minute of the meridian of the

---

Gellibrand's discovery see Stephen Pumfrey, " 'O tempora, O magnes': A Sociological Analysis of the Discovery of Secular Magnetic Variation in 1634," *British Journal for the History of Science,* 1989, *22*:181–214

[28] E.g., Anthony Jenkinson, "A Proposal for a Voyage of Discovery to Cathay, 1565," BL Cotton Galba D9, fols. 4–5; "Advise of William Borrowe for the Discerning of the Sea and Coast Byyonde: Perhaps Whether the Way Be Open to Cathayia, or Not" (ca. 1568), BL Lansdowne 10, fols. 132–133b; Humphrey Gilbert, BL Add. 4159, fol. 175b; Richard Hakluyt, "The Chiefe Places Where Sundry Sorts of Spices Do Growe in the East Indies, Gathered out of Sundry the Best and Latest Authors," Oxf. Bodl. MS Arch. Selden 88, fols. 84–88; William Bourne, *A Regiment for the Sea* (London, 1592), fol. 76; William Barlow, *The Navigator's Supply Conteining Many Things of Principall Importance belonging to Navigation* (London, 1597), fol. b1a; Sir Dudley Digges, *Of the Circumference of the Earth: or, A Treatise of the Northeast Passage* (London, 1612); and Henry Briggs, "A Treatise of the NW Passage to the South Sea, to the Continent of Virginia and by Fretum Hudson" (London, 1622), published in Samuel Purchas, *Hakluytus Posthumus; or, Purchas His Pilgrimes* (London, 1625), Vol. II, pp. 848–854 (see note 43 for the publication history of Purchas's work).

[29] John Dee, *General and Rare Memorials Pertayning to the Perfect Arte of Navigation* (London, 1577); see also Waters, *Art of Navigation* (cit. n. 15), p. 210. William Borough describes a map depicting two compass needles in *Variation of the Cumpas* (cit. n. 22), fol. F2b.

paradoxicall chart whose degrees dothe increase from the pole towardes the equinoctiall."[30]

Just as the plane chart proved totally inaccurate for polar sailing, it was soon found wanting for sailing at most latitudes away from the equator. The basic problem with a plane chart was that it was drawn as if the lines of latitude and longitude were a constant distance apart. Since this was approximately true for lines of longitude at only a very limited distance from the equator, in northern or southern waters distortion rapidly became a problem. Thus a straight-line course on a plane chart would not have corresponded to a straight course on the globe, and a great circle course could only have been approximated by a series of steps across the map.

Mathematical geographers throughout the late sixteenth century had been intrigued by the fact, first discovered in 1537 by Pedro Nuñez, a Spanish geographer, that loxodromes or rhumb lines formed a spiral on the globe. Although this demonstration had no real practical application, the elegance of the proof appealed to the English mathematical geographers and they delighted in repeating it.[31] This proof provided an intellectual challenge and aesthetic delight for these mathematical geographers, as did the creation of map projections, demonstrating that the theoretical dimension of their investigation was as compelling as the practical one. It would thus be misleading to view these new map projections as merely practical or technological improvements, since they also satisfied the aesthetic requirement of neatness of fit so important to the theoretician.

In 1569 Gerard Mercator published his now-famous map projection, which allowed sailing courses and rhumb lines to be drawn as straight lines. It was not until the end of the century, however, that the precise technique of using geometric progression for mapmaking was enunciated mathematically in print. In 1597 William Barlow, in *The Navigator's Supply,* presented a graphical method for creating a Mercator projection that encouraged navigators to draw their own charts. Barlow described all the instruments that were necessary and useful for the navigator, beginning with the compass and ending with Mercator's map projection. His method for creating this projection did not entail any insight into the mathematics involved, merely an ability to use a quadrant and ruler and to follow his somewhat confusing directions. The second, more significant, explanation of Mercator's work came in Edward Wright's *Certaine Errors in Navigation* (1599). Wright provided an elegant Euclidean proof of the geometry involved in this map projection. He pictured the globe as a cylinder, in which every parallel was equal to the equinoctial, and proved that the rhumb lines "must likewise be streight lines." He also published a table of meridian parts for each degree, which enabled cartographers to construct accurate projections of the meridian network, and offered straightforward instructions on map construction.[32] His work was the

---

[30] Captain John Davis, *The Seaman's Secrets* (London, 1595); see also Waters, *Art of Navigation,* p. 201. George Waymouth, *The Jewelle of Artes* (1604), BL Add. MS 19,889, fol. 78b. Both Davis and Waymouth had been on northern expeditions. Davis's travels are recorded in Purchas, *Hakluytus Posthumus; or, Purchas His Pilgrimes* (1625), Vol. II, pp. 463–470; Weymouth's, *ibid.,* pp. 809–813.

[31] E.g., Borough talks about this proof, citing Pedro Nuñez de Medina, in *Variation of the Cumpas* (cit. n. 22), fol. G1b; Barlow, *Navigator's Supply* (cit. n. 28), fols. I1b–2a, and Blundeville, *His Exercises* (7th ed., London, 1636), p. 693, also refer to the proof, but not specifically to Nuñez.

[32] Barlow, *Navigator's Supply,* fols. A1b, K4b; and Wright, *Certaine Errors in Navigation* (Lon-

first truly mathematical rendering of Mercator's projection and placed English mathematicians, for a time, in the vanguard of European mathematical geography.

English mathematical geographers were thus interested in problems that were current, sophisticated, and usually practical. Grounded in the mathematical framework of Ptolemy and attempting to manipulate their image of the world into a geometrically satisfying design, mathematical geographers developed a research program, unspoken though it might have been, that was driven by practical problems and an attempt to use mathematics to improve England's standing in the world. The questions of longitude determination, navigation in northern waters, and map projection were all of paramount importance to England in her drive for new trading routes and new colonies. As well, problems relating to the determination of longitude and map projection had to be solved before geographers could understand the magnetic globe in a geometrical manner. Mathematical geography was thus equally a theoretical and a practical investigation.

DESCRIPTIVE GEOGRAPHY

Descriptive geography developed as a subdiscipline quite distinct from mathematical geography. Less rigorous, less exacting, less arcane, it attracted more casual investigators than did its mathematical counterpart. It also provided important information for English merchants, politicians, and scholars. Descriptive geography helped the English identify themselves as different from the rest of the world. With the incremental accumulation of information about Europe, patterns of national behavior and trading relations could emerge. Descriptions of the four corners of the world confirmed English sentiments of superiority and otherness. This study taught them how to exploit their world and gave them license to do so. Descriptive geography thus emerges as a study that was less theoretical than mathematical geography, but that taught people a great deal about the world around them and was fundamentally useful—socially, politically, and economically.

This branch of geography developed from ancient roots, just as mathematical geography had done. Strabo defined geography as a science that was the concern of the philosopher, while maintaining that it must be first and foremost a *useful* enterprise. For Strabo, the utility of geography "is manifold, not only as regards the activities of statesmen and commanders but also as regards knowledge both of the heavens and of things on land and sea, animals, plants, fruits, and everything else to be seen in various regions—the utility of geography, I say, presupposes in the geographer the same philosopher, the man who busies himself with the investigation of the art of life, that is, of happiness."[33] Strabo cast his net as far afield as possible in his search for geographical information, eschewing the specialized investigation inherent in mathematical studies (which he deemed of lesser importance). The geography he developed was an inductive exploration

---

don, 1599; rpt. Amsterdam: Theatrum Orbis Terrarum, 1974), fols. C4a, D3a–E4a. See also Taylor, *Late Tudor and Early Stuart Geography* (cit. n. 20), p. 76; and Taylor, *Mathematical Practitioners* (cit. n. 3), biographies no. 95, 99.

[33] *The Geography of Strabo*, trans. Horace L. Jones (Loeb Classical Library) (London: William Heinemann, 1917), 1.1.1, p. 3.

of interesting detail, with utility of information (in the widest sense) as the arbitrator of what was to be examined and reported.

Many sixteenth-century geographers followed Strabo's lead; they developed a subdiscipline that owed much to this anecdotal and rambling source.[34] Descriptive geography became a more or less rigorous investigation of the physical features of the globe, including mountains, rivers, cities, and climates, as well as political institutions, different peoples and cultures, and the ever-varying flora and fauna. The most popular aspects of this descriptive branch in the sixteenth and seventeenth centuries might somewhat anachronistically be labeled physical and political geography, though practitioners took occasional forays into the fields of what would now be called cultural geography and anthropology.

The most significant contribution made by English descriptive geographers was the collection of tales of exploration and adventure. Although this genre was first seriously developed by Europeans such as Giovanni Battista Ramusio, whose *Delle navigationi et viaggi* (Venice, 1554–1559) recorded Marco Polo's adventures and Columbus's discoveries, it might be said to recall the earlier, if fanciful, accounts of the supposed Englishman Sir John Mandeville.[35] In a genre dominated by such giants as Ramusio, Peter Martyr, Joseph de Acosta, Jan Huygen van Linschoten, and Theodore and Johann Theodore de Bry, Richard Hakluyt and Samuel Purchas take their place as significant and popular contributors to the European field of collections of voyages and new lands.[36]

When Thomas More encountered Raphael Hythlodaeus and questioned him concerning that hitherto unknown island of Utopia, he fancifully entered into the

---

[34] Strabo's *De situ orbis* was published in five Latin folio editions in the fifteenth century: Rome, 1469 (trans. G. Veronensis and G. Tifernas); Venice, 1472; Treviso, 1480; Venice, Jan. 1494; Venice, Apr. 1494. Versions in the sixteenth century: one Greek (Venice: Aldus, 1516); three Greek and Latin (Basel, 1549; Basel, 1571 [ed. G. Xylander]; Geneva, 1587 [ed. Isaac Casaubon]); three Latin (Venice, 1510; Basel, 1523; Lyons, 1559 [16mo]); and one Italian (Venice/Ferrara, 1562–1565 [4to]). It was published only once in the seventeenth century, in Greek and Latin (Paris, 1620), and thereafter not until the nineteenth century. See Aubrey Diller and Paul Oskar Kristeller, "Strabo," in *Catalogus translationum et commentariorum: Medieval and Renaissance Latin Translations and Commentaries*, ed. P. O. Kristeller and F. Edward Cranz (Washington, D.C.: Catholic Univ. Press, 1960–), Vol. II (1971), pp. 225–233.

[35] *The Voyages and Travels of Sir John Mandeville* was written in about the middle of the fourteenth century; see *The Travels of Sir John Mandeville and the Journal of Friar Odoric*, with introduction by Jules Bramont (Everyman's Library) (London: J. M. Dent, 1928), p. vii. It is apparently an imaginary account of travels through the Middle East to India and Africa. Its putative author claims to have made the pilgrimage to Jerusalem and then an extended journey eastward. From this book come tales of dog-faced men, men with heads in their chests (cited along with cannibals in *Othello* 1.3.143–145), and people with one large umbrella foot. Mandeville was very popular in the late Middle Ages and preserved that popularity well into sixteenth-century Europe. See C. W. R. D. Moseley, "The Availability of *Mandeville's Travels* in England, 1356–1750," *Library*, 5th Ser., 1975, *30*:125–133. The *Travels* went through numerous editions in the sixteenth and seventeenth centuries as well as appearing in epitome in Purchas, *Hakluytus Posthumus; or, Purchas his Pilgrimes* (1625), Vol. II, pp. 128–157.

[36] Petrus Martyr Anglerius, *De orbe novo decades* (Compluti, 1530); Joseph de Acosta, S.J., *De natura novi orbis, libri duo, et de promulgatione evangelii, apud barbaros . . . sive de procuranda Indorum salute libri sex* (Salamantica, 1589); Jan Huygen van Linschoten, *Itinerario: Voyage ofte Schipvaert, van J.H.v.L. naer Ooost ofte Portugaels Indien* (Amsterdam, 1595–1596), also published in Latin as *Navigatio ac itinerarium J. H. Linscotani in Orientalem sive Lusitanorum Indiam* (The Hague, 1599); Theodore de Bry, *America*, Pts. 1–13 (Frankfurt, 1590–1634); Johann Theodore de Bry (son), *India orientalis*, Pts. 1–10 (Frankfurt, 1598–1613). For a discussion of Hakluyt's place in the ranks of these great compilations see G. B. Parks, "Tudor Travel Literature: A Brief History," in *The Hakluyt Handbook*, ed. D. B. Quinn, 2 vols. (2nd Ser., 144) (London: Hakluyt Society, 1974), pp. 97–132.

vocation of Richard Hakluyt. Hakluyt, a Master from Christ Church, Oxford, and a diplomat, spy, and churchman, spent his life interviewing mariners, navigators, and travelers and collecting stories of new countries, hair-raising adventures, and sea dramas. His great work, *The Principal Navigations, Voyages, Traffiques, and Discoveries of the English Nation,* enumerated the voyages and discoveries of Englishmen in the Americas and the East.[37] Hakluyt's collection, seen with Edmund Spenser's *Faerie Queene* and William Camden's *Britannia* as the cornerstone of both the mature English language and nascent English patriotism, encouraged Britons to see themselves as leaders in the exploration of and trade with the wider world. Rather than offering tales of the valor of the Spanish or Dutch, *Principal Navigations* supplied the English with reflections of themselves. Hakluyt used the words of people who had been there, a style of reporting that lent great verisimilitude to his stories and allowed his readers to see the real passion and poetry, as well as the hard-nosed business sense, of England's travelers. His book let the English mariner or merchant develop a self-consciousness of his role in the world and encouraged him to risk life and limb for the glory of queen, country, and purse. Thus Hakluyt's book provided an ideological foundation for a descriptive study of the wider world. It combined an energetic and often dramatic literary style, and patriotic and pragmatic pride, with a huge portion of fascinating descriptive and navigational information.

Hakluyt's great work encouraged the English to see themselves as separate from the Continent and the rest of the world in two different ways. First, he stressed the primacy of English exploits and contacts, beginning with Arthur's voyage to Britain, and including the trade of Britons in the Mediterranean "before the incarnation of Christ" and the "ancient trade of English marchants to the Canarie Isles, Anno 1526," among others.[38] Second, Hakluyt stressed the dissimilarities between the English and other peoples by describing strange customs and practices.[39] By stressing odd and foreign attributes, he drew a distinction between the "other" and the homely English reader.

---

[37] Richard Hakluyt, *The Principal Navigations, Voiages, and Discoveries of the English Nation* (London, 1589), was largely concerned with explorations of America. His later work, too much enlarged and revised to be seen as an edition of the first, dealt with exploration of the whole world: *The Principal Navigations, Voyages, Traffiques, and Discoveries of the English Nation,* 3 vols. (London, 1598–1600). Subsequent references in this article will be to this version of the work.

The authoritative biography of Hakluyt, though now old, continues to be George Bruner Parks, *Richard Hakluyt and the English Voyages* (New York: American Geographical Society, 1930). For more modern treatments of this important geographer see E. G. R. Taylor, "Richard Hakluyt," *Geographical Journal,* 1947, *109*:165–174; Taylor, *The Original Writings and Correspondence of the Two Richard Hakluyts,* 2 vols. (London: Hakluyt Society, 1935); Edward Lynam, ed., *Richard Hakluyt and His Successors: A Volume Issued to Commemorate the Centenary of the Hakluyt Society* (2nd Ser., 93) (London: Hakluyt Society, 1946); and Quinn, ed., *Hakluyt Handbook.*

[38] Richard Hakluyt, *Principal Navigations,* Vol. I, p. 1; Vol. II, Pt. 1, p. 1, taken from Camden's work; Vol. II, Pt. 2, p. 3. See also Vol. II, Pt. 1, p. 96.

[39] E.g., the Lappians and "Scrickfinnes" "are a wilde people who neither know God, nor yet good order . . . they are a people of small stature, and are clothed in Deares skinnes, and drinke nothing but water, and eate no bread but flesh all raw": *ibid.,* Vol. I, p. 233; "The king of Persia (whom here we call the great Sophy) is not there so called, but is called the Shaugh. It were there dangerous to cal him by the name of Sophy, because that Sophy in the Persian tongue, is a begger, & it were much as to call him, The great begger": *ibid.,* pp. 397–398; concerning China, "So great a multitude is there of ancient and grave personages: neither doe they use so many confections and medicines, nor so manifold and sundry wayes of curing diseases, as wee saw accustomed in Europe. For amongst them they have no Phlebotomie or letting of blood: but all their cures, as ours also in Japon, are atchieved by fasting, decoctions of herbes, & light or gentle potions": *ibid.,* Vol. II, Pt. 2, pp. 88 ff.

Hakluyt's book also reveals an interesting movement toward a "Baconian" or collecting methodology in the human and descriptive sciences. The collection of useful facts—insignificant when taken individually, but amalgamated into a complete world view that was greater than the sum of its parts—corresponds to Francis Bacon's later methodology of tabulation. The many lists of foreign phrases, useful for the traveler and the trader, demonstrated an early consciousness of comparative languages, while descriptions of natives and their customs introduced concepts of anthropology.[40] Botany, zoology, and natural history were not forgotten, with descriptions of native plants and animals, and Hakluyt could even be said to have contributed to economic theory with his faithful rendering of merchants' reports and commodities pricing.[41] These human sciences are found in nascent form in this huge tabulation of geographical description. Hakluyt's massive work can thus be seen as helping to develop a methodology of data collection, as well as an ideology that helped establish this technique of collection and its product as morally neutral. In other words, by collecting information in a putatively objective manner (simply recording what others reported), Hakluyt presents the information itself as neutral.[42] This was clearly not the case, since the very essence of Hakluyt's work was embued with values of the supremacy of England and Protestantism and of the power of the Old World over the New. *Principal Navigations* thus provides an early example of, and perhaps an inspiration for, the methodology of inductive tabulation and the hidden ideology of power that went hand in hand with such a technique.

Samuel Purchas, Hakluyt's literary and spiritual successor, developed a very different emphasis in his great collection of English voyages. Purchas worked as Hakluyt's assistant during the latter years of Hakluyt's life and purchased Hakluyt's unpublished manuscripts after his death. He used these manuscripts and other geographical research in two great works: *Purchas His Pilgrimage* (1613) and *Hakluytus Posthumus; or, Purchas His Pilgrimes* (1625).[43] Both were highly

---

[40] E.g., on languages: "Divers Words of the Language Spoken in New France, with the Interpretation Thereof": *ibid.*, Vol. III, pp. 211, 231; "The Interpretation of Certeine Words of the Language of Trinidad Annexed to the Voyage of Sir Robert Duddeley": *ibid.*, pp. 577–578. For anthropological material, e.g.: "Certaine Letters in Verse, Written out of Moscovia, by M. George Turbervile, Secretary to M. Randolfe, Touching the State of the Countrey, and Maners of the People": *ibid*, Vol. I, p. 384; and "Observations of the Sophy of Persia and of the Religion of the Persians": *ibid.*, p. 397.

[41] E.g., on native plants and animals: "A Testimony of Francis Lopez de Gomara, concerning the Strange Crook-Backed Oxen, the Great Sheepe, and the Mightie Dogs of Quivira," *ibid.*, Vol. III, p. 308; and "A Notable Description of Russia—the Native Commodities of the Countrey," *ibid.*, Vol. I, pp. 477–479. And on topics of economic interest, e.g.: "The Letters of the Queenes Majestie Written to the Emperour of Russia, Requesting Licence and Safe-Conduct for Anthonie Jenkinson," *ibid.*, p. 338; and "A Note of All the Necessary Instruments and Appurtenances Belonging to the Killing of the Whale," *ibid.*, p. 413.

[42] *Ibid.*, Vol. I, preface, p. xxiv.

[43] *Purchas His Pilgrimage; or, Relations of the World and the Religions Observed in All Ages and Places Discovered, from the Creation unto This Present* went through four editions: 1613, 1614, 1617, and 1626. The 1626 edition is often seen as Volume V of *Hakluytus Posthumus; or, Purchas His Pilgrimes, Contayning a History of the World, in Sea Voyages, and Lande Travels, by Englishmen and Others* (London, 1625), but is actually a completely separate work. *Hakluytus Posthumus* was published in four volumes and has been printed in twenty volumes in the definitive Glasgow edition (1905); it will hereafter be cited as Purchas, *His Pilgrimes*. His third work, *Purchas His Pilgrim or Microcosmos; or, The Historie of Man* (1619), a moral rather than geographical treatise, was highly condemnatory of the degeneracy of man and was never reprinted. For an analysis of the relationship

moralistic, almost Puritan in tone, and were aimed at the armchair traveler and island-bound country gentleman rather than at the practical navigator or merchant.[44] This change in intended audience reflects the growing popular interest in travel literature, as well as the increased stress on religious controversies as the century progressed. Purchas told tales of new discoveries and described peoples, customs, and natural settings in great detail; he tended, however, to dwell on "Mans diversified Dominion in Microcosmicall, Cosmopoliticall, and that spirituall or heavenly right, over himselfe and all things, which the Christian hath in and by Christ" and "the diversities of Christian Rites and Tenents in the divers parts of the world," rather than dealing with rates of exchange or the feasibility of a Northwest passage, as Hakluyt had done. He also stressed the treachery of other races toward the English and the victories over these lesser people by his superior Protestant countrymen.[45] Purchas's pedantic prose is far less captivating than Hakluyt's more lively renditions, and he succeeds best where he follows Hakluyt in allowing the explorers and travelers to speak for themselves. While Purchas continued to encourage the self-promotion of the English nation, this movement relied less on his original work than on the impetus already supplied by Hakluyt. What Purchas contributed to English descriptive geography was a growing Protestant bias and an increasing belief in the ability and need of the English to achieve a Protestant hegemony over the pagan and Catholic world.

Most original English descriptive geography, aside from these collections and translations, consisted of travel literature. Briefly, the vast majority of such English travel texts described the New World, especially Virginia. Many were promotional, attempting to attract colonists and investors to the new settlement; others were apologias for the strife-ridden colony.[46] The second and smaller group of travelers' tales described Europe. These were largely political in nature, commenting on government and mores in the various countries visited. The most frequently described country was Russia, significant for English mercantilism at this time. After the first disastrous voyage of Sir Hugh Willoughby and the somewhat more successful one of Sir Richard Chancellor, the English maintained significant contacts with the court of Ivan the Terrible, and through him were often able to circumvent the Mediterranean trade embargo of Suliman the Magnificent. These adventures were frequently described in highly evocative prose.[47]

---

between Hakluyt and Purchas see C. R. Steele, "From Hakluyt to Purchas," in *Hakluyt Handbook,* ed. Quinn (cit. n. 36), pp. 74–96.

[44] Purchas, *His Pilgrimes,* Vol. I, preface, fol. ¶5a. Purchas claimed that travel might be injurious to naive Protestant youths, while they could read about other countries with no danger.

[45] Ibid., Vol. I, bk. 1, pp. 6, 147 (quotations). On treachery and triumph see, e.g., "A True and Briefe Discourse of Many Dangers by Fire, and Other Perfidious Treacheries of the Iavans": *ibid.,* bk. 3, pp. 167–170; see also pp. 156, 179, 206, 251–253, and bk. 10, p. 1853.

[46] The former included John Brereton, *A Briefe and True Relation of the Discoveries of the North Part of Virginia . . . Made This Present Year 1602,* with annexed: *A Treatise Containing Inducements for Planting,* by Edward Hayes (London, 1602); Robert Johnson, *Nova Britannia: Offering Most Excellent Fruites by Planting in Virginia* (London, 1609); and Captain John Smith, *A Description of New England* (London, 1616). The latter included John Smith, *A True Relation of Such Occurrences as Hath Happened in Virginia* (London, 1608); William Symonds, ed., *The Proceedings of the English Colonie in Virginia* (Oxford, 1612); and Rev. Alexander Whitaker, *Good Newes from Virginia* (London, 1613).

[47] Accounts of these Russian exploits included Jerome Horsey, "Coronation of the Emperor of Russia" (printed in Hakluyt, *Principal Navigations,* Vol. I, pp. 466–470, and in Purchas, *His Pil-*

The Middle East also held out lures of rich trade, exotic culture, and traditional peregrination routes; this region provided a third area of emphasis for original descriptive geography. Many a traveler encountered significant hardship as well as stunning beauty, as reported by such travelers as Sir Anthony Sherley, George Sandys, and that unlucky Scot, William Lithgow.[48] Travel literature by and for English readers and writers thus stressed the New World, Europe (including Russia), and the Middle East, areas where England had significant financial involvement.

Descriptive geography, as these examples show, was based largely on personal experience, an emphasis in which it differed from mathematical geography. Descriptive geography was also much more accessible than mathematical geography, lending itself to the enthusiastic amateur, rather than merely to the highly trained specialist. In fact, descriptive geography, in line with its emergence as a "collecting science," rejected specialization in favor of the work of generalists.

Descriptive geography helped the English to identify themselves as separate from the Continental unrest they saw before them, especially in the French wars of religion and the Dutch revolt. It provided them with a model of inductive methodology, which would become more important as the human sciences developed. It encouraged them to regard the world as an endless source of wondrous tales and new goods, thereby creating a mentality that would condone and encourage the exploitation of foreign peoples and resources. In the short term, this study provided court polish, skill in vernacular languages, economic information, and political comparisons. In the long run, it set in place a mentality of separateness and exploitation that would encourage the growth of the English empire.

CHOROGRAPHY OR LOCAL HISTORY

Chorography, the third branch of geography studied in the late sixteenth century, was the most encompassing of the geographical arts, in that it provided the specific detail to make concrete the other general branches of geography. Chorography was essentially useful, since it emphasized the surveying of estates close to

---

grimes, Vol. II, pp. 70–74); Edward Webbe (servant at Moscow to Anthony Jenkinson), *The Rare and Most Wonderful Thinges which Edward Webbe an Englishman Borne, Hath Seene and Passed in His Troublesome Travailes* (London, 1590); and Giles Fletcher, *Of the Russe Commonwealth* (London, 1591), BL Lansdowne MS 60. There is some doubt about the date of Fletcher's work. According to Taylor, *Late Tudor and Early Stuart Geography* (cit. n. 20), p. 202, it was "written 1583 and temporarily suppressed" until 1591. According to the *Dictionary of National Biography* (cit. n. 22), Vol. VII, p. 301, the book was "suppressed and partially printed only in Hakluyt and Purchas." Both Horsey and Fletcher were reprinted in *Russia at the Close of the Sixteenth Century*, ed. Edward A. Bond (London: Hakluyt Society, 1856).

[48] Sir Anthony Sherley traveled to Venice, where he received the title "Count," then proceeded to Persia; he acted as envoy for the Shah to Czar Boris in Russia, to Rudolph II, and to Clement VIII. This was reported in several works, including his own *True Report of Sir A. Shierlies Journey Overland to Venice* (London, 1600); Sherley, *Relation of His Travels into Persia* (London, 1613); William Parry, *A New and Large Discourse of the Travels of Sir A. Sherley, Kt, by Sea and Overland to the Persian Empire* (London, 1601); and Robert Cottington, *A True Historicall Discourse of Muley Hamets Rising and the Three Kingdoms of Morocco, Fes and Sus . . . the Adventures of Sir Anthony Sherley, and Divers Other English Gentlemen . . .* (London, 1609). George Sandys's travels were recorded in *A Relation of a Journey Begun An. Dom. 1610* (London, 1615); while William Lithgow's appeared in print in *A Most Delectable and True Discourse, of an Admired and Painful Peregrination from Scotland to the Most Famous Kingdomes in Europe, Asia and Affricke* (London, 1614).

home, the compilation of genealogies of local families, and the description of local attractions and commodities. Yet there was a theoretical thread running through it, especially in the emphasis on the mapping of time, chronology, which was related to genealogy and local description but was founded on an attempt to develop mathematically a theoretical definition of absolute and relative time. The political motivations for and manifestations of the study of chorography were more overt than were those for mathematical and descriptive geography. By describing local places, chorography implicitly called to mind descriptions of the wider world, thereby encouraging the English to define themselves as self-contained and separate from their Continental counterparts. The improvement of surveying techniques and practices was motivated by attempts at political and economic aggrandizement on the part of middling gentry and merchant families. Even more important, local genealogies were drawn up in response to a search for ancient and respectable roots that would allow this group to enter the ranks of the armigerous gentry, as well as identifying them with their county or geographical area. Finally, the study of chorography, more perhaps than that of the other two subdisciplines of geography, encouraged the development of an inductive and public spirit in the human sciences. Chorographers classified time and place in an effort to reduce their surroundings to a manageable and controllable object. The methodology of incremental fact gathering was fundamentally important to chorographers in the development of their subdiscipline and, when combined with the political and economic motivations that drove the study, provided a science that accepted public scrutiny while placing a high value on thoroughness, tabulation, classification, and the development of a community of like-minded scholars.

Chorography, like mathematical geography, owed its early definition to Ptolemy's *Geographia*. By the sixteenth century the term *chorography,* or local history as it might be called, was applied to any study of local places or people. This study often, though not always, focused on one town or county, where, through its detailed enumeration of local sights, marvels, and commodities, it would engender local pride and loyalty.

The Renaissance study of chorography originated in Italy.[49] Flavio Biondo's *Italia illustrata* (1482) influenced an entire generation of local historians. A Renaissance humanist, Biondo applied the techniques he had mastered in dealing with classical texts to the evaluation of the Italian countryside, proceeding province by province, town by town. This investigation of the local setting as text spread throughout Europe in the fifteenth and sixteenth centuries, influencing Konrad Celtis, Georgius Braun, and Francis Hohenberg in Germany, and even Jean Bodin in France.[50] Its influence came late to England, reaching British

---

[49] My description of chorography owes much to Mendyk, "*Speculum Britanniae*" (cit. n. 10); and Mendyk, "Early British Chorography," *Sixteenth Century Journal,* 1986, *17*:459–481. Mendyk describes well the work of sixteenth- and early seventeenth-century British chorographers, although I take exception to his claim that the introduction of Baconian ideals into this discipline, occurring with the foundation of the Royal Society in the 1660s, changed chorography into natural history. Rather, these "Baconian" tendencies existed long before the Civil War and themselves helped to shape the ideology of both the Lord Chancellor and the Royal Society.

[50] Konrad Celtis, *Germana illustra; or, De Origine situ moribus et institutis Norimbergae libellus incipit* (Nuremberg, 1502); Georgius Braun and Francis Hohenberg, *Civitates orbis terrarum* (Cologne, 1597); Jean Bodin, *Methodus, ad facilem historiarum cognitionem* (Paris, 1566); and Bodin,

shores only in the sixteenth century. There it mingled with a much older native chronicle tradition to form a chorographical study that was uniquely British and helped to develop pride in county and to breed familiarity with a rapidly developing inductive approach to the natural world.

The native English tradition owed its clearest articulation to Ranulf Higden, whose *Polychronicon* (ca. 1350) was a means by which sixteenth-century antiquaries became acquainted with earlier chorographies. Higden encompassed and superseded earlier English works. He changed the emphasis from chronicle to chorographical description by presenting physical and political details on much of England, as well as including antiquarian information such as the original location of Roman roads and ancient place names, linguistic history, and accounts of legal terminology and organization.[51]

This medieval chorographical tradition represented by Higden combined with the work of Continental local historians, inspired by a reevaluation of classical literature and methods, to produce a new English chorography in the sixteenth century. This study was distinct from its medieval antecedents but was not simply a copy of its Continental sources of inspiration. Rather, sixteenth-century English chorography united a medieval belief in the ordered development of time and place with a more modern interest in classifying and dividing that same time and place. As economic gains could be won by a more accurate accounting of disputed property, as social advancement could be attained through careful genealogical analysis, as religious controversy necessitated the establishment of God's time, rather than simply local or relative time, so chorography developed to answer these concerns.

The geographical subdiscipline of chorography developed in response to these problems in three separate yet clearly related directions: surveying and cartography supplied a geometrical and legal emplacement of local lands and county seats;[52] chronology attempted, using a rigorously mathematical method, to set the world and its history, both natural and local, into a single schema of time measurement;[53] and antiquarianism, genealogy, and local history more generally

---

Les six livres de la Republique (Paris, 1576). See also Denis Hay, *Annalists and Historians* (London: Methuen, 1977), p. 109.

[51] Ranulph Hidgen, *Polychronicon*, written ca. 1350; first published in complete form as *Cronica Ranulphi cistrensis monachi (the book named P. Proloconycon [sic]) . . . Compiled by Ranulph Monk of Chrestre . . .* ([Westminster: W. Caxton], 1482, rpt. 1495 (both folio)). Selections of Higden's work were also printed in 1480 as *The Descrypcyon of Englonde* and rpt. 1497, 1502, 1510, 1515, 1528 (all folio); rpt. in *Rerum Anglicarum scriptorum veterum*, ed. H. Savile, 3 vols. (London, 1596) (folio). On Higden see Mendyk, "Early British Chorography" (cit. n. 49), pp. 462, 463.

[52] E.g., Valentine Leigh, *The Most Profitable and Commendable Science of Surveying* (London, 1577); Leonard Digges, *A Geometrical Practise, Named Pantometria . . . Lately Finished by Thomas Digges His Sonne* (London, 1571); Edward Worsop, *A Discoverie of Sundrie Errours* (London, 1582); Cyprian Lucar, *A Treatise Named Lucarsolace* (London, 1590); as well as Christopher Saxton, *Atlas* (London, 1579); and John Norden, *Speculum Britanniae* (London, 1592). See Sarah Tyacke and John Huddy, *Christopher Saxton and Tudor Map Making* (British Library Series, 2) (London: British Library Reference Division, 1980), for a full discussion of Christopher Saxton's life, times, and work.

[53] For an interesting treatment of this question of time see Anthony Grafton, "Joseph Scaliger and Historical Chronology: The Rise and Fall of a Discipline," *History and Theory*, 1975, *14*(2):156–185; and Grafton, "From *De die natali* to *De emendatione temporum:* The Origins and Setting of Scaliger's Chronology," *Journal of the Warburg and Courtauld Institutes*, 1985, *48*:100–144. For the larger picture see Donald Wilcox, *The Measure of Times Past* (Chicago: Univ. Chicago Press, 1987). The most famous English chronologer was James Ussher, author of *Annales veteris testamenti a prima*

provided new techniques drawn from classical sources and fueled both a curiosity concerning local and ancient sites, as well as natural history, and a social drive for antiquity or gentility. The study and practice of surveying and cartography, mathematical in technique and in application, developed in part as a response to the break with Rome and the resultant dissolution of the monasteries, leading to a race to survey and claim various unidentified and neglected plots of land formerly belonging to the religious foundations.[54] The study of chronology, a descendant of both medieval chronicles and the *computus* tradition of calendar reform, became imbued in the sixteenth century with a new sense of historical perspective and a passion for objective mathematical accuracy. The main branch of chorography, which included genealogy and antiquarianism, was perhaps motivated by notions of the desirability of objective description through the incremental collection of facts, leading to a categorization of individuals and of places that would allow the discovery of God's order. As well, it gained much of its momentum from a desire on the part of its practitioners to establish their own categories of gentility and social value and from an increasing sense of local and national pride, shown by a tendency for local inhabitants implicitly to identify themselves with their country and increasingly with their county.[55]

The most important figure of this "collecting" study in Elizabethan England was William Camden. Camden's *Britannia,* first published in 1586 and enjoying a seventh Latin edition by 1607, amalgamated the volumes of chorographical information collected over the preceding twenty years into one huge work. Just as Hakluyt's collection of voyages forever changed descriptive geography, so Camden defined and stabilized the genre of local history. In *Britannia* Camden brought together the study of all aspects of human habitation: history, locale, linguistics, genealogy, and etymology. He greatly expanded his work in succeeding editions, but never lost sight of the goal of local description.[56] He carefully defined the etymologies of each place-name, in a manner reminiscent of medieval encyclopedists, but with a new sense of textual and critical rigor. He quoted ancient and medieval authorities, but with a highly critical eye. He described

---

mundi origine deducti (London, 1650). For chronological manuscripts see, e.g., Thomas Lydiat, Oxf. Bodl. MS Bodl. 666; Oxf. Bodl. MS Add. C 297, fols. 9–10b; MS Bodl. 313, fols. 28–29, 68a–b; Robert Chambers, Oxf. Bodl. MS Add. C 297, fols 104a–b; and Robert Vaughan, Oxf. Bodl. MS Add. A 380.

[54] Mendyk claims that the English chorographical movement was a result of the break with Rome because the Church of England needed to establish its own historical precedents: *"Speculum Britanniae"* (cit. n. 10), p. 8. Rather, I believe the relationship of chorographical interests to Henry's "Great Matter" was personal and economic—an attempt on the part of land-hungry gentry to take over more land with as little cost to themselves as possible. See, e.g., Peter Eden, "Three Elizabethan Estate Surveyors; Peter Kempe, Thomas Clerke, and Thomas Langdon," in *English Map Making, 1500–1650,* ed. Sarah Tyacke (London: British Library, 1983), pp. 68–84, esp. p. 76.

[55] Richard Helgerson, "The Land Speaks: Cartography, Chorography, and Subversion in Renaissance England," *Representations,* 1986, 16:51–85, contends that the proliferation of county maps in the seventeenth century and their changing iconography indicate a movement away from the crown toward county identification and loyalty. Mendyk also sees patriotism driving the investigations of such men as Leland, Camden, and Lambarde: *"Speculum Britanniae,"* pp. 38–56.

[56] William Camden, *Britannia, sive Florentissimorum Regnorum Angliae, Scotiae, Hiberniae et insularum adiacentium ex intima antiquitate chorographic descriptio* (London, 1586). This publication was followed by seven more Latin editions: London, 1587; London, 1590; Frankfurt, 1590; London, 1594; London, 1600; London, 1607; Frankfurt (2 vols.), 1616. There were also two English editions: London (2 vols.), 1610; London, 1637. For a biography of Camden see F. J. Levy, "William Camden as a Historian" (Ph.D. diss., Harvard Univ., 1959); see also Levy, "The Making of Camden's *Britannia"* (cit. n. 10).

monuments, especially Roman inscriptions, churches, the countryside, and country seats. In keeping with his heraldic position as Clarencieux King of Arms, Camden carefully included genealogies of prominent English families. Anything involving the British human condition, especially of the middle or upper classes, came under his purview. In *Britannia,* a new standard of critical treatment of the sources combined with a long-standing interest in the wealth and antiquity of Britain to create a book that set the pace for British local history in the century that followed. It helped the English define themselves by their setting and history in a way that aided burgeoning national consciousness and individual self-expression.[57]

Influenced by such geographers as Abraham Ortelius and Mercator, Camden relied on a huge network of compilers, including William Lambarde, Humphrey Lhuyd, Sampson Erdeswick, Richard Carew, Sir Robert Cotton, and Reginald Bainbrigg.[58] Most of these men contributed their own descriptions to the genre, as well as aiding Camden in his great work. William Lambarde, for example, the first Englishman to produce a chorographical study of a single county, described his own county of Kent in a work published in 1576 that was copied by most chorographers up to the Civil War. This work grew out of Lambarde's interest in Old English law. It was a detailed description of notable sights and people of the county and exuded a sense of Lambarde's genuine pleasure in the depiction of native antiquities and genealogies. Lambarde was also interested in compiling a complete description of Britain, but abandoned the task when he discovered his friend Camden to be working on an identical project. Instead, Lambarde supplied Camden with material on Kent and read Camden's manuscript for him.[59]

Camden also drew on material concerning Staffordshire gathered by Sampson Erdeswick. Erdeswick's *View of Staffordshire,* compiled from 1593 until his death in 1603, contained a wealth of local historical detail, some of which he had previously supplied to Camden and other sections of which were incorporated into the *Britannia* in its later editions. Erdeswick wrote in an informal and conversational style, rather as if writing a letter to a friend. He described Staffordshire by following the courses of the major rivers and detailing each town, manor, or other habitation or marvel that lay along their paths. This was a clever organizing principle, allowing readers to set out vicariously on such a journey. It was this style, indeed, that would soon lead to the geographical novel, similarly constructed to simulate a journey through space and time.[60]

---

[57] William Camden, *Britannia, 1695: A Facsimile of the 1695 Edition,* with an introduction by Stuart Piggott (Newton Abbot, Devonshire: David & Charles Reprint, 1971), preface, fol. e1b. See also the introduction, p. 9.

[58] See, e.g., Oxf. Bodl. MS Smith 71, for letters from Camden to Cotton concerning antiquities; Oxf. Bodl. MS Smith 86, fol. 97, for collections from Dr. Dee; BL Lansdowne MS 121, fols. 160–164b, "Instructions for the Pictes Wall Sent by Mr. Reginald Bainbrigg to Mr. William Camden"; and Oxf. Bodl. MS Smith 74, for letters from Camden to various foreign scholars, including Paul Merula and Abraham Ortelius.

[59] William Lambarde, *A Perambulation of Kent: Contayning the Description, Hystorie, and Customes of That Shyre* (London, 1576). See Oxf. Bodl. Rawl 4to 263 for Lambarde's copy with corrections and additions for the second edition. Many of Lambarde's editorial notes involved additions to genealogical trees or new marvels and details of various places. On Lambarde's help to Camden see Mendyk, *"Speculum Britanniae"* (cit. n. 10), p. 49.

[60] Sampson Erdeswick, *A View of Staffordshire Containing the Antiquities of the Same Country,* compiled from 1593 to 1603. I have discovered two manuscript exempla: Oxf. Bodl. MS Gough

Richard Carew, a student with Camden at Christ Church, supplied another source of local material for the *Britannia* as well as producing his own eloquent and comprehensive work. Carew's *Survey of Cornwall,* first circulated in manuscript form and only published in 1602 at Camden's behest, was a model of chorographical eloquence. In it he described the fish, fowl, and animals of Cornwall, including a dog who had altruistically supplied his blind mastiff companion with daily meals. Probably more important to Carew and to Cornwall was his lengthy description of Cornish tin and tin mining, subjects corresponding to Dee's earlier definition of chorography.[61] He also discussed legal and political problems in the county, as well as devoting some time to the more usual etymologies of place-names and genealogies of prominent families. Still, Carew's work begins to approach natural history, with its relative stress on flora and fauna, rather than remaining merely local history. It also demonstrates that an inductive methodology had been introduced into English chorography well before the Civil War.

John Speed, too, was an important chorographer. His hugely popular book, *The Theatre of the Empire of Great Britain* (1611), was largely cartographical, but combined with his *History of Great Britain* (1611) to provide a comprehensive work of chorography. Speed produced a concrete view of the counties of England, both in verbal and cartographical terms, and in so doing implicitly encouraged the English to conceive of their counties in the highly abstract format of geometrical maps, while identifying their own lives (as Speed described them) with those maps and counties.

All of these local descriptions seem to express a belief in the power of categorizing; by naming the parts of their world chorographers perhaps felt that they could establish its natural order. Just as the divine presence and purpose could be uncovered through the study of natural law in the political arena, so too could the proper classification of the countryside, that is, the establishment of a natural order based on names and measurement, reveal the divine hand in English local affairs. This revelation of order and therefore purpose was, in essence, the strength of this nascent inductive method. It gained currency in chorography because of its political and economic implications; chorographers, like lawyers in political philosophy, saw themselves empowered with certain privileges over nature by right of their taxonomic knowledge. Thus the often tedious recitation of places, names, and dates, seen in every local history text, was essential in establishing objective criteria for the natural order of the world and humankind's control over it.[62]

Staffordshire 4 and BL Harley MS 1990. For evidence of Erdeswick's contribution to Camden consider the following: "There is in the other Church-yard a Monument with Saxon caracters (as I take 'em) whereof I caused Worley to take a noat and send the same to Mr. Cambden to Westminster": Oxf. Bodl. MS Gough Staffordshire 4, p. 48. Concerning the geographical novel see Percy G. Adams, *Travel Literature and the Evolution of the Novel* (Lexington: Univ. Press Kentucky, 1983).

[61] Richard Carew, *The Survey of Cornwall* (London, 1602), fols. ¶4a, 113a (the mastiff), 7b–18b (tin mining). The last indicates the close relationship between chorography and the emerging technology of mining and minerals, as exemplified in such works as Agricola's *De re metallica* (Basel, 1556). See also Mendyk, "*Speculum Britanniae*" (cit. n. 10), pp. 77–79.

[62] Donald Wilcox sees the sixteenth century as the beginning of this trend of looking, historically, for *secondary* causes, rather than concentrating on divine ones: *The Measure of Times Past* (cit. n.

What Happened to Occult Qualities in the Scientific Revolution?

By Keith Hutchison

IN THIS ESSAY I seek to re-evaluate current conceptions of the role of occult qualities in the Scientific Revolution. In Renaissance science "occult" qualities were commonly characterized as *insensible*, as opposed to "manifest" qualities, which were directly perceived. Christian Aristotelianism tended to deny the existence of occult qualities, and when it did allow that such a quality was real, it insisted that it was unintelligible, because *scientia* in the medieval tradition was restricted to entities within the range of the human senses. This attitude constituted a major epistemological impasse not surmounted until the seventeenth century. At that time occult qualities became fully and consciously accepted in natural philosophy, just as it became recognized that no qualities were ever directly perceived.

Existing secondary literature, however, tends almost universally to claim that the Scientific Revolution produced a scientific outlook that rejected these occult qualities. The misunderstanding seems to result principally from overlooking significant changes in the connotations of the word "occult" since the year 1600. For if their writings are closely examined, many leaders of the Scientific Revolution can be seen to be explicitly urging the acceptability of occult entities. When they *appear* to be recommending the abandonment of occult qualities, close examination reveals that they are instead objecting to the earlier thesis that the occult is unintelligible, to the use of substantial forms as causal explanations, or to the extremely idiosyncratic occult causes posited by some writers. With the acceptance of insensible agencies into the scope of natural philosophy, the word "occult" lost its connotation of "insensible" and henceforth referred solely to unintelligibility. The Scientific Revolution culminated in a good deal of dispute over occult causes because different philosophies differed in their estimation of the intelligibility of the world. But these disputes have little to do with the original application of the word "occult," and hence must not affect our judgment of what happened to those properties of bodies which were declared occult by the orthodox before the seventeenth century.

WHAT WAS AN "OCCULT" "QUALITY"?

I do not pretend to present here any definitive semantic history of "occult," but even my preliminary analysis will suffice to give us important insights not available if we insist on using the word only in its elusive modern sense. In fact, the current misunderstanding of the term "occult" is compounded by an important ambiguity in seventeenth-century usage of the term "quality." "Quality" was at that time indeed used in its modern sense, to refer to the properties, attributes, or features of an object, but it was also used in a technical Peripatetic sense to refer to the causes of those attributes: the forms or hypostatical *qualitates*, which had a real existence in the ontology of Christian Aristotelianism and related philosophies and served as the explanation of the attributes of bodies. This causal theory was widely rejected by seventeenth-century philosophers, and causal occult *qualitates* were banished *a fortiori*. But it was through their being real *qualitates* that they were thus banished, and not (I shall show) through their being occult: *qualitates* that were not occult were rejected in precisely the same manner. However, these philosphers often used occult qualities as examples when they wished to attack the theory of real causal *qualitates*, creating the impression that they were attacking the existence of the occult effects of those *qualitates*.[1]

At the beginning of the seventeenth century, furthermore, "occult" was part of the technical Peripatetic terminology used to distinguish qualities which were evident to the senses from those which were hidden. In this context it was the antonym of "manifest." Typical manifest qualities were tastes and colors, because they could be immediately apprehended by the senses. Typical occult qualities were planetary influences, the magnetic virtue (apparently unrelated to the perceptible qualities of a piece of rock), or the purported abilities of certain chemicals to effect specific medical cures. If a drug like aspirin, for example, manages to relieve a headache, it does so by virtue of qualities which are imperceptible, and its effect is no direct or indirect reflection of its being a silent, white powder of bitter taste and medium density. We can observe the effects of aspirin, but we cannot observe what it is in aspirin which achieves those effects. As Daniel Sennert put it early in the seventeenth century:

> Qualities are divided in respect of our knowledg into *Manifest* and *Occult*. The manifest are those, which easily evidently and immediately, are known to, and judged by the Senses. So light in the Stars, and Heaviness and Lightness. . . . But occult or hidden Qualities are those, which are not immediately known to the Sences, but their force is perceived mediately by the Effect, but their power of acting is unknown. So we see the Load-Stone draw the Iron, but that power of drawing is to us hidden and not perceived by the Sences. . . . So we perceive with our senses the evacuation caused by purgative medicaments; but we do not perceive that quality by which the purging medicaments do work that effect. After the same manner, we perceive with our Senses the symptoms which Poysons do stir up in our Bodies; but the qualities whereby they cause the said symptoms we perceive not by the sense. By our Senses . . . we perceive Heat in the Fire, by means whereof it heats: but it is not so in those operations which are performed by occult qualities. We perceive the Actions but not the qualities whereby they are affected.[2]

---

[1]This distinction is particularly clear in 16th-century disagreements over the nature of the sacraments. Both Luther and Calvin reject the idea that there is an occult *qualitas* in, e.g., baptismal water, which renders that water effective, but they do not deny the effect of the water. They simply attribute the effect to God rather than to an inherent virtue. See Martin Luther, *Luther's Works*, Vol. I, ed. J. Pelikan (St. Louis: Concordia, 1958), pp. 95–96, 227–228, and Jean Calvin, *Institutes of the Christian Religion*, ed. John T. McNeill, trans. Ford L. Battles, 2 vols. (London: S. C. M. Press, 1961), Vol. II, pp. 1289, 1292 (= 4.14.14, 17). For examples in a more "scientific" context, compare the section heading with the text of René Descartes, *Principia philosophiae*, Pt. IV, §187; *Oeuvres de Descartes*, ed. C. Adam & P. Tannery, 13 vols. (Paris, 1897–1913), Vol. VIII, pp. 314–315; and see Robert Boyle, *The Works of the Honourable Robert Boyle*, ed. T. Birch, 6 vols. (London, 1772), Vol. III, p. 44.

[2]Daniel Sennert, *Thirteen Books of Natural Philosophy*, apparently a translation by N. Culpepper & A. Cole of the 1632 *Epitome naturalis scientiae* (London, 1661), pp. 29, 431.

Today we accept such powers as a matter of course, as my superficially anachronistic aspirin example indicates, very simply and without the need for sophisticated argument, and we have accepted such powers continuously since the seventeenth century.

Substantial evidence that occult qualities were fully accepted by other seventeenth-century philosophers than Sennert will be presented later. More importantly, the same evidence indicates that these philosophers saw their acceptance of such occult qualities as one of the marks of the superiority of their new philosophy over then-orthodox systems of thought. They saw Aristotelianism as unable to handle occult qualities because it placed too much emphasis on the importance of sensation, and failed to solve the central epistemological paradox posed by occult qualities: How can a science based on sense perception handle agencies which by very definition are insensible? Montaigne for one had explicated this paradox late in the sixteenth century, when he attacked the Aristotelian thesis that our senses are complete:

> I make a question whether man be provided of all naturall senses, or no. I see divers creatures that live an entire and perfect life, some without sight, and some without hearing; who knoweth whether we also want either one, two, three, or many senses more: For, if we want any one, our discourse cannot discover the want or defect thereof. It is the senses priviledge to be the extreme bounds of our perceiving. There is nothing beyond them that may stead us to discover them: No one sense can discover another. . . . Who knowes whether . . . by this default the greater part of the visage of things be concealed from us? Who knowes whether the difficulties we find in sundry of Natures workes proceede thence? . . . The proprieties which in many things we call secret [*occultes*] . . . is it not likely there should be sensitive faculties in nature able to judge and perceive them, the want whereof breedeth in us the ignorance of the true essence of such things?[3]

Occult agencies are likely to exist then, says Montaigne, but if they do they will be unknowable. Later natural philosophers agreed that they exist, but found acceptable methods of knowing at least something about them. William Gilbert showed, for example, that even though one could not perceive the magnetic virtue (he believed that its cause was some kind of living soul), the effects of magnetism could be reliably studied by experiment.

OCCULT QUALITIES IN CHRISTIAN ARISTOTELIANISM

Many Aristotelians shared Montaigne's view that occult properties, even when real, were methodologically unstudyable.[4] Indeed, the intellect was seen, in Peri-

---

[3] Michel de Montaigne, *The Essayes of Michael Lord of Montaigne*, trans. J. Florio (1603), ed. H. Morley (London, 1886), p. 302.

[4] I use the terms "Peripatetic" and "Aristotelian" rather loosely, attaching a philosopher to this tradition if he roughly adhered to a doctrine of immanent *qualitates*. In Lynn Thorndike's *History of Magic and Experimental Science*, 8 vols. (London: Macmillan, 1923; New York: Columbia Univ. Press, 1934–1958), there is no sustained discussion of epistemological issues, but my proposals are supported through numerous scattered instances: see Vol. I, pp. 377–379, 431, 644, 646, 778; Vol. II, pp. 8, 29–31, 131, 135, 144, 160–161, 166, 220, 281, 299, 336, 363, 387, 408, 508ff., 535, 545, 555, 573, 603–604, 632, 652–653, 701, 733–734, 769, 789, 829, 837, 886, 891, 893; Vol. III, pp. 157–158, 408, 577, 582; Vol. IV, pp. 118, 170–171, 208, 225, 229, 313; Vol. V, pp. 109, 117–118; Vol. VI, pp. 391, 432. See also David Knowles, *The Evolution of Medieval Thought* (London: Longmans, 1962), pp. 101–102; Armand A. Maurer, *Medieval Philosophy* (New York: Random House, 1964), pp. 183–184, 198, 221, 237–238, 282; John Herman Randall, *The Career of Philosophy*, Vol. I (New York: Columbia Univ. Press, 1962), pp. 31, 33–34, 104, 263, 305; William A. Wallace, *Galileo's Early Notebooks: The Physical Questions* (Notre Dame, Ind.: Notre Dame Univ. Press, 1977), p. 297.

patetic psychology, as operating by means of abstracted sense images, and since only the effects of occult virtues could be sensed, the causes of these effects were outside the range of man's intellect. Occult qualities could thus be detected experimentally, but could not be studied scientifically, since *scientia* in the Aristotelian tradition was, above all, a knowledge of causes. Built upon foundations laid by Plato and Augustine, mainstream medieval thought incorporated a large measure of skepticism and denied that man's reason was capable of achieving extensive knowledge, except when granted divine aid. With the accommodation of Aristotelian realism in the thirteenth century, the demarcation between reason and revelation was established around the level of sense perception: if an entity could not be sensed, then it was unlikely that God wished ordinary men to understand that entity. Occult agencies furthermore were widely regarded as unreliable in operation. Spurious experiential reporting and failure to isolate the precise preconditions of the operation of these agencies commonly led to the cause and effect relations involved being perceived as irregular. This perceived irregularity strengthened the refusal of Aristotelians to classify knowledge of the occult as a branch of science, since *scientia* was seen as dealing only with universal necessary causes. Accordingly, supernatural revelation was widely regarded as the path to a knowledge of occult virtues, and the occult was closely associated with mysticism and demonism. Being outside the province of *natural* philosophy, and dependent on a *supernatural* epistemology, occult powers were excluded from official science, just as their namesakes are today, now that the originals have been fully accepted.

To pretend that these extremely general remarks apply to the whole of that vast and heterogeneous field, medieval and Renaissance philosophy, would be to claim somewhat too much. But the philosophy of this era exhibited a very strong tendency to dismiss the occult, and furthermore, (as we shall see below), the innovators of the seventeenth century perceived this inability to handle the occult as an important fault in the philosophy they were supplanting. No doubt their view of this philosophy was somewhat warped, but the presumptive evidence provided by these opponents can fortunately be supported by strong, though scattered, direct evidence from the Aristotelians themselves. Numerous examples exist of medieval philosophers either failing to recognize that insensible entities can be corporeal, or declaring that what is insensible can only be known imperfectly.

Insensibility a Token of Incorporeality. Perhaps the most telling illustration of the failure to recognize the possibility of insensible matter is Aquinas's declaration that no animals can exist below the threshhold of our senses. "It is not possible," he writes in his commentary on the *Physics*, "that there should be certain parts of flesh and bone which are non-sensible because of smallness."[5] This stance had theological significance, because to accept the existence of animals that man could not sense would seem to lead to a clash with Genesis 2:19-20, where Adam is said to have given names to all the animals in a parade.[6] Genesis suggests further that the whole of creation functions to serve man: the stars are described as "adorn-

---

[5]Thomas Aquinas, *Commentary on Aristotle's Physics*, trans. R. J. Blackwell *et al.* (London: Routledge & Kegan Paul, 1963), p. 34. See also *Aristotle's De anima in the Version of William of Moerbeke and the Commentary of St. Thomas Aquinas*, trans. K. Foster and S. Humphries (London: Routledge & Kegan Paul, 1951), p. 490, and Galileo's discussion in Wallace, *Early Notebooks*, pp. 208-209, 224-225.

[6]See Aquinas, *Summa theologiae*, 1a. 94. 3.

ment," and man is said to have command over all creatures. This view, widely held until the seventeenth century, runs counter to the more modern idea that God has filled his universe with objects that make no impact on the human senses. Elsewhere Aquinas suggests in passing that the sense faculties of fallen men are inferior to those of the original creation: this would allow a fuller corporeal nature to have been accessible to Adam when he gave the animals their names. Augustine certainly includes epistemic impairment as part of God's punishment after the fall.[7]

Yet another symptom of the reluctance of medieval philosophers to accept the possibility of material entities that cannot be seen is their common tendency to use, if only in passing, terms such as "invisible" to refer to spiritual entities. Aquinas, indeed, lumps invisible material things together with darkness, and hence argues that such things are in fact perceived by sight: "Sight perceives both the visible and the invisible, the invisible being darkness, which is apprehended by sight." Aquinas does, however, classify the sun as invisible, because it is so bright that it overpowers the eye. Accordingly the owl was characterized as the animal with the *weakest* eyesight, since it could not even bear normal levels of illumination. Though it was recognized that some animals could see in the dark, such vision was often not explained through increased sensitivity to light but rather seen as evidence for the theory of extramission.[8] As these examples indicate, medieval philosophy had great difficulty in accommodating the existence of anything too "small" to be sensed.

Aquinas does in fact accept that there are some insensible actions in the corporeal world, like magnetic attraction, but he cites such attraction as an "occult virtue which man is not capable of explaining." Further, he insists that many actions which seem to be natural, like magnetism, are in fact supernatural. He rejects, for example, the claim that saintly relics have an occult curative virtue, and insists that since the cures performed by such relics are only performed selectively and do not succeed with every patient, they must be performed by angelic intervention.[9]

Aquinas's relegation of some insensible operations to the realms of the supernatural accords with a standard medieval and Renaissance view of magic, that it was not the magician who performed wonders but rather demons, who were summoned, implicitly or explicitly, by the magician. Such a theory of magic implies either that the magician's paraphenalia does not have occult powers, or that if it does, it is the demon rather than the magician who can deploy the powers. In Augustine's view, demons were aided in tapping such powers by the fact that they had keener senses than men, and Aquinas endorsed this idea, albeit ambiguously. Late in the Renaissance this view met an important competitor when the idea of a natural magic, which proceeded without supernatural intervention, was promulgated, but such a magic continued to be viewed with suspicion by the

[7]See Peter Brown, *Augustine of Hippo* (London: Faber, 1967), pp. 261–262; Aquinas, *Summa theologiae* 1a.99.1, 1a.101. See also Henry Power, *Experimental Philosophy* (1664; New York: Johnson, 1966), preface; George Atwell, *An Apologie, or Defence of the Divine Art of Natural Astrologie* (London, 1660), p. 59; Alexander Ross, *The Philosophical Touch-stone* (London, 1645), pp. 2, 56–57; *Luther's Works*, Vol. I, p. 62.

[8]See, e.g., Augustine, *Soliloquia* 1.3; Aquinas, *Summa theologiae* 1a.64.1, 2a.2ae.171.3; Aquinas, *Commentary on the Metaphysics of Aristotle*, trans. J. P. Rowan, 2 vols. (Chicago: Regnery, 1961), Vol. I, p. 118; Aquinas, *Commentary on De anima*, pp. 301, 305, 317; David C. Lindberg, *Theories of Vision from Al-Kindi to Kepler* (Chicago: Univ. Chicago Press, 1976), pp. 53, 88, 160.

[9]Aquinas, *Summa theologiae* 2a.2ae.96.2; Aquinas, "On the Occult Works of Nature," in J. B. McAllister, *The Letter of Saint Thomas Aquinas De occultis operibus naturae* (Washington: Catholic Univ. Press, 1939), pp. 20, 22. See also Thorndike, *History*, Vol. IV, p. 208.

orthodox. Moreover, magic was not learned by the normal processes of human investigation, but from another magician who in turn learned from another magician and so on back to a magician who learned by demonic revelation.[10]

Insensibility a Token of Unintelligibility. As this conception of the epistemics of magic suggests, medieval thought also had great difficulty accepting the intelligibility of the insensible. Central to this difficulty was Aristotelian psychology, which required the distinction outlined by Sennert between occult and manifest qualities. When an object became known, according to this psychology, it became known through its sense image.[11] As it was sensed, its manifest qualities entered the imagination without the matter composing the object. The forms in the imagination were identical to the sensible forms in the object, and the *modus operandi* of the human intellect was the "sifting" of these forms to abstract the universal and essential forms from the accidental and singular. That process just could not occur in the absence of a sense image, and an occult quality was *a fortiori* outside the scope of the human intellect. As Aquinas comments, "all the objects of our understanding are included within the range of sensible things existing in space. . . . Whenever the intellect actually regards anything, there must at the same time be formed in us a phantasm [i.e., sense image]"; and elsewhere, "Man is not competent to judge of interior actions that are hidden [*qui latent*] but only of exterior motions that are manifest [*qui apparent*]."[12] Such a position could well be used to deny that God is knowable, since he is the prime example of an occult cause, and both Aquinas and Scotus consider the argument when examining the bounds of human reason. This view had the attraction of implying a major limitation on reason as opposed to faith and supporting the traditional skepticism of Christian theology, but both Aquinas and Scotus wish to establish that a measure of natural knowledge of God is possible. So Aquinas does allow some epistemic access to insensible causes, but he insists that such knowledge, acquired from sensed effects of the cause, is defective knowledge, nonquidditive in character.[13]

That Aquinas did not see these epistemic problems as restricted solely to the arena of theology appears from his opinion that magnetism was beyond human comprehension. This pessimism about understanding the nature of magnetism was very common, and persisted up to the end of the sixteenth century and beyond. In 1597, for example, William Barlow contrasted the marvelous-but-explicable behavior of gunpowder with the truly inexplicable behavior of the magnet.[14] Similarly, Augstine cited the *occultissimi* characteristics of quicklime, characteristics that cannot be directly sensed yet can be "experienced" (*sed compertus experimento*) in the sense that they have sensible effects, as a parallel in the material world to the miracles of Christian tradition. Hence he implied that the behavior of quicklime, which grows hot when mixed with the cold element water, yet remains cool when mixed with inflammable oil, is beyond man's understanding. Twelve hundred years later Augustine's example was still being used as a specimen of a

[10]Augustine, *Contra academicos* 1.7.20; Augustine, *De civitate dei* 9.22, 10.8–11, 21.6; Aquinas, *Summa theologiae* 1a.57.4, 1a.110.4, 1a.114.4; Aquinas, *Summa contra gentiles* 3. 101–107.

[11]See, e.g., Aquinas, *Summa theologiae* 1a.85.1; Randall, *Career of Philosophy*, Vol. I, pp. 31–36.

[12]Aquinas, *Commentary on De anima*, p. 456; *Summa theologiae* 1a.2ae.91.4.

[13]Duns Scotus, *Philosophical Writings*, ed. and trans. A. Walter (London: Nelson, 1963), pp. 14–33, Aquinas, *Summa theologiae* 1a.84.7; *Summa contra gentiles* 1.3. On God as an occult cause, see Aquinas, *Summa theologiae* 1a.64.1; Calvin, *Institutes*, Vol. I, pp. 52, 209 (=I.v.1, I.xvi.9).

[14]William Barlow, *The Navigators Supply* (London, 1597), page opp. p. B.

natural marvel "that man's understanding . . . may not apprehend," but could only be known through experience.[15] Cornelius Agrippa's discussion of occult virtues reflects this same general epistemic attitude:

> There are . . . vertues in things, which are not from any Element, as to expell poyson, to drive away the noxious vapours of Minerals, to attract Iron, or any thing else; and this vertue is a sequell of the species, and form of this or that thing; whence also it being litle in quantity, is of great efficacy; which is not granted to any Elementary quality. For these vertues having much form, and litle matter, can do very much; but an Elementary vertue, because it hath more materiality, requires much matter for its acting. And they are called occult qualities, because their Causes lie hid [from our senses], and mans intellect cannot in any way reach, and find them out. Wherefore Philosophers have attained to the greatest part of them by long experience [and conjecture], rather then by the search of reason.[16]

Agrippa makes the distinction, already met in Augustine, between *sensing* an entity and *experiencing* it, occult qualities being within the realm of experience, but outside the realm of sense. The fact that Aristotelianism emphasized the dependence of natural philosophy upon sense images is often regarded as evidence of the "empirical" nature of scholastic thought. But to insist on direct sensation as the foundation of one's epistemology is to devalue all other forms of experience. Thus in the seventeenth century Henry Power could castigate the Aristotelians for being "Sons of Sense" while himself recommending an experiential philosophy.[17] Scholastic *scientia* was reluctant to deal with entities which could *only* be experienced, and hence this philosophy must be regarded as having viewed experience as a poor basis for knowledge. Since experience normally indicates effects separated from their causes, it did not seem to supply the causes required by the Aristotelian conception of *epistêmê*.[18] Furthermore, even the effects themselves were commonly thought to be in doubt, for, as Hobbes put it, "to remember all the circumstances that may alter the success is impossible." This philosophical attitude, a remnant of Aristotelianism, explains why Kepler adopts a markedly defensive tone when he insists that experience is ultimately reliable, and that the old wives' tales polluting contemporary knowledge of occult actions can indeed be eliminated:

> Some lovers of nature . . . have found there are attributed to the stars effects that are certainly not fabricated, but that through protracted empirical experience are attested as regards some general consistency [*convenientia*]. Similarly, the physician first derives from experience that some herb, collected between two [festive] days . . . is supposed to be good for this or that specific ailment; now, since a very great number of such observations, certainly false, have nothing to do with the matter . . . such as the festive-days in themselves, such a herb is used effectively and curatively because of its own nature, or because of a quality that it has in common with many other herbs. . . . Therefore, in the case of materia medica, experience is not suspect, but diligent physicians know how to cultivate this empirical knowledge so that it is no longer mere empiricism or old wives' lore, but something true, reliable. In every way it is also like this with astrological experience. . . . Thus, just as there is little cause to exclude

---

[15] Augustine, *Civitate dei* 21.4–5; *The Book of Secrets of Albertus Magnus*, ed. M. R. Best and F. H. Brightman (ca. 1550; Oxford: Clarendon Press, 1973), pp. 82, 104.

[16] Heinrich Cornelius Agrippa of Nettesheim, *Three Books of Occult Philosophy*, trans. J. French (London, 1651), p. 24 (= 1.10) (inserting material from p. 34).

[17] Power, *Experimental Philosophy*, preface. Cf. *Secrets of Albertus Magnus*, pp. 82–83.

[18] See, e.g., Aquinas, *Commentary on the Metaphysics*, Vol. I, p. 13; Thorndike, *History*, Vol. I, p. 585; Vol. II, pp. 71, 508–509, 769; Vol. VI, p. 358.

medicine from the number of the arts by reason of false or defective experience, so there is as little cause to demand this of the entire and perfect astrology. . . . In its enquiry into the kinds and properties of herbs, medicine initially knew nothing of necessary and certain causes, but has finally learnt of these through diligence and rational conjecture, and it is to some extent still seeking. . . .[19]

Sensibility and the Four Elements. Another important idea introduced in the passage from Agrippa is the classification of a quality as occult if it cannot be accounted for in terms of the four elements of Aristotelian sublunar cosmology. This deficiency was commonly held to define an occult quality.[20] Since the four elements functioned as the basic principles of "perceptible body" in Aristotelian physics, the definition is effectively equivalent to that formulated in the quotation given from Sennert. Aristotle himself presented the four-element theory in the course of analyzing the sensible qualities of matter. Aquinas also relates the four-element theory to a theory of sensation, using it to argue that our senses are complete: qualities that might be sensed by any hypothetical additional sense, he seems to argue, would require that there be additional elements beyond the traditional four. He dismisses the idea abruptly.[21]

Idiosyncrasy of Insensible Actions. Apart from confirming that occult qualities could not be handled by human reason, consigning them outside the four-element system also supported the view that occult qualities were not universally distributed in nature. Since the Peripatetic ideal of *scientia* dealt only with causes which were universal (or near-universal), this was yet another ground for excluding occult qualities from the province of scientific knowledge. Indeed, an occult quality was often referred to as a *property* or *idiosyncrasy*, technical terms used to indicate that it was peculiar to a relatively narrow class of individuals, as opposed to the manifest qualities, which reflected universal characteristics of the four elements present in all terrestial bodies. Every individual body in the sublunary world was similar to every other body by virtue of its being composed of the elements, but it was also a unique body to the extent that it had an individual composition, shared to some extent with other bodies of its species and genus. The occult properties of the body were seen as attached to some entity representing this individuality, such as the substantial form of the body, or its "complexion," or "temperament," or the "whole substance," or the mathematical proportions of the elements.[22]

We have already seen that Aquinas explained the fact that saints' bones performed cures which were selective in nature by proposing that such cures were in fact performed supernaturally, but a naturalistic explanation could also be given by proposing that saints' bones have an occult curative virtue that applies only to particular individual patients. The most famous proponent of this approach to occult virtues is Paracelsus, who rejected the prevailing theory of disease as primarily a disorder of the whole body generated by an imbalance of the four

[19]Thomas Hobbes, *Leviathan* (London: Dent, 1973), p. 22; Johannes Kepler, *Gesammelte Werke*, Vol. IV (Munich: Beck, 1941), pp. 163–164 (my translation).
[20]Aquinas, "Occult Works," p. 21. See also Thorndike, *History*, Vol. II, pp. 664, 667, 892–893; Vol. III, pp. 114, 130–139, 156, 240–245, 395, 408, 414, 440–441, 449, 483; Vol. IV, p. 34.
[21]Aristotle, *On Generation and Corruption*, 328b25–330a30; Aquinas, *Commentary on De anima*, pp. 352–353.
[22]Sennert, *Natural Philosophy*, pp. 432, 436, 439; Thorndike, *History*, Vol. I, p. 643; Vol. II, pp. 209–210, 535, 565–566, 854–855, 906, 910; Vol. III, pp. 245–246, 395, 415, 429, 440–441, 448, 499, 543; Vol. IV, pp. 190–191, 208, 532; Vol. IV, pp. 369, 371; *Secrets of Albertus Magnus*, pp. 75–76.

humors in favor of a conception of disease as a specific affliction of a specific section of the body. As such it was not to be attacked by universal remedies aimed at restoring bodily equilibrium through the manifest qualities of the four elements, but rather by specific chemical or natural agents with a special capacity to cure the particular affliction in question. These curative virtues were so specific that they were even subject to a variability both in time and between individual specimens, just as no two human bodies are identical:

> If the physician is to understand the correct meaning of health, he must know that there are more than a hundred, indeed more than a thousand kinds of stomach; consequently, if you gather a thousand persons, each of them will have a different kind of digestion, each unlike the others. One digests more, the other less, and yet each stomach is suitable to the man it belongs to. . . . It follows that no one drinks the same amount as another, that no one has the same thirst as another. . . .[23]

Precisely how he thought one could come to know such radically individual properties Paracelsus leaves unclear. He often recommends experience, yet reliance on experience presupposes some stability in the virtues being examined. The same applies also to his adoption of a doctrine of "signatures," according to which nature has so arranged things that the occult curative virtue of a plant or chemical will be indicated by some manifest external "sign," just as a man's internal character is revealed by his external physiognomy. Thus the "*Siegwurz* root is wrapped in an envelope like armour; and this is a magic sign showing that like armour it gives protection against weapons. And the *Syderica* bears the image and form of a snake on each of its leaves, and thus, according to magic, it gives protection against any kind of poisoning."[24] Perhaps the truly idiosyncratic virtues in nature could only be recognized by the suprarational intuitions of individual adepts. Only dependence on experience remained current as a solution of Montaigne's epistemological impasse at the end of the seventeenth century, but the survival of that solution required exiling occult virtues that were not universal in scope.

All Actions Ultimately Sensible in Kind. Some Aristotelians recognized the force of the argument from experience, and acknowledged that there were significant insensible actions in the material world. But to reconcile this acceptance with their theoretic commitment to sensibility, they resorted to what might be called a "manifestization" of occult qualities. Although, for example, Agrippa delineates occult virtues as those which exceed the elemental powers, a good deal of his discussion prior to the passage quoted gives elemental accounts of many properties of objects that others would typically have classified as occult. Agrippa claims, for example, that the strange behavior of quicklime noted by Augustine does in fact "follow the nature, and proportion of the mixtion of the . . . vertues" of the elements. Such manifestization of occult qualities was reasonably common.[25] It accepts that certain actions in nature may be insensible and does not interpret this insensibility as evidence that the actions are supernatural, but it insists that the insensibility is more or less incidental, and that the actions are really sensible in kind. The

---

[23]Paracelsus, *Selected Writings*, ed. Jolande Jacobi, trans. Norbert Guterman (London: Routledge & Kegan Paul, 1951), p. 161. See also pp. 102–103, 152–153, 170–171, 203.
[24]*Ibid.*, p. 197.
[25]Agrippa, *Occult Philosophy*, pp. 22–23; cf. *Secrets of Albertus Magnus*, pp. 78–79; Sennert, *Natural Philosophy*, pp. 433–438. See also Thorndike, *History*, Vol. II, pp. 564, 908; Vol. III, pp. 481–483, 531; Vol. IV, p. 228; Vol. VI, p. 358.

approach accepts occult qualities as effects to be accounted for, and attempts to account for these effects in terms of qualities which are preeminently intelligible. But other philosophers insisted that some qualities were genuinely occult. Sennert, for example, argued that if poisons really did act by cold, then ice would be a poison par excellence: and no combination of hot, cold, moist, and dry would ever produce magnetic attraction, just as no mixture of pigments would ever produce anything but a color, even when mixed by the most skilled painter. The philosopher had no right to expect that every effect he finds in nature would be readily intelligible, and he had to accept effects as he found them, whether explicable or not:

> . . . it is a ridiculous thing to deny that which is manifest by Experience, because we cannot tel the reason thereof. As if it were impossible any thing might happen in Nature of whose cause we are ignorant. We are ignorant of most things. And therefore they that would in Natural Philosophy find out the Truth, and not fal into wild and sophistical Opinions, they must begin with things known to the Sense, and so proceed to the Causes and having found them rejoyce in the Works of Nature; and not finding them, confess their own ignorance; but by no means deny things that are manifest. For it is less shameful having found out the effect to be ignorant of the Cause, which is frequently hid from the most expert Philosophers, than together with the cause to be ignorant of the effect.[26]

OCCULT QUALITIES IN THE NEW PHILOSOPHIES

The Aristotelians whom Sennert attacks are often "praised" in secondary literature for their "modernity" in denying the existence of occult agencies, but this is a very dubious judgment. Apart from the question whether there is much point in the historian's distributing such laurels, it is hardly a modern position to insist that the sole natural actions in the world are hot, cold, moist, and dry qualities. Despite their various differences, all adherents of the new science of the seventeenth century were at least agreed that actions beyond these four pervaded the universe, and that such "occult" actions were within the scope of the human intellect. Furthermore, such agreement was not merely implicit in their work, visible only to the retrospective gaze of the historian, but it was explicit and self-conscious. These innovators openly argued that the ability to accommodate occult qualities was one of the signs of the superiority of their new science.

Insensibility No Token of Incorporeality: Descartes. Just after his long discussion of the cause of magnetism in the *Principles*, Descartes announces his confidence that similar mechanical explanations will eventually be found for all other occult qualities: these have finally been brought within the scope of science: ". . . there are no qualities which are so occult, no effects of sympathy or antipathy so marvelous or so strange, nor any other thing so rare in nature (granted that it is produced by purely material causes destitute of thought and free will), that its reason cannot be given by [the principles of the mechanical philosophy.]"[27] Unlike the Aristotelians, Descartes does not have to posit an unknowable *qualitas* behind each occult quality. Instead he can give an explanation based on an insensible mechanism. Furthermore, he does exactly the same with manifest qualities. There are no *qualitates* behind them either, and the apparently sensible qualities of

---

[26]Sennert, *Natural Philosophy*, p. 435; cf. Boyle, *Works*, Vol. III, pp. 294, 297–301.
[27]Descartes, *Principia philosophiae*, Pt. IV, §187 (*Oeuvres*, Vol. IX, p. 309).

bodies are also generated by insensible mechanisms. There remains thus no strict distinction in Descartes's philosophy between the occult and the manifest. All qualities have become occult, for there are no properties of bodies that directly enter the intellect in the manner of the sensible forms of the Peripatetics. In Descartes's view, the function of our perceptions is not to give us a direct picture of reality, but simply to safeguard our bodies. It is then manifest qualities, not occult ones, that Descartes rejects.

This rejection of manifest qualities is in fact commonly recognized as an important feature of the Scientific Revolution, though the fact that it is usually referred to in Lockean terminology, as "the distinction between primary and secondary qualities," obscures its connection with the problem of occult causes. To insist, as adherents to this distinction did, that one's psychological perception of a sensible quality is of a different order of reality from the physical cause of that quality is tantamount to declaring that cause occult. So accepting Locke's distinction is equivalent to denying the existence of manifest qualities, and on this point all proponents of any form of the mechanical philosophy were agreed, though few expressed it this way. On the contrary, many retained the terminology of the Aristotelian distinction, but reinterpreted that terminology, perhaps not too consciously, to accord with their own philosophical outlook. But the persistence of the old terminology should not prevent us from recognizing that the new philosophy did not allow that bodies had attributes that were manifest in the Aristotelian sense. The only attributes that bodies have are those which satisfy the Aristotelian criterion for being occult.

This rejection of manifest qualities is implicit in most of Descartes's work. *Le Monde* begins with a direct attack on manifest qualities, and Descartes constantly reiterates his rejection of them in the *Meditations,* where he particularly wishes to deny that there is anything especially intelligible about the sensible, since the aim of the work is to reverse Peripatetic conceptions of the relative strengths of natural theology and natural philosophy. Descartes wishes to show that natural reasoning alone can lead to a knowledge of God superior to the knowledge it gives us of the sensible world.[28]

Not only does Descartes attack the prevailing belief in the especial intelligibility of the sensible, but he also consciously insists on the existence of insensible entities. He rejects the idea that lack of a sense image of these things prevents us from understanding them and uses his mechanical philosophy to explain how it is that such things do not register on our senses:

> ... [M]any men are unable to believe that there is any substance unless it is imaginable and corporeal and even sensible. . . . [T]hey persuade themselves . . . that there is no body which is not sensible. . . . I consider that there are many particles in each body which cannot be perceived by our senses, and this will perhaps not be approved by those who take their senses as a measure of the things they can know. . . . [I]t should not be wondered at that we are unable to perceive very minute bodies, for the nerves which must be moved by objects in order to cause us to perceive, are not very minute . . . and thus cannot be moved by the minutest of bodies.[29]

---

[28]Descartes, *The Philosophical Works of Descartes*, trans. E. S. Haldane & G. R. T. Ross (Cambridge: Cambridge Univ. Press, 1931), Vol. I, pp. 133–134.

[29]*Ibid.*, pp. 209, 251, 297. Cf. Boyle, *Works*, Vol. I, p. 516; Charleton, *Physiologia Epicuro-Gassendo Charltoniana* (1654; New York: Johnson, 1966), pp. 113–116; Francis Bacon, *Works*, trans. and ed. J. Spedding, R. L. Ellis, and D. D. Heath, 14 vols. (London, 1858–1861), Vol. IV, p. 26.

Sensibility No Token of Intelligibility: Charleton. The most explicit discussion that I have come across of the idea that bodies do not have manifest attributes occurs in Walter Charleton's *Physiologia Epicuro-Gassendo-Charltoniana*. This book contains an illuminating chapter entitled "Occult Qualities Made Manifest," though "Manifest Qualities Made Occult" would perhaps be a more accurate description. In this chapter Charleton attempts, like Descartes, to give a "scientific" treatment of occult qualities by drafting mechanical explanations for them. Charleton begins by explicitly rejecting not the *existence* of occult qualities, but rather the Peripatetic distinction between the occult and the manifest. *All* qualities, redness just as much as magnetism, he argues, are occult, for the causes of what the Aristotelians see as a simple act of sense perception are really quite complex, and dependent upon the hidden mechanical structure of matter:

> . . . the Schools . . . too boldly praesuming, that all those Qualities . . . which belong to the jurisdiction of the senses, are dependent upon Known Causes, and deprehended by Known Faculties, have therefore termed them *Manifest*: and as incircumscriptly concluding, that all those Proprieties of Bodies, which fall not under the Cognizance of either of the Senses, are derived from obscure and undiscoverable Causes, and perceived by Unknown Faculties; have accordingly determined them to be *Immanifest* or *Occult*. Not that we dare be guilty of such unpardonable Vanity and Arrogance, as not most willingly to confess, that to *Ourselves all the Operations of Nature are meer Secrets*; that in all her ample catalogue of Qualities, we have not met with so much as one, which is not really Immanifest and Abstruse, when we convert our thoughts either upon its Genuine and Proxime Causes, or upon the Reason and Manner of its perception by that Sense, whose proper Object it is: and consequently, that as the *Sensibility* of a thing doth noe way praesuppose its *Intelligibility*, but that many things, which are most obvious and open to the Sense, as to their *Effects*, may yet be remote and in the dark to the Understanding, as to their *Causes*. . . .[30]

To say that Charleton and Descartes rejected the existence of manifest qualities but accepted occult ones is not, of course, to say that they accepted the existence of every agency put into this classification by one or another of their opponents. Accordingly, most of Charleton's chapter is an attack on the notions of sympathies and antipathies, "windy terms" (as Charleton calls them) referring, not to real actions at a distance, but to the mere visible effects of insensible mechanism:

> The means used in every common and Sensible Attraction . . . of one Bodie by another, every man observes to be Hooks, Lines, or some such intermediate Instrument continued from the Attrahent to the Attracted; and in every Repulsion . . . there is used some Pole, Lever, or other Organ. . . . Why therefore should we not conceive, that in every Curious and Insensible Attraction of one bodie by another, Nature makes use of certain slender Hooks, Lines, [and] Chains . . . and likewise . . . in every Secret Repulsion. . . . Because, albeit those Her Instruments be invisible and imperceptible; yet are we not therefore to conclude, that there are none such at all. . . . [F]or us to affirm, that nothing Material is emitted from the Loadstone to Iron . . . only because our sense doth deprehend nothing . . . is an Argument of equal weight with that of the Blind man, who denied the Being of Light and Colours, because He could perceive none.[31]

Similarly, if a viol tuned with some strings of sheep gut and some of wolf gut refuses to play in perfect consonance, the reason is *not* an occult *antipathy* be-

---

[30]Charleton, *Physiologia*, pp. 341–342.
[31]*Ibid.*, p. 344.

tween sheep and wolves but rather an occult *mechanism*, "the aer be[ing] unequally percussed and impelled by [the two strings, so that] the sounds created by one . . . confound and drown the sounds resulting from the other."[32]

The fact that Charleton, in common with many of his fellow mechanical philosophers, rejects sympathies and antipathies, is evidence that might be used to support the description of the seventeenth century as rejecting the occult. In the modern sense of the term, this description is probably quite accurate, but it misrepresents Charleton's real attitude. Another possible piece of evidence is a clear attack by Charleton on what he terms "that ill-contrived sanctuary of ignorance, called occult qualities." The Aristotelians who founded this sanctuary, he says,

> thought it a sufficient Salvo for their Ignorance, simply to affirme all such Properties to be *Occult*; and without due reflection upon the Invalidity of their Fundamentals they blushed not to charge Nature Herself with too much Closeness and Obscurity, in that point, as if she intended that all Qualities, that are *Insensible*, should also be *Inexplicable*. . . . [I]nstead of setting their Curiosity on work to investigate the Causes [of a difficult problem], they lay it in a deep sleep, with that infatuating opium of Ignote Qualities: and yet expect that men should believe them to know all that is to be known, and to have spoken like Oracles . . . though at the same instant, they do as much confess, that indeed they know nothing at all of its Nature and Causes. For, what difference is there, whether we say, that such a thing is Occult; or that we know nothing of it.[33]

In this passage Charleton might *seem* to be rejecting occult qualities, but close scrutiny reveals a subtly different attitude: what Charleton really objects to is a *doctrine* of occult qualities used as an intellectual refuge, as the termination rather than initiation of an enquiry. Insensible agencies certainly exist, in Charleton's view, and the natural philosopher has to do a lot more than simply designate them: he must investigate and explain them. Charleton is attacking not occult qualities but the Aristotelians. The same applies to many other apparent attacks on occult qualities in the seventeenth century. Hobbes, for example, makes almost the same point as Charleton:

> in many occasions [the Aristotelians] put for cause of Naturall events, their own Ignorance; but disguised in other words . . . as when they attribute many Effects to *occult qualities;* that is, qualities not known to them; and therefore also (as they thinke) to no Man else. And to *Sympathy, Antipathy, Antiperistasis Specificall Qualities,* and other like Termes, which signifie neither the Agent that produceth them, nor the Operation by which they are produced.[34]

Charleton's remarks also point to another major theme of the Scientific Revolution: the recognition that nature is permeated with *secrets* to which man has reasonable access. Like the closely related existence of occult properties, this is accepted as commonplace today. Because it has become a hackneyed metaphor, the intellectual advance it represents is often unappreciated, but the Scientific

---

[32] *Ibid.*, p. 357.
[33] *Ibid.*, pp. 342–343.
[34] Hobbes, *Leviathan*, pp. 371–372. Another well-known "attack" is that in Galileo, *Dialogue Concerning the Two Chief World Systems*, trans. Stillman Drake (Berkeley: Univ. California Press, 1953), pp. 445, 462, where Salviati says that he "cannot bring himself to give credence to such causes [of the tides] as lights, warm temperatures . . . , occult qualities, and similar idle imaginings." To argue from this that Galileo saw occult qualities as "idle imaginings" is no more valid than to argue that he also saw light and heat as "idle imaginings." Galileo is in fact rejecting *all* celestial influences, occult and manifest, on the tides.

Revolution, with its emphasis on thoroughness and active experimentation in place of uncritical passive observation, depended on such a recognition. Virtually all seventeenth-century scientists draw attention to this issue. Thus Galileo, discussing the unexpected results he has discovered in his study of the strengths of materials and the strange effects of scale, observes "how conclusions that are true may seem improbable at a first glance, and yet when only some small thing is pointed out, they cast off their concealing cloaks [*le vesti che le occultavano*] and, thus naked and simple, gladly show off their secrets." Bacon similarly writes of the need "to penetrate into the inner and further recesses of nature," and criticizes existing "speculation" for ceasing "where sight ceases. . . . Hence all the workings of the spirits enclosed in tangible bodies lies hid and unobserved . . . unless these . . . things . . . be searched out and brought to light, nothing great can be achieved in nature. . . ." To get at the truth Nature must be interrogated under torture and *forced* to reveal her secrets. Hooke urged a study of the "many excellent Experiments and Secrets" of the mechanical arts, and in the *Opticks* Newton also writes of the search for "the more secret and noble works of nature." It is, I suggest, important not to discount these passages as mere rhetoric. They are symptoms of a new approach to nature, new at least among men whom we classify as natural philosophers rather than magicians. Though there were numerous "books of secrets" circulating before the seventeenth century, these generally had a poor reputation, and were not part of official science. One of the most widely known of such works, the *Book of Secrets of Albertus Magnus*, for example, explicitly connects itself with the "science of magic" and declares that it deals with marvels "in which we know no reason."[35]

Sensibility No Token of Effectiveness: Boyle. Like Descartes and Charleton, Robert Boyle took a philosophical stance that assumes no ultimate distinction between the occult and seemingly manifest. But Boyle often avoided emphasizing this consequence of his adopting the mechanical philosophy; he frequently used the old terminology of occult and manifest qualities without constantly reminding his reader that he did not believe manifest qualities were really manifest, just as he did not constantly remind his reader that he did not believe in *qualitates*. He was, he said, interested in things, not words, and he was frequently content to use an old form of words, so long as the conceptions to be attached to these words were not the old misconceptions. Yet on occasion he did confront the issue as to what his terminology should be taken to mean, and he then endorsed the ideas we have already met above. In the *Origin of Forms and Qualities*, he explicitly denied the existence of manifest qualities—"there is no distinct quality in [a] pin answerable to what I am apt to fancy pain"—while the whole of the *Sceptical Chymist* can be interpreted as an elaborate argument against the existence of these qualities. In this work Boyle rejected the four-element theory by showing how impossible it was to use that theory to account for observed effects: even colors, paradigmatic manifest qualities, could not be accommodated. So Boyle showed that all qualities exceed the powers of the elements and thus, like Descartes and Charleton, effectively demolished the Aristotelian distinction between the occult and the manifest by arguing that all qualities are occult.[36]

[35]Galileo, *Two New Sciences*, trans. Stillman Drake (Madison: Univ. Wisconsin Press, 1974), p. 14; Bacon, *Novum organum*, I.18, 50, 98; Robert Hooke, *Posthumous Works*, ed. R. Waller (1705; New York: Johnson, 1969), pp. 27, 36, 43; Isaac Newton, *Opticks* (1730; New York: Johnson, 1952), p. 262; *Secrets of Albertus Magnus*, pp. xi, 3, 82.

[36]Boyle, *Works*, Vol. II, pp. 83–96, Vol. III, pp. 23–26, 41, 292–293, Vol. IV, p. 340.

Such arguments, explicit and implicit, against the existence of manifest qualities Boyle supplemented with overt support for occult qualities. Perhaps his fullest discussion of these qualities takes place in a medical context, in a review of the controversial Paracelsan theory of specific cures. Boyle does not reject Paracelsus's idea of occult curative virtues but specifically commends the mechanical philosophy as being able to accommodate such ideas. "Among the several kinds of occult qualities," he writes, "[those] afforded by the specific virtues of medicines . . . appear to be of much greater importance, than . . . commonly thought . . . because divers learned physicians do . . . disfavour the corpuscular philosophy [because] they think it cannot be reconciled to the virtues of specific remedies. . . ." Indeed not only does Boyle see occult properties as reconcilable with the mechanical philosophy, but he explicitly attacks the Aristotelians for refusing to recognize such virtues. He ascribes their refusal to an outworn theoretical commitment to the view that manifest effects can only be produced by manifest agencies:

> [The reason] physicians are wont to reject, if not deride, the use of such specificks, as seem to work after a secret and unknown manner, and not by visibly evacuating peccant humours (or by other supposedly manifest qualities) [is] generally this; that they see not, how the promised effects can well be produced by bodies, that must work after so peculiar and undiscerned a manner. . . . [T]he naturalists may do much towards the removal of this impediment by shewing . . . as strange operations, as are ascribed to these specificks, are not without example in nature; and consequently ought not to be rejected, barely as being impossible. And indeed the physiology . . . [of] the schools, has done . . . no small disservice, by accustoming [physicians] to gross apprehensions of nature's ways of working. Whence it comes to pass, that not a few even learned doctors will never expect, that any great matter should be performed in diseases, by such remedies, as are neither obvious to the sense, nor evacuate any gross, or at least sensible matter. Whereas, very great alterations may be wrought in a body, especially if liquid, as is the blood and peccant humour, without the ingress or egress of any visible matter, by the intestine commotion of the parts of the same body acting upon another. . . . How much an unperceived recess of a few subtile parts of a liquor may alter the nature of it, may be guessed at, by the obvious change of wine into vinegar; wherein upon the avolation (or perhaps but the misplacing) of so little of the spirituous and sulphureous part, that its presence, absence, or new combination with the other parts is not discernible to the eye, the scarce decreased liquor becomes of a quite differing nature from what it was. . . . That . . . invisible corpuscles may pass from amulets, or other external remedies, into the blood and humours, and there produce great changes, will scarce seem improbable to him, that considers, how perspirable . . . a living body is. . . .

And to demonstrate that the mechanical philosophy can accommodate such actions, Boyle argues that even in *ordinary* machines, it is quite common for manifest effects to have hidden, or at least tiny, causes:

> The faint motion of a man's little finger upon a small piece of iron, that were no part of an engine, would produce no considerable effect; but when a musket is ready to be shot off, then such a motion being applied to the trigger by virtue of the contrivance of the engine . . . throws out the ponderous leaden-bullet, with violence enough to kill a man at seven or eight hundred foot distance.

And the same is true of the human body,

> . . . that scarce sensible quantities of matter, having once obtained access to the mass of blood . . . may . . . give such a new and unnatural impediment or determination to the motion of the blood, as to discompose . . . its texture . . . (as a spark of fire reduceth a whole barrel of gunpowder . . .) need be manifested by nothing, but the

operations of such poisons, as work not by any of those (which physicians are pleased to call) *Manifest Qualities*. For though I much fear, that most of those, that have written concerning poisons, supposing that men would rather believe than try what they relate, have allowed themselves to deliver many things more strange than true; yet the known effects of a very small quantity of opium, or of arsenick, of the scarce discernable hurt made by a viper's tooth, and especially of the biting of a mad dog, (which sometimes, by less of his spittle than would weigh half a grain, subdues a whole great ox into the like madness, and produceth truly wonderful symptoms both in mens bodies and beasts) are sufficient to evince what we proposed.[37]

In this discussion of Boyle's we can observe a repetition of the idea that we encountered in Agrippa, that occult agencies produce disproportionately large effects. Francis Bacon took such "inequality" between cause and effect as one of the defining characteristics of magic. It is then the poor handling of such *instantiae magicae* by Aristotelianism that Boyle is comparing unfavorably with the ease of their accommodation by the mechanical philosophy.[38]

No Tokens of Intelligibility At All: Constructive Skepticism. Boyle, like Charleton, was a figure active in achieving a reconciliation between the new natural philosophy and the skepticism of the late sixteenth century. This skepticism was a continuation of the medieval debates over the roles of revelation and reason, and the problem of delineating the domain of competence of the human mind. Late sixteenth-century skeptics had maintained that the human mind was totally incompetent, leaving revelation as the only source of knowledge. In response to this paralyzing stance, the doctrine of "mitigated" or "constructive" skepticism was developed by Mersenne, Gassendi, and their English followers, while the Cartesians retreated to a new dogmatism. Accepting that the Peripatetic ideal of *epistêmê* was unattainable, the mitigated skeptics settled for an "inferior" science of appearances and effects, in which the search for definitive knowledge about ultimate reality was abandoned. Sensations were accepted (apparently on the theological grounds that God is no deceiver) as being generally reliable, and capable of effective self-correction in cases of illusion, but attempts to glimpse the *Ding an sich* behind these perceptions were seen as futile. For things other than internal sensations, a doctrine of "degrees of certainty" was adopted, and assent was only to be granted partially, in proportion to the evidence available.[39]

Superficially, it might seem that this constructive skepticism would have been hostile to the occult. It denied that we would ever know the ultimate secrets of nature, and in denying further that anything but the immediate sensation is certain, it seemed to support the Peripatetic contention that the insensible is unintelligible. But the skeptics argued that everything else was equally unintelligible, and hence again put the manifest into the same basket as the occult: the cause of redness was just as unintelligible as the cause of magnetism, and the effects of magnetism were just as sensible as the effects of redness.[40] As soon as the negative side of skepticism was sidestepped, the occult became acceptable through the process we have

---

[37]*Ibid.*, Vol. II, pp. 170–171, 175, 183; Vol. V, p. 77.

[38]Bacon, *Novum organum*, II.51. Cf. Thorndike, *History*, Vol. III, p. 441.

[39]See Richard H. Popkin, *The History of Scepticism from Erasmus to Spinoza* (Berkeley: Univ. California Press, 1979), esp. pp. 129–150; Henry G. Van Leeuwen, *The Problem of Certainty in English Thought 1630–1690* (The Hague: Nijhoff, 1963).

[40]See Joseph Glanvill, *Scepsis scientifica* (London, 1885), pp. 145–148; Sennert, *Natural Philosophy*, p. 431.

witnessed: the destruction of the Aristotelian distinction between the occult and the manifest, and the abandonment of the idea that bodies have genuinely manifest qualities. The range of human intellect is thus paradoxically extended by an intellectual movement stressing its impotence, through a reduction in the standards of what constitutes rational thinking. The inconclusiveness of the hypothetico-deductive method, for example, ceased to be a barrier against its use in science, and the method was self-consciously adopted as a means of exploring the insensible realm of nature. Furthermore, the skeptical arguments were used to refute Peripatetic objections to such occult phenomena as actions at a distance: it is beyond the power of man's reason to know that these are impossible, the skeptics say, so they may well exist. Thus Glanvill, one of the leading constructive skeptics in the Royal Society, writes:

> . . . to shew how rashly we use to conclude things *impossible*; I'le instance in some reputed *Impossibilities*, which are only strange and difficult performances. . . . That Men should confer at very distant removes by an *extemporary* intercourse, is . . . a reputed *impossibility*; but yet there are some hints in Natural operations, that give us probability that it is feasible, and may be compast without unwarrantable correspondence with the people of the Air. That a couple of *Needles* equally touched by the same *magnet*, being set in two Dyals exactly proportion'd to each other, and circumscribed by the Letters of the *Alphabet*, may effect this *Magnale*, hath considerable authorities to avouch it. . . . Now though this pretty contrivance possibly may not yet answer the expectation of inquisitive *experiment*; yet 'tis no despicable item, that by some other such way of *magnetick efficiency*, it may hereafter with success be attempted, when *Magical* History shall be enlarged by riper inspections. . . .[41]

Even Descartes enlisted skepticism as an ally in the fight to achieve acceptance of occult entities in natural philosophy. When we investigate the remoter regions of nature, he says, we do not need to insist on rigorous demonstration. The certainty to be required of such explanations as that he has given for magnetism is only moral certainty, comparable in kind to that of the man who manages to decipher a code by trial and error. Other explanations may well exist in both cases, but the philosopher has done his duty when he has found a possible explanation.[42]

The view proposed above of the leaders of the mechanical philosophy, Boyle, Charleton, and Descartes, that they accepted the importance of occult qualities in natural philosophy and criticized the Aristotelians for failure to give a wide enough recognition to occult agencies, does not accord with prevailing descriptions of the seventeenth-century scientific movement. Even recent studies of the "hermetic" component of the Scientific Revolution have not confronted this view, for they have emphasized the survival and influence of seemingly irrational attachments to pre-seventeenth-century belief in occult qualities, and this emphasis has tended to obscure the essential soundness of the occult qualities themselves. Yet it was a consequence of the accommodation of occult qualities by official science that these "irrational" trappings could be dispensed with, for rational techniques to deal with the insensible had finally become available. Many historians have pointed out the affinities between natural magic and post-seventeenth-century science, but the prevailing misunderstanding of the role of occult virtues in the Scientific Revolution has led to the erroneous view that belief in these virtues on the part of the natural magician marks an irreconcilable difference between the two

---

[41]Glanvill, *Scepsis scientifica*, pp. 171–176: cf. Van Leeuwen, *Problem of Certainty*, p. 88.
[42]Descartes, *Principia philosophiae*, IV.204–205; *Philosophical Works*, Vol. I, pp. 300–301.

systems of thought. But in fact the two systems have in common a willingness to deal with occult qualities and a refusal to accept that insensibility implies spirituality: it is within natural magic that we can find precedents for the confidence with which seventeenth-century philosophy insisted that the insensible realms of nature could be profitably entered by human thought. Only in the case of Newton has there been significant recognition that something like occult agencies eventually achieved acceptance in the course of the Scientific Revolution. But even here the Newtonian position is strongly contrasted with the earlier mechanical philosophy, and many historians do not in any case accept the description of gravity as occult. If, however, my evidence is accepted, then all these descriptions clearly require modification.

Unintelligibility No Token of Noneffectiveness: The Dispute over Gravity. The one important obstacle to recognizing that the Scientific Revolution accommodated occult qualities is the dispute between the Newtonians and the Cartesians over gravity, in which the Cartesians claimed that gravity is occult. If the mechanical philosophy could openly accept occult agencies, why did this accusation apparently have force? To resolve this dilemma, we must recall the drift in meaning that the word "occult" has suffered since the late sixteenth century, the drift I have already labeled as responsible for much of existing misunderstanding of the role of occult qualities in the Scientific Revolution. For the disputes over gravity reveal that a significant part of this drift actually took place within the seventeenth century, so that when a Cartesian in 1700 refused to accept universal gravitation on the grounds that it was occult, he almost certainly did not mean the same thing by this accusation as might have been meant some half century earlier. When the seventeenth century opened, "occult" had the double connotation of "insensible" and "unintelligible," the two ideas being bound together by the belief that natural reason could not accommodate the insensible. Over the course of the Scientific Revolution, the intelligibility of many insensibles was recognized, and the distinction between the sensible and the insensible lost most of its earlier force, so the connotation "insensible" became somewhat vacuous. Accordingly, the bond between the two ideas was broken, and "occult" lost the connotation of "insensible," to retain only that of unintelligibility.

The most evident symptom of this drift is the fact that the dispute over gravity was clearly about intelligibility, not about sensibility: everyone agreed that gravitation acted insensibly. But the Cartesians were willing to introduce occult qualities in the old sense of the word into their science only on condition that they were not occult in the new sense, that is, that mechanical explanations could be framed for them. To the Newtonians, on the other hand, intelligibility was not essential, and they were happy to deal with occult entities they could not understand, so long as those occult entities satisfied other criteria, notably that they had been reliably detected, and that they were free of the idiosyncrasy so commonly attached to occult qualities in the Aristotelian era.

Thus the Newtonians did *not* maintain that they had banished occult entities, in either the old or the new sense of the word, but only that they had banished objectionable features of earlier approaches to such entities. Echoing Sennert and Charleton, Newton's spokesman Samuel Clarke insists that observed effects must be accepted even if their causes are unknown. He replies to Leibniz's charge that gravity is a "chimerical thing, a scholastic occult quality," with a rhetorical question that *allows* the possibility that gravity may have an occult cause: "[Is] a manifest quality to be called . . . *occult* because the immediate efficient cause of it (perhaps) is occult?" Newton himself describes gravity and other "active Principles" as "manifest Qualities [whose] Causes only are occult." John Keill sees the successful Newtonian philosophy as an eclectic one, based on borrowings from the other main philosophers: what it has borrowed from Aristotelianism is the idea of a quality. "If the true causes be hid from us," he asks, "why may we not call them occult Qualities?"[43]

Although it is somewhat uncertain what Newton and Clarke meant by a manifest quality here, it is quite evident that neither of them had any objection to Newtonian gravitation's having a cause that might be called occult. But they did insist on an epistemic separation between a discussion of effects and a discussion of causes, and they maintained that one can detect effects reliably, whether or not one understands causes. This methodological point was by no means original with Newton (indeed we have already seen Sennert and Kepler argue to the same effect in earlier defenses of occult causes), but Newton showed more than anyone else how powerful the new method could be. As the attitudes of the opponents of Newton, Kepler, and Sennert indicate, this was a real intellectual advance.

Not only did Newton disapprove of the Cartesian reluctance to endorse the manifest effects of causes which are occult, but he disapproved of the way the Cartesians dealt with the occult causes themselves. Although Descartes rejected the Aristotelian thesis that the insensible was outside philosophy, his attempts to reduce all occult qualities to the effects of peculiar combinations of extension and motion had ended in patent fabrication, and it was impossible to feel confidence in the reality of the speculative mechanisms his imagination had devised. In Newton's view part of the reason for this failure was that Descartes's explanations had been devised individually, with a new mechanical cause postulated for each new effect:

> Could all the phaenomena of nature be deduced from only thre or four general suppositions there might be great reason to allow those suppositions to be true: but if for explaining every new Phaenomenon you make a new Hypothesis if you suppose y^t ye particles of Air are of such a figure size and frame, those of water of such another, those of Vinegre of such another, those of sea salt of such another, those of nitre of such another. . . . If you suppose that light consists in such a motion pressure or force & that its various colours are made of such & such variations of the motion & so of other things: your Philosophy will be nothing else than a system of Hypotheses. And what certainty can there be in Philosophy wch consists in as many Hypotheses as there are Phaenomena to be explained.

Precisely the same objection could be raised against the idiosyncratic virtues of the Aristotelian era:

> To tell us that every Species of Things is endow'd with an occult specifick Quality by which it acts and produces manifest Effects, is to tell us nothing.[44]

Here is a sense in which it might be said that Newton banished occult qualities, but

---

[43]*The Leibniz-Clarke Correspondence*, ed. H. G. Alexander (Manchester: Univ. Press, 1956), pp. 94, 118; Newton, *Opticks*, p. 401; John Keill, *An Introduction to Natural Philosophy* (London, 1745), p. 4.

[44]Isaac Newton, Cambridge University Library MS. Add. 3970.3, fol. 479, quoted from Richard S. Westfall, *Force in Newton's Physics* (London: MacDonald, 1971), p. 386; Newton, *Opticks*, p. 401.

it is not their occultness that he objects to. Rather, it is the earlier practice of positing individual qualities—or even mechanisms—to explain individual effects. To the Peripatetics this was reasonable because the *qualitates* were seen as "real" and separate from the effects they produced, and to attribute the effect of a drug, for example, to a "soporific virtue" served the far from trivial task of locating the cause of drowsiness in the drug itself rather than in some supernatural agency summoned by the drug. To the moderns, by contrast, the seating of the cause of drowsiness *within* the drug was not the only alternative to supernatural causation. The action of the drug, to them, represented some special relationship between the mechanical properties of the drug and the frame of the human body, so that to locate it in the drug itself was mere nominalism, an acceptable way of speaking, but no causal explanation. Furthermore, even if it were true that the action of the drug was supernatural in origin, as Newton at times thought gravity might be, such nominalism allowed one to continue to speak of the action as attached to the drug, and one could study its effects exactly as one would study the effects of nonsupernatural actions, so long as they were regular. So the automatic positing of a *qualitas* behind each observed power was pointless, because such descriptions could only be generally true in a nominalistic sense. And given that each *qualitas* was an isolated individual, no explanatory reduction to general laws was even effected. Occult qualities were certainly banished in this sense, but only because they were real and individual. Their being occult was quite irrelevant here: it was just as unacceptable to Newton to explain individual colors through manifest qualities.[45]

As an alternative, Newton sought "two or three" *universal* occult causes, as exemplified in the gravitational force he discovered and in the chemical and optical forces he continually searched for. Not only do such causes have real explanatory powers, even if interpreted nominalistically, but their existence can be soundly confirmed by the accumulation of evidence. Though the seventeenth century saw removed any objection of principle to occult virtues, the same century also saw abandoned many occult virtues previously believed in, because sound evidence for these particular virtues could not be accumulated. The skepticism pervading the seventeenth century imposed new standards of evidence upon claimants to the title of established fact. Science became intolerant of events which could not be widely observed, and following Bacon's lead, rejected the idea of "unlevel wits," men whose subjective experiences were more valid than others. Experiments were expected to be repeatable, or else the evidence provided by them would be too weak to command significant assent.[46] Totally idiosyncratic occult virtues in the Paracelsan mold could not be accepted into science, because it was impossible to accumulate evidence for them. Universal occult actions such as Newton's gravity, by contrast, could be repeatedly detected by anyone, and evidence for them could be substantial. The less specific a virtue is, the more assent it can command, and the more it can explain. Occult virtues are acceptable to the constructive skeptic, but only after they have been shaved by Ockham's razor.

---

[45] A. I. Sabra, *Theories of Light from Descartes to Newton* (London: Oldbourne, 1967), pp. 290, 294.

[46] Joseph Glanvill, *Essays on Several Important Subjects* (1676; New York: Johnson, 1970), pp. xv, 49; John Locke, *An Essay Concerning Human Understanding* (London, 1690) 4.15–16; Bacon, *Works*, Vol. IV, p. 26; Jacques Rohault, *A System of Natural Philosophy*, trans. Samuel Clarke, 2 vols. (1723; London: Johnson, 1969), Vol. I, pp. 13–14. See Paolo Rossi, *Francis Bacon: From Magic to Science*, trans. S. Rabinovitch (London: Routledge & Kegan Paul, 1968), pp. 27–35.

Although the success of the Newtonian program partially eclipsed constructive skepticism, with Newton himself giving much support to this new dogmatism, there remains a large measure of skepticism in Newton's attitudes, even his conscious ones. Like his skeptical predecessors, Newton insisted that fundamental truth is beyond our reach, since God has the freedom and power to produce the sensible appearance of the world through any of a variety of unknowable means.[47] Newton adopted the notion of different levels of verification, and he accepted effects without understanding their causes. It is unclear whether he regarded gravity as beyond understanding or simply as not yet understood. His voluntarism would have allowed him to accept gravity's incomprehensibility, while his attempting to devise mechanisms for it suggests he thought it within the grasp of reason. But for a skeptically inclined mind the issue is not urgent: it is the effect rather than the cause that takes priority. The Cartesians interpreted the Newtonians' willingness to describe the cause of gravity as occult as a declaration that gravity *could not* be understood, that it was some sort of primary quality imposed directly by divine participation. It was this type of occultness that Leibniz objected to, not occultness in general:

> . . . the ancients and moderns who avow that gravity is an *occult quality*, are right if they mean thereby that there is a certain mechanism unknown to them, by which bodies are impelled toward the center of the earth. But if their notion is that this transpires without any mechanism, by a simple *primitive property*, or by a law of God which brings about this effect without using any intelligible means, then it is a senseless occult quality. . . .[48]

It was thus Newton's voluntarism, and the attached skepticism, or perhaps caricatures of these attitudes, that the Cartesians attacked under the banner of occult qualities. Unlike Newton, the Cartesians refused to base their philosophy upon any entities that were less than perfectly intelligible, and for them, or others who shared their insistence on intelligibility, the word "occult" could be applied in its new sense as a term of abuse. Some who shared this insistence did not agree that the Cartesian or Leibnizian ideas were as perfectly intelligible as their proponents made out. To them, the Leibnizian inherent activity of matter, or the basic mechanism of the Cartesian system, the impact interaction, could be just as occult as Newtonian forces were to a Cartesian.[49] Considerable dispute thus emerged from the seventeenth century as to what was to be counted as intelligible, that is, as to what constituted the reference of the word "occult." But there was widespread agreement over its sense of "beyond understanding." More importantly, there was universal agreement that the Aristotelian criterion for intelligibility—sensibility—was inadequate. The abandonment of this criterion and the exploitation of the epistemological ideas that lay behind this abandonment were undoubtedly major components of the Scientific Revolution.

[47]See, e.g., Isaac Newton, *Mathematical Principles of Natural Philosophy*, trans. A. Motte (1729), rev. F. Cajori, 2 vols. (Berkeley: Univ. California Press, 1966), Vol. II, p. 546; and Newton, *Unpublished Scientific Papers*, ed. and trans. A. R. and M. B. Hall (Cambridge: Cambridge Univ. Press, 1962), pp. 138–145.
[48]As quoted by Cajori in Newton, *Mathematical Principles*, Vol. II, pp. 668–669.
[49]See, e.g., Leonhard Euler, *Opera omnia*, Series II, Vol. III, ed. C. Blanc (Leipzig: Teubner; Zürich: Füssli, 1948), p. 50; Berkeley and Maupertuis, as cited and discussed on pp. 159–160 of Thomas Hankins, *Jean d'Alembert* (Oxford: Clarendon Press, 1970); *Leibniz-Clarke Correspondence*, p. 116; and Peter van Musschenbroek, *The Elements of Natural Philosophy*, 2 vols., trans. J. Colson (London, 1744), Vol. I, preface.

Galileo, Motion, and Essences

By Margaret J. Osler

THE SCIENTIFIC REVOLUTION of the sixteenth and seventeenth centuries produced changes in the foundations of science as well as in the specific content of the several sciences. One important characteristic of the revolution was a shift from an epistemological position which may be called essentialism to a position of nonessentialism. In this paper I shall attempt to characterize these two positions and to argue that the development of Galileo's new concept of motion and his assertion of the doctrine of primary and secondary qualities are both reflections of the move from essentialism to nonessentialism which characterized this period of intellectual history. The discussion will center on certain questions concerning the nature of matter, theories of knowledge, and their interconnections, questions which go to the heart of the nature of science.

Essentialism is the view which maintains that science is concerned with and is able to discover facts about the inner natures or real essences of things. It is based on the assumption either that the world in some way corresponds to our conceptions or that our observations in some manner reveal the inner natures of objects to us. Essentialism is, in this sense, contrary to the view that science is concerned with and is able to discover only the operational or phenomenological attributes of things but can never come to grips with their real essences. There had been nonessentialist philosophies of science before the seventeenth century, most notably in the instrumentalism that followed from Plato's instructions to the astronomers to "save the phenomena" using uniform circular motions, without concerning themselves with the real machinery of the heavens.[1] Plato had not doubted the existence of real essences; he simply denied the possibility of discovering them empirically. In spite of this important counterexample, most science—other than astronomy—and many writings on scientific method before the scientific revolution expressed an essentialist viewpoint. One outcome of the scientific revolution was a shift in emphasis from an essentialist to a nonessentialist philosophy of science.

Deeply enmeshed in the mechanical philosophy, the fundamental assumptions of the new science came to deny the possibility of any knowledge of essences. In any case, such knowledge was regarded as beside the point: what counted, for formulating

---

An earlier version of this paper was presented at the meeting of the West Coast Lazzaroni in San Francisco in March 1972. I wish to thank my colleague Tad Beckman for his helpful suggestions for revising the paper.

[1] Pierre Duhem, *To Save the Phenomena: An Essay on the Idea of Physical Theory from Plato to Galileo,* trans. Edmund Doland and Channah Maschler (Chicago: University of Chicago Press, 1969). Further discussion of the essentialist and nonessentialist positions can be found in L. L. Laudan, "The Idea of a Physical Theory from Galileo to Newton: Studies in 17th Century Methodology" (Ph.D. dissertation, Princeton, 1966); see esp. pp. 4–12.

mathematical laws and for applying science to practical needs, was knowledge of appearances. Galileo himself expressed such a view in the *Two New Sciences* with an eloquence comparable to Newton's *hypotheses non fingo* which was aimed at much the same point:

> The present does not seem to be the proper time to investigate the cause of the acceleration of natural motion concerning which various opinions have been expressed by various philosophers, some explaining it by attraction to the center, others by the decreasing amount of medium to be penetrated, while still others attribute it to a certain stress in the surrounding medium which closes in behind the falling body and drives it from one of its positions to another. Now all these fantasies, and others too, ought to be examined; but it is not really worth while. At present it is the purpose of our Author merely to investigate and to demonstrate some of the properties of accelerated motion (whatever the cause of this acceleration may be)....[2]

Galileo's statement stands in marked contrast to the Aristotelian philosophy of science which had sought the ultimate causes or principles of things as its goal.[3]

I want to argue that the development of Galileo's thought from *De motu* to his mature mechanics—in particular his change of outlook regarding the nature of motion—represents a move from an essentialist to a nonessentialist epistemology. In particular, I maintain that just as the doctrine of primary and secondary qualities asserted in *Il saggiatore* signifies a nonessentialist philosophy of science in contrast to the essentialist Aristotelian analysis of qualities, so too Galileo's new inertial conception of motion (insofar as it was inertial) represents a move away from a traditional essentialist position. I shall attempt to support this assertion in the remainder of the paper.

In his early treatise, *De motu* (1590), Galileo expressed many views that are consistent with traditional essentialism. These views can be summed up in the assertion that a body's state of motion or rest reveals something of its specific nature. Just as Aristotle believed that from the observable characteristics of substances we can learn something of their inner essences, so the young Galileo regarded the motions of a body as indicative of its inner nature. I will support this assertion by considering, in turn, the implications of several important aspects of the physics of *De motu*.

---

[2] Galileo Galilei, *Dialogues Concerning Two New Sciences*, trans. Henry Crew and Alfonso de Salvio (New York: Dover, 1914), pp. 166–167. I have used a version of this translation as modified by Richard S. Westfall, *Force in Newton's Physics* (New York: American Elsevier, 1971), pp. 45–46. Galileo asserted a nonessentialist position as early as Dec. 1612 in the Third Letter on Sunspots, where he made the following statement: "in our speculating we either seek to penetrate the true and internal essence of natural substances, or content ourselves with a knowledge of some of their properties. The former I hold to be as impossible an undertaking with regard to the closest elemental substances as with more remote celestial things.... I know no more about the true essences of earth or fire than about those of the moon or sun, for that knowledge is withheld from us, and is not to be understood until we reach the state of blessedness." *Discoveries and Opinions of Galileo*, trans. Stillman Drake (Garden City, N.Y.: Doubleday, 1957).

[3] "When the objects of an inquiry, in any department, have principles, conditions, or elements, it is through acquaintance with these that knowledge, that is to say, scientific knowledge, is attained. For we do not think that we know a thing until we are acquainted with its primary conditions or first principles, and have carried our analysis as far as its simplest elements. Plainly, therefore, in the science of Nature, as in other branches of study, our first task will be to try to determine what relates to its principles." Aristotle, *Physics*, 184a9–16. Aristotle proceeds to argue that science aims toward knowledge of the nature of things, and that an important meaning of the word "nature" is the form of the thing in question.

In *De motu* Galileo assumed that all bodies possess heaviness, to one degree or another, and that this heaviness is the cause of all motion. He began the treatise with the assertion "all natural motion, whether upward or downward, is the result of the essential heaviness or lightness of the moving body."[4] He proceeded to argue that there is no such thing as absolute lightness or natural upward motion, strictly speaking, since effective heaviness and lightness are simply relative qualities determined by specific gravity, and upward motion proceeding from relative lightness is simply the result of the extruding action of the heavier medium.[5] Although all bodies possess heaviness, some are heavier than others. In a void, where there is no ambient medium to interfere with the motion caused by its heaviness, a body experiences its *natural* motion;[6] that is, in a void a body of any given heaviness will fall with a fixed, determinate speed characteristic of its heaviness. Now, since heaviness, at least insofar as motion is concerned, is *the* essential quality of a body, the natural motion of a body in a void will reveal the essence of that body. By implication, if we know the relative weight of the medium, we can likewise determine the degree of essential heaviness accruing to a body by observing its motion through the medium. Therefore, in a purely operational manner, it is possible to discover something about the essences of moving bodies. In fact, much of the discussion in *De motu* hinges on the essences of bodies insofar as it concerns the way in which heaviness produces different kinds of motions.

Galileo explained projectile motion in *De motu* by appeal to a self-expending impetus, in the tradition of Philoponus and Avicenna. A naturally heavy body, one which would move downward by its nature, is caused to move upward when a lightness is temporarily impressed upon it. This lightness or upward motive force is a quality which temporarily deprives a body of its heaviness and thus renders it light.[7] Impetus is a quality which is impressed upon the projectile. But within a fundamentally Aristotelian ontology, qualitative changes alter, if only temporarily, the essences of the things changed. Galileo stated as much himself:

> But, this force, since it is lightness, will indeed render the body in motion light by inhering in it.... I would not say that a stone after its [upward] motion has become [permanently] light. I would say rather that it retains its natural weight, just as the hot glowing iron is devoid of coldness but, after the heat [is used up], it resumes the same coldness that is its own. And there is no reason for us to be surprised that the stone, so long as it is moving [upward] is light. Indeed, between a stone in that act of motion [upward] and any other

---

[4] Galileo Galilei, *On Motion and On Mechanics*, trans. I. E. Drabkin and Stillman Drake (Madison:University of Wisconsin Press, 1960), p. 13; hereafter called *De motu*.

[5] *Ibid.*, pp. 22, 116.

[6] In *De motu* Galileo measured the speed of motion by the relationship $V = F - R$, where F is the heaviness of the body and R is the heaviness of the medium. Thus in a void $R = 0$ and $V = F$. Moreover, he assumed that once the upward impetus of a body is consumed, the body will attain a uniform downward speed, and in a void this terminal speed is the "natural speed" of the body in question. See pp. 100–105.

[7] *Ibid.*, pp. 78–79. Although Galileo's appeal to the impetus theory of Philoponus and Avicenna is a departure, strictly speaking, from Aristotle's explanation of projectile motion in the *Physics*, nevertheless the assumption that continued motion requires the action of a continued cause is Aristotelian. The ontology of motion in the theory of impetus (whether self-expending or not) remains consistent with the Aristotelian analysis of qualities and essences. For further discussions of the background to *De motu* see Ernest A. Moody, "Galileo and Avempace," *Journal of the History of Ideas*, 1951, *12*:163–193, 375–422, and also Edward Grant, "Aristotle, Philoponus, Avempace, and Galileo's Pisan Dynamics," *Centaurus*, 1965, *11*:79–95.

light body it will not be possible to assign any difference. For since we call light that which moves upward, and the [projected] stone does move upward, the stone is therefore light so long as it moves upward.[8]

To explain more fully just how such qualitative change takes place, Galileo drew an analogy between the motive force impressed on a projectile and the sonority acquired by a bell when it is rung:

> Do you wonder what it is that passes from the hand of the projector and is impressed on the projectile? Yet you do not wonder what passes from the hammer and is transferred to the bell of a clock, and how it happens that so loud a sound is carried over from the silent hammer to the silent bell, and is preserved in that bell when the hammer which struck it is no longer in contact. The bell is struck by the striking object; the stone is moved by the mover. The bell is deprived of its silence; the stone of its state of rest. A sonorous quality is imparted to the bell contrary to its natural silence; a motive quality is imparted to the stone contrary to its state of rest. The sound is preserved in the bell, when the striking object is no longer in contact; motion is preserved in the stone when the mover is no longer in contact. The sonorous quality gradually diminishes in the bell; the motive quality gradually diminishes in the stone.[9]

The unnatural upward motion of a projectile, then, indicates that a change in the body's essence has taken place. The change is only temporary, as the unnatural motion eventually dies out. Nevertheless, the fundamentally noninertial and Aristotelian assumption that every motion, as long as it continues, requires the continuing action of a cause combined with the essentialist ontology of qualities (i.e., qualitative change is the result of and reveals some inner, essential change) as applied to impetus leads to the conclusion that motion, in *De motu*, reveals something of the essence of the body moved.

In his maturity Galileo radically changed his views both of the nature of motion and of the ontology of qualities. Both of these changes reflect a weakening of the grip of essentialism on Galileo's assumptions about the foundations of science. The inertial properties of motion in his mature mechanics imply that from a body's motions we can tell nothing of its inner nature; and the doctrine of primary and secondary qualities, as he stated it in *Il saggiatore*, embodies a new ontology of qualities according to which the appearances a body presents to us do not reveal its essence.

Galileo's mature physics, as presented in the *Dialogues Concerning Two New Sciences* (1638), contains an entirely new concept of motion, a concept which represents a fundamental departure from the assumptions of *De motu* and medieval physics. Motion is inertial,[10] and as a consequence, I will argue, the motions of a body alone no longer permit us to deduce facts about its essence.

The first significant result that Galileo derived in the *Two New Sciences* is the law of free fall, to wit, that for *any* body undergoing naturally accelerated motion, the distance increases as the square of the time,

$$s \propto t^2$$

---

[8] *Ibid.*, pp. 80–81.
[9] *Ibid.*, pp. 79–80.
[10] The fact that Galileo's principle of inertia seems to apply only to bodies moving on spherical surfaces about centers such as the center of the earth does not seriously detract from my argument. Moreover, the correct formulation of the principle of inertia was stated by Descartes in 1644. Consequently, the theme of increasing non-essentialism appears in the science of mechanics, in spite of the vestiges of heaviness and circles that remained with Galileo until his death.

By implication (and with Galileo, by prior definition), the velocity increases in direct proportion to the time elapsed from the beginning of the motion,

$$v \propto t$$

This relationship holds true for all terrestrial bodies (all of which happen to possess the quality of heaviness, endowing them with the inclination to undergo naturally accelerated motion). Thus, in direct contrast to the results of *De motu* we can infer *nothing* of a body's nature from its speed of fall. All bodies in free fall undergo the same acceleration. From their motions alone, we are entirely unable to distinguish between a cannon ball and a Siamese cat. Free fall is a phenomenon we can describe with mathematical precision, but it gives us no insight into the specific natures of individual bodies. Free fall, however, does reveal that all bodies are heavy, since heaviness is the property giving them the inclination to fall. Galileo did not address this question explicitly, but it seems clear that for him heaviness, like certain geometrical properties, is one of the primary qualities of matter. From the motions of bodies we may indeed be able to detect the primary qualities of bodies in general; but since in important respects their motions are all alike, we cannot distinguish the specific natures of individual bodies from their motions. Having abandoned the dynamical viewpoint of *De motu*, Galileo abandoned the essentialist implications of the traditional doctrine of motion.

A second important aspect of Galileo's mature mechanics is his statement of the principle of inertia, which he used to derive the parabolic trajectory of projectile motion. Although he never isolated the principle as a special proposition, he stated it in the midst of discussions of specific problems in several places in the text. For example,

> ... any velocity once imparted to a moving body will be rigidly maintained as long as the external causes of acceleration or retardation are removed, a condition which is found only on horizontal planes; for in the case of planes that slope downwards there is already present a cause of acceleration, while on planes sloping upward there is retardation; from this it follows that motion along a horizontal plane is perpetual; for, if the velocity be uniform, it cannot be diminished or slackened, much less destroyed.[11]

From the principle of inertia it follows that a body is entirely indifferent to the motions impressed on it by external forces. Its state of motion or rest expresses nothing of its inner nature or essence. If a body is accelerating or decelerating, we can reason to the action of some accelerating or decelerating force external to the body. If it is at rest or in a state of uniform motion on a horizontal plane, we can only reason that no external forces are acting on it. In either case we have been able to make inferences only about factors external to the body: its motions reveal nothing of its essence. Now this claim is not entirely true for Galileo. It is well known among historians of science that by a "horizontal plane" he meant a spherical surface whose center coincides with the center of the earth. For Galileo, naturally accelerated motion was still "natural" motion, and bodies still possessed heaviness. He had not entirely freed himself from the shackles of circles and the cosmos. Nevertheless, as I have argued in connection with the law of free fall, the heaviness that bodies reveal, though an essential property of all bodies, is not specifically essential and tells us nothing about the natures of individual bodies. According to Galileo's mature concept of motion, we cannot reason from observed

---

[11] *Two New Sciences*, p. 215.

motions to specific natures. And if Galileo himself did not state the principle of inertia entirely correctly, his successors Descartes and Newton surely did. We observe in the transition from the motion of *De motu* to that of the mature Galileo a giant step toward the abandonment of essentialist implications of mechanics.

In his own thinking about the nature of matter and qualities, Galileo similarly moved from the essentialism that was implicit in *De motu* to a nonessentialism which is explicit in the doctrine of primary and secondary qualities expressed in *Il saggiatore* (1623). In this polemical work Galileo was led to state a position on the nature of qualities which was very much the view that dominated science for at least another century.

> I do not believe that for exciting in us tastes, odors, and sounds, there are required in external bodies anything but sizes, shapes, numbers, and slow or fast movements; and I think that if ears, tongues, and noses were taken away, shapes and numbers and motions would remain, but not odors or tastes or sounds. These, I believe, are nothing but names, apart from the living animal—just as tickling and titillation are nothing but names when armpits and the skin around the nose are absent.[12]

Having argued for this position from the relativity of all qualities but the mathematical ones, Galileo abandoned the Aristotelian, essentialist ontology that we saw in *De motu*. In fact, to illustrate his new position, he considerately provided an analysis of sonority, just the quality he had used in the earlier work as a paradigm of a real quality:

> Sounds are created and are heard by us when—without any special 'sonorous' or 'transonorous' property—a rapid tremor of the air, ruffled into very minute waves, moves certain cartilages of a tympanum within our ear. External means capable of producing this ruffling of the air are very numerous, but for the most part they reduce to the trembling of some body which strikes upon the air and disturbs it; waves are thereby very rapidly propagated, and from their frequency originates a high pitch, or from their rarity a deep sound.[13]

No longer is sonority a real quality that somehow modifies, if only temporarily, the naturally silent essence of the bell; it is the result of the impact of certain mechanical phenomena on our sense organs. No longer can the essential nature of the object be determined from its phenomenal attributes. Qualities, like motions, no longer reveal anything of the real essences of bodies.

By way of conclusion, then, I would like to say that certain patterns in Galileo's own intellectual development parallel what appears to be a characteristic feature of science during the scientific revolution. Whereas traditional physics and philosophy of science had been dominated by the search for real essences, the new science was content to know the appearances with a fair degree of probability. As philosophers later in the century came to reflect on the nature of scientific knowledge (e.g., Gassendi and Locke), they offered theoretical reasons why empirical knowledge must in principle be confined to appearances and probabilities.[14] Galileo's work, which reflects the transition from medieval to modern science in so many ways, also represents a transition from essentialist to nonessentialist assumptions about the nature of scientific inquiry.

---

[12] Galileo Gailiei, *The Assayer*, in *The Controversy on the Comets of 1618*, trans. Stillman Drake and C. D. O'Malley (Philadelphia: University of Pennsylvania Press, 1960), p. 311.

[13] *Ibid.*, p. 311.

[14] For further discussion of this point see my article "John Locke and the Changing Ideal of Scientific Knowledge," *J. Hist. Ideas*, 1970, *31*: 3–16.

Science and Patronage

Galileo and the Telescope

By Richard S. Westfall

SOMETIME LATE IN 1610, probably near 11 December, Galileo received a letter from his disciple Benedetto Castelli:

> If the position of Copernicus, that Venus revolves around the sun, is true (as I believe), [Castelli wrote], it is clear that it would necessarily sometimes be seen by us horned and sometimes not, even though the planet maintains the same position relative to the sun. . . . Now I want to know from you if you, with the help of your marvellous glasses, have observed such a phenomenon, which will be, beyond doubt, a sure means to convince even the most obstinate mind. I also suspect a similar thing with Mars near the quadrature with the sun; I don't mean a horned or non-horned shape, but only a semicircular and a more full one.[1]

How readily the passage summons up familiar images of Galileo, the Copernican polemicist, who turned the telescope on the heavens, if not first, surely first in an effective manner, and with his discoveries forever transformed the terms of the debate. Some twenty years after Castelli's letter, in the Fourth Day of the *Dialogue Concerning the Two Chief World Systems* (1632), Galileo summed up what he considered the most convincing arguments in favor of the Copernican system: first, the retrograde motions of the planets and their approaches toward and recessions from the earth (a reference primarily to Venus and Mars); second, the rotation of the sun on its own axis; third, the tides.[2] Half of the first argument and all of the second and third arguments were Galileo's own work. The theory of the tides did not depend on the telescope, of course, but the arguments from the rotation of the sun and the approaches of the planets could not have existed without it. For all that, we must not allow the *Sidereus nuncius*

---

[1] Castelli to Galileo, 5 Dec. 1610; *Opere di Galileo Galilei,* ed. Antonio Favaro, 20 vols. (Florence: G. Barbera, 1890–1909), Vol. X, pp. 480–483. (Unless otherwise specified, all translations are mine.) Two versions of this letter exist among Galileo's papers, both apparently in Castelli's hand, one labeled (by Castelli) "copy," though its wording differs modestly from the other's. Castelli first dated the original 5 Nov. 1610 but altered it at some time to 5 Dec. I am strongly inclined to treat the November date as a slip and to believe that Castelli made his correction at the time rather than later, which might call the December date into question. For other interpretations, which I find unconvincing, see Raffaello Caverni, *Storia del metodo sperimentale in Italia,* 6 vols. (Florence, 1891–1900), Vol. II, pp. 359–360; Antonio Favaro, "Galileo Galilei, Benedetto Castelli e la scoperta delle fasi di Venere," *Archeion,* 1919, *1*:283–296.

[2] Galileo, *Dialogue Concerning the Two Chief World Systems,* trans. Stillman Drake, (2nd ed., Berkeley: Univ. California Press, 1967), p. 462; cf. pp. 349–355. The argument from the approaches and recessions of Venus embodied its phases (see p. 321), which furnish the only rigorous part of the argument.

and Galileo's early discoveries with the telescope to dazzle us. Before the arrival of Castelli's letter, Galileo does not appear to have thought out a serious program of observation with his new instrument to settle, or to attempt to settle, the Copernican question. Quite the contrary, his attention appears to have focused almost exclusively on the telescope's capacity to insure his own future. The episode of the Castelli letter tells us something about Galileo's commitment to Copernicanism, but it tells us a great deal more about the system of patronage and the material circumstances under which Galileo pursued his career in science.

We can best understand the Castelli letter if we look first into its background. Let us start with Galileo's father, who was a distinguished musician but hardly an economic success. As a consequence, though Galileo was descended from an old Florentine patrician family, he found himself upon the death of his father, which came when Galileo himself was approaching thirty, heir to some sizable obligations but to no material means worth mentioning. To make matters worse, he would soon have no income; although he held a chair (with a miserable salary, to be sure) at the University of Pisa, he had offended powerful people and thus insured that the appointment would not be renewed in 1592. From that time his primary asset would be his wits, a formidable asset indeed, but not one directly negotiable in the marketplace. In the late sixteenth century, the system of patronage was one of the principal devices by which to convert wits (I refer of course to wits of Galileo's caliber) into the material necessities of life.

An appointment as professor of mathematics at the University of Padua in the autumn of 1592 provided an income. In the mind of a twentieth-century reader, such a position does not raise the suggestion of patronage. In the late sixteenth century, however, different considerations determined university appointments, and without the effective intervention of Guidobaldo del Monte, Galileo would never have occupied the Paduan chair.[3] Moreover, the university appointed him for a limited period of four years with guaranteed extension of two more. Although he continued in 1598 and in 1604 to teach and to be paid after two six-year appointments had expired, the university did in the end explicitly reappoint him both times. Not only was reappointment not guaranteed, but Galileo always wanted an increase in salary. That is, he had continuing need of patrons in Padua, and he took care that he did not lack them. The great majority of the friends he cultivated in Venice were members of the highest ranks of Venetian nobility.[4] The word *friend* carries special connotations within a context of pa-

---

[3] Guidobaldo was a recognized authority in mathematics whose friendship with Galileo rested almost entirely on their shared interest in mathematics. Their correspondence concerning Galileo's appointment in Padua (*Opere*, Vol. X, pp. 26–54), however, convinces me that his being the Marchese del Monte, brother of the Cardinal del Monte, was the significant factor there.

[4] For Galileo's correspondents during the Paduan years see *Opere*, Vol. X, pp. 55–256; for his private students see Antonio Favaro, *Galileo e lo studio di Padova* (Florence: Le Monnier, 1883), Vol. II, pp. 184–192. Galileo's initial salary at Padua was 180 florins, which according to my calculations somewhat more than doubled his previous salary at Pisa. For the Paduan salary, actually calculated in lire, at the rate of five lire per florin, see *ibid.*, p. 142. For the silver in these lire, see Nicolò Papadopoli, *Le monete di Venezia*, 4 vols. (Venice/Bologna: Ongania, 1893–1907), Vol. II, pp. 393–422. For the silver equivalent of the Pisan salary, see Giuseppe Parenti, *Prime ricerche sulla rivoluzione dei prezzi in Firenze* (Florence: C. Cya, 1939), p. 58. My calculations from the exchange rates as found in José-Gentil da Silva, *Banque et crédit en Italie aux XVIIe siecle*, 2 vols. (Paris: Klincksieck, 1969), Vol. I, pp. 296, 320, gave a similar result (1:2.13, as against 1:2.4). Experts advise that such calculations involve many pitfalls.

tronage; authorities on patronage distinguish what they call instrumental friendship from emotional friendship. Galileo's "friends" in Venice appear to have understood that the "friendship" entailed the use of their connections and influence on his behalf; certainly Galileo expected as much. In the late summer of 1599, a year after his appointment had expired, the *riformati*, the highest authorities of the university, who were appointed by and responsible to the Venetian government, took his reappointment under consideration. Antonio Quirini, who had himself received commendations from three men of influence, visited Leonardo Donato, one of the *riformatori*, on Galileo's behalf. Giovanfrancesco Sagredo visited all three of the *riformatori*, one of them three times, and he indicated that the nephews of this man, Zaccaria Contarini, had also been working on him. The sticking point was the raise that Galileo wanted. Sagredo, who was clearly tiring of the exercise, wanted to be sure that Galileo understood he had fulfilled his duty as a patron. "Since I have already satisfied abundantly enough the friendship I hold for you, the obligations to you which I acknowledge, and the favor and help that true gentlemen try to extend to the qualified who deserve it," he wrote, he thought he might now honorably desist.[5] One would be hard pressed to find a better example of the language of patronage. The net result of this marshalling of influence was a raise from 180 to 320 florins *per annum*, effective in 1598, when the original appointment had expired. In the fall of 1605, when the second reappointment was taken up, Galileo was able to call on no one less than the grand duke of Tuscany, who did indeed intervene and was apparently instrumental in obtaining a further raise to 520 florins.[6]

By that time Galileo had already begun to cast his gaze beyond Padua, where he continually had to direct a choir of patrons to sing his praises, and where the officials frankly told him that he would need to supplement his salary by private lessons. This he had resorted to, in addition to using his home as a hostel and manufacturing instruments for sale. He could have been writing of his own situation when, soon afterwards, he described to Grand Duchess Cristina the plight of his friend Fabrizio, who, "finding himself . . . scarcely able to endure the continual labors he needs to undertake every day to serve his many friends and patrons, and hence wanting very much to find a little quiet, both to sustain his life and to bring some of his works to a conclusion," hoped to enter the service of the grand duke.[7] Galileo had similarly concluded that serving a single patron might be easier than serving many. In the spring of 1604, after instructing Vincenzo Gonzaga, the duke of Mantua, in the use of his geometric and military compass, he presented the duke with a compass and received in return gifts worth more than a year's salary at its current rate. This was suggestive indeed; without further ado he undertook negotiations to enter Gonzaga's service. The gift turned out to have been misleading, however. Gonzaga was only a duke, whereas Galileo had a princely stipend in mind. Galileo decided to stay on in Padua for the time.[8]

Almost at once Galileo earnestly began to woo the Medici rulers of his native

---

[5] Quirini to Galileo, 24 Aug. 1599; Sagredo to Galileo, 1 Sept. 1699; *Opere,* Vol. X, pp. 76, 77.
[6] Barbolani to Ferdinand 1, 29 Oct. 1605, 10 June 1606; Saracinelli to Galileo, 26 May 1606; unsigned to Vinta, 12 Aug. 1606; *ibid.,* pp. 147–161.
[7] Galileo to Cristina, 8 Dec. 1606; *ibid.,* pp. 164–166.
[8] See Galileo's accounts, *ibid.,* Vol. XIX, p. 155, line 194; Galileo to Gonzaga, 22 May 1604; *ibid.,* Vol. X, pp. 106–107.

Tuscany. He had prepared his instructions for the use of the geometric and military compass for publication as a pamphlet, and in the spring of 1605 he formally sought permission to dedicate it to the crown prince, Cosimo. Galileo parlayed acceptance of the dedication into an invitation to instruct the prince in mathematics during his summer vacation and did not thereafter relent in his quest. He wrote the prince the flattering letters that an absolute ruler expected of a client, declaring himself "one of his most faithful and devoted servants," and proclaiming his desire to demonstrate "by how much I prefer his yoke to that of any other Master, since it seems to me that the suavity of his manner and the humanity of his nature are able to make anyone desire to be his slave."[9] The terms of Galileo's address, jarring to twentieth-century ears, would not have seemed sycophantic to Galileo's contemporaries. Almost no one challenged the legitimacy of a hierarchically ordered society, the precondition of the patronage that supported Galileo in an economically unproductive occupation. Even if Cosimo was a dull student incapable of profiting from the instruction he received, Galileo's words spoke to his position, not to his person, and expressed the practical necessities that had to be faced.

Summer instruction became an annual affair. In the fall of 1607, when magnetism seized the fancy of the prince, the court consulted Galileo as its authority on such matters. He informed them that he owned a lodestone that weighed about half a pound and was quite strong but not well formed. Like everything he owned, he assured them, it was at the prince's command. However, he knew of another, far better, one that weighed about five pounds. It was owned by a friend of his—he referred to Sagredo—who was ready to part with it at a fair price. The friend had refused an offer of two hundred gold scudi from a representative of the emperor; Galileo suggested that four hundred scudi sounded about right. It quickly transpired that four hundred scudi was more than the Tuscan court intended to spend to pamper the ephemeral fancy of even the crown prince, and with some embarrassment Galileo ascertained that Sagredo would accept two hundred after all.[10] The negotiations and arrangements stretched out over a period of six months, during which Galileo experimented extensively with the lodestone's capacity. It was even better than he had originally determined. It would support over twice its own weight, but it would do so only in the hands of one who knew how to apply the weights properly to the poles. To insure that the prince not be disappointed, Galileo decided to send the lodestone with the weights in place. Not just any weights would please a crown prince of Tuscany, however, and seizing upon the fable of a lodestone strong enough to raise an anchor, he had two little anchors of the proper weight made of iron. Indeed he traveled to Venice more than once to supervise the artisans at work on them. Meanwhile, as he kept learning about the lodestone, his imagination embroidered his knowledge into ever more elaborate images. Because the lodestone, far from exhausting itself, appeared to gain strength as it held the weight, Galileo proposed that on the support for the whole device one could inscribe the motto, *Vim facit amor*, "love produces strength." This motto suggested, he added, "the dominion of God conferred upon the just and legitimate prince over his subjects, which should be such that with loving vio-

---

[9] Giugni to Galileo, 4 June 1605; Galileo to Cosimo, 29 Dec. 1605; *ibid.*, pp. 144, 153–154.

[10] Galileo to Picchena, 6 Nov. 1607; *ibid.*, pp. 185–186. Correspondence about the lodestone occupies most of pp. 185–213.

lence it draws to itself the devotion, loyalty, and obedience of the subjects."[11] Galileo, taken with the figure and doubtful that the secretary of state, to whom he had sent it, had passed it on to the ruling family, repeated the conceit to Grand Duchess Cristina in the fall on the occasion of Cosimo's marriage, when it became doubly suitable. By now his imagination had embroidered it still further. Obviously the lodestone represented the prince. The ancient symbol of the Medici embodied its shape. Moreover, the earth was known to be a great magnet, and the prince's name, Cosimo (or Cosmo) was a synonym for *Mondo,* or Earth. Hence it was possible, he concluded, "through the most noble metaphor of a globe of lodestone to indicate our great Cosimo."[12] Such evidence reveals how the subtle alchemy of patronage transmuted an object of science into an *objet d'art* to amuse and flatter a prince.

Meanwhile, well before the prince's marriage, Galileo had sent the lodestone to Florence on 3 May 1608. Florentine officials, concerned that it not be injured in transit, had instructed him to deliver it to the Tuscan resident in Venice for shipment via their own courier, but he had sent it instead by the common courier. Three weeks later, having received no word of acknowledgment, Galileo was almost beside himself with anxiety.[13] After another week had passed with still no word, convinced that his use of the common courier in defiance of explicit orders had given offense, he composed a long letter of explanation. The scene he painted is not one commonly associated with Galileo; we might search some time, however, to find a better description of the mores of patronage. Galileo had spent all of the first three days of May in Venice supervising the final completion of the device. (We ought to remember those three days when we hear Galileo complain about the demands that his life in Padua placed on his time.) On the third day, a Sunday, the festival of Santa Croce, he had forced two artisans to work on the anchors against their will. When night came with the work not yet completed, he sent a note to the resident asking the hour at which the Tuscan courier would leave. Word came back that he could deliver the package as late as the fourth hour of the night, and the artisans worked on until the deadline approached. Galileo called for a gondola, which he had trouble finding because of the hour and the rain. They set out, the gondolier grumbling every stroke of the way, and found the general area of the resident's home. Alas, it was so dark and rainy that they could not locate his house. They knocked on various doors, and received in reply either silence or words Galileo thought it better not to repeat in his letter. Finally, in desperation, determined that his offering not wait for the next official courier, he took his box to the master of the couriers.[14] By the time he wrote the explanation, it was no longer necessary; a letter had been sent from Florence informing him that the lodestone had arrived safely.

Soon it appeared that his diligent labor to provide an occasional amusement for the crown prince had been most prudently expended. Nine months later, Ferdinand I died, and Cosimo succeeded him as grand duke of Tuscany. Galileo wrote to him, of course, offering the customary mixture of condolences on his father's death and congratulations on his succession. "I supplicate Your Most

---

[11] Galileo to Vinta, 3 May 1608; *ibid.*, pp. 205–209.
[12] Galileo to Cristina, Sept. 1608; *ibid.*, p. 222.
[13] Galileo to Vinta, 23 May 1608; *ibid.*, p. 209.
[14] Galileo to Vinta, 30 May 1608; *ibid.*, p. 212.

Serene Highness," he concluded, "that as you have been established by God as the ruler of all your most devoted subjects, do not disdain now and then to turn the favorable eye of your grace toward me, one of your most faithful and devoted servants, for which grace I devotedly entreat while I bow before you in all humility and kiss your hand."[15] To an official of the court he wrote a much longer letter that frankly laid out his aspirations:

> Having labored now twenty years, the best ones of my life, in dispensing at retail, as the saying goes, at the demand of everyone, that little talent in my profession that God and my own efforts have given me, my desires would truly be to obtain enough leisure and quiet as would enable me before I die to complete three great works that I have in hand in order to be able to publish them, perhaps with some praise for me and for whoever has helped me in the business. . . . It is not possible to receive a salary from a Republic, however splendid and generous, without serving the public, because to get something from the public one must satisfy it and not just one particular person; and while I remain able to teach and to serve, no one can exempt me from the burden while leaving me the income; and in sum I cannot hope for such a benefit from anyone but an absolute prince.[16]

He specified his current income and assured the grand duke once more that the prince who patronized him could expect to receive more reflected glory than most clients delivered. A reply came back that Cosimo would write when he was able. It was the classic formula of evasion, the seventeenth-century equivalent of a promise to call for lunch, and indeed silence followed the letter. Galileo, who had spent his vacations in the summer with the ducal family, instructing Cosimo the crown prince in mathematics in 1605, 1606, and 1608 (with 1607 omitted at his own request), did not receive an invitation from Cosimo the grand duke for 1609. The weeks stretched into months. Though Galileo could not have known it in the summer of 1609, the months would stretch into a year, and who knows how long Galileo might have waited had nothing else intervened. Perhaps his efforts in wooing the prince had been in vain after all; perhaps he was destined to stay on in Padua wearing himself out peddling his wares at retail.

But something did intervene. Call it fate. In the spring of 1609 a man from Flanders appeared in Venice with a device that magnified the images of things and enabled one to see objects at a distance. Galileo may have seen the instrument firsthand; he certainly heard it described by some who had. Within a short time he was able to reproduce it and, what was far more important, to improve upon it.[17] Whereas the instrument from Flanders magnified three times, by late August Galileo had one that magnified eight or nine times. It caused a sensation in Venice. Aged senators struggled to the top of the campanile to see ships approaching the harbor two hours before they became visible to the naked eye. The Flemish adventurer had offered his eyeglass to the Senate for 1,000 *zecchini*, very nearly four times Galileo's annual salary. With supreme insight, Galileo presented his to the Doge. He made the presentation before the College, a sort of council of ministers, which forthwith ordered the *riformatori* of the uni-

---

[15] Galileo to Cosimo, 26 Feb. 1609; *ibid.*, pp. 230–231.
[16] Galileo to "S. Vesp." (Geraldini?), Feb. 1609; *ibid.*, pp. 231–234.
[17] See Albert Van Helden, *The Invention of the Telescope* (Transactions of the American Philosophical Society, 67. 4) (Philadelphia: APS, 1977); and Van Helden, "The Telescope in the Seventeenth Century," *Isis*, 1974, 65:38–58.

versity to renew his contract for life at a salary of 1,000 florins. "Knowing how hope has wings that are very slow and fortune wings most swift," Galileo reported to his brother-in-law, "I said that I was content with how much it pleased His Highness."[18]

Since a great deal has been made of that 1,000 florins, it is useful to put it briefly into perspective. During his entire preceding career in Padua, Galileo had lived, in terms of salary, in the shadow of Cesare Cremonini, the Aristotelian philosopher who was at once his rival and his friend. Cremonini had gone to Padua in 1590 at a salary of 200 florins; Galileo's initial salary had been 180. In 1599, when Galileo's salary had increased to 320 florins, Cremonini's had risen to 400. In 1601, Cremonini's salary had increased again to 600 florins. Only in 1606 had Galileo's reached 520, and two years later Cremonini's had risen once more, this time to 1,000. What the college offered to Galileo in 1609 was, in fact, not an unprecedented salary but one merely equal with Cremonini's. Moreover, the lifetime contract would have effectively foreclosed any future negotiations for more. Cremonini's contract did not, and his salary did increase further, to 1,400 florins in 1616, then to 1,800 in 1623. Ultimately it reached 2,000 florins. As for the indignation in Venice when Galileo chose to leave, it is worth noting that the original action of the college, to make the new salary effective for the current year, was rescinded by the senate.[19] Since he left for Florence before the next year began, Galileo never received the salary of 1,000 florins.

Meanwhile, the Tuscan court was as interested in the new device as everyone else. Six days after the presentation of the instrument to the doge, exactly the time ordinarily required for communications between Venice and Florence, a Tuscan official wrote to tell Galileo how much the grand duke would like to have one of the eyeglasses, and three weeks later they sent him pieces of glass to work into lenses for it.[20] There is no evidence, however, that the grand duke received one at this time. The Venetian government had ordered its servant to make twelve more telescopes and not to reveal their secret. Galileo needed no encouragement on the latter score. Already in September others were offering for sale telescopes like the one that had appeared during the summer, and by November every ordinary maker of spectacles in Venice could produce them.[21] Galileo's prestige rested on the fact acknowledged by everyone who tried them that his telescopes were better than any others, and he kept them that way by making them himself in private. Early in 1610 he stated that he had already made over sixty telescopes.[22] Galileo was usually inclined to exaggerate to his own advantage, but there is no reason to doubt that he spent much of the autumn of 1609 making telescopes and further improving them. By late autumn, as a result of his own efforts, he had a twenty-power instrument.

---

[18] Galileo to Landucci, 29 Aug. 1609; *Opere*, Vol. X, pp. 253–254. Cf. the independent account in Priuli's chronicle, cited in Edward Rosen, "The Authenticity of Galileo's Letter to Landucci," *Modern Language Quarterly*, 1951, 12:482.

[19] Rosen, "Authenticity," p. 481; on Cremoni see Favaro, *Galileo e Padova*, Vol. II, pp. 424–425.

[20] Piccolomini to Galileo, 29 Aug., 19 Sept. 1609; *Opere*, Vol. X, 255, 258–259.

[21] Bartoli to Vinta, 26 Sept., 7 Nov. 1609; *ibid.*, pp. 259–260, 267. In strict usage, the word *telescope* is an anachronism before April 1611, when it was put into currency at the banquet in Rome that Federigo Cesi gave in honor of Galileo; see Edward Rosen, *The Naming of the Telescope* (New York: Schuman, 1947).

[22] Galileo to Vinta, 7 May, 19 Mar. 1610; *Opere*, Vol. X, pp. 350, 301. In the first draft of the second letter, the number was more than 100 (*ibid.*, p. 298).

Sometime during the autumn, probably near the end of November, Galileo found time to turn his telescope on the heavens.[23] He was not the first to do so. He was not even the second, though he had no knowledge of earlier observers at the time.[24] Having the best instrument, he discovered things no one before him had seen, and having a clear idea of what he might do with such discoveries, he succeeded in attaching his name to the new celestial world forever. No records of his observations before 7 January 1610 have survived. The *Sidereus nuncius*, which offers the principal account of them, seems to suggest that initially Galileo merely looked at the most obvious celestial objects—the moon and then the stars, including the Milky Way. From the first moment, however, he was aware that he was taking a historic step. As he wrote to Antonio de' Medici on 7 January 1610, after describing his discoveries, "None of the observations mentioned above are seen or can be seen without an exquisite instrument; hence I can believe that I am the first in the world to observe so close and so distinctly these features of the heavenly bodies." Since the letter is long and carefully composed and uses some phrases and comparisons that would make their way into the *Sidereus nuncius*, we can well believe that Galileo had already given thought to the advantages of a publication that would announce what had been observed and, of course, who the observer was. That he had not yet begun to compose the *Sidereus nuncius* appears from the reference in the first paragraph to discoveries made after 7 January: Galileo waited until he had a purpose associated with patronage.[25]

On the night he wrote the letter, Galileo had turned his telescope toward Jupiter, and he reported to Antonio that among the new stars he had seen in the heavens were three near Jupiter. Why had he chosen Jupiter as the first planet to observe? Like Saturn, it circles far outside the earth in both the Ptolemaic and the Copernican systems, and it seems one of the least likely candidates to reveal anything of decisive importance to an astronomer. Why look at Jupiter? For no better reason than the fact it was there. The records of Galileo's observations seem to reveal a man who, like many, preferred staying up in the evening to rising before dawn. On 7 January 1610, only Jupiter and Saturn appeared

---

[23] In the *Sidereus nuncius*, composed in early February 1610, Galileo spoke of observations made over the past two months; see Galileo, *The Starry Messenger*, in *Discoveries and Opinions of Galileo*, trans. Stillman Drake (Garden City, N.Y.: Doubleday, 1957), p. 31. Guglielmo Righini, in *Contributo alla interpretazione scientifica dell'opera astronomica di Galileo* (Istituto e Museo di Storia della Scienze, monografia 2) (Florence: Istituto e Museo di Storia della Scienza, 1978 [1980]), concludes that Galileo probably began observing early in October. His argument, also set forth in Righini, "New Light on Galileo's Lunar Observations," in *Reason, Experiment, and Mysticism in the Scientific Revolution*, ed. M. L. Righini Bonelli and William R. Shea (New York: Science History, 1975), pp. 59–76, is challenged by Owen Gingerich, "Dissertatio cum Professore Righini et Sidereo nuncio," pp. 77–88. More recently, Ewan A. Whitaker, by comparing modern photographs of the moon with Galileo's drawings, convincingly places the date at the end of November, in agreement with Galileo's assertion; see Whitaker, "Galileo's Lunar Observations and the Dating of the Composition of *Sidereus nuncius*," *Journal for the History f Astronomy*, 1978, 9:155–169.

[24] On a report published in 1608 mentioning that the new eyeglass revealed stars invisible to the naked eye, see Edward Rosen, "Stillman Drake's *Discoveries and Opinions of Galileo*," *Journal of the History of Ideas*, 1957, 18:446. On Thomas Harriot's observations of the moon in England by August 1609, see Van Helden, *Invention of the Telescope*, p. 27.

[25] Galileo to Antonio de' Medici, 7 Jan. 1610; *Opere*, Vol. X, p. 277; cf. Galileo, *Sidereus nuncius*, in *Opere*, Vol. III, Pt. 1, p. 17. Favaro guessed that the letter's recipient, not named, was Antonio de' Medici. In an article filled with information, Stillman Drake opines that the recipient was instead Enea Piccolomini; see Drake, "Galileo's First Telescopic Observations," *Journal for the History of Astronomy*, 1976, 7:153–168; cf. Drake, "Galileo and Satellite Prediction," ibid., 1979, 10:75–95.

in the evening sky, and Saturn was so low that half an hour after sunset it stood only a few degrees above the horizon and was probably invisible. Jupiter was high in the eastern sky. When he pointed his telescope at Jupiter, Galileo did nothing more premeditated than to look at the only planet readily available for observation.

For the second time fate intervened, offering with Jupiter a prize there could have been no reason to expect. Galileo was even fortunate in the time when he happened to look at Jupiter. As he wrote Antonio de' Medici that night, he was just completing a new telescope that would magnify thirty times, nearly the ultimate level to which he brought his instrument, and he had it available for the observations that followed.[26] As before, he was not slow to seize the bounty that fate held out. The three new stars he saw near Jupiter must have sparked some premonition. Not only did he mention them to Antonio de' Medici; not only did he make a record of the observation, the first such one that survives; but he returned to observe the three objects on the following night. The ninth was cloudy, but he observed them again on the tenth, increasingly intrigued by their changing positions relative to Jupiter. By the eleventh he was certain about what he discovered: "there are around Jupiter three other wandering stars that have been invisible to everyone until this time."[27] Two days later his cup ran over; he identified a fourth. Four was a happy number; Cosimo was one of four brothers. Now Galileo was sure he had found what he wanted, a ticket to Florence.

By the end of the month he had prepared the *Sidereus nuncius,* a message from the stars to be sure, but a message composed with the grand duke always in mind. He wrote to Belisario Vinta, the Florentine secretary of state, from Venice, where he had gone to have it printed, about his observations: "as they are amazing without limit, so I render thanks without limit to God who has been pleased to make me alone the first observer of things worthy of admiration but kept secret through all the centuries." He ran through his discoveries—the surface of the moon, the new stars, the nature of the Milky Way. "But that which exceeds all the marvelous things, I have discovered four new planets."[28] Vinta answered in the least possible time. Immediately upon the arrival of Galileo's letter, he had taken it to the grand duke, who was rendered "stupified beyond measure by this new proof of your almost supernatural genius." Assured that the prey was taking the bait, Galileo sprang the trap. He was willing, he told Vinta, to publish his observations only under the auspices of the grand duke, in order that "his glorious name live on the same plane with the stars." As the discoverer of the new planets, it was his privilege to name them. "However, I find a point of ambiguity, whether I should consecrate all four to the Grand Duke alone, calling them with his name the *Cosmici,* or whether, since they are exactly four in number, I should dedicate them to the group of brothers with the name of *Medicean Stars.*"[29] In fact, Galileo did not entertain any large doubt as to which alternative would be preferable, and he proceeded to print the pamphlet

---

[26] Galileo to Antonio de' Medici, 7 Jan. 1610, p. 277. In the *Sidereus nuncius,* Galileo stated that he was already using the new instrument on 7 Jan.; Galileo, *Starry Messenger,* trans. Drake, p. 51. Drake argues instead that Galileo used his twenty-power instrument; Drake, "Galileo's First Observations," pp. 158–159.
[27] Galileo, *Osservazioni,* in *Opere,* Vol. III, Pt. 2, p. 427.
[28] Galileo to Vinta, 30 Jan. 1610; *ibid.,* Vol. X, p. 280.
[29] Vinta to Galileo, 6 Feb. 1610; Galileo to Vinta, 13 Feb. 1610; *ibid.,* pp. 281, 282–284.

using the name *Cosmica Sydera,* in the heading on the first page and elsewhere. Alas, someone in the court observed that since *Cosimo* derived from the Greek *kosmos,* people might mistake the name *Cosmici* for a reference, not to the grand duke, but to the nature of the bodies, and the *Sidereus nuncius* came out with the name *Medicean Stars* on a slip pasted over the original.[30]

The book, rushed into print, as Galileo explained to Vinta, lest someone forestall him both in the discoveries and in the right to name them, duly appeared in March.[31] The dedication to the grand duke explained how through the ages mankind had attempted to preserve the memory of distinguished men by erecting statues of them and by attaching their names to columns, pyramids, and even cities. In the end, however, every human invention perishes; the heavens alone are eternal. Hence men have given to the brightest stars the names of those "whose eminent and godlike deeds have caused them to be accounted worthy of eternity in the company of the stars."[32] He received for his pains a gold chain worth four hundred scudi plus a medal of the grand duke to hang on it.

Galileo's discoveries have not suffered from neglect, but in assessing them from the perspective of nearly four centuries, one should not underestimate the excitement they aroused in their own age. Italian dukes, German princes, the queen of France, the Holy Roman Emperor, half the cardinals in Rome wrote to Galileo asking for one of the instruments that made the celestial wonders visible.[33] Correspondents and commentators vied with each other in plundering mythology and history to find an appropriate symbol for the man who had uncovered such marvels. Giovanni Battista Manso of Naples likened him to Atlas in relation to the new heavenly spheres. Orazio del Monte, Guidobaldo's son, compared the discovery of new planets to the discovery of the new world and assured Galileo that he would "compete in glory with Columbus." In England, Sir William Lower opined that "my diligent Galileus hath done more in his threefold discoverie than Magellane in opening the streights to the South Sea."[34] More than the other revelations of the telescope, the new "planets" and Galileo's masterstroke of naming them for the Medicean ruler of Florence fired men's imaginations. Should he discover another planet, he heard from Paris, and name it for the French king, not only would he do a just thing but he would "make himself and his family rich forever."[35] During the following years, a French supporter of the proposition that sunspots were in reality planets near the sun named them for the Bourbons, while a Belgian proponent of the same position named them for the Hapsburgs.[36] Obviously both imitated what Galileo had done first.

In a word, Galileo had raised himself with one inspired blow from the level of an obscure professor of mathematics at the University of Padua to the status

---

[30] Galileo, *Sidereus nuncius, ibid.,* Vol. III, Pt. 1, p. 9, cf. p. 46; Vinta to Galileo, 20 Feb. 1610; *ibid.,* Vol. X, pp. 284–285.

[31] Galileo to Vinta, 19 March 1610; *ibid.,* p. 298.

[32] Galileo, *Starry Messenger,* trans. Drake, p. 23.

[33] See among the letters in *Opere,* Vol. X, p. 318, to Vol. XI, p. 208.

[34] Manso to Galileo, 18 Mar. 1610; Del Monte to Galileo, 16 June 1610; *ibid.,* Vol. X, pp. 296, 372; and Lower to Harriot, 21 June 1610; quoted in John Roche, "Harriot, Galileo, and Jupiter's Satellites," *Archives internationales d'histoire des sciences,* 1982, *32*:16.

[35] Galileo to Giugni, 25 June 1610; *Opere,* Vol. X, p. 381.

[36] John North, "Thomas Harriot and the First Telescopic Observations of Sunspots," in *Thomas Harriot: Renaissance Scientist,* ed. John W. Shirley (Oxford: Clarendon Press, 1974), p. 134.

of the most desirable client in Italy. He had no intention of allowing his opportunity to slip by. With a copy of the *Sidereus nuncius*, he sent the grand duke the very telescope with which he had made the discoveries. Already he had indicated his concern that people unfamiliar with the telescope might not succeed in observing the Medicean stars, and he had resolved to follow his gift to Florence during the Holy Week vacation, which would last most of April, to insure that the grand duke not be disappointed. "Because of my having been able to demonstrate how much I am a very devoted servant of my Lord in so exotic a manner," he wrote to Vinta, himself a client who would understand, "and because I could never hope for another chance like it, this occasion is so important to me that I do not want to be disturbed by any difficulty or obstacle."[37] The upshot was that he did spend approximately two weeks with the court in Pisa, and correspondence makes it clear that they came to general agreement on his entering the grand duke's service. Almost immediately upon his return, he wrote a long letter that spelled out his current financial situation and stated his desires. He was able, he assured the Florentine court, to double his salary of one thousand florins by giving private lessons and taking students into his house, but they consumed his time. "Hence if I am to return to my native land, I desire that the primary intention of His Highness shall be to give me leave and leisure to draw my works to a conclusion without my being occupied in teaching." Arrangements were quickly made final. Galileo would return to Tuscany with the titles of professor of mathematics at the University of Pisa, without any obligation to teach or to reside in Pisa, and philosopher and mathematician to the grand duke. More than once the court explicitly acknowledged its primary intention of giving Galileo leisure to complete his works. The stipend was set at one thousand scudi in Florentine money. He departed from Padua on 7 September, after a final observation of the Medicean stars before dawn, and arriving in Florence on 12 September, entered the service of the grand duke. Henceforth, as Galileo himself put it, he would earn his bread from his books, "dedicated always to my lord."[38]

Two letters that Galileo received soon after the move to Florence suggest both the values and aspirations that animated the system of patronage and the status that Galileo had achieved within it. Cardinal del Monte was delighted that the grand duke had called Galileo back home with an honorable title and a noble provision, "which action is truly worthy of such a Prince, who has shown himself to be like Augustus in favoring the exceptional." For his part, Giuliano de' Medici expressed his pleasure that Galileo had received from the grand duke "that recognition which corresponds to his quality."[39] If the grand duke demonstrated his magnificence by favoring excellence, Galileo in his turn could not afford to assume that his excellence had been established once and for all. Here

---

[37] Galileo to Vinta, 13 March 1610; *Opere*, Vol. X, p. 389.

[38] Galileo to Vinta, 7 May 1610; *Opere*, Vol. X, pp. 350–351; as trans. by Drake in *Discoveries and Opinions*, p. 62. If calculations are made like those in note 4 above, the stipend from the grand duke appears a bit less than 50% higher than Galileo's final promised salary at the University of Padua. Galileo gave up the additional income earned from private lessons and taking in students, an income he probably exaggerated in the letter to Vinta. To insure his economic security, he abandoned his common-law wife in Padua, and he protected himself from the threat of dowries by placing his two daughters in a convent.

[39] Del Monte to Galileo, 9 Oct. 1610; Giuliano to Galileo, 6 Sept. 1610; *Opere*, Vol. X, pp. 444, 427.

was a new dilemma, one that he could not avoid once he had gained the summit of the pyramid of patronage—to wit, those at the top must fight to stay there.

Galileo began to face the issue even before he arrived in Florence. His letter of 7 May had promised the grand duke "many discoveries and such as perhaps no other prince can match, for of these I have a great many and am certain I can find more as occasion presents itself."[40] But the opportunity was also a threat. Thus the thought occurred that other "planets" like the Medicean stars, which had made his fortune, might be waiting out there to be discovered. What if someone else beat him to them? What if someone else conferred another name on them so that Galileo's gift to the grand duke ceased to be unique and thereby lost its value? On 18 June 1610, in the letter that accepted the terms offered by the Grand Duke, Galileo assured Vinta that he had looked for satellites around Mars and Saturn many times as he observed them in the morning before dawn. Nearly a month before, in May, Jupiter had become invisible in the western sky at sunset as it approached conjunction. Mars and Saturn both stood high in the morning sky, though later evidence suggests that the observations Galileo alluded to were more perfunctory than his words implied. A week later, in a letter to Vincenzo Giugni, also a member of the court, he was more explicit about the object of central concern. After quoting the letter from Paris that urged him to name any further "planet" after the French king, he went on to assure the grand duke he had reserved that honor for him alone, for many observations and investigations had convinced him he would not find any more.[41]

Meanwhile, the Medicean stars did not yet prove to be his private possession. Discovering them was one thing; defining their periods would be something else. In the *Sidereus nuncius*, Galileo stated that he had not yet been able to determine the periods. Letters to Florence during the following months indicated that he was working hard on the problem, and the record of his observations confirms as much.[42] From the moment he discovered them, they became the primary object of his attention in the heavens. Almost no records of other observations survive, and there is no reason to think they ever existed. When he went to Rome in 1611, he set his telescope up every night as he traveled south, in order to measure the positions of the satellites, and he did the same every clear night in Rome. He even recorded observations for the nights of Cesi's banquet and the gathering at the Collegio Romano, at both of which he was the guest of honor.[43] He would continue to return to the satellites until 1619, always working to determine their periods more accurately, always concerned to maintain his priority. There was an issue of practical importance involved, to be sure: if he could define the periods with sufficient accuracy to permit tables of the satellites' positions to be calculated, he would have the crucial ingredient of a method to determine longitude at sea. Galileo made more than one effort to peddle this idea to the seafaring nations, and he would undoubtedly have reaped a bounteous harvest had he brought the method to perfection. More was at stake than a practical device, however. His acknowledged position as the messenger from

---

[40] Galileo to Vinta, 7 May 1610; *ibid.*, p. 351; as trans. by Drake in *Discoveries and Opinions*, p. 62, modified slightly (esp. in rendering *invenzioni* as *discoveries*).
[41] Galileo to Vinta, 18 June 1610; Galileo to Giugni, 25 June 1610; *Opere*, Vol. X, pp. 374, 382.
[42] Galileo, *Starry Messenger*, trans. Drake, p. 56; Galileo to Vinta, 19 Mar, 7 May 1610; *Opere*, Vol. X, pp. 299, 352.
[43] Galileo, *Osservazioni*, in *Opere*, Vol. III, Pt. 2, pp. 442-444.

the heavens was threatened. His friend Cigoli reported that Magini had said that the discovery of the satellites was nothing and that all the praise would belong to the man who established their periods. Magini, the report continued, was bending every effort to be that man.[44] When Kepler pronounced the task to be almost impossible, and the Jesuits in Rome agreed with him, the glory to be won increased all the more.[45] Galileo did succeed in defining the periods in 1612. Quite unmindful of the possibility that he was giving away a lucrative advantage in regard to navigation, he rushed his discovery into print in a work to which they had no relation.[46]

Well before he had untangled the periods, his pursuit of the satellites had led him to another discovery. On 25 July 1610 Jupiter became visible to him again after its conjunction with the sun. Perforce he had to observe it before dawn, when Saturn was also visible high in the western sky. Whatever Galileo's fine words a month earlier about his many observations of Saturn, he cannot have looked at it closely before this time, for now he immediately saw, or thought he saw, exactly what he had solemnly assured the grand duke was not there. He had discovered, he wrote in excitement to Vinta, "another most fantastic marvel," which he wanted the ducal family to know about although he asked them to keep it secret until he could publish it. "I have wished to give an account of it to their Most Serene Highnesses," he continued, "so that if others chance upon it they [their Highnesses] will know that no one observed it before me."[47] Saturn, he had discovered, consists of three bodies, a central sphere plus two smaller ones immediately adjacent on either side, Galileo's interpretation of the rings as they appeared through his telescope. "Behold," he wrote elsewhere, "I have found the court of Jupiter and the two servants of this old man, who help him walk and never leave his side."[48] "I have found"—one cannot miss the significance of the personal pronoun. To protect his priority, he concealed the discovery in a cipher and sent it to Prague, where it thoroughly excited the emperor. When Galileo sent the solution to the cipher in the fall, without waiting for publication, he received word in return that it had given the emperor "no less pleasure than amazement" and that the emperor had immediately set Kepler the task of confirming it.[49] Although we do not have any explicit evidence, there is no reason to think the new observation aroused any less pleasure and amazement in Florence than it did in Prague. Thus scarcely a month before he formally entered the grand duke's service, he began to fulfill his promise of many more discoveries, the first installment against the never-ending demand to justify his status as a client likely to confer glory upon his patron.

Exactly three weeks and one day after Galileo sent the solution of the Saturn cipher to Prague, Benedetto Castelli sat down in Brescia to write the letter quoted at the opening of this article, asking about the appearances of Venus and

---

[44] Cigoli to Galileo, 23 Aug. 1611; *ibid.*, Vol. XI, p. 175.
[45] Galileo to Vinta, 1 Apr. 1611; *ibid.*, p. 80.
[46] Galileo, *Discourse on Bodies in Water*, trans. Thomas Salusbury, ed. Stillman Drake (Urbana: Univ. Illinois Press, 1960), pp. 1–2. See also the excellent account of this considerable achievement in Drake, "Galileo and Satellite Prediction," (cit. n. 25). Measured by our present figures, Galileo's periods, published in 1612, were accurate for each of the four satellites to well less than one part in a thousand; see Righini, *Interpretazione scientifica* (cit. n. 23), p. 56.
[47] Galileo to Vinta, 30 July 1610; *Opere*, Vol. X, p. 410.
[48] Galileo to Giuliano de' Medici, 13 Nov. 1610; *ibid.*, p. 474.
[49] Hasdale to Galileo, 19 Dec. 1610; *ibid.*, p. 491.

Mars. Six days later, on 11 December 1610, Galileo sent a second cipher to Prague, "of another thing just observed by me which involves the outcome of the most important issue in astronomy and, in particular, contains in itself a strong argument for the Pythagorean and Copernican system."[50] The cipher concealed the sentence, "The mother of love emulates the figures of Cynthia," that is, Venus reveals phases like those of the moon. Galileo sent the cipher not only to Giuliano de' Medici in Prague but to at least three others.[51] At the end of December he wrote two long letters, one to Father Clavius in Rome and one to Castelli, describing in detail the observations of Venus that he had been making over the past three months—its appearance, round and very small, as it emerged in the western sky after its superior conjunction with the sun, its steady increase in size, its loss of roundness as it approached maximum elongation, and its assumption of a crescent shape as it passed maximum elongation, where it was standing at the date of his letter. He went on to predict its future changes of appearance as it went through inferior conjunction and proceeded through the rest of a complete cycle.[52] Towards Castelli he assumed an avuncular tone:

> O how many consequences and ones of such import have I deduced, my Master Benedetto, from these and from my other observations. You almost made me laugh when you said that with these manifest observations the obstinate could be convinced. Well then, don't you know that to convince those capable of reason and anxious to know the truth the other demonstrations already produced were enough, but to convince the obstinate who care only for the empty applause of the stupid and dull crowd, the testimony of the stars themselves, come down to earth to discuss themselves, would not suffice? Let us then endeavor to learn something for ourselves and rest satisfied with this alone, but as for advancing ourselves in popular opinion or gaining the assent of philosophers in books, let us give up the desire and the hope.

Castelli's letter and Galileo's activities during December raise several questions. Had Galileo received Castelli's letter from Brescia before he sent off the cipher on 11 December? We do not know, but it was easily possible. Brescia is about eighty kilometers east of Milan and about the same distance, via Bologna, from Florence as Venice is. Repeatedly, when he was in Padua, Galileo answered letters from Florence six days after they were written, and officials in Florence answered him after the same intervals. There are instances of letters making the trip in five days. Galileo himself, no longer a young man, spent six days on the journey when he moved from Padua to Florence in September. Castelli had indicated on another occasion that Brescia did not have regular mail service to Florence; that time he had sent his letter with a friend to Milan for posting.[53] In the silence surrounding the episode, we can readily imagine any number of possible scenarios, such as a friend's setting out for Bologna, or even Florence. This would have put the letter in Galileo's hands on 11 December. The coincidence of Galileo's cipher leaving on exactly the day that Castelli's letter would most likely have arrived makes it difficult to believe that Galileo wrote his cipher in ignorance of Castelli's communication.

---

[50] Galileo to Giuliano de' Medici, 11 Dec. 1610; *ibid.*, p. 483.
[51] See the references to the cipher in Santini to Galileo, 25 Dec. 1610; Magini to Galileo, 28 Dec. 1610; and Gualdo to Galileo, 29 Dec. 1610; *ibid.*, pp. 495, 496, 498.
[52] Galileo to Clavius, 30 Dec. 1610; Galileo to Castelli, 30 Dec. 1610; *ibid.*, pp. 499–500, 502–504.
[53] Castelli to Galileo, 27 Sept. 1610; *ibid.*, p. 436.

A more important question concerns the truth of Galileo's assertions to Clavius and to Castelli at the end of December, as well as to others later, that he had been observing Venus for three months, that is, from the beginning of October, long before Castelli wrote. Again we cannot answer with assurance. Aside from Galileo's claims, there is no evidence whatever for such observations, but silence does not yield a definite answer. However, there is considerable evidence that casts doubt on Galileo's statements. At the beginning of October Galileo was still completing his transfer to Florence. He did not systematically resume his interrupted observations of the satellites of Jupiter until the middle of November, though he made three isolated observations on 25 October and 4 and 5 November. It is true that Jupiter presented a problem that he did not experience with Venus. Jupiter was then visible in the east in the morning, and the house in which he lived temporarily until the end of October did not, as he complained, have a view to the east.[54] In contrast, Venus stood well up in the western sky in the autumn of 1610, so there was not any obstacle, as far as we know, to seeing it. Not only is there no mention of such observations, however, but in the middle of November, together with his translation of the cipher about Saturn, Galileo informed Giuliano de' Medici that he did not have any new discoveries about the other planets to report.[55] If he was in fact observing Venus at that time, a time when its gibbous appearance would have been in flat contradiction to Ptolemaic expectations, it is hard to reconcile his words with the excitement of the new cipher composed less than a month later.

Possibly relevant here is another statement by Galileo. In his letter in late December to Castelli about the observations of Venus, he wrote that he had been observing Mars for four months. We might note in passing what this dating implies about the truthfulness of his assertion to the grand duke the previous June that he had observed Mars many times. Be that as it may, in the autumn of 1610 Mars stood high in the morning sky, observable when Jupiter and Saturn were observable, and therefore possibly subject as well to the same obstacles that interrupted observation of them through September and October. I say "possibly subject" because Mars was west of the meridian in the morning, and if Galileo was moved to get up before dawn to look at Mars, the lack of an eastern view would not have presented a problem. I should add that Galileo's exact words were "four months," which take us back to the end of August, when he did record observations of Jupiter made in Padua as late as 7 September, the morning he left. He could easily have looked at Mars at the same time. In assessing the truth of his claims, moreover, we must also bear in mind the ease with which he could set up his telescope for the sort of qualitative observations in question here. Recall that in order to observe the satellites of Jupiter, he set up his telescope every night as he journeyed south to Rome in March of 1611. The domestic upheaval of the move to Florence and the house that lacked a proper view may have posed obstacles to observations during the early fall of 1610, but they were not insuperable obstacles. Nevertheless, the preponderance of evidence, especially the statement to Giuliano de' Medici in mid November that he had not made any new discoveries about the other planets, decidedly inclines me to doubt the truth of Galileo's assertions.

[54] Galileo to Giuliano de' Medici, 1 Oct. 1610; *ibid.*, p. 441.
[55] Galileo to Giuliano de' Medici, 13 Nov. 1610; *ibid.*, p. 474.

The evidence thus suggests that at the time Galileo began his celestial observations, he had not formulated a program of systematic observations designed to settle the Copernican issue. Rather, as I have asserted, he saw the telescope more as an instrument of patronage than as an instrument of astronomy. When Galileo, having seized what the moon and stars could quickly offer, had turned his telescope on the next brightest object in the evening sky, Jupiter, early in January, Venus was visible in the predawn sky. For a Copernican, Venus was in a critical part of its orbit, past maximum elongation, approaching superior conjunction, and thus exhibiting a shape incompatible with the Ptolemaic system. As we have seen, however, Jupiter had offered something quite different, an incomparable present to the grand duke, and Galileo had not paused to look further. We have two accounts of his early telescopic discoveries, by informed friends, apparently written before publication of the *Sidereus nuncius* and thus independent of it.[56] They bring up all the major discoveries announced by the book; they do not, however, mention observations of Venus, thereby tending to confirm the reading of Galileo's own silence as due to his not having noticed the planet's phases. In fact, the *Sidereus nuncius* was not totally silent about Venus. Galileo mentions it once, in his discussion of irradiation, his theory of an optical phenomenon within the eye that makes stars appear larger than they are. He suggests that his readers observe how small stars look when they first emerge from the twilight at sunset. "Venus itself, when visible in broad daylight," he adds, "is so small as scarcely to appear equal to a star of the sixth magnitude."[57] If he paused to consider its apparent shape at that time, he made no mention of it. By the time he completed the *Sidereus nuncius*, Venus was too close to the sun to be visible.

When Venus reappeared in the western sky near the end of July (judging by the angular distance from the sun when Jupiter was visible to Galileo), a few days after Jupiter reappeared in the east before dawn, Galileo was intently observing the satellites of Jupiter so that he could define their periods. At this time he also made his discovery about Saturn. He never claimed that he observed Venus during the month it was visible to him before he left Padua. The move to Florence with its attendant upheaval accompanied the steady rise of Venus in the western sky. When Galileo was finally ready to resume regular observations, he riveted his attention once more on the satellites of Jupiter; his records show almost daily observations of them beginning in the middle of November, that is, predawn observations while Venus was visible in the evening sky. All during this period Galileo seems to have used his telescope to further his advancement rather than Copernicanism.

Galileo probably read Castelli's letter on 11 December 1610, against the background I have described of eighteen months dominated by the telescope and the new possibilities it had opened to him. He had been attacked by enemies and challenged by rivals. It was Benedetto Castelli, however, his devoted student and disciple, who unintentionally delivered the unkindest cut of all, by pointing

---

[56] Sarpi to Leschassier, 16 Mar. 1610; Manso to Beni, Mar. 1610; *ibid.*, pp. 272, 291–296. Paolo Beni lived in Padua; Manso's letter replied to his account of the observations and discussed them in great detail.

[57] Galileo, *Starry Messenger*, trans. Drake, p. 46. An astronomer friend assures me that such an observation in broad daylight is entirely possible.

out that in his pursuit of celestial novelties to dazzle the grand duke, Galileo had neglected a phenomenon of supreme importance, which had been there all the time, virtually asking to be observed. Galileo must have understood instantly the significance of Castelli's query, a significance not confined to astronomy. Less than a month later, as he deciphered the anagram for Giuliano de' Medici, he explained the significance as he saw it. "From this marvelous experience we have a sensible and sure proof of two major suppositions which have been doubted until now by the greatest minds in the world"—namely, that the planets are dark by nature, and that Venus and Mercury revolve around the sun as Copernicus believed.[58] The subordinate clause in the sentence is no less important than the main one. The premise on which Galileo's position in Florence rested was his status as the one who revealed things the greatest minds in the world had not known. Was he now to admit that someone else, indeed one of his students, had understood what he had not? It appears evident to me that he instantly decided no. With good fortune but also with much effort, he had won the position he desired. At that point he had held it exactly three months, and he could well have felt insecure. The cipher was his announcement that he did not intend to surrender it lightly.

As we know, his tactics were completely successful. The very existence of Castelli's letter was not known for another two centuries and has been generally ignored since it was published. Castelli, who was not only a devoted follower of Galileo but his client, who hoped to exploit Galileo's new prominence to move himself from the monastery in Brescia into a life in science, was not one to raise claims. When Galileo wrote his first letter on sunspots in May 1612, he described the phases of Venus "discovered by me about two years ago."[59] His addition of three more months to his priority was of no great import. No one came forward to dispute any aspect of his claim.

I have suggested that the episode casts light on Galileo's commitment to Copernicanism. It is frequently asserted—indeed I think it is the majority opinion—that Galileo remained unconvinced of the truth of the Copernican system until the telescope revealed new evidence to him.[60] I see good reason to doubt this proposition. The *Sidereus nuncius* was an explicitly Copernican work, but when he wrote it, Galileo was not yet aware of any of the telescopic observations that he would later point to as decisive evidence. Castelli's letter, written by a student who had left Padua in 1607, well before Galileo's involvement with the telescope, is the expression of one Copernican who understood that he was writing to another. If my analysis of the episode is correct, Galileo composed the cipher asserting that Venus displays phases on the sole basis of his prior commitment to the Copernican system, which allowed him to predict what he had not then seen. His letter to Castelli did, after all, insist that even without Venus there was plenty of evidence to convince those capable of reason.[61]

The cipher he sent on 11 December had a double advantage. It protected his priority, but it concealed the statement he was making in case he might need to

---

[58] Galileo to Giuliano de' Medici, 1 Jan. 1611; *Opere*, Vol. XI, p. 12.

[59] Galileo, *Letters on Sunspots*, in *Discoveries and Opinions*, trans. Drake, p. 93; on Castelli's attitude see, e.g., Castelli to Galileo, 24 Dec. 1610; *Opere*, Vol. X, pp. 493–494.

[60] See, e.g., Willy Hartner, "Galileo's Contribution to Astronomy," in *Galileo: Man of Science*, ed. Ernan McMullin (New York: Basic Books, 1968), pp. 178–194.

[61] Cf. Galileo to Giuliano de' Medici, 1 Jan. 1611; *Opere*, Vol. XI, p. 11.

withdraw it. Before Galileo wrote to Clavius and Castelli at the end of December, he had two and a half weeks in which to observe. Fate had returned to his side once more. During those two and a half weeks Venus was going through a critical portion of its orbit, in which at maximum elongation it gradually changed from a slightly gibbous shape to a thick crescent. At no point during December was its shape compatible with the Ptolemaic system. Galileo could also call upon the single observation of Venus mentioned in the *Sidereus nuncius*, when he had seen the planet small and fully round as it approached superior conjunction, and seizing its significance now, fit it into the Copernican pattern. By the end of December he was sure enough to send the full argument, not only to Castelli, but to Clavius in Rome. He had not earlier committed himself so far as to send Clavius the cipher. I should add, in partial support of the accepted position, that Galileo's sudden eagerness in January to carry the Copernican argument to Rome, the first serious step down the path that would lead him, two decades later, into the dungeon of the Inquisition, may very well have stemmed from his observation of Venus's phases.[62]

My primary interest in the episode focuses, however, on the light it casts on the system of patronage. As I realize all too well, even to bring the matter up is to court hostility. So well have we defended the pantheon of science from any suggestion of stain that only after I had pursued this question nearly to its conclusion did I discover it had been raised once before, nearly a century ago, by an Italian scholar, Raffaello Caverni. Caverni, who wrote during the springtime of Italian unity, was summarily drummed right out of the Italian learned community for casting a shadow on the name of a national hero.[63] Since I have no desire to suffer a similar fate, I trust that I am far enough removed from the seat of such emotions. I do wish to emphasize that it is not my purpose in any way to call Galileo's position in the history of science into question. If my analysis is correct, it does take away one small part of his reputation, though nothing that pertains to his major accomplishment. I reject, however, any implication that similar detailed analyses will dissolve away the whole. It appears to me that some degree of disillusionment is likely to accompany most examinations of the social setting of science. By their nature, they lead us out of the context of justification and into the context of discovery, where we see the play of human motives instead of the finished products of reason. To show that Galileo was no plaster saint does not discredit him. The *Dialogue* remains the *Dialogue*, the *Discourses* remain the *Discourses*, and nothing in the episode of Venus even begins to undermine the validity of his enormous achievement.

Nevertheless, we do face the question of why Galileo acted as he appears to have done. Why did a man of so many accomplishments and achievements strive compulsively to monopolize all discoveries with the telescope? In different circumstances with different details and more justice, he would before long repeat his performance in regard to the discovery of sunspots. Why did Galileo feel

---

[62] See Galileo to Vinta, 15 Jan. 1611; *ibid.*, p. 27.

[63] Caverni, *Storia* (cit. n. 1), Vol. II, pp. 357–361. See also Favaro's passionate repudiation of Caverni, "Galileo e la scoperta delle fasi"; and Giorgio Tabarroni's appreciative introduction to Caverni (New York: Johnson Reprints, 1971), Vol. I, pp. vii–xxii. Even a sympathetic observer such as Tabarroni regards Caverni's passage on the phases of Venus as a "malicious and unfounded accusation" (p. xviii).

that he had to take all the credit? I suspect no single explanation suffices to answer these questions. Surely Galileo's aggressive personality must enter into consideration. I do want to argue, however, that part of the answer has to do with patronage. The ultimate truth about what I have been calling the system of patronage is that it was not a system at all. It was a set of dyadic relations between patrons and clients, each of them unique. Patronage had no institutions and little if any formal structure. It embodied no guarantees. The relation between patron and client was voluntary on both sides and subject always to disintegration. Past performance counted only to the extent that it promised more in the future. A client's only claim on a patron was his capacity to illuminate further the magnificence of the man who recognized his value and encouraged him. From our perspective, Galileo was secure beyond possible challenge, and without loss to himself could have acknowledged Castelli's critical role in the discovery of the phases of Venus. Thus, in his second letter on sunspots, Galileo did credit Castelli with the method of drawing the spots that he described, and he did not suffer any consequent loss of reputation.[64] No doubt Galileo's perspective differed from ours, and living in early seventeenth-century Italian society, he may have understood its imperatives better than we do. In 1610 the *Dialogue* and the *Discourses* existed only as aspirations. His one serious accomplishment was the telescope plus the *Sidereus nuncius* it had produced. How far could he ride that horse? Consider the reply Galileo received from Antonio Santini, one of the men to whom he sent the cipher about Venus. Accepting the promise of the cipher without question, Santini assured Galileo that he "was born to honor our century with new discoveries and with the perfection of your industry to enrich the world of knowledge with so many noble and hidden objects that I am persuaded you walk out there among those lights."[65] Such a perception, spread abroad through the world of culture, was the very rock on which Galileo's position as a client stood, and Santini's letter is by itself evidence that the phases of Venus contributed to sustaining that perception. As John North has said of Galileo, "He gave glory to senates and princes, in the expectation of rewards, and it was important that he not only reveal new truths, but also master the art of establishing priority in his discoveries."[66] Such an argument does not excuse Galileo's apparent dishonesty in the matter of Venus. It may help us to understand what drove him to it.

Patronage cannot provide the universal key to an understanding of the social history of the Scientific Revolution. Some figures in the movement were not sustained by patronage, and it is not yet clear how many were so supported. Nevertheless, patronage was perhaps the most pervasive institution of preindustrial society; it animated the worlds of art, music, and literature.[67] Only now are scholars beginning to chart its course in the science of the age, and we have every reason to expect that it will prove to be very important there as well. I would like to suggest, that patronage, together with other practices that the age

---

[64] Galileo, *Letters on Sunspots*, trans. Drake, p. 115.
[65] Santini to Galileo, 25 Dec. 1610; *Opere*, Vol. X, p. 495.
[66] John North, "Thomas Harriot and the First Telescopic Observations of Sunspots," p. 130.
[67] See Werner L. Gundesheimer, "Patronage in the Renaissance: An Exploratory Approach," in *Patronage in the Renaissance*, ed. Guy F. Lytle and Stephen Orgel (Princeton: Princeton Univ. Press, 1981), pp. 3–23.

itself reveals to us, may be the avenue most likely to lead us into a fruitful social history of the Scientific Revolution, a movement to which the present generation of scholars has devoted itself extensively. In our investigations, it appears to me, we have allowed ourselves to be dominated excessively by concepts derived in the nineteenth century which are more applicable to that century and our own than to Galileo's age. Efforts to impose them on the seventeenth century have appeared forced and largely barren, and I want to propose, not as a new dogmatism, but as a topic for discussion, the possibility that we need to come at the problem from a different angle, using seventeenth-century categories instead of nineteenth-century ones. Patronage was certainly a seventeenth-century category. I hope that the episode of Galileo and the phases of Venus reveals to others as it has to me the promise that patronage offers as one device by which we can probe the social dimension of the Scientific Revolution.

The Telescope in the Seventeenth Century

By Albert Van Helden

THE AIM OF THIS PAPER IS TWOFOLD: first, to discuss the role played by science in the development of the telescope, and second, to attempt to assess the influence of the telescope on scientific ideas in the seventeenth century. Much of what I have to say is not new, but it is not my purpose to present new information. I want to try to interpret, or rather reinterpret, the available information. I believe that the development of the telescope has not been properly understood, and that the role of the instrument in seventeenth-century science has not been fully appreciated. The traditional treatment of the telescope is replete with optical diagrams, explanations of spherical and chromatic aberration, and references to the theoretical investigations of Galileo, Kepler, Descartes, Huygens, and Newton, with the result that one is left with the impression that the telescope was an instrument which, if not invented through science, was at any rate turned into the sophisticated instrument it became by science —the science of optics.[1] This view I shall examine. To place the telescope in its proper perspective in seventeenth-century science is a rather more difficult task; the second part of the paper will therefore be more speculative, and for this I beg the reader's indulgence.

THE DEVELOPMENT OF THE TELESCOPE IN THE SEVENTEENTH CENTURY

It is a pity that Cornelis de Waard's book on the invention of the telescope (*De Uitvinding der Verrekijkers*) has never been translated into a language somewhat more accessible than the original Dutch. Although de Waard left all the quotations in the

---

This paper was first delivered at a joint session of the Society for the History of Technology and the History of Science Society, Chicago, December 28, 1970. My thanks are due to Dr. Wilbur Applebaum, the late Dr. C. Doris Hellman, Dr. Derek de Solla Price, Dr. Stanley Jaki, Dr. Charles Garside, Jr., Dr. A. Rupert Hall, Dr. John Olmsted, and Dr. Stephen Straker for their helpful comments.

[1] This view is sometimes expressed explicitly, e.g., E. J. Dijksterhuis, *The Mechanization of the World Picture*, trans. C. Dikshoorn (Oxford: Clarendon Press, 1961), p. 390: "The construction of telescopes and microscopes, which was the fruit of the development of geometrical optics, was a factor of eminent importance for the growth of classical science." A seemingly more balanced but equally misleading point of view is expressed in M. Caspar, *Kepler*, trans. C. D. Hellman (London: Abelard-Schuman, 1959), p. 190: "From the very beginning, theory and practice joined hands and found it necessary to work together to achieve abundant success." See also W. P. D. Wightman, *The Growth of Scientific Ideas* (Edinburgh: Oliver & Boyd, 1951), pp. 57, 59.

original languages, his researches have not had as much impact as they could have had.[2] It seems best to assume, based on the research presented by de Waard, that the telescope was invented toward 1600 in Italy but remained practically unknown until, because of a patent application in 1608, its usefulness became common knowledge.

We are usually treated at this point to the passage from Galileo's *Sidereus nuncius* in which he explains his construction of the instrument after he had heard about it. (I shall not here go into the possibility that Galileo may actually have seen a telescope before he made his own.[3]) I quote from Stillman Drake's translation in *Discoveries and Opinions of Galileo:*

> A few days later the report [that a certain Fleming had constructed a spyglass] was confirmed to me in a letter from a noble Frenchman at Paris, Jacques Badovere, which caused me to apply myself wholeheartedly to inquire into the means by which I might arrive at the invention of a similar instrument. This I did shortly afterwards, my basis being the theory of refraction [*quam paulo post, doctrinae de refractionibus innixus assequutus sum*].[4]

Before Drake's translation the only English version of the tract was Carlos' translation of 1880, in which the last line of the quotation reads "through deep study in the theory of refraction."[5] Drake's translation is much more faithful to the original, but we are still left with the impression that Galileo somehow constructed a telescope after having figured out its theory. In *Il saggiatore* (1623) he tries to maintain this impression

---

[2] Cornelis de Waard, *De Uitvinding der Verrekijkers* (The Hague, 1906). De Waard summarized his findings in French in "L'invention du Télescope," *Ciel et Terre*, 1907, *28*:81–88, 117–124. A. Favaro wrote a lengthy article on the subject in 1907, prompted by de Waard's publication: "La Invenzione del Telescopio secondo gli ultimi Studi," *Atti del Reale Istituto Veneto di Scienze, Lettere ed Arti*, 1907, *66* (2):1–54. A. Danjon and A. Couder also dealt with de Waard's work in their classic *Lunettes et télescopes* (Paris: Editions de la Revue d'optique théorique et instrumentale, 1935), pp. 583–604, giving references to both de Waard's book and article, as well as Favaro's article. Vasco Ronchi has discussed de Waard's work in *Galileo e il Cannocchiale* (Udine: Casa editrice Idea, 1942), pp. 132–134, and has referred the reader to it in his more recent works, e.g., *Histoire de la lumière*, trans. J. Taton (Paris: Armand Colin, 1956), p. 83. Although Sir Harold Spencer Jones refers to de Waard's conclusions in his foreword to H. C. King's *The History of the Telescope* (London: Charles Griffin, 1955), p. vii, King himself seems unaware of them. Stillman Drake does not agree with de Waard's conclusions; see his *Galileo Studies. Personality, Tradition, and Revolution* (Ann Arbor: University of Michigan Press, 1970), pp. 155–156.

[3] The charge that Galileo had in fact *seen* a telescope before he set about making his own was made by several of his contemporaries. In his article "Did Galileo Claim He Invented the Telescope?" (*Proceedings of the American Philosophical Society*, 1954, *98*:304–312) E. Rosen shows that Galileo in fact never claimed to have been the *first* inventor of the instrument, and Rosen also concludes that there is no reason to doubt his statement that he constructed his first telescope on the basis of rumors and reports by others. I entirely agree with this conclusion, but it should be pointed out that telescopes were exhibited in Milan, Padua, and Venice during the summer of 1609 and that in all likelihood Galileo did speak to several men who had seen one (see, e.g., Stillman Drake, *Isis*, 1959, *50*:245–254). It seems thus likely that Galileo constructed his first telescope in much the same way as Simon Marius did, who had the instrument described to him by Hans Philip Fuchs von Bimbach. Fuchs had seen a telescope in the hands of a Dutch peddler at the Frankfurt fair in August or September 1608 (S. Marius, *Mundus Jovialis*, [Nuremburg, 1614], Praefatio ad candidum lectorem). The evidence also suggests that Sacharias Janssen, Hans Lipperhey, and Jacob Metius made their first telescopes after having seen or heard about the instrument (de Waard, *Uitvinding, passim*).

[4] Stillman Drake, *Discoveries and Opinions of Galileo* (Garden City: Doubleday, 1957), pp. 28–29.

[5] E. S. Carlos, *The Sidereal Messenger of Galileo Galilei and a Part of the Preface to Kepler's Dioptrics* (London, 1880), p. 10.

when refuting Sarsi's accusation that he had made the telescope into his "foster child."[6] Galileo admits that he had not invented the telescope (indeed, he never claimed that he did)[7] but argues that unlike the "simple maker of ordinary spectacles" who first invented the instrument, he had "discovered the same by means of reasoning."[8] He then goes on to explain what this reasoning consisted of:

> My reasoning was this. The device needs either a single glass or more than one. It cannot consist of one alone, because the shape of that one would have to be convex ... or concave ..., or contained between parallel surfaces. But the last named does not alter visible objects in any way, either by enlarging or reducing them; the concave diminishes them; and the convex, while it does indeed increase them, shows them very indistinctly and confusedly. Therefore a single glass is not sufficient to produce the effect. Passing next to two, and knowing as before that a glass with parallel faces alters nothing, I concluded that the effect would still not be achieved by combining such a one with either of the other two. Hence I was restricted to trying to discover what would be done by a combination of the convex and the concave, and you see how this gave me what I sought.[9]

Aside from the fact that the combination of the concave and convex lenses does not necessarily follow from what goes before, it is clear that any "simple maker of ordinary spectacles" would have been capable of similar reasoning. In fact, the instrument was so simple that any lens grinder could have made one after hearing a cursory description of it. For this reason the Council of the State of Zeeland advised the Estates General of the Netherlands, in connection with Lipperhey's patent application, that in their opinion the art could not be kept secret "because after it is known that the art exists attempts will be made to duplicate it, especially after the form of the tube has been seen and from it the reasons could have been understood to some extent."[10]

We may be sure that Galileo was familiar with the theoretical optics of his day. Drake has written on this subject, "His knowledge of perspective was at least equal to that of the ordinary professor of mathematics in any Italian university of the time, and his knowledge of refraction equaled that of any other professor of astronomy."[11] But when he made his first telescope, in the summer of 1609, and even when he made his famous thirty-powered instrument in January 1610, there still was no adequate theory of image formation through lenses and lens systems. It was therefore quite impossible to arrive at the construction of a telescope starting from the theoretical optics of the day. Not until 1611, when Kepler published his *Dioptrice*, was there any kind of theoretical foundation for the study of lens systems. Kepler did in fact have a knowledge of optics around 1610 which enabled him to do what Galileo claimed to have done: investigate the *theory* of the telescope. And although Kepler's treatment could not be quantitative and definitive in the absence of the sine law of refraction, his investigations did lead him to the conclusion that one could also construct a telescope by combining two convex lenses.[12] Here is a telescope whose theoretical description pre-

---

[6] H. Grassi, *On the Three Comets of the Year 1618* (Rome, 1619), in Stillman Drake and C. D. O'Malley, *The Controversy of the Comets of 1618* (Philadelphia: University of Pennsylvania Press, 1960), p. 81.
[7] Rosen, *op. cit.*
[8] Galileo, *Il saggiatore* (Rome, 1623), in Drake and O'Malley, *Controversy of the Comets*, pp. 212–213.
[9] *Ibid.*, p. 213.
[10] De Waard, *Uitvinding*, pp. 173–174.
[11] Drake, *Galileo Studies*, p. 142.
[12] Johannes Kepler, *Dioptrice* (Augsburg, 1611). Kepler not only points out that objects

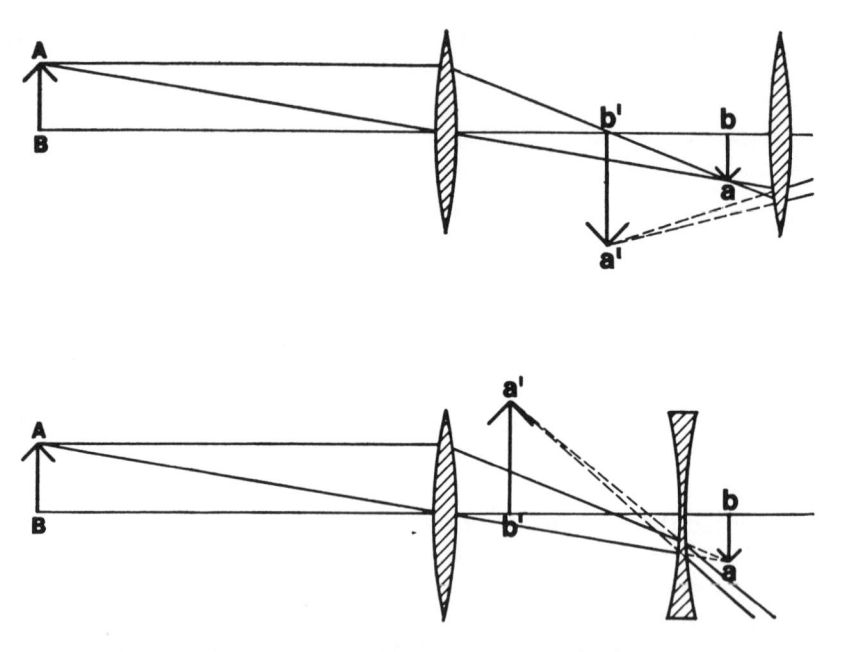

Figure 1. Top: astronomical telescope; bottom: Galilean telescope. In each case AB represents the object, ab the image formed by the objective lens alone, and a'b' the image as seen through the eyepiece.

ceded its actual construction, unless we believe Francesco Fontana, a telescope maker from Naples, who claimed in 1646 that this particular instrument had been invented by him as early as 1608.[13] Fontana's claim has never been substantiated.

But we are now faced with the fact that this configuration of lenses, the so-called astronomical telescope, did not come into widespread use until the 1640s, that is, more than thirty years after Kepler had first described it. The advantages of the astronomical telescope are that it has a larger field of view than the Galilean telescope and that it has a positive focus which allows one to introduce measuring devices into the instrument (see Fig. 1).[14] It was thus a more suitable instrument for celestial observations, yet it was virtually ignored for thirty years. The probable explanation of this delay lies in the fact that both these advantages could only be discovered through the actual use of such a telescope; they were not predicted by theory. A person reading Kepler's comments on this configuration of lenses would therefore see no advantages but one obvious and major *disadvantage*: the inverted image. It is true that the inverted image of the Keplerian or astronomical telescope presents no particular problems to modern observers

---

can be enlarged and made more distinct by using two convex lenses, although the image will be inverted, but he also shows that such a configuration will *project* an erect image on a screen, while the image seen through a telescope consisting of two convex lenses can be erected by the introduction of a third convex lens. Johannes Kepler, *Gesammelte Werke* (Munich:C. H. Beck, 1937—), Vol. IV, pp. 387–389.

[13] Francesco Fontana, *Novae coelestium, terrestriumque rerum observationes* (Naples, 1646), p. 7.

[14] Neither of these advantages was known to Kepler, who never put his own suggestion into practice.

of the heavens, but to observers in the early part of the seventeenth century it appears to have been a serious disadvantage. The instinctive preference for an erect image was difficult to overcome.

Moreover, we tend to think that the full potential of the telescope as a research instrument was immediately apparent to Galileo and his contemporaries; this was not the case. During the first twenty-five years of its existence the telescope remained primarily an instrument for terrestrial use, usually for naval or military purposes.[15] When one turned such an instrument to the heavens, it was usually to see for oneself the discoveries of Galileo or to work on specific projects which readily suggested themselves.[16] Routine telescopic observations of heavenly bodies to see if anything new could be discovered did not come until later. Thus, although Pierre Gassendi made telescopic observations from 1618 onward, these were limited to Jupiter's satellites, solar and lunar eclipses, and transits, until 1633.[17] Only then did he begin to investigate other areas.[18] Furthermore, although Gassendi had access to a variety of telescopes through his patron Nicolas Claude Fabri de Peiresc, the instruments with which he worked during this early period were of very poor quality.[19] The story is much the same for Ismael Boulliau,[20] Johannes Hevelius,[21] and Giovanni Battista Riccioli.[22] In the case of Gassendi, Hevelius, and Riccioli the telescope came into its own through efforts to map the moon.

There are few other astronomical studies with the telescope to which we can point before the late 1630s.[23] One of these is Christoph Scheiner's exhaustive study of sun-

---

[15] Let us not forget that Lipperhey's presentation to Prince Maurice and Galileo's presentation to the Doge stressed the military usefulness of the new instrument.

[16] Obvious areas for detailed work were mapping the moon, observing the satellites of Jupiter in the hope that eventually eclipse tables would allow the determination of longitude at sea, observing eclipses, sunspots, and transits of Venus and Mercury across the sun.

[17] Pierre Gassendi, *Opera omnia* (Lyons, 1658), Vol. IV, pp. 81–82, 84–88, 90–93, 99, 101–109.

[18] *Ibid.*, pp. 110–166. In this year Gassendi began keeping elaborate notebooks for his observations. His first known comments on and sketches of Venus (p. 141) and Saturn (p. 142) date from this year.

[19] For Peiresc's telescopes see Gassendi's life of Peiresc, *The Mirrour of true Nobility & Gentility being the Life of the Renowned Nicolaus Claudius Fabricius Lord of Peiresk, Senator of the Parliament at Aix written by the Learned Petrus Gassendus, . . . Englished by W. Rand* (London, 1657), p. 144. Gassendi remarks of a telescope which he used in 1619 and 1620 to observe the separation of Jupiter's satellites from their parent body: "Memini Telescopium, quo tunc utebar haud valde magnum, exquisitumque fuisse; ut proinde neque praedicta Diametrorum designatio pro summe exquisita non sit" (*Opera omnia*, Vol. IV, p. 82).

[20] MSS Collection Boulliau, Bibliothèque Nationale, F.F. 13058, *passim*.

[21] Johannes Hevelius, *Selenographia* (Gdansk, 1647). The earliest telescopic observations reported by Hevelius date from 1642 (e.g., p. 34). On p. 42 he writes: "Memini namque quod ipsum [Saturnum], mense Septembri & Octobri, Anni 1642. plane rotundum conspexerim. . . ." Apparently he did not bother to make notes on all his telescopic observations made in 1642! See also p. 46.

[22] G. B. Riccioli, *Almagestum novum* (Bologna, 1651). The earliest telescopic observations made by Riccioli himself that are reported here date from 1643 (pp. 484, 487, 488).

[23] Besides the work by Gassendi already mentioned, and by Scheiner and Fontana discussed below, Hortensius used a telescope for his observation of the transit of Mercury in 1631 (Martin Hortensius, *Dissertatio de Mercurio in sole viso et Venere invisa*, Leyden, 1631, p. 6). Wilhelm Schickard missed this event because he did not use a telescope, but rather followed Kepler's suggestion that a *camera obscura* be used for such observations: the aperture of this *camera obscura* was too large to show Mercury on the sun (W. Schickard, *De Mercurio sub vole viso*, Tübingen, 1632, pp. 14–16). Langrenus (M. F. van Langren) drew his first known moon map in 1644 and produced a more elaborate engraving the following year. O. Van de Vyver has recently dealt with some of the confusion regarding Langrenus'

spots. Apparently Scheiner used Kepler's configuration of lenses as early as 1617 (which, in all probability, was about a decade before Fontana began using it.) The reason for this is probably that when one looks through an astronomical telescope the image is inverted, but when one *projects* the image onto a screen it is erect. The reverse is true for the Galilean telescope.[24] It is therefore not surprising that the actual use of an astronomical telescope should first be mentioned in connection with the study of sunspots, the subject of Scheiner's major work, *Rosa ursina*, published in 1630.[25] Although it is fairly certain that Fontana had used this configuration of lenses since at least 1625,[26] and although Scheiner discussed its advantages in print as early as 1630, still the astronomical telescope did not immediately catch on.[27]

The telescope remained first and foremost a terrestrial instrument, used for spying on the enemy from a distance or identifying a ship when it was far away. Since good telescopes were very expensive, very few astronomers had good telescopes.[28] During

---

moon maps ("Original Sources of Some Early Moonmaps," *Journal for the History of Astronomy*, 1971, *2*:97). His monograph "Lunar Maps of the Seventeenth Century" (*Vatican Observatory Publications*, 1971, *1*, No. 2), is a definitive review of this subject. Gassendi's moon maps date from the middle 1630s (A. Tiberghien, "Cartes lunaires peu connues; II C. Mellan 1634-5," *Ciel et Terre*, 1932, *48*:106–111, 122–132). Throughout this period Galileo was making observations of Jupiter's satellites for the purpose of using their eclipses for determining longitude at sea (*Le Opere di Galileo Galilei*, Edizione Nazionale, Florence, 1890–1909, Vol. III, Pt. 2).

[24] This had already been pointed out by Kepler in his *Dioptrice* of 1611 (*Gesammelte Werke*, Vol. IV, pp. 389, 396). Scheiner dealt with images projected through two convex lenses in his *Oculus hoc est fundamentum opticum* (Ingolstadt, 1619), pp. 176 ff., and treated them in his major work, *Rosa ursina* (Bracciano, 1630), fol. 129v.

[25] *Rosa ursina*, fol. 130r. Scheiner mentions here also for the first time the advantages of the Keplerian or astronomical telescope: "Si similes duas lentes eodem modo optaveris in Tubum, oculumque debite applicaveris, videbis everso quidem situ, sed magnitudine, claritate, et amplitudine incredibili obiecta quaecunque terrea. . . ." He goes on to say that the disadvantage of the inverted image is not important for astronomical observations. The fact that these advantages are mentioned here for the first time and not in his 1619 publication seems to indicate that Scheiner first used the Keplerian configuration of lenses just for projecting images (as early as 1617, as he indicates on this same page) and only later began making astronomical observations *through* such a telescope.

[26] The bulk of the observations presented by Fontana date from 1645 and 1646, but a few go back to the early 1630s. The earliest reported observation (of the moon) dates from Oct. 31, 1629 (*Novae coelestium*, p. 80). The book also contains statements by G. B. Zuppo and H. Sirsalis that they saw astronomical telescopes belonging to Fontana in 1614 and 1625, respectively (*ibid.*, pp. 3, 5).

[27] There is no evidence that Gassendi and Hortensius used astronomical telescopes for their transit observations. As late as 1639 both Jeremiah Horrocks and William Crabtree still used the Galilean configuration for their observations of the transit of Venus (Jeremiah Horrox, *The Transit of Venus over the Sun*, in A. B. Whatton, *Memoirs of the Life and Labors of the Rev. Jeremiah Horrox*, London, 1859, pp. 125, 198).

[28] Even Peiresc, a man of means, did not have telescopes good enough to show "Jupiter or Saturn or Venus entirely denuded of their rays . . . ," something which was, especially in the case of Venus, of course, quite impossible with anachromatic telescopes. He asked Galileo to send him one of his famous telescopes (Peiresc to Galileo, Jan. 26, 1634, Galileo, *Opere*, Vol. XVI, p. 28). Galileo did send a telescope which was evidently much better than all the instruments Peiresc and Gassendi had used up to that point (Peiresc to Galileo, Feb. 24, 1637, Vol. XVII, pp. 34–35). Gassendi used this telescope for his observations until his death in 1655 (*Opera omnia*, Vol. IV, *passim*). In 1637 Hortensius wrote to Galileo on the subject of observing Jupiter's satellites for the purpose of longitude determination; "we have found none in Holland up to now which can be polished to the precision which is required for these observations. For the best usually show the disc of Jupiter hairy and poorly defined, whence the Jovial stars in its vicinity are not perceived well" (Hortensius to Galileo, Jan. 26, 1637, Galileo, *Opere*, Vol. XVII, p. 19).

this early period the superior instruments were designed for seeing things on earth and were only occasionally turned to the heavens. This situation did not change very much until the telescope for celestial purposes parted company with the telescope designed for use on earth, a process which began slowly in Italy in the late 1630s and spread gradually until 1645. In that year Antonius Maria Schyrle de Rheita published a book entitled *Oculus Enoch et Eliae*, in which an instrument which is usually referred to as the terrestrial telescope is mentioned for the first time. This telescope consisted of four convex lenses, an objective, a field lens, an erector lens, and an ocular.[29] Now this instrument was found to be infinitely superior to the Galilean telescope for terrestrial purposes, the great disadvantage of the Galilean telescope being of course its small field of view. The telescope used by Galileo, the cracked objective of which is mounted and still preserved in Florence, had a field of view of about 15 minutes of arc—usually large enough to show all four of Jupiter's satellites at once, although only about one quarter of the moon could be seen through it at one time. But for terrestrial purposes this was not very satisfactory. It means that at a distance of 500 yards the field of view covered an area only about 2 yards wide—indeed a severe handicap. This new instrument, besides having an erect image, had a field of view much larger than the Galilean telescope; some telescope makers claimed it was thirty times as large![30] It is not surprising that these telescopes replaced the older Galilean ones.

But when one of these new telescopes was turned to a bright heavenly body, the image was so severely affected by aberration that the instrument was useless. In fact, Rheita does not refer to its astronomical use at all.[31] Aberration could be reduced to a nonobtrusive level, however, if the field lens and the erector lens were removed, leaving only the convex objective and the convex ocular. And although this combination did produce an inverted image, this was found not to be such a great inconvenience for celestial observations after all, especially if one got into the habit of drawing and engraving the figures of the planets, sun, and moon inverted. This practice was adopted gradually after 1645. And of course there was the larger field of view: for example, the 23-foot telescope made by Christiaan Huygens in 1656 had a field of view of 17 minutes of arc at a magnification of almost 100,[32] whereas such a magnification in a Galilean telescope would have resulted in a field of view of less than 5 minutes. Starting in about 1645, therefore, the astronomical telescope could and did follow a course which was independent of that of the various telescopes made for terrestrial purposes.

---

[29] A. M. Schyrle de Rheita, *Oculus Enoch et Eliae seu radius sidereo-mysticus* (Antwerp, 1645). Actually this telescope, consisting of four lenses, was not described in detail here. After stating that a telescope with three convex lenses gives an erect image, and that such an instrument had been used by him for terrestrial purposes, he gives his *secretum*, an easily solved cryptogram which mentions yet a fourth convex lens in the eyepiece. But Rheita does not give further information about it.

[30] T. H. Court and M. von Rohr, "New Knowledge of Old Telescopes," *Transactions of the Optical Society*, 1930–1931, *32*:119. The authors have reproduced here a manuscript which contains a price list of Johann Wiesel's telescopes. Rheita in fact claimed that the telescope with three lenses which he used for terrestrial purposes had a field of view about one hundred times as large as that of a Galilean telescope (*Oculus Enoch et Eliae*, p. 356). He also claimed that a 15-ft astronomical telescope had as large a field of view as a Galilean telescope of 1 ft (p. 352).

[31] Of the telescope consisting of three convex lenses he states specifically: "quali tubo pro terrestribus nos utimor..." (*Oculus Enoch et Eliae*, p. 356).

[32] *Oeuvres Complètes de Christiaan Huygens* (The Hague: Martinus Nijhoff, 1888–1950), Vol. XV, pp. 15–16, 56, 60, 350–351.

The work of Descartes has been pointed to as a scientific influence on the development of the telescope. It is true that Descartes gave a full treatment to the geometry of lens systems in his *Dioptrique* of 1637, after which the cause, or at least one of the causes, of the diffuseness of images seen through lenses with highly curved surfaces became generally known.[33] Previously the problems had been ascribed to imperfect grinding of spherical lenses,[34] but it was now realized that even if a lens could be ground with perfect spherical curvature, this would not eliminate the diffuseness of the images. If one took Descartes' advice, one would try to grind lenses with hyperbolic surfaces. Besides the fact that such lenses would in fact not have eliminated the problem of *chromatic* aberration (not treated until 1672), lens grinders in those days were not up to such a task—indeed, they never were in the seventeenth century. Thus, Descartes' study caused telescope makers to spend many fruitless hours trying to grind lenses with hyperbolic surfaces.[35] To my knowledge there was not a single successful attempt in the seventeenth century. Acknowledging both Descartes' work in optics and the practical considerations of lens grinding, Rheita advised his readers to try to make hyperbolic lenses of large radii of curvature in the hope that the error would be so small or actually so advantageous as to make these lenses superior to any lens with spherical curvature.[36]

It is fair to say that between 1610 and 1660 the improvements made in the telescope were due almost solely to improvements in lens grinding techniques. A good example of this is found in a comparison of the lenses made by Galileo with those made by Giuseppe Campani, by far the best telescope maker of the second half of the seventeenth century. The telescopes ascribed to Galileo, preserved in the Museo di Storia della Scienza in Florence, were tested by Ronchi and Abetti in 1924. The best objective, the cracked lens mounted in the stand, has a useful aperture of 38 millimeters and an actual resolving power of perhaps 10 seconds at best.[37] A Campani objective dating from about 1660 was tested in Utrecht around the turn of the twentieth century. For a useful aperture of 42 millimeters its actual resolving power was found to be 3.7

---

[33] René Descartes, *La Dioptrique* (1637), in *Oeuvres de Descartes*, publiées par Charles Adam et Paul Tannery (Paris, 1897–1913), Vol. VI, pp. 81–228. Descartes was not the first to treat the problem of spherical aberration. In his *Ad vitellionem paralipomena* (1604) Kepler had dealt with the subject of focusing parallel rays through a dense medium (*Gesammelte Werke*, Vol. II, pp. 143–183), and in his *Dioptrice* (1611) he took the subject up again as it applied to the telescope, showing that through a convex lens with spherical curvature the rays farthest from the optical axis are refracted more than those nearer the axis, so that they will not all come together in one point. He then states that in order for parallel incident rays to be combined at one point the lens surface must be hyperbolic (Vol. IV, pp. 371–372).

[34] The objective lenses used by Galileo in 1610 were already stopped down drastically in order to obtain a less diffuse image.

[35] Gaston Milhaud has written on this subject: "Toutes les figures théoriques que nous apporte la *Dioptrique* pour les instruments grossissants restent des figures, et si l'auteur de la *Dioptrique* a espéré qu'ils seraient construits un jour d'après ses indications, il s'est trompé: ses efforts sont restés en marge des progrès continus réalisés par les physiciens et les astronomes dans la confection des lunettes et des télescopes; tout s'est passé, comme s'ils n'avaient pas existé" (*Descartes savant*, Paris: Librairie Félix Alcan, 1921, p. 196). See also M. Daumas, *Les instruments scientifiques aux XVIIe et XVIIIe siècles* (Paris: Presses Universitaires de France, 1953), p. 45.

[36] Rheita, *Oculus Enoch et Eliae*, p. 353.

[37] G. Abetti, "I Cannocchiali di Galileo e dei suoi Discepoli," *L'Universo*, Sept. 1923, 685–692; V. Ronchi, "Sopra i Cannocchiali di Galileo," *L'Universo*, Oct. 1923, 791–804. See also G. Abetti, *The History of Astronomy*, trans. B. Burr Abetti (London: Sidgwick & Jackson, 1954), pp. 94–95.

seconds.[38] Since the aperture determines the theoretical resolving power,[39] and since the apertures of these lenses are almost the same, the difference in actual resolving power represents a great improvement in the quality of lenses between 1610 and 1660. Both Galileo and Campani used Venetian glass, and since it is unlikely that the quality of glass improved that much in fifty years, we must conclude that the improvement in resolving power was due to a great improvement in lens grinding techniques. One of the men who tested the Campani lens mentioned that the lens was so perfectly ground that it would hardly be possible today (i.e., 1898) to grind a nonachromatic lens with spherical curvature which would be more perfectly shaped than this Campani lens.[40] This is not to say that the lens did not have defects—it most certainly did. Although there were few flaws in the glass, the images, especially of stars, suffered heavily from aberration.[41] But the point of this example is that great improvements were to be gained from grinding better *spherical* lenses, quite apart from any improvements which might be gained from the much more difficult task of grinding lenses with hyperbolic curvature.

This introduces the problem of the length of seventeenth-century telescopes. It is usually postulated that after Descartes' analysis of lenses efforts were made to minimize the curvature of lens surfaces, which led to an increase in the length of telescopes. It is true that telescopes became longer after 1645, but as pointed out above, the realization that increased curvature led to more diffuse and colored images was a matter of practical experience which preceded Descartes' theoretical work. Moreover, the Galilean telescope imposed an upper limit on magnification because of its restricted field of view, and therefore not until about 1645, when the Galilean telescope began to be replaced rapidly by the astronomical telescope, do we see telescopes with lengths of more than about 8 feet.[42] It is thus difficult to credit Descartes with the impetus behind the increase in length of the telescope.

From about 1645 to 1650 the length of good telescopes increased from 6–8 feet to 10–15 feet, while magnifications increased from about 30 to 40 diameters.[43] After about 1650 lengths increased steadily but not rapidly. There has been some confusion on this subject recently. Both King and Bedini ascribe a telescope of 123 feet to Christiaan Huygens in 1656.[44] This is an error; the telescope made by Huygens in 1656 was

---

[38] H. J. Klein, "Eine Pruefung alter Fernrohrobjektive von Huygens und Campani," *Sirius*, 1899, *32*:278. The superiority of Campani's lenses over those of his contemporaries is well illustrated in this article. Albertus Antonie Nijland tested a Huygens objective with an aperture of 52 mm as well as the Campani lens. For the Huygens lens theoretical and actual resolving powers were 2.3″ and 3.8″, respectively, while for the Campani lens they were 2.8″ and 3.7″. Campani thus came much closer to the predicted theoretical resolution for a given aperture than did Huygens.

[39] According to the formula $\theta = 1.22\lambda/A$, where θ is the theoretical resolution in radians, λ the mean wavelength of white light, and A the aperture.

[40] Klein, *op. cit.*, pp. 277–278.

[41] *Ibid.*, pp. 278–279.

[42] In Italy magnifications and lengths of telescopes probably increased somewhat more rapidly, starting with the astronomical telescopes of Fontana *c.* 1635.

[43] Rheita gives tables of aperture ratios and focal lengths for astronomical telescopes. In each case the magnification is 40 diameters (*Oculus Enoch et Eliae*, p. 351).

[44] King, *History of the Telescope*, p. 51. S. A. Bedini, "The Aerial Telescope," *Technology and Culture*, 1967, *8*:398–399. The well-known 123-ft telescope made by Constantijn Huygens, Christiaan's brother, bears the date June 4, 1686. It was given to the Royal Society by Constantijn in 1692 (Huygens, *Oeuvres*, Vol. XV, pp. 23–24). Further confusion may result from the error in the description of Fig. 23 in the second volume of R. Taton, *A General History of the Sciences* (London: Thames and Hudson, 1963–1966), Vol.

23 feet long.[45] By 1660 the average length of a good telescope was about 25 feet, and by 1670 this length had increased to 40 or 50 feet. Only after 1670 did a more rapid increase take place which, in the early 1680s, necessitated eliminating the tube between the objective and the eyepiece. This form of telescope is called the aerial telescope.[46] The lengths of telescopes could now increase even further, and before the end of the century telescopes of well over 200 feet had been made.

I realize that a host of exceptions can be pointed out. Eustachio Divini made a 35-foot telescope in 1649,[47] and Hevelius erected a 140-foot telescope on the beach near Gdansk in the early 1670s.[48] But I have been discussing the lengths of *good* telescopes. The telescopes made by Campani are a case in point: they were consistently much shorter than the telescopes made by others at the same time, and they were consistently better. The point of this discussion of the length of telescopes is one of accent. The history of the telescope in the seventeenth century has been too much the history of fancy and useless instruments. The fact that very few, if any, astronomical discoveries were made with telescopes which, by contemporary standards, were very long seems to have escaped writers on this subject. If one wishes to place the telescope in its proper perspective in the history of astronomy this fact will have to be taken into consideration. Between 1650 and 1700 practically all useful astronomical work was done with telescopes of very modest lengths compared with other telescopes made at the same time, and we should not be far wrong if we put the upper limit of usefulness of seventeenth-century telescopes at about 35 feet.[49] Beyond this point the law of diminishing returns set in very rapidly, and we may properly consider the very long telescopes of the last few decades of the century as white elephants. It is important, therefore, to find out more about telescopes such as the one supplied to Riccioli by

---

II, pp. xii–xiii. The figure (opposite p. 288) shows Hevelius' 140-ft telescope; this well-known illustration is taken from *Machinae coelestis pars prior* (1673), *not* from *Dissertatio de nativa Saturni facie* (1656).

[45] Christiaan Huygens, *Systema Saturnium* (1659), *Oeuvres*, Vol. XV, p. 238; see also pp. 15, 16.

[46] In commenting on the cover design in "The Aerial Telescope," Bedini writes: "Inasmuch as the tube of the telescope shown appears to be made of wood, however, the artist probably intended to delineate one of the smaller aerial telescopes of the seventeenth century" (p. 395). The term "aerial telescope" has since its inception in the seventeenth century usually referred to a telescope in which the objective and eyepiece were not connected by a tube.

[47] Divini mentions this telescope in an advertising sheet published in 1649. The sheet is reproduced in G. Govi, "Della Invenzione del Micrometro per gli Strumenti Astronomici," *Bullettino di Bibliografia et di Storia delle Scienze matimatiche e fisiche*, 1887, *20*, facing p. 614.

[48] Hevelius, *Machinae coelestis pars prior* (Gdansk, 1673), pp. 403–419. Reproductions of the engraving showing this telescope can be found in a number of books, e.g., Taton, *General History*, opposite p. 288; King, *History of the Telescope*, p. 52.

[49] Of all the celestial discoveries made between 1650 and 1700 only Cassini's discoveries of two new satellites of Saturn in 1684 were made with telescopes longer than 35 ft: "Ils ont premierement été découverts au mois de Mars 1684, par deux objectifs excellens, de 100 & de 136 pieds, & ensuite par deux autres de 90 & de 70 pieds que M. Campani avoit tous travaillez & envoyez de Rome à l'Observatoire Royal par ordre de sa Majesté" (J. D. Cassini, "Nouvelle decouverte des deux satellites de Saturne les plus proches, fait à l'Observatoire Royal," *Mémoires de l'Académie des Sciences, contenant les Ouvrages adoptés par cette Académie avant son Renouvellement en 1699*, Paris, 1730, *10*: 701). Cassini goes on to explain that all four glasses were used in the form of aerial telescopes (pp. 701–702). These discoveries were in fact the only contributions made to astronomical knowledge by the aerial telescope. That these Campani glasses were of comparatively modest focal lengths for those days is illustrated by the fact that at the same time Hartsoeker glasses of 155 and 250 ft were in use at the Paris observatory (*ibid.*, p. 702). See also C. Wolf, *Histoire de l'observatoire de sa fondation à 1793* (Paris, 1902), p. 164.

Johann Wiesel, with which Francesco Maria Grimaldi made his lunar observations,[50] and the 17- and 34-foot efforts by Campani,[51] rather than to elaborate further on the 140-foot telescope of Hevelius or the 200-foot-plus telescopes of the Huygens brothers and Nicolaas Hartsoeker.[52]

To return briefly to the role of science in the improvement of the telescope, we must consider Christiaan Huygens. The invention of the Huygenian eyepiece is usually considered a scientific contribution.[53] Huygens was a man who had all the manual dexterity of a Galileo and the theoretical interest in optics of a Kepler. In 1656 he used an eyepiece consisting of two contiguous plano-convex lenses,[54] the advantages of which are not obvious to me. But given this configuration of the eyepiece, it became possible by means of trial and error, or experiment if one prefers, to notice the improvement of the image obtained by moving the lenses apart to certain distances. In doing so, Huygens invented an eyepiece which corrects for transverse chromatic aberration. In 1662, after having found this combination, Huygens stated that its advantages were these: it gives a larger field of view, the image is less deformed, and the lack of homogeneity in the glass is less of a nuisance.[55] But as to the relationship governing the distance between the two lenses in the eyepiece, Huygens remarked, "it would be difficult to apply theoretical precepts to this subject, because the consideration of colors cannot be reduced to geometrical laws...."[56]

In fact, in 1662, when Huygens discovered his eyepiece, chromatic aberration was a subject which could not be treated scientifically because there was as yet no theoretical foundation for it. This only came with Newton's paper of 1672.[57] Furthermore, this kind of "research" could be done by men who were by no means as skilled in theoretical optics as Huygens was. The three-lens eyepiece of the terrestrial telescope was invented by Wiesel and Rheita,[58] and in the 1660s Campani developed an eyepiece con-

---

[50] Riccioli, *Almagestum novum*, p. 204. Riccioli mentions that for their lunar observations he and Grimaldi had used telescopes by Galileo, Fontana, Torricelli, and Manzini, either gifts or on loan, and another sold to him by a "Bavarian craftsman." I believe that this craftsman was Johann Wiesel in Augsburg, whose reputation had spread to all parts of Europe by this time. Riccioli writes about this telescope that it served best for lunar observations, "not so much because of its length, although this was 15 feet, as because of the combination of lenses, both convex, so favorably associated that although they show at least the whole lunar disc at Apogee at once, they nevertheless amplify it and its individual parts so that in these parts the smallest particles are disclosed to sight, which with the others we had either not been able to see or had neglected."

[51] When he went to Italy in 1668 Adrien Auzout compared his new 35-ft telescope with one of equal length made by Divini and with a 17-ft effort by Campani which was in the possession of Cassini. He found his own equal to the Divini telescope, but definitely inferior to the Campani one (Huygens, *Oeuvres*, Vol. VI, p. 300). Cassini made a number of discoveries with this instrument during the 1660s. The objective of the 34-ft Campani telescope, delivered to the newly built Paris observatory in 1672, is still preserved in that building. It has been described by Danjon and Couder (*op. cit.*, pp. 645–647).

[52] The objective glass with the longest focal length made by the Huygens brothers was a 210-ft glass made by Constantijn in 1686 (Huygens, *Oeuvres*, Vol. XV, p. 23). Hartsoeker made one objective with a focal length of 330 ft (Vol. IX, pp. 15, 25), and, according to Philippe de La Hire, made lenses with focal lengths of up to 1,200 ft (Vol. IX, p. 114).

[53] E.g., A. E. Bell, *Christian Huygens and the Development of Science in the Seventeenth Century* (London: Edward Arnold, 1950), pp. 51–52, 170–171.

[54] *Systema Saturnium* (1659), *Oeuvres*, Vol. XV, p. 230. See also Vol. II, p. 362.

[55] *Ibid.*, Vol. XIII, pp. L, 256–257, 462–463. See also Vol. IV, pp. 242–243.

[56] *Ibid.*, Vol. XIII, pp. 264–265.

[57] *Phil. Trans.*, Feb. 19, 1671/1672, 80:3075–3087.

[58] Rheita, *Oculus Enoch et Eliae*, pp. 339–340. It is not clear exactly who invented this eyepiece, Rheita, Wiesel, or Gervasius Mattmüller.

sisting of three lenses which was superior for terrestrial purposes to the Huygenian eyepiece (combined with an erector lens).[59] It is fair to say then, as did the editors of Huygens' *Oeuvres complètes*, that "it must be admitted that experience [or experiment] played the greatest role in the invention of his eyepiece."[60] This is not to say that Huygens did not work on the theory of the telescope; he most certainly did. His theoretical investigations, however, did not directly lead to improvements in the telescope during the seventeenth century.

In view of the above, I believe that the influence of science on the development of the telescope has been overestimated. Except for one clear example—Kepler's description of the astronomical telescope—practice preceded theory, and in the case of Kepler there was no immediate impact. I have not dealt with the reflecting telescope here because it was of no importance in the seventeenth century, but I might point out that Newton's reflector is a clear example of a scientifically conceived telescope, perfect in principle, that could not be realized on a useful scale by contemporary craftsmen.[61]

Nevertheless, we can say that science, the science of astronomy, had something to do with the improvement of the telescope. Astronomers encouraged telescope makers to produce better and better telescopes in the firm belief that better instruments would add to astronomical knowledge.[62] Scientists were in constant touch with these craftsmen and worked closely with them,[63] which resulted in a generation of telescope makers who were quite unlike the simple artisans of the beginning of the century. Indeed, in some cases these instrument makers made new astronomical discoveries with their telescopes and even published these discoveries.[64] But although the telescope was certainly instrumental in bringing scientist and artisan closer together, the theoretical researches of the scientist into optical systems at no point in the seventeenth century led directly to demonstrable improvements in the telescope.

THE INFLUENCE OF THE TELESCOPE ON SCIENCE IN THE SEVENTEENTH CENTURY

I now turn to the second part of this discussion: the influence of the telescope on scientific thought and endeavor in the seventeenth century. First, it should be obvious that the telescope influenced the science of optics profoundly. There is no question

---

[59] Huygens, *Oeuvres*, Vol. IV, pp. 266–267; Vol. VI, p. 48.

[60] *Ibid.*, Vol. XIII, p. L.

[61] After the initial excitement caused by Newton's small reflecting telescope, it was soon found that there was "the great obstacle in this manner of Telescopes, vz. the softnesse of the metal in comparison of Glasse, wherefore it doth not receive so perfect a polishing, neither is it able to keep it, so that I hope but little of it, for practice" (Christiaan Huygens to Constantijn Huygens [father], Aug. 17, 1674, trans. Constantijn Huygens, *ibid.*, Vol. VII, p. 392).

[62] Very often the tasks which astronomers set telescope makers were very difficult indeed. In 1668 Hevelius ordered a telescope of "40, 50, 60 or 70 feet" from England, through the offices of Henry Oldenburg. After a year and one failure he finally received lenses which were deemed good enough by the Fellows of the Royal Society. *Correspondence of Henry Oldenburg*, ed. A. Rupert Hall and Marie Boas Hall (Madison: University of Wisconsin Press, 1965—). Vol. IV: *1667–1668* (1968), pp. 444, 447; Vol. VI: *1669–1670* (1969), pp. 165–174.

[63] During his visit to England in 1661, Christiaan Huygens missed the coronation of King Charles II in order to observe a transit of Mercury. For this purpose he spent the day in the workshop of John Reeves, a telescope maker (*Oeuvres*, Vol. III, pp. 271, 279).

[64] E.g., G. Campani, *Ragguaglio di due nuove osservazioni* (Rome, 1664), and *Lettere intorno all'ombre delle Stelle Medicee* (Rome, 1665).

that this science went through a revolution in the seventeenth century. Spectacles had been invented late in the thirteenth century, but while scientists knew of their existence, lenses were not considered a proper field of study. After the invention of the telescope, however, the study of lenses became an important part of optics. Ronchi has written on this subject:

> After more than three centuries of empirical life among the artisans and under the ban of science, lenses acquired a mathematical theory and became scientific instruments.
>
> This was one of the most momentous and catastrophic revolutions recorded in the history of science. It is really amazing that so stupendous an event is practically unknown. For it meant the establishment of a new faith, which radically altered the attitude of the scientist and research worker toward observational instruments. Formerly the skeptic was unwilling to look through them for fear of being deluded by appearances. Now the insatiable investigator pushes a device's potentialities to the limits, seeking to obtain from it information, even fragmentary and deceptive information, about the macrocosm and microcosm.[65]

The discoveries made by Galileo gave lenses a respectability which they had never had before, and lenses gradually became an integral part of the tool kit of the investigator into the nature of light. But lenses and lens systems brought with them a whole new set of problems. As pointed out above, problems encountered in attempts to improve the telescope resulted in the formulation of spherical aberration and later in the discovery of chromatic aberration. All these studies go back to Kepler, who is rightly called the father of modern optics. Whereas in his *Supplement to Witelo* of 1604 Kepler had mentioned lenses only briefly, explaining how they correct nearsightedness and farsightedness,[66] in his *Dioptrice* of 1611 he gave them a full treatment. The subtitle of this work mentions that its writing was occasioned by the invention of the telescope,[67] and all the work for it was done during a few weeks in August and September of 1610.[68] Kepler's work was continued by his successors, and one has only to read the beginning of Descartes' *Dioptrique* to see that this whole treatise is concerned primarily with the telescope.[69] The earliest studies in the field of optics of both Huygens and

---

[65] Vasco Ronchi, *Optics: The Science of Vision*, trans. Edward Rosen (New York: New York University Press, 1957), p. 47. It should be pointed out that not all historians of science agree with Ronchi's views here presented. In their recent article "The Sense of Vision and the Origin of Modern Science" (Allen G. Debus, ed., *Science, Medicine and Society in the Renaissance*, London: Heinemann, 1972, Vol. I, pp. 29–45), David C. Lindberg and Nicholas H. Steneck argue persuasively against Ronchi's claims that the sense of sight was mistrusted before Porta, Galileo, and Kepler.

[66] *Gesammelte Werke*, Vol. II, pp. 180–183.

[67] *Ibid.*, Vol. IV, p. 329. The preface deals with Galileo's latest discoveries, especially the phases of Venus and the tricorporeal appearance of Saturn. See also Carlos, *Sidereal Messenger*.

[68] Caspar, *Kepler*, p. 198.

[69] Descartes, *Oeuvres*, pp. 81 ff. After relating the invention of the telescope (which he wrongly attributes to Jakob Metius of Alkmaar), Descartes writes about the Dutch telescope: "Et c'est seulement sur ce patron, que toutes les autres qu'on a veües depuis ont esté faites, sans que personne encore, que ie sçache, ait suffisanment determiné les figures que ces verres doivent avoir. Car, bien qu'il y ait eu depuis quantité de bons esprits, qui ont fort cultivé cete matiere, & ont trouvé a son occasion plusieurs choses en l'Optique, qui valent mieux que ce que nous en avoient laissé anciens, toutefois, a cause que les inventions un peu malaysées n'arrivent pas a leur dernier degré de perfection du premier coup, il est encore demeuré assés de difficultés en celle cy, pour me donner suiet d'en escrire" (p. 82). Therefore, he plans to explain light and light rays first, then to describe the parts of the eye and to explain vision, after which he will show how vision can be rendered more perfect and how it can be altered by telescopes and microscopes (pp. 82–83).

Newton were on lenses and telescopes,[70] although their particular interests took these two men in different directions. Unquestionably the telescope and the problems connected with its improvement were the focal point of the science of optics during the first three quarters of the seventeenth century and therefore influenced this science greatly.

The impact of the telescope on astronomical thought is somewhat more difficult to deal with. Not that I am suggesting that there was no impact, but rather that this very profound influence is difficult to realize fully. I am referring to man's change in outlook an the cosmos. But first let us look at the obvious.

The discoveries made by Galileo, Scheiner, Johannes Fabricius, Simon Marius, and Thomas Harriot (and probably some others) were bombshells indeed. Galileo is credited with most of them because of his priority in publication. In his *Sidereus nuncius* he announced three discoveries—the earth-like nature of the moon, the multitude of stars, and the satellites of Jupiter.[71] In 1610 he also discovered the curious appearances of Saturn[72] and the phases of Venus,[73] while in 1611 he first observed sunspots.[74] Now much has been made of the conservative opposition to these discoveries, but I should like to suggest that in view of the circumstances, the time it took Galileo to convince all reasonable men was astonishingly short. The reason why I say this is that the importance of the new discoveries was tremendous, the opposition was powerful, and the instruments were very poor. We tend to forget these facts in discussing Galileo's discoveries. When Galileo published the *Sidereus nuncius* he singlehandedly took on the scholastic establishment which was in many cases part of the Church. And the argument was not about angels dancing on the head of a pin: it struck at the roots of Aristotelian physics and metaphysics, which was the official natural philosophy of the Church. Yet, only one year later, he had convinced the very men who were the guardians of the faith in matters of astronomy—the mathematicians of the Collegio Romano, as well as the head of the college, Cardinal Bellarmine.[75] Father Clavius, the senior member of the mathematicians, had condemned the Copernican system as being absurd in his commentary on the *Sphere of Sacrobosco*, published in 1590, and remained an adherent of the Ptolemaic system until he was shaken in his belief by Galileo's discoveries.[76] In 1610, when he was seventy, he allegedly said that in order for Galileo to see the things he claimed to see through the

---

[70] In Dec. 1653 Huygens had finished a manuscript tract "Tractatus de refractione et telescopijs" (Huygens, *Oeuvres*, Vol. XIII, pp. 1–269). Newton started his letter to Oldenburg which contained his new theory of light and colors: "Sir, To perform my late promise to you, I shall without further ceremony acquaint you, that in the beginning of the Year 1666 (at which time I applyed my self to the grinding of Optick glasses of other figures than *Spherical*,) I procured me a triangular glass-Prisme, to try therewith the celebrated *Phaenomena of Colours*" (*Correspondence of Isaac Newton*, Vol. I, Cambridge: Cambridge University Press, 1959, p. 92).

[71] Galileo, *Sidereus nuncius*, in Drake, *Discoveries*, pp. 31–58.

[72] Galileo, *Opere*, Vol. X, pp. 409–410;
Kepler, *Gesammelte Werke*, Vol. IV, pp. 344–346; Carlos, *Sidereal Messenger*, pp. 88–91.

[73] Galileo, *Opere*, Vol. X, pp. 481–483; Kepler, *Gesammelte Werke*, Vol. IV, pp. 346–351; Carlos, *Sidereal Messenger*, pp. 94–102.

[74] Drake, *Discoveries*, pp. 87–144.

[75] G. de Santillana, *The Crime of Galileo* (London: Heinemann, 1958), pp. 23–24. See also Galileo, *Opere*, Vol. XI, pp. 89–90, 102, but esp. 92–93, which is the letter from Clavius, Grienberger, Maelcote, and Lembo to Cardinal Bellarmine, saying that they have verified Galileo's observations (Apr. 24, 1611).

[76] A. Pannekoek, *A History of Astronomy* (London: Allen & Unwin, 1961), p. 224; Drake, *Discoveries*, pp. 153, 195.

telescope, he first had to put them in the telescope.[77] Yet in 1611 he and his colleagues honored Galileo at the Collegio Romano,[78] and in the same year Christoph Scheiner postulated the existence of new satellites (of the sun this time) to explain sunspots.[79]

Futhermore, these early telescopes were not at all good instruments. Galileo's great contribution to the telescope was his improvement of the instrument. His contemporaries who were not lucky enough to be able to look through a telescope made by Galileo might very well decide that the evidence presented to their senses was not nearly as conclusive as Galileo claimed it to be. An example will illustrate this. In 1612, Jacob Christmann, professor of logic at the University of Heidelberg, published a little work entitled *Nodus Gordius*, in which he described some of his observations made with the telescope:

> On the 22nd of October of last year, at half past seven p.m., Saturn crossed the meridian. His body showed itself in three distinct scintillations through the smaller rod. Through the larger rod, on the other hand, he was perceived split into four fiery balls. Through the uncovered rod [two lenses not in a tube], Saturn was seen to be one long single star. Therefore it appears to be the case that two, three, or four companion stars have been detected about Saturn with the aid of the new sight. But from this it does not follow that two or three servants have been assigned to Saturn by Nature, who like bodyguards watch over him and march with him constantly. How this fantasy can arise is evident from the above....
>
> At the beginning of the month of December, by means of either radius cylinder, the body of Jupiter appeared in three distinct scintillations and exhibited two shimmering diameters, or rather, the body of Jupiter was seen completely on fire, so that it appeared separated into three or four fiery balls, from which thinner hairs were spread in a downward direction, like the tail of a comet.[80]

With telescopes such as the ones in the possession of Christmann it must have been tempting indeed to reject Galileo's discoveries if one were conservative. It seems therefore fair to say that except for a few cranks, scientists acted like reasonable men, no matter how conservative they were on the issue of the Great Debate, and it is in fact surprising how little resistance they mounted.

None of the discoveries of Galileo and his contemporaries proved in the legalistic sense of Bellarmine[81] that the earth moves. They did, of course, show that the perfec-

---

[77] De Santillana, *Crime of Galileo*, p. 23; Drake, *Discoveries*, p. 75.

[78] Drake, *Discoveries*, p. 75.

[79] Christoph Scheiner, *Tres epistolae de maculis solaribus scriptae ad Marcum Velserum* (Augsburg, 1612). The letters are dated Nov. 12, Dec. 19, and Dec. 26, 1611. Scheiner explains his hypothesis in the third letter. The pages are not numbered.

[80] J. Christmann, *Nodus Gordius ex doctrina sinuum explicatus. Accedit appendix observationum quae per Radium artificiosum habitae sunt circa Saturnum, Iovem & Lucidiores stellas affixas* (Heidelberg, 1612), pp. 41–42.

[81] De Santillana, *Crime of Galileo*, p. 92: "The trouble with these leaders' [i.e., higher church officials] minds, so subtly logical on points of law, was that they stopped functioning as soon as they dealt with a diagram or with this kind of 'new stuff, paradox to the philosophical vulgar,' as Ciàmpoli called it." De Santillana also quotes Cardinal Bellarmine's letter to Foscarini: "If there were a real proof that the Sun is in the center of the universe, that the Earth is in the third heaven, and that the Sun does not go round the Earth but the Earth round the Sun, then we should have to proceed with great circumspection in explaining passages of Scripture which appear to teach the contrary, and rather admit that we did not understand them than declare an opinion to be false which is proved to be true. But, as for myself, I shall not believe that there are such proofs until they are shown to me. Nor is it proof that, if the Sun be supported at the center of the universe and the Earth in the third heaven, everything works out the same as if it

tion of the heavens could no longer be maintained. The satellites of Jupiter went a long way toward demolishing the crystalline spheres (if anyone still believed in them), and the phases of Venus gave the *coup de grace* to the Ptolemaic system. By this time most conservative astronomers had gone over to the system of Tycho Brahe, and after 1610 the rest followed suit.

In a way the telescope was a blessing to conservative astronomers, because it established an area of research within astronomy in which one's religious and astronomical convictions did not particularly matter. In the difficult climate of Italy after Galileo's condemnation one could still work on new projects such as mapping the moon or making tables of the motions of Jupiter's satellites without running the risk of getting into trouble with the Inquisition. Some of the best observational astronomy in the seventeenth century was, in fact, done by astronomers who either rigidly adhered to a geocentric system or never came out publicly with their opinion. I am thinking particularly of Giovanni Battista Riccioli, the Jesuit astronomer of Bologna, whose *Almagestum novum* of 1651 was one of the important astronomical books of the seventeenth century,[82] and who proposed the system of nomenclature to distinguish lunar features which we still use today.[83] But it is also true of Christoph Scheiner, who was the most dedicated student of sunspots in the seventeenth century, Giovanni Domenico Cassini, whose numerous celestial discoveries included four new satellites of Saturn. the rotational period of Jupiter, and the division in Saturn's ring system which is still known as Cassini's division, and a number of others. It may be said that perhaps some of these men, like Cassini, were secret Copernicans; that may be true. But it does not take away from the argument.[84] The telescope allowed orthodox astronomers, especially in Italy where they were very much under the scrutiny of the Inquisition, to do good work in astronomy.

Clearly, the meaning of the word "astronomy" changed with the introduction of the telescope. A whole new dimension was added to this study. But the change was not immediate; as pointed out above, the telescope broke into the arsenal of the professional astronomer only gradually, and for almost a generation its true potentials were not recognized. Some uses of the instrument were obvious: mapping the moon so that eclipses could be timed more accurately for purposes of determining longitudes, observing the satellites of Jupiter and preparing tables of their motions in an attempt to solve the problem of longitude at sea, and observing transits of Mercury and Venus to try to determine astronomical constants. But aside from these endeavors, Scheiner's

---

were the other way around. In case of doubt we ought not to abandon the interpretation of the sacred text as given by the holy Fathers" (pp. 99–100).

[82] This book is a review of all of astronomy up to 1650. As the title suggests, it was designed to be a complete compendium of astronomy. It is, in Montucla's words, "un vrai trésor d'erudition et de savoir astronomique..." (J. E. Montucla, *Histoire des mathématiques*, Paris, 1799–1802, Vol. II, p. 340).

[83] *Almagestum novum*, pp. 204½ ff. Riccioli's proposed nomenclature was of course not the first. Langrenus had proposed a system based on the names of royalty and philosophers, and Hevelius had suggested a system based on geographical names transposed from earth.

[84] The interesting question is surely not whether Cassini was a confirmed Tychonian or a Copernican. The important point is that he could go through life without ever expressing an opinion on the subject (to the best of my knowledge) and still be one of the greatest astronomers of his age!

study of sunspots was the only use of the instrument for serious scientific work.[85] When other astronomers became interested in observational astronomy, it was in areas where there was an obvious problem, such as the appearances of Saturn,[86] or an obvious use for their studies, such as Langrenus', Hevelius', and Gassendi's work on the moon.[87] In fact, there appears to have been a feeling that Galileo had discovered all there was to be discovered with the telescope—a feeling which was probably at least partly due to the scarcity of good telescopes during the early period.[88] Apparently even astronomers of the stature of Gassendi and Riccioli did not have telescopes which surpassed those Galileo had used around 1610 until well after 1640![89]

The first inkling of a change in this attitude comes in the late 1630s,[90] and only after 1646 do any new discoveries—in this case surface markings on Jupiter—become widely

---

[85] The observations of Jupiter's satellites did not bear fruit until over half a century after they were begun by Galileo. Determining longitude at sea by means of tables of the eclipses of Jupiter's satellites never became a practical reality. But Cassini's tables of 1668, only the second set of tables published, were sufficiently accurate to allow Ole Rømer to determine the speed of light approximately, in 1676. Although in 1612 Scheiner thought that sunspots were satellites of the sun, he subsequently changed his mind, and in his major work, *Rosa ursina*, he put them squarely on the sun's surface. Good moon maps did allow a more precise comparison of eclipse data, resulting in a more accurate determination of the difference in longitude between various observatories.

[86] Galileo discovered the tricorporeal appearance of Saturn in 1610 (*Opere*, Vol. X, pp. 409-410); in 1612 he saw that Saturn had lost its lateral bodies (Drake, *Discoveries*, pp. 143-144); in 1616 he first observed the so-called "handled" appearance (*Opere*, Vol. XII, p. 276, 276 n.1). He never developed an adequate theory to explain these phenomena, and he seems to have lost interest in the problem. Not until the edgewise appearance of the ring in 1642 were any sustained observations made by others. Between that time and the next edgewise appearance in 1655/1656, regular observations were made by several astronomers, and thus by 1656 sufficient information was available to allow a number of coherent theories to be proposed, of which Huygens' ring hypothesis was one. See *Systema Saturnium* (1659), *Oeuvres*, Vol. XV, pp. 270-295.

[87] Hevelius, *Selenographia*. Gassendi's moon maps were published in 1636 (Tiberghien, *loc. cit.*). Langrenus' moon maps were apparently not published until 1645, although he began making them much earlier (Niesten, *loc. cit.*).

[88] When Galileo was informed about the new type of telescope made by Fontana which enlarged things more, he wrote: "Quanto all'ingrandire gli oggetti più de gli altri telescopii nostrali e più corti, e verissimo: e circa all'ingrandire la luna e mostrarla maggiore del mercato di Napoli, questo è un parlare del volgo, argomento della poca intelligenza del Napolitano artefice, che ne ha dato relazione a esso Padre [Benedetto Castelli]. Del vedervisi infinite differenze è vero, ma sono le medesime chi si veggono co i telescopii nostri, ma alquanto più conspicue mercè dell'ingrandimento; ma non è già che vi si scorgano cose nuove e differenti dalle prime scoperte da me e poi riconosciute da molti altri," and again, "Tutto questo è stato osservato, nè di novo ci si vede altro che un maggiore ingrandimento, mercè di questi novi telescopii più lunghi" (Galileo, *Opere*, Vol. XVIII, pp. 18, 19). In 1658 Christopher Wren began his tract on the appearances of Saturn as follows: "The incomparable Galileo, who was the first to direct a telescope to the sky . . . so overcame yielding nature, that all celestial mysteries were at once disclosed to him. . . . His successors are envious because they believe that there can scarcely be any new worlds left, about which they can boast, and believe that only to succeeding Lyncei is it granted to add to the discoveries of Galileo" (Christopher Wren, *De corpore Saturni*, trans. Albert Van Helden, in "Christopher Wren's *De Corpore Saturni*," *Notes and Records of the Royal Society of London*, 1968, 23:219).

[89] Gassendi used the telescope sent to him by Galileo until his death in 1655 (see n. 28 above). In 1653 he obtained a 12-ft Divini telescope through the offices of Sir Kenelm Digby (Gassendi, *Opera omnia*, Vol. IV, pp. 474, 480). Riccioli does not distinguish between the telescopes by Galileo, Fontana, Torricelli, and Manzini. His praise is reserved for the one made by the Bavarian craftsman (see n. 50 above). Assuming this craftsman to be Wiesel, we can be fairly certain that Riccioli learned of Wiesel's telescopes through Rheita's book, published in 1645 (see n. 58 above).

[90] At this time references to the new telescopes by Fontana begin to creep into the Galileo correspondence, and there are a number of comments about new phenomena seen through them (*Opere*, Vol. XVII, pp. 192, 308, 363-364, 375, 383-384).

known.[91] Between 1650 and 1685 there was a second wave of discoveries,[92] and during this period astronomers finally came to realize the full potential of the new instrument. In the second half of the seventeenth century the telescope became a full-fledged research instrument which was used not only for discovery but also for measurement.[93]

But we must come to terms with the most difficult of all aspects of the telescope—also the aspect which is probably most obvious—the impact of the telescope on man's perception in general and on the perception of astronomers in particular. Although I shall not deal with the microscope, it is an integral part of this change in perception. The telescope was an extension of man's senses—the first real extension in history. In one stroke a whole new dimension of the cosmos had been opened up. Men who had reached maturity by the time Galileo published his first discoveries (and those include Galileo himself) were not affected in their basic conception of the cosmos,[94] but those who grew up with the telescope and were taught what it revealed formed a different fundamental conception of the cosmos from that of their predecessors. The Aristotelian duality between the sublunary and superlunary regions had been the common-sense notion for a long time: the only permanence was found in the heavens, and all bodies above the moon, with the exception of the sun, looked alike. They were all stars; some were fixed, and some wandered. But after the telescope had been directed to the heavens, a new type of common sense emerged. On the one hand the planets were shown to have many things in common with the earth, while on the other the fixed stars now appeared much farther away than had ever been dreamed before.[95]

---

[91] Fontana, *Novae coelestium*, pp. 110–125. This is the first mention of these discoveries in print. After this time it became customary to include figures of observations, not merely verbal descriptions with an occasional small sketch, in printed works. Fontana's book contains figures of the moon in various phases (pp. 32–87), Mercury, showing several phases (pp. 88–91), Venus in various phases (pp. 92–103), Mars with something which could signify a central spot but could also be a diffraction effect, also showing a phase (pp. 104–106), Jupiter, showing two and sometimes three equatorial belts (pp. 107–125), and Saturn in a variety of appearances (pp. 126–141). The book thus contains a large number of planetary appearances which had never before appeared in print.

[92] These include five satellites of Saturn, many phenomena connected with Saturn's rings, and surface markings on Jupiter and Mars, from which their periods of rotation were deduced accurately.

[93] Through the work of Huygens and Auzout the micrometer (already invented by William Gascoigne in the early 1640s but unknown up to that point) came into general use in the 1660s (Pannekoek, *op. cit.*, p. 258). Picard and Flamsteed used telescopes mounted on astronomical measuring instruments and improved the accuracy of positional astronomy tremendously (*ibid.*, pp. 259–260), and Rømer further refined this technique with his meridional telescope (p. 278). Many of these developments had already been foreshadowed earlier, e.g., by Horrocks, Gascoigne, and Crabtree, but they did not bear fruit until after 1660. The watershed in positional astronomy was the debate between Hooke and Hevelius about the practicability of telescopic sights on measuring instruments, after the publication of Hevelius' *Machinae coelestis pars prior* in 1673. See E. F. MacPike, *Hevelius, Flamsteed and Halley* (London: Taylor & Francis, 1937).

[94] It is certainly true, as Koyré states, that "the conception of the infinity of the universe is, of course, a purely metaphysical doctrine that may well—as it did—form the basis of empirical science; it can never be based on empiricism" (Alexandre Koyré, *From the Closed World to the Infinite Universe*, Baltimore: Johns Hopkins Press, 1957, p. 58). Koyré goes on to show that deep down Kepler was bound by the tradition of Aristotle (p. 87). When the telescope became known, Kepler was in his late thirties and Galileo in his middle forties. Neither one changed his basic conception of the universe, although they altered it somewhat in detail.

[95] Before Copernicus the distance to the fixed stars was usually thought to be about 20,000 earth radii. Copernicus himself gives this distance as *indefinita*. Kepler thought it to be 34,077,067 earth radii before the telescope came into use and revised this figure to 60,000,000 after 1610. Even the men who believed in a geocentric system in the seventeenth century had to increase their estimates. Riccioli believed the figure to be at least 100,000 earth radii at first and then revised

THE TELESCOPE IN THE SEVENTEENTH CENTURY 151

Thus, the planets were brought closer to the earth and the stars became farther removed from the earth and planets. The new common sense made it natural to think in terms of a new concept: *the solar system*.[96] We are on a body which is part of a system (whether it is in the center or not) which is itself almost insignificant in terms of the distances to the fixed stars suggested by the telescope.

Against this background of the change wrought by the telescope, the Great Debate palls somewhat. Even the men who had grown up with the telescope but who for intellectual reasons still believed in a geocentric universe conceptualized their universe in a manner quite different from their sixteenth-century predecessors. Every new discovery brought the planets closer to the earth, and every improvement of the telescope showed the stars to be even farther away. Thus, the Great Debate becomes more and more irrelevant as the century progresses. Cassini did just as much to supply Newton with his information as did John Flamsteed.[97] There is something pathetic about Riccioli's *Almagestum novum*. The frontispiece shows the goddess Astraea holding up a balance from which the Copernican system and Riccioli's own modification of the Tychonic system are suspended. The balance is tipped decisively in favor of the Tychonic system, while the Ptolemaic system lies discarded at the bottom. In the upper corners the appearances of the planets and moon are shown as revealed by the telescope held by Argus.[98] In the book Riccioli discussed the arguments for and against the Copernican system and did his best to prove that it could not be right.[99] But he also gave detailed telescopic observations of the moon and planets.[100] His suggested system of nomenclature gave the names of philosophers—earthlings—to lunar features, thus making the moon even more earth-like.[101] While fighting the battle for orthodoxy he

---

it to at least 210,000. Rheita thought it to be about 60,000,000 (Riccioli, *Almagestum novum*, p. 419). After Newton the distance from the solar system to the nearest fixed star was commonly assumed to be about 20,000,000,000 earth radii (e.g., John Locke, *Elements of Natural Philosophy*, Berwick upon Tweed, 1754, p. 12).

[96] The first use of the term "solar system" in the English language, to the best of my knowledge, is in Locke's *Elements of Natural Philosophy* of 1706, an edition which I was unable to consult. The term is used in the 1754 edition which I used, e.g., pp. 8 ff. In the first (Latin) edition of his *Elements of Physical Astronomy* (Oxford, 1702), David Gregory used the terms "systema Solis & Planetarum primariorum" (p. 62) and "systema Solis & Planetarum primariorum & secundariorum" (p. 69). I was unable to consult the first English edition (1716), but the second English edition (London, 1726) does contain "solar system," e.g., Vol. I, p. 132.

[97] For his astronomical data Newton usually has reference to Flamsteed or Cassini.

[98] Riccioli, *op. cit.*, frontispiece.

[99] M. Delambre, *Histoire de l'astronomie moderne* (Paris, 1821), Vol. I, pp. 672–681, is an excellent summary of Riccioli's arguments. Delambre apparently believed that Riccioli was a secret Copernican: "On pourrait pencher indifféremment pour l'une ou l'autre hypothèse, sans la témoignage de l'Ecriture qui tranche la question; d'où l'on peut conclure que si Riccioli n'est pas copernicien, c'est qu'il est jésuite" (p. 680). "Enfin, quand on compare les éloges que Riccioli donne à l'hypothèse qu'il combat, à la foiblesse des raisons qu'il lui oppose, on croit voir un avocat chargé malgré lui de plaider une cause qu'il sait mauvaise, qui n'apporte que des argumens pitoyables, parce, qu'il n'y en a pas d'autre, et qui sait lui-même que sa cause est perdue" (p. 681). Dorothy Stimson agrees with Delambre: "His conclusions seem to show that only his position as a Jesuit restrained him from being a Copernican himself" (Dorothy Stimson, *The Gradual Acceptance of the Copernican Theory of the Universe*, New York, 1917, p. 83. The frontispiece of this book is a reproduction of the frontispiece of the *Almagestum novum*. On the page facing it there is an explanation of the frontispiece).

[100] *Almagestum novum*, Vol. I, pp. 204–204½, 484–488, 712, 724.

[101] The system of nomenclature is not biased against Copernicans. The crater Copernicus is surely equal in prominence to the crater Tycho, while Kepler, Boulliau, Gassendi, and Galileo merit craters as prominent as any of their Ptolemaic or Tychonic counterparts. This impartiality was probably an important reason why Riccioli's proposed system was accepted.

could not help but aid the enemy! It can in fact be argued that the Tychonic astronomer of 1660 had more in common with the Copernican astronomer of 1660 than with the Tychonic astronomer of 1600.

During the seventeenth century the telescope was instrumental in the growth of the idea that the laws of nature apply everywhere equally in our solar system and, by implication, everywhere in the universe. This was, of course, a natural concomitant of the growth of the concept of a solar system. As the affinity between the earth and the other planets increased with every telescopic discovery, men's minds started, almost subconsciously, to extrapolate earthly attributes to the planets. The subject of extraterrestrial life changed from the enthusiastic speculation of a Giordano Bruno[102] to a scientific possibility soberly entertained by a Huygens.[103] Astronomers actually began to anticipate discoveries. Thus, the discoveries of the rotation of Mars and Jupiter were verifications of predicted planetary behavior analogous to the behavior of the earth.[104] If the earth rotates, why shouldn't the planets also? Although these discoveries were greeted with enthusiasm, they did not cause great surprise. (The only exception to this was the planet Saturn, whose strange appearances had been a constant source of puzzlement, and when Huygens published his ring hypothesis in 1659 it was greeted with surprise and skepticism.[105]) It was this growing conviction of the universality of nature's laws—impossible in the two-tiered system of Aristotle—which allowed men to identify earthly gravity with the virtue which keeps the planets tied to the sun and satellites to their parent bodies,[106] a notion which Newton quantified and thus raised to the status of a physical law.

I should mention the development of the telescope as an astronomical measuring instrument briefly here. Although this was rather a late development, it bore some important fruit in the seventeenth century. The application of telescopic sights to measuring instruments such as quadrants resulted in an increase in accuracy of an order of magnitude in measuring angular separations, from Tycho's 4 minutes of arc to Flamsteed's 10 seconds of arc.[107] The most important immediate result of this increase in accuracy was the measurement of the parallax of Mars by the simultaneous observations of Cassini in Paris and Jean Richer in Cayenne.[108] This gave astronomers

---

[102] Giordano Bruno, *On the Infinite Universe and Worlds* (1584), in D. W. Singer, *Giordano Bruno, His Life and Thought* (New York: Henry Schuman, 1950), pp. 323–327.

[103] Christiaan Huygens, *Cosmotheoros* (1698), *Oeuvres*, Vol. XXI, pp. 677–821. Huygens' first sentence of the tract is (in the French translation by the editors): "Il n'est guère possible, mon cher frère, qu'un adepte de Copernic, considérant la Terre que nous habitons comme une des Planètes en mouvement autour du Soleil et recevant de lui toute leur lumière, ne se figure parfois qu'il n'est pas déraisonnable d'admettre que, de même que notre Globe, les autres aussi ne soient pas dépourvus de culture et de parure, ni peut-être d'habitans" (p. 680).

[104] Rheita had already assigned periods of rotation to all the planets in 1645 (*Oculus Enoch et Eliae*, p. 277).

[105] His contemporaries had a difficult time coming to terms with the idea that such a huge structure could be suspended around Saturn without any support, especially since Huygens insisted that the ring was a thick solid structure (e.g., Boulliau to Huygens, Nov. 21, 1659, *Oeuvres*, Vol. III, pp. 510–511; Riccioli, *Astronomiae reformatae tomi duo*, Bologna, 1665, pp. 367–368).

[106] E.g. G. A. Borelli (1660), Huygens, *Oeuvres*, Vol. III, pp. 161–162.

[107] For a convenient comparison of the accuracies of observations by Tycho, Hevelius, and Flamsteed, see Delambre, *Histoire*, Vol. II, p. 476.

[108] J. W. Olmsted, "The Scientific Expedition of Jean Richer to Cayenne (1672–1673)," *Isis*, 1942, *34*:124–126. See also Flamsteed's determination of the parallax of Mars, *Phil. Trans.*, July 21, 1673, *96*:6000.

for the first time a relatively accurate absolute distance between two heavenly bodies, and from this the distances between the planets and the sun, as well as the size of our solar system, were determined to within 10 per cent of their modern values. Furthermore, the introduction of measuring devices into the telescope itself allowed astronomers to measure the apparent diameters of planets. Coupled with the knowledge of the distances to the various planets, this gave the actual size of the planets and the sun. Thus, by about 1675 the solar system had been given its modern form and dimensions. It was *this* system, shaped by the telescope, that Newton quantified, not the cosmos of Copernicus.

Descartes on Refraction

Scientific Versus Rhetorical Method

By Bruce Stansfield Eastwood

I. THE *DIOPTRICS:* PROOF, DISCOVERY, OR PERSUASIVE DESCRIPTION

OUT OF THE EXTENSIVE BODY OF ANALYSIS of Descartes's controversial *Dioptrics,* the works of two modern scholars, A. I. Sabra and Gerd Buchdahl, best set the stage for an adequate appreciation of Descartes's assumptions and intentions in that work. Transcending the long history of antagonism toward the Cartesian account of refraction (which runs from Huygens and Leibniz in the seventeenth century through Ernst Mach in the twentieth), Sabra and Buchdahl approach the *Dioptrics* with sympathy, attempting to understand the meaning of the text by taking it at face value.[1] Their analyses succeed more completely than most critics', but they require significant modification if we are to discern clearly why Descartes chose this particular path to the law of refraction.

Sabra attempts to trace the evolution of Descartes's account of refraction, painstakingly analyzing all the relevant texts—from *Cogitationes privatae* (1619–1621) through the *Regulae* (1628) and *Le monde* (1632) to the *Dioptrics* of 1637. Sabra brings together from each of these texts, as well as from letters, Descartes's many statements that the two foundations for knowledge of refraction are scientific explanation (especially through analogy) and the instantaneous transmission of light. Their centrality in the *Dioptrics* is undeniable. But concentrating on the relationship between the *Dioptrics* and the writings that preceded it obscures the relative importance in it of explanation and instantaneous propagation.

The idea of instantaneous propagation (or transmission) is generally taken as one of the hallmarks of Descartes's physics. Yet that emphasis seems to be more the critics' than Descartes's. The idea emerges in his letters and *Le monde* as well as the *Dioptrics,* but in the *Dioptrics* the treatment of refraction transcends the idea. For Descartes, the assumption that transmission is instantaneous is the simpler assumption in an account that *could* have assumed a finite velocity for

---

I owe many thanks for the extensive, detailed, and often challenging comments made by an anonymous reviewer for *Isis,* from which I profited greatly, although I have ungraciously ignored some of his criticisms. Students of Descartes are not always the most rational creatures.

[1] See A. I. Sabra, *Theories of Light from Descartes to Newton* (London: Oldbourne, 1967), Ch. 4 (which develops from and partly depends upon Chs. 1–3); and Gerd Buchdahl, *Metaphysics and the Philosophy of Science* (Cambridge, Mass.: MIT Press, 1969), pp. 118–147, esp. 136–147. For an account of the historical antagonism, see Sabra, *Theories,* pp. 101–105.

light and still have been worded in a similar way.[2] But although Descartes's mode of presentation eliminates time from the refraction of light, most modern authors insist on reintegrating time into the formulation, thereby obscuring the formulation's original purpose.[3]

As early as 1619–1621 (in *Cogitationes privatae*) Descartes speculated about refraction in the following way:

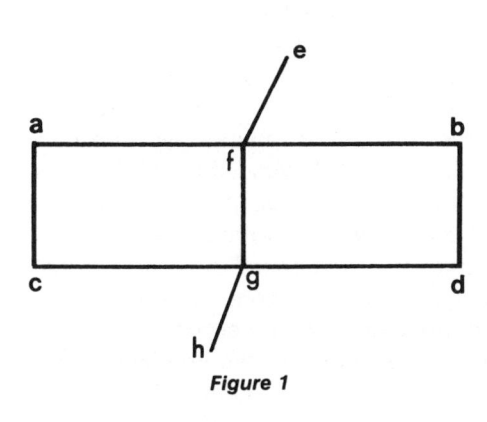

Because light cannot be generated except in matter, where there is more matter light is more easily generated, other things being equal. Therefore it penetrates a denser medium more easily than a rarer. Thus it happens that refraction occurs away from the perpendicular in the latter case and toward the perpendicular in the former. Moreover the greatest refraction of all would be along the perpendicular itself, were the medium extremely dense, from which the exiting ray would pass again along the same angle [as before entering the dense medium]. Let there be *a b c d* [see Fig. 1], an extremely dense medium, and the ray *e f* will cross perpendicularly into *g h*, so that *b f e* and *c g h* are equal angles.[4]

Figure 1

This youthful speculation clearly does not prefigure the law of refraction. Descartes conceives of refraction as contradicting any established proportionality between incident and refracted rays. What does appear at this early point, however, is a sense of the usefulness (to him) of the terminology of facility (ease of passage) rather than of speed. Descartes has found his basic vocabulary of refraction. Within the next seven years he was to formulate a law of refraction as well; but we have no statement of the law from his own hand, whether complete or partial, until the *Dioptrics*. Given this gap, we must be wary of assuming that the account of 1637 faithfully represents the discovery of 1626 or earlier, as Sabra does.[5]

---

[2] The wording of Descartes's account of refraction in the *Dioptrics* does depend on instantaneous propagation; see René Descartes, *Oeuvres de Descartes*, ed. Charles Adam and Paul Tannery, rev. by Bernard Rochot with Pierre Costabel, 13 vols., Vol. VI (Paris: Vrin, 1965) (hereafter *Oeuvres*), Discourse 2, pp. 93–105 (Discourse 1 appears on pp. 81–93.) For a survey of Descartes's writings containing the doctrine of instantaneous propagation, see Sabra, *Theories*, pp. 48–60.

[3] See, e.g., Sabra, *Theories*, pp. 105–135, esp. 110–113; Buchdahl, *Metaphysics*, pp. 141–147; and the discussion in Sections VIII–IX below.

[4] Descartes, *Oeuvres*, Vol. X, pp. 242–243; see Sabra, *Theories*, pp. 105–106.

[5] Sabra ignores the latter half of the quoted statement ("Moreover . . . equal angles") and speculates that Descartes may have "possess[ed] the sine law when he wrote the above passage in 1619–21"; *Theories*, p. 106. Sabra carefully reviews the evidence for Descartes's originality as well as earliest date for his discovery of the law; *ibid.*, pp. 101–103, 106 n. 33. The evidence that he had discovered it by 1626 rests on a manuscript in Mersenne's hand that copies material that the mathematician Claude Mydorge sent him in or after 1631. Mydorge demonstrates that the work he described in an earlier letter to Mersenne, written in 1626, was based on a primitive form of the law of refraction. See Marin Mersenne, *Correspondance du P. Marin Mersenne, réligieux minime*, ed. Cornelis de Waard and Rene Pintard (Paris: Beauchesne, 1932), Vol. I, pp. 404–415. For a recent reconstruction of the discovery of the sine law, see John A. Schuster, "Descartes and the Scientific Revolution, 1618–1634: An Interpretation" (Ph.D. dissertation, Princeton University, 1977), Ch. 4, esp. pp. 299–308 and 329–354.

During those intervening years, Descartes seems to have consciously held yet not put to full use the doctrine of instantaneous transmission. In *Le monde* (1632) he preserves his vocabulary of "facility" for the instantaneous speed of light: "When they [light rays] enter obliquely a medium where they can propagate more or less easily than in the medium from which they depart, they should bend at the point of this change and suffer refraction."[6] Rather than a doctrine of instantaneous propagation in terrestrial physics, what we have is a concept of varying facilities of transmission; this concept, as Descartes admits in his letters, translates into the notion of instantaneous velocity.

To understand the *Dioptrics* we must accept the idea of instantaneous transmission in the exact sense in which it is used in that text (to be fully discussed below). We must also carefully observe the form of explanation developed in that text. It is tempting to ignore that form and look beyond the *Dioptrics* for the principles behind its explanation of the transmission of light. Sabra does this rather persuasively, even looking at the *Meditationes* (1641) and *Principia philosophiae* (1644). He argues that the principles are to be found in *Le monde* and the general reasoning process in the *Regulae* (which mentions the use of analogies), and that the *Dioptrics* assumes the principles and proceeds through the use of analogies to a nonrigorous and elliptical resolution of the problem.[7] If we are committed to integrating the *Dioptrics* fully into the methodological and physical-metaphysical doctrines of Descartes's other works, then Sabra's approach is a reasonable and well-stated one. However, the *Meditationes* and *Principia* appeared four and seven years after the *Dioptrics*, respectively, and the relevant works that precede it, the *Regulae* and *Le monde*, were not published until 1684 (in Holland) and 1664, respectively. Surely Descartes's sense of his audience was not so dull that he would have published a treatise requiring a knowledge of works that were unavailable to the reader.[8] More important, the immediate intention of the author of the *Dioptrics* seems best determined by internal analysis of that text rather than by references to his other writings.

Just what Descartes meant to accomplish in his optical work is the question. Sabra, relying on the notion of proof, suggests one possibility: "Descartes gives no other *proof* of the laws of optical reflection and refraction than the one in which he compares these phenomena to the mechanical reflection and refraction of a ball. Instead of providing better *proofs* in the *Traité de la lumière*, as we might be led to expect, he simply refers to the *proofs*, by comparisons, that are given in the *Dioptric* [emphases added]."[9] Sabra's remark points to the importance of the comparisons, or analogies, in Descartes's optics, a point that both he and Buchdahl stress. Going beyond Sabra's analysis, Buchdahl points to

---

[6] Descartes, *Oeuvres*, Vol. XI, p. 103.
[7] Sabra, *Theories*, pp. 24–33.
[8] Descartes's sensitivity to his audience appears in the famous statement of Nov. 1633—made to a churchman—that he has no desire to publish a work disagreeing with the church (*Oeuvres*, Vol. I, pp. 270–271) and in the letter of 28 Jan. 1641, where he confides his hope that the readers of the *Meditationes* "will insensibly get used to my principles, and recognize their truth before they learn that they destroy those of Aristotle" (*Oeuvres*, Vol. III, pp. 297–298; quoted in Sabra, *Theories*, p. 25, n. 22). Descartes also speaks of writing for a broad audience—women, as well as subtler minds—in the *Discours de la méthode* (*Oeuvres*, Vol. I, p. 560) and for an audience of laymen, especially in the *Dioptrics*.
[9] Sabra, *Theories*, p. 28; Sabra's discussion of how light can add momentum in a denser medium also depends on the notion of "proof"; *ibid.*, p. 114.

three different types of analysis in the Cartesian writings themselves: mathematical (a method of proof), hypothetico-deductive (also a method of proof), and resolutive (a method either of proof *or* of discovery). With regard to resolutive analysis and its application to optics, Buchdahl writes: "The interest attaching to the example from optics is . . . [in part] that it indicates the sense in which this method functions as a method of discovery."[10] Like Sabra, Buchdahl believes there is a direct methodological connection between the *Regulae* (primarily rules 8 and 9) and the *Dioptrics,* but he elects to see in the latter work a method of discovery rather than, like Sabra, a method of proof. In Buchdahl's account, the use of analogies constitutes the relationship between the two treatises. Thus he moves from the mention of analogies in *Regulae* 8 and 9 to the actual analogies in *Dioptrics* 1 and 2, taking Descartes's usage in the explanation of refraction as an example of the procedure laid down in the earlier methodological treatise. Buchdahl sets forth this procedure and its exemplification as follows:

> Descartes is . . . perfectly aware that he has no direct evidence for, and hence cannot be certain of, the actual nature of the fundamental physical process involved. He expressly remarks that it may be that when we are trying to develop the theory *deductively,* we are in fact "unable to determine straightforwardly the nature of the action of light." In that case, we should run over the various possibilities which earlier inspection of the variety of natural powers has already suggested. . . . Descartes . . . suddenly switches to the remarkable position that a knowledge of one of the other alternatives may "help us to understand it [*sc.* the action of light], *at least by analogy.*" The intended analogy must be . . . that of a body in local motion . . . , moving with *finite* velocity.
>
> We see, then, that at this point . . . the deductive line is snapped and original restrictions are relaxed; the reasoning thereafter proceeds in accordance with a series of *models*. The models of the *Dioptrics* are in fact on Descartes' own admission employed with the utmost abandon and disregard for mutual consistency in their actual physical action, though founded on the basic paradigm of the mechanical laws of matter and motion. We might describe all this as a procedural tightening of the qualitative hypothetical situation, which was initially introduced to begin the movement from the data to the colligating formula.[11]

Neither Sabra's account of the *Dioptrics* as an attempt at proof nor Buchdahl's as a recounting of the method of discovery seems to me to describe it adequately. Of the two I prefer Buchdahl's, which seems most appropriate in its concern with the nature of the analogies used. He finds that Descartes argues in a noncausal manner when moving from physical theory to specific actions of light, conducting his argument by means of analogues that do not "reproduce the theoretical *structure* as such, . . . [for each one] is only a kind of formal 'analogue.'" Being formal, an analogy establishes no more than a structural parallel. This sense of Descartes's analogical argument—that it is only structurally descriptive and not any kind of proof—leads us toward a more accurate understanding of the presentation of refraction of the *Dioptrics*.[12]

---

[10] Sabra, *Theories,* pp. 29–33; Buchdahl, *Metaphysics,* p. 128–130, 137 (quoted).

[11] Buchdahl, *Metaphysics,* pp. 141–142.

[12] *Ibid.,* quoting p. 144. Buchdahl's account is, however, somewhat marred by an extraneous concern; see *ibid.,* p. 139; and esp. Buchdahl, "The Relevance of Descartes' Philosophy for Modern Philosophy of Science," *British Journal for the History of Science,* 1963, *1*:227–249, esp. p. 244.

The exposition below interprets Descartes's actual presentation of refraction in the *Dioptrics* 1 and 2 as an exercise in a distinctive rhetoric, cast in the mode of persuasive instruction for an idealized artisan. While Descartes seems to have a physical theory in the back of his mind, the account we are dealing with does not present it. Revealing the law of refraction through easily understood analogies between light and common observables, his narrative instructs and persuades in everyday terms. What he offers is surely an explanation but just as surely not a proof. What he explains is a way to understand a refraction diagram. His vocabulary and methods of explanation, which employ a distinct type of analogical reasoning, make it clear that persuasive description, not proof or discovery, is the purpose of the account. The *Dioptrics* is an essay in educational, and thereby rhetorical, method rather than an example of scientific method.

II. DESCARTES'S AUDIENCE AND EDUCATIONAL ASSUMPTIONS

Descartes published his *Dioptrics* (1637) as an essay in the method laid out in the *Discours de la méthode* published in the same volume. Whether or not the method of the *Discours* is exemplified by the *Dioptrics,* it is fruitful to analyze the optical treatise as an entity in itself, working from the inside out.[13] This was certainly the perspective of Descartes's contemporaries, for in 1637 they did not have the benefit of knowing the *Regulae* or *Le monde,* and only a few were privileged to know his thoughts through letters. Michael Mahoney has recently pointed out the significance of this lack for Pierre Fermat's objections to the Cartesian presentation of the law of refraction. Mahoney even claims that knowledge of the two earlier works is necessary if the arguments in the *Dioptrics* are to be understood, since "Fermat's critique . . . focused precisely on the points that required the fuller context."[14]

But what could one understand from the *Dioptrics* by reading it alone? One could begin by noting its scope: the ten discourses, which take up 153 pages in the original edition, deal successively with light, refraction, the eye, the senses in general, images in the retina, vision, means of perfecting vision, shapes for lenses, telescopes, and the method for cutting lenses. Refraction and the nature of light may be more interesting to the historian of scientific theory, but they are only two of the ten topics covered. The first six sections, the more theoretical parts, constitute less than half of the total work. The last four sections all relate to the telescope, the invention that Descartes mentions as his point of departure. After reviewing the invention of the telescope in Holland by "a man who had never studied," he says that "many good minds" have recently been

---

Sabra proposes with much insight an alternative conjecture to explain Descartes's actual means of discovery of the law of refraction; see *Theories*, p. 116.

[13] On the *Dioptrics* as exemplifying the method, see Descartes, *Oeuvres*, Vol. VI, p. xiii; see also Vol. I, p. 559; and Sabra, *Theories*, p. 33 n. 43: "All we can say (and all Descartes wants us to believe) is that the *Dioptric* contains results which he has discovered by his method without showing how this has in fact been done."

[14] Michael Sean Mahoney, *The Mathematicial Career of Pierre de Fermat (1601–1665)* (Princeton: Princeton Univ. Press, 1973), pp. 375–390, quoting p. 378. My discussion in this study in part explores just how far Mahoney's claim is justified. Alan E. Shapiro, "Light, Pressure, and Rectilinear Propagation: Descartes' Celestial Optics and Newton's Hydrostatics," *Studies in History and Philosophy of Science*, 1974, 5:242, takes Descartes's *caveat lector* at face value and observes that it would be a mistake to look into the *Dioptrics* for an understanding of or even an introduction to the physics of light.

prompted to study the science of lenses. Still, he maintains, there is room for his own contribution:

> Because somewhat difficult inventions do not reach their final degree of perfection on the first try, there still remain enough difficulties in this one to give me something to write about. And, inasmuch as the execution of the things that I shall say must depend on the work of artisans, who ordinarily have not studied at all, I shall attempt to make myself intelligible to everyone, and to omit nothing, nor to assume anything that one might have learned from the other "sciences."[15]

Whether conceived as an apology or as a sincere reflection of his intentions, this statement begins to define limits for what can be expected in the ensuing treatise. First, the *raison d'être* of the work is the telescope: the nature of the telescope, the educational level of its makers, and the problems in understanding how to improve it provide the guidelines for Descartes's procedure in the *Dioptrics*. It is a work of education—not just a manual of technical instructions—for makers of lenses. Descartes wishes to assume nothing from other sciences, yet to omit nothing and to make himself intelligible to everyone. This goal does not require that he state how he has arrived at the knowledge he chooses to impart; instead, it suggests that he will proceed as he thinks best in order to communicate that knowledge to others, including lens-grinders.

The educated layman who had been to a *collège* would have perceived that Descartes was revising the foremost procedures of contemporary education. In the Jesuit *collèges* of his day the *modus parisiensis*, or Parisian method, was the framework for learning. Adopted by the Jesuits in the mid-sixteenth century and canonized in its Jesuit form by the *Rationes studiorum* published in 1586 and 1599, the *modus* emphasized ordered study, disciplined memory, and a disputatious frame of mind. The medieval scholastic methods of *quaestiones, disputationes,* and so forth, had been expanded to encompass not only higher education but the whole range of secondary studies as well.[16]

An implicit critique of that contemporary education underlies the *Dioptrics*. Descartes's rhetorical stance is based on certain assumptions: that the currently accepted academic sequence for presenting scientific knowledge is not the best order; that lens-grinders will generally not possess the prescribed academic knowledge for optics; that the *Dioptrics* is aimed at the intelligent lens-grinder—in fact, a hypothetical and idealized lens-grinder; and that the *Dioptrics* can make its argument without the academic presuppositions that the *collège* required for optics (even though the text gives a reasoned account of optical facts that the academics cannot explain). The reader of the *Discours* would have already encountered Descartes's desire for reform not only in "the body of the sciences" but also in "the established order of teaching them in the schools."[17]

---

[15] Descartes, *Oeuvres*, Vol. VI, pp. 82–83. The detail and seriousness of the final book of the *Dioptrics*, devoted completely to the means of constructing a lens-cutting machine, gives further support to Descartes's claim to be writing for a nonacademic audience. Albert Van Helden, "The Telescope in the Seventeenth Century," *Isis*, 1974, 65:38–58, on p. 45, notes that 17th-century instrument makers were aware of the *Dioptrics*. The work ultimately proved useless for the contemporary practice of telescope making.

[16] On the adoption of the *modus parisiensis* by the Jesuits see Gabriel Codina-Mir, *Aux sources de la pédagogie des Jésuites: Le "modus parisiensis"* (Rome: Institutum Historicum S.I., 1968), esp. pp. 99–150.

[17] Descartes, *Discours*, 2, in *Oeuvres*, Vol. VI, p. 13.

The seemingly innocuous phrase quoted above, "nor to assume anything that one might have learned from the other 'sciences,'" contradicts the order of study practiced at La Fleche and Clermont. From 1606 to 1626 the philosophy curriculum placed mathematics with physics in the middle year of study, preceded by a year of logic and followed by a year of metaphysics. After 1626 mathematics and metaphysics changed places. Optics came under mathematics and presumed a sizable body of prior knowledge.[18] But Descartes preferred to introduce his readers to the science of optics through everyday-language hypotheses and familiar analogies and to assume no prior knowledge of conic sections in his discussion of lenses. To initiate his teaching on the best way to make telescopic lenses, he planned to "begin with the explanation [*explication*] of light and of its rays"[19] for those "who ordinarily have not studied."

III. DESCARTES'S FORMAL ANALOGIES AND THEIR DIAGRAMS

Throughout his exposition in Chapters 1 and 2, Descartes uses analogies as his explicit building blocks for the models of direct, reflected, and refracted light. Each analogue for light—a cane, wine in a perforated vat, a tennis ball and racket—has readily understood characteristics upon which Descartes can depend, thereby avoiding a general discussion of mechanical principles. Rather than discussing incompressibility theoretically, he can refer to the blind man's cane and say, "It is like this." With the wine, and even more with the tennis ball analogues, he depends not only on easily understood characteristics (such as the incompressibility of the wine or the bounce of the ball) but also on illustrative diagrams, which help restrict the number of characteristics of these analogues that the reader need consider. In order to say the minimum needed about the mechanics of light, Descartes first tutors the reader to conceive of direct, reflected, and refracted transmission in the most ordinary situations: in air, on level surfaces, or across common media such as water or glass. He then uses already structured complex situations, such as the tennis ball crossing two media, and takes from the complex as little as possible, using only those elements he needs to expound what is happening. Thus, in a very real sense, the notion of light imparted here by Descartes is little more than a carefully delimited and only partially explained notion of how a cane or wine or a tennis ball acts.

The type of analogy that best supports Descartes's purpose is formal analogy.

---

[18] See René Descartes, *Discours de la méthode,* ed. Etienne Gilson (Paris: Vrin, 1925), pp. 128, 184–185; Etienne Gilson, *La liberté chez Descartes et la théologie* (Paris: Alcan, 1913), pp. 9–10; Camille de Rochemonteix, *Un collège de Jésuites aux XVIIe et XVIIIe siècles: Le Collège Henri IV de La Flèche* (Le Mans: Leguicheux, 1889), Vol. IV, pp. 21–49; Joseph Emmanuel Sirven, *Les années d'apprentissage de Descartes (1596–1628)* (Albi: Impr. Coopérative du Sud-Ouest, 1928), p. 33. The actualities of instruction in mathematics, and thereby in optics, show that it was poorly supported and not widely or deeply studied before the second half of the 17th century. See François de Dainville, "L'enseignement des mathématiques dans les Collèges Jésuites de France du XVIe au XVIIIe siècle," *Revue d'histoire des sciences et de leur application,* 1954; 7:6–21, 109–123, esp. 9–15. Like the municipal *collèges* of 16th-century France, the Jesuit *collèges* offered occasional courses of lectures for the public in mathematics (unlike other courses). I am indebted to George Huppert, Professor of History, University of Illinois at Chicago Circle, for information on 16th-century municipal colleges.

[19] Descartes, *La dioptrique,* Discourse 8 (on elliptical, parabolic, and hyperbolic lenses), in *Oeuvres,* Vol. VI, pp. 165–196; quoting Discourse 1, ibid., p. 83.

Formal analogy expresses the same explicit structure for each of the objects compared, making the objects (light and cane, for example) parallel, rather than convergent or divergent, conceptions. Rectilinearity, incompressibility, and instantaneous transmission thus become understood and explicit as characteristics of the cane, and it is these characteristics alone that are carried over to light. With the tennis ball analogue for reflection and refraction, Descartes directs the reader to imagine for light only the characteristics of the moving ball that he chooses to use. He emphasizes this limitation by tying the chosen characteristics to a geometrical illustration. Using perpendiculars and radii in a circular diagram, he introduces his distinction between motion and determination for the moving ball, then adroitly transforms these to action and determination for light. The reader's acceptance of the diagrammatic representation of the moving ball permits Descartes, without any theoretical discussion, to carry over to light only those aspects of the ball that he needs. The shift from motion (the ball) to action (light) is permissible because of the earlier analogy of light with a cane. Similarly, the transition from speed of the ball to ease of passage of light depends

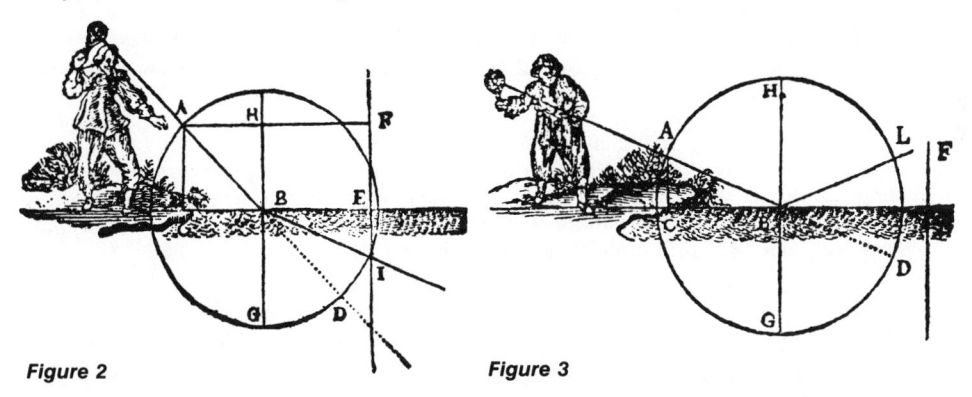

Figure 2

Figure 3

in part on the prior analogies (instantaneous action) and ultimately on the claim that the ease of passage for light entering water is analogous to the speed of the ball when struck by a racket as the ball passes through a penetrable surface. Why this should be so Descartes never says. The assumptions needed he takes from the now thoroughly contrived analogue of a ball struck again as it crosses a surface. Constructed of very familiar images from experience, this analogue is easy for readers to understand.

Descartes's strategy, then, is to use images from common experience to construct analogies with light, to make these images and the analogies more precise by employing diagrammatic illustrations, and to use the diagrammed analogies in place of a full discussion of mechanical principles. Thus the formal analogy establishes a limited structural parallel, shutting off any importation of attributes beyond those explicitly allowed in the construction. Formal analogy shows the ways in which the compared objects are the same, and only the sameness is relevant to the discussion. A carefully controlled, persuasive picture of *how to describe* the path of refracted light is the outcome of Descartes's approach—an approach for teaching the uninitiated, not for establishing mechanical foundations. Another result, of course, is that more general, and perhaps more revealing, theoretical questions have no place in the discussion.

The nature of the diagrams is subtle but important. Up to the point where

Descartes is ready to discard the tennis ball and to speak only of refracted light, his diagrams have all been specific images of his analogues for light: a vat of grapes, a candle, a tennis player with racket and ball, a river bank with a river's surface (see Figs. 2 and 3). But with the transition to a description fully appropriate to light, Descartes exhausts the analogue of the ball and ceases using visually suggestive images. In both words and images, Descartes says that tennis balls are only analogues and introductions to light. The diagrams for tennis balls are geometrical but not simply geometry, while the remaining diagrams in the second discourse, which refer directly to light and not to its analogues, are simple geometrical figures. Thus the limitations of the analogues, recalled as much by the diagrammatic images as by the text, no longer apply.[20]

In the second paragraph of the *Dioptrics*, Descartes briefly presents this broader use of analogies: that the analogues relate to certain macroscopic properties of light but may also suggest other, less easily observed properties. In his own words:

> Not having here any other occasion to speak of light except to explain how its rays enter the eye, and how they can be turned aside by the different bodies they encounter, I need not undertake to explain its true nature. And I believe it will suffice that I use two or three comparisons which help to conceive it in the manner which seems to me the most convenient to explain all its properties which experience makes us aware of, and then to deduce all the others which cannot be so easily observed; imitating in this the Astronomers, who, although their assumptions [F. *suppositions;* L. *hypotheses*] are almost all false or uncertain, nevertheless, because they refer to various observations which they have made, never cease to draw many very true and very firm conclusions from them.[21]

Descartes claims that he need not explain the "true nature" of light. Instead of searching for truth, he finds it sufficient to introduce analogies. These can be seen as hypotheses; like astronomical hypotheses, they summarize data and suggest properties that are to be assumed, even though they are unobservable.[22] These properties, such as light's greater ease of passage in denser media, find explicit expression through the analogies. Since Descartes is explicitly concerned only with how light rays pass through media, a full discussion of the "true nature" of light is beyond the scope of his undertaking in the *Dioptrics*.

---

[20] Descartes does once again use the analogue of the tennis ball; appropriately, he appends a diagram with an image of a watery surface, not simply a representative line (as in the schematics for light); Descartes, *Oeuvres*, Vol. VI, p. 103. Paul J. Olscamp's translation in *Discourse on Method, Optics, Geometry, and Meteorology* (Indianapolis: Bobbs-Merrill, 1965), pp. 65–83, reproduces the more suggestive visual images in Discourse 1 but shifts to strictly schematic constructions beginning with the discussion of reflection at the inception of Discourse 2. Possibly he considered the argument to be purely geometrical from that point on, contrary to my interpretation.

[21] Descartes, *Oeuvres*, Vol. VI, p. 83. French allows only *lumière* for light, but Latin offers two options, *lux* and *lumen*. The Coimbra commentary on *De caelo* speaks of light in bodies as *lux* and of light radiating as *lumen;* see Etienne Gilson, *Index scolastico-cartésien* (Paris, 1912), pp. 159–163. In the Latin translation of the *Dioptrics* by Etienne de Courcelles (*Dioptrice*, 1644), *lumière* becomes *lux vel lumen* (*Oeuvres*, Vol. VI, p. 585); thereafter *lumen* is used generally, but there is no clear distinction in the *Dioptrice*.

[22] For the hypothetical nature of many of Descartes's physical models, see *Discours*, Pt. 5, where he says that his cosmology and his physiology are suppositions, or hypotheses; see also *Discours*, Pt. 6, and *Meteorology*, Discourse 1. In the generally perceptive "Science et hypothèses chez Descartes," *Archives internationales d'histoire des sciences*, 1974, 24:319–339, M. Martinet labels Descartes's three analogues (stick, vat of wine, tennis ball) as "suppositions fausses" (p. 325). I would modify this label to "instructive fictions."

IV. DIRECT TRANSMISSION OF LIGHT RAYS

The first analogy used in the *Dioptrics* is the blind man's stick. This analogy was a commonplace in the ophthalmological literature from as far back as Ḥunain, who used it, and Galen before him, who rejected it. Descartes uses it here, however, to relate his argument to concrete experience. In the introduction he says: "It has doubtless sometimes happened to you, in walking at night without a torch through difficult places, that it became necessary to use a stick." He then sharpens the analogue by asking the reader to imagine its significance for a blind man. Here the stick becomes the blind man's equivalent of the reader's eyesight. The comparison between light and the stick is consummated through the willingness of the reader to "think of light as nothing else, in bodies that we call luminous, than a certain motion or action, very responsive and quick, which passes to our eyes through the medium of the air and other transparent bodies, *in the same manner* [emphasis added] that either the motion or the resistance of bodies encountered by this blind man passes to his hand through the medium of his stick."[23]

If there are ambiguities in the meaning either of stick or of light, such uncertainties are to be ignored, for Descartes intends no more than a strict parallel between his two relationships. Acceptance of his limited definition of light as a "motion or action" passing "through the medium" is necessary, Descartes tells the reader, "in order to draw a comparison" between light and the blind man's stick. Just as light is defined as a motion or action, the stick is felt as motion or resistance. Thus the analogy is carefully framed in terms of perception. The stick tells us something about light in a medium, only because they are both means of perception.

Insofar as light is an action, the stick analogue suggests to us that light "can extend its rays in an instant from the sun to us."[24] Here we can compare the "action" (not motion) from the sun to us with the "resistance," or pressure, felt through the stick. Since we understand clearly the example of the stick, we can now imagine with some clarity how, descriptively, light may act similarly over a much greater distance. The comparison demands that both actions be conceived in terms of rigid incompressible media, but the purpose of the stick analogue is to suggest and illustrate, not to justify.

Descartes proceeds further to suggest that we explore other aspects of visual perception with the blind man's stick in mind. Surely the stick tells the blind man nothing about colors and light tells us nothing about, for example, the viscosity of substances. In both cases, however, action through a medium permits the perception of properties of bodies at a distance. No substance need pass from a jar of honey at one end of the stick to the hand at the other end, nor from the green leaf through the intermediate air to the distant eye, at least, says Descartes, "there is no need to *assume* [emphasis added] that something material passes" in order for the perception to occur. His message is clear enough: we do not need to understand how light effects sensations such as color in the eyes. Here Descartes skillfully offers both a criticism of traditional physics and

---

[23] Descartes, *Oeuvres,* Vol. VI, pp. 83, 84.
[24] *Ibid.,* p. 84.

suggestions for a better and more immediate way to introduce a student to the properties of light. He goes on to say that the reader will "easily be able to decide the question that is current among them [the philosophers] concerning the origin of the action that causes the sensation of sight."[25]

Pressing the analogy a bit further, Descartes points out that the blind man can act against the stick when it is in contact with bodies. From this one can assume that the eye, too, can be a source of the action that causes vision. In keeping with the analogy, we must consider this action that originates in the eye to be light. As Descartes says:

> Whereas this action is nothing else but light, we must note that it is only those who can see in the darkness of the night, such as cats, in whose eyes this action is found; and that, as for the ordinary man, he sees only by the action which comes from objects, since experience shows us that these objects must be luminous or illuminated in order to be seen, and not that our eyes must be luminous or illuminated in order to see them.[26]

Descartes allows the extension of the analogy to support an emission theory of vision, but then he divides specifically the last conclusion of the analogical reasoning. The emission theory is limited to cats, who, according to common belief at the time, were able to see in the dark. In any case, an understanding of vision as *we* experience it renders an emission theory superfluous, for it is evident that objects we see are lit either by themselves or by another light source. Since external light is always present in *our* experience of vision, any explanation of vision through light from the eye is pointless. Descartes develops the hypothesis, here a common-sense analogy, only insofar as it accords with experience. Such descriptive, illustrative analogies are an essential tool in his pedagogical rhetoric.[27] They do not, however, serve either for discovery or demonstration.

To deal with further aspects of the transmission of light, Descartes introduces the analogue of a vat filled with grapes and wine. In the bottom of the vat there are two holes, *A* and *B*, through which the wine can freely flow. This new analogue will enable us to imagine the medium through which the action of light passes. While resistance transferred through a blind man's stick describes the action of light, a fluid (wine) is a closer analogue for the actual medium for transmission of light; almost all philosophers agree that nature abhors a vacuum, and therefore porous bodies must be filled with a universal and very subtle fluid matter. This analogy likens the solid parts of observed bodies to the grapes in the vat and the subtle matter to the wine. Given random points *C*, *D*, and *E* on the surface of the liquid in the vat and holes *A* and *B* in the bottom, Descartes makes the following comparison (see Fig. 4):

> Just as the parts of this wine that are for example at *C* tend to go down in a straight line through the hole *A* at the same instant that it is opened, and at the same time through hole *B*, and as those which are at *D* and at *E* tend also at the same time to go down through these two holes, without any one of these actions being impeded

---

[25] *Ibid.*, p. 85.
[26] *Ibid.*, p. 86.
[27] For the argument that such analogies are more characteristic of pedagogy than of discovery, see Pierre Duhem, *The Aim and Structure of Physical Theory*, trans. P. P. Wiener (New York: Athenaeum, 1962), pp. 75, 93–99, and *passim*.

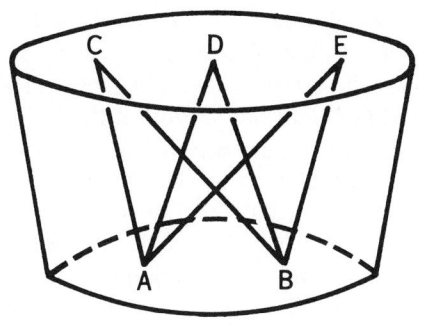

Figure 4

by the others or by the resistance of the grapes in this vat . . .—so all of the parts of the subtle matter which touches the side of the sun towards us tend in a straight line to our eyes at the very instant we open them, and this without these [subtle] parts impeding each other or even being impeded by the heavier particles of transparent bodies between the two [sun and eye], whether these bodies are moved in other ways, like the air, which is almost always agitated by some wind, or whether they are without motion, perhaps like glass or crystal.[28]

The only thing new here is the notion that transmission of light can be understood as nothing more than rectilinear action through a fluid, implicitly incompressible. Lest the point be lost, Descartes explicitly tells us that "it is necessary to distinguish between motion and the action or inclination to move." For the wine in the vat, the rectilinear and unimpeded character of the transmission follows only from such a distinction. Likewise for the other side of the analogy: "And thus considering that it is not so much the motion as the action of luminous bodies that must be taken for their light, you should judge that the rays of this light are nothing else but the lines along which this action tends."[29] The second analogy complements the first, and the two analogies reinforce each other since each suggests instantaneous transmission. The first is better for suggesting rigidity and incompressibility; the second, for suggesting multiplicity of directions and unimpeded intersection in transmission. The analogies do not exhaust the properties of light.

V. THE TENNIS BALL AS ANALOGUE

The reader of the *Dioptrics* has considered vision first as mechanical action and next as mechanical action through a fluid pervading any medium. A third analogy, using the tennis ball, presents transmission across more than one medium. Light rays crossing successive media "are subject to deflection or weakening by them in the same way as the motion of a ball or of a rock thrown in the air is deflected by those [media] it encounters." Again the reader finds Descartes preserving strict parallelism in his analogy. The light rays compare with the motion of a ball across different media, requiring the reader to accept that "it is very easy to believe that the action or the inclination to move, which I have said should be taken for light, must follow *in this* [emphasis added] the same laws as motion."[30] First of all, the statement "it is very easy to believe" signifies only that there is a persuasiveness—not certitude, not necessity—in the doctrine. Second, Descartes carefully circumscribes the doctrine that the actions of light and of moving balls follow the same laws. When he says "in this," he limits the comparison of luminous action and moving bodies to the specific case of deflected transmission across media.

---

[28] Descartes, *Oeuvres*, Vol. VI, pp. 87–88.
[29] *Ibid.*, p. 88.
[30] *Ibid.*, pp. 88–89.

In developing the tennis ball analogy, Descartes classifies any medium met by a ball as soft, hard, or liquid. Soft bodies, such as linen sheets, sand, or mud, absorb the motion of the ball. Hard bodies immediately deflect the ball in some other direction, depending on the deflecting surface. And the motion of the ball, we are reminded, is dual. The ball travels rectilinearly. It can also have spin. Black bodies are for Descartes like soft bodies; just as a soft body absorbs the motion of a ball, a black body breaks up a light ray, absorbing all its force.[31] The main point is that each of the three types of body met by a ball in motion has its analogue among types of body met by light, and the effects on the two sides of the analogy are comparable. In finishing the first section of the *Dioptrics,* Descartes says that the third type of body, liquid, has its analogue in a transparent body through which light passes either "more or less easily" than it did through the previous medium. In the case of the ball and in the case of the ray, we find that deflection is consequent to oblique incidence.

On the qualitative level, Descartes has gone as far as he needs. Through his three analogies he has explored three characteristics of the transmission of light in vision. In order to deal with quantitative questions, Descartes retreats temporarily from refraction and opens his discussion in the second discourse with the subject of reflection, again using the ball in motion.

VI. THE PROPAEDEUTICAL USE OF REFLECTION

The discussion of reflection in the second discourse of the *Dioptrics* is an incomplete one, and, according to Descartes himself, its function is simply preparatory. Taking up again the analogue of the moving deflected ball, the narrative does not mention the light ray until the last sentence devoted to reflection, which begins characteristically with "likewise, if a ray. . . ."[32] The discussion of the reflection of a ball serves primarily as an entrée to the topic of refraction, and the text goes on to develop carefully the analogy between moving ball and refracted light ray in the same parallel manner as in the first discourse.

The understanding of reflection develops from a continuity of the analogue introduced in the last section of the first discourse. A moving ball travels at constant speed and strikes a rigid flat surface from which it reflects at the same speed. Relevant to the analogy are only those characteristics of the moving ball in reflection "which occur in the action of light."[33] Initially the moving ball is shown to have two distinct powers (*puissances*): motive force and directional force. The motion is a result of the force (*force*) imparted by a racket, and it is

---

[31] *Ibid.*, p. 91. Descartes consistently uses the word *force* in connection with light rather than *motion. Force* appears 13 times in *Dioptrics:* French text: pp. 91, 94 (4 times), 97, 98, 100 (2), 132, 204, 207, 208; Latin text: pp. 589, 590 (3), 591, 592, 593 (2), 607, 640 (2). The French text is thoroughly consistent in using this single word where the sense is needed; the Latin text uses *vis, impetus,* or *robor.* In connection with a body in local motion, such as the ball, *force of motion* is a typical phrase. In connection with light, *force* is used without *motion,* and Descartes clearly means to evade the sense of motion, with the exception of a single partial slip on p. 129 (French), p. 605 (Latin). The language Descartes employs in connection with force is intimately related to his language for ease of passage in connection with light. The analogous term for *ease of passage* is *speed,* which he employs only in connection with actual motions. I discuss this at greater length in Sections VII–VIII.
[32] *Ibid.*, p. 96.
[33] *Ibid,* p. 94.

clear, says Descartes, that the same force should be imparted in any direction. On the other hand, the orientation (*situation*) of the racket gives the directional power, or determination, to the ball independent of the motive force imparted. This distinction of forces offers a rationale for the assumption that reflection of a ball brings no change in its speed, only a change of direction. While that assumption prescribes a simplification only when dealing with a ball, the analogy with light, especially in refraction, needs this simplification in order to explain deflection of an instantaneously transmitted action. The distinction of motive force and directional force also permits Descartes to dispense with an element of Peripatetic physics, the *quies media,* or medial rest. According to many Aristotelians, a body moving in one direction must cease moving for an identifiable instant before resuming motion in another direction. Such a notion makes no sense to Descartes, for it would require a discontinuity of motion, which would then need a new cause for resumption. In his radical separation of motive force and determinative force, there is only change of direction in reflection. When refraction arises there will still be no change in the force of action of light, and the force of motion of the analogous tennis ball will be adjusted to account for this. Motion, therefore, is preserved unchanged.

In the case of reflection the actual determination AB divides into two determinations, AC and $AH,$ which are perpendicular and parallel to the reflecting plane (see Fig. 5). To Descartes, this is the only resolution of the determination AB that makes sense. Pierre Fermat, however, objected to the apparent arbitrariness of the choice of component directions, whose only virtue seemed to be their suitability for Descartes's conclusions. Fermat's approach requires the assumption that the construction proposed by Descartes, or any alternative to it, is purely imaginary. Given this imagined construction, we must find that the angle of reflection does not equal the angle of incidence. But Descartes's original account does not admit Fermat's reconstruction.[34]

Descartes introduces his resolution of the determination AB with the words "we can easily imagine," by which he intends to emphasize the ease of comprehending such an approach.[35] The approach is in no sense a construction of fantasy; it is based upon the concrete reality conceived by the ordinary man. Descartes's intention in describing reflection is clear enough. The elements to be reckoned with are (a) perpendicular forces of determination, related to the reflecting surface; (b) unchanging speed along the actual path of motion; and (c) the reference line $FE,$ which represents the vertical determination of motion, placed along the horizontal in relation to the change in the horizontal component of motion (not in reflection). The reference line FE is the crucial diagrammatic element; to understand its placement is to have adequate insight for the description of reflection.

Throughout his description of a reflecting ball Descartes is preparing for the construction of a model for refraction. The lack of any adequate description of reflection of light, following the extended discussion of a reflected ball, shows how unimportant reflection as such is for him. Completed analogies appear in the previous discussions of the properties of light transmission and appear again

---

[34] See Fermat to Mersenne, Apr. or May 1637, and the rather strongly worded rebuttal in Descartes to Mersenne, 5 Oct. 1637, *Oeuvres,* Vol. I, pp. 358–359; 452–453.

[35] *Ibid.,* Vol. VI, p. 95.

in the succeeding discussion of refraction. Only in reflection does Descartes depart from the use of fully developed formal analogy.

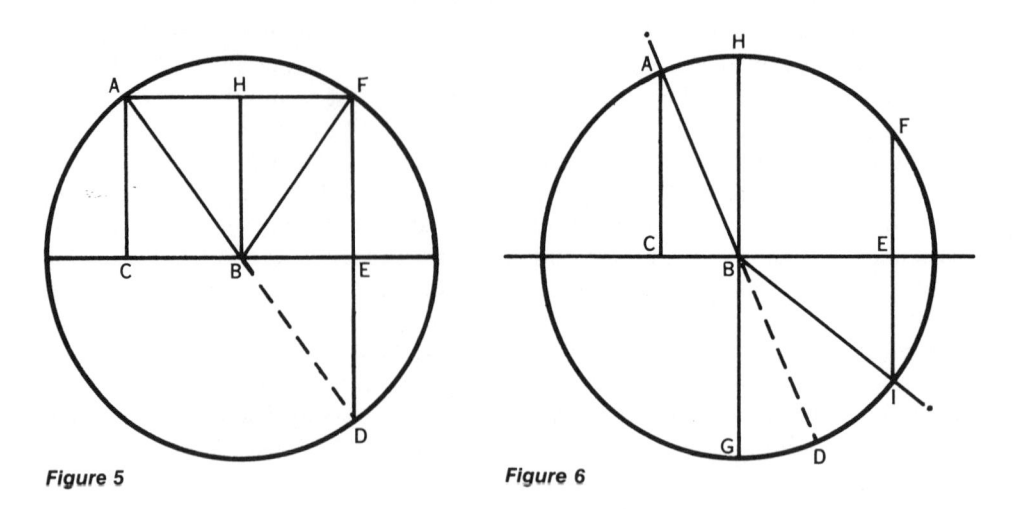

Figure 5

Figure 6

VII. A REFRACTION DIAGRAM FOR A TENNIS BALL

Now we come to refraction. In reviewing the nature of reflection we learned the utility of two ideas. First, there was the assumption that the reflected ball maintained constant speed and implicitly that the reflected light ray did as well. This simplification aided in replacing mechanical elements of reflection with the geometry of reflection. Second, there was the primary importance of the reference line *FE* in Figure 5, which is much more than just another perpendicular in the construction. The placing of the reference line *FE* on the basis of the change in the determination is the essential step in determining the angle of reflection, and it remains the essential step for refraction. Descartes's focus on the placing of *FE* is consistent with the character of his formal analogies in the descriptions already given. The reader should not expect a causal explanation of refraction. Where mechanical notions appear, as in the discussion of the force of the ball (or light), the mechanical elements translate into geometrical magnitudes and interrelate diagrammatically rather than mechanically.

Refraction is first presented by means of the common-sense image of a ball striking a penetrable cloth. Descartes emphasizes that this is the same tennis ball as in the analogy for reflection, only now it is striking "no longer the surface of the earth but a cloth."[36] As in reflection, we again find a determination of the ball toward the right and a determination vertically downward. The thin cloth surface directly affects only the latter direction. On the basis of a hypothetical diminution by one half, we construct again a circle with three perpendiculars (see Fig. 6). One perpendicular rises from the center, the point where the ball is to contact the cloth; another drops from the point of the ball's entrance into the constructed circle; the third is the reference line *FE*. By the time the ball leaves the circle, the duration of the horizontal determination has doubled. Therefore, our reference line *FE* must be placed twice as far from the point of

[36] *Ibid.*, p. 97.

the ball's impact (*B*) as the previous perpendicular (*AC*), which marked the ball's entry into the circle. Before the impact, the *time* for motion along *AB* is one unit, and the determination to move in direction *CB* is one unit. After impact, the *time* for motion along *BI* (or any radius below the horizontal) is two units, and the duration of the determination to move in direction *BE* also doubles.

In Figure 6, *AB* and *BI* are equal distances traveled but in different times. Likewise, *CB* and *BE* represent a unit distance and its double traversed in a unit time and its double. Because of the postulated lack of change in the horizontal determination when the ball strikes *CBE*, we know that the horizontal distance is doubled. By construction—not by analysis—Descartes describes a diagram containing only units of distance. No time units appear in the design. What is represented directly in the diagram is the placing of the reference line *FE* to show the change in path of motion, not the change in the speed of motion. The absence of any direct representation of time elements in the figure will be convenient when the analogy is made with light, for Descartes has no desire to introduce time into the explanation of light transmission. His first step in the analysis of refraction has already eliminated the element of time *from the diagram,* calling the reader's attention primarily to the location of the perpendicular *FE*. Descartes will complete the strategy of this construction by eliminating the element of time *from the vocabulary* accompanying the diagram when he shifts to the other side of the analogy—the transmission of light.

Continuing to deal with the moving tennis ball, Descartes asks that we now "imagine" (*pensons*) it to meet the surface of water rather than a thin cloth. Again the speed is cut in half at the surface, "the rest being assumed as before." The purpose of the diagram and the exposition is descriptive; they prove nothing. Descartes tells us to assume that the medium of water has no mechanical property continuously affecting the ball's speed. He also tells us to assume that the slowing by the medium is a surface phenomenon, and so we must follow our teacher's mental image from one surface (cloth) to another surface (water), ignoring what really happens to a ball in water. The imagined cloth-penetrating ball is now a water-penetrating ball, and we move from one situation to the next by means of a diagram—an image of transmission. Descartes then makes a pointed assertion: "I say that this ball must pass from *B* in a straight line, not towards *D,* but towards *I*. For, first of all, it is certain that the surface of the water must deflect it towards the latter *in the same way* as did the cloth. Given that it takes from the ball just as much of its *force* and opposes it from the same direction [emphases added]."[37] Noting the shift from "speed" to "force" in describing the ball's motion, we now find the time element adroitly removed from the vocabulary accompanying the diagram.

Fermat's objections to the procedures in this part of Descartes's exposition touch an important distinction. Yet Fermat does not appear to understand what Descartes intends to do, that is, what he really means by the term *determination*. There is irony in his misunderstanding, for Fermat's view seems to derive from his great familiarity with mathematics and not from ignorance. As he says in his elaborate attempt to destroy the Cartesian analysis, "Archimedes and other an-

---

[37] *Ibid.*, p. 98.

cients" understood composition of motions in the same way (as Fermat), and so Fermat immediately assumes the traditional approach upon seeing the Cartesian diagrams and reading the Cartesian language of division into perpendicular components of motion. Were Fermat less a mathematician, perhaps he might see and accept more willingly the innovation Descartes proposes. Instead, Fermat lays out in detailed fashion the paralogisms in which determination means no more than direction and yet varies quantitatively apart from speed. For Fermat, determination means direction. While Descartes does not directly define determination, he introduces the term as the directional power of motion.[38]

If Fermat cannot decipher Descartes's meaning, can we expect anyone else to? That is, what assumptions does the hypothetical artisan need in order to understand Descartes? We could begin by excluding the assumptions of mechanics to which Fermat refers. Ironically, Fermat could more profitably have adopted the Cartesian ideology of assuming as little as possible "learned in the other sciences" (as Descartes said at the beginning of the *Dioptrics*), for Descartes never conceived his *Dioptrics,* or any part of it, as a set of rigorous demonstrations. His educational approach emphasizes analogies, simple geometrical constructions, and description rather than strict demonstrative reasoning and causal analysis. Fermat's objections, however, were based on a mathematician's sense of rigor and a mathematician's categories of mechanical analysis.[39] But Descartes uses his own approach throughout the book. In the eighth discourse, upon introducing the ellipse, he does not define it mathematically but speaks of "a curved line . . . which I have . . . seen used sometimes by gardeners in partitioning of their flower beds, where they describe it in a manner which, although it is really very clumsy and inexact, seems to me to render its nature more comprehensible than the section of a cylinder or of a cone. They plant two pickets in the earth—for example, the one at point *H,* the other at point *I*—and having tied the two ends of a string together, they put it around them in the way that you see here." Descartes then proceeds to draw the familiar design he has described and to provide a further diagram, a geometrical figure, which he uses to discuss the properties of elliptical lenses. Not only does he show little interest in geometrical demonstration; he does not even deal in geometrical construction in the rigorous sense. For hyperbolic lenses he begins the same way, referring to a gardener's construction. The text continues with geometrical

---

[38] See Fermat to Mersenne, Nov. 1637, *Oeuvres,* Vol. I, pp. 464–474; that Fermat considers *determination* to mean no more than *direction* appears from pp. 465–466; he focuses on Descartes's supposed error of analysis on pp. 468–473. But Descartes's more complex use of *determination* is clear from *Oeuvres,* Vol. VI, pp. 64 (French), 590 (Latin). For an extensive analysis of Fermat's view and Descartes's intention, see Sabra, *Theories of Light,* pp. 116–127; on Descartes's *determination* and its conceptual difficulties, see Alan Gabbey, "Force and Inertia in the Seventeenth Century: Descartes and Newton," *Descartes: Philosophy, Mathematics and Physics,* ed., Stephen Gaukroger (Totowa, N.J.: Barnes & Noble, 1980), pp. 230–320, esp. 250, 256, 258, 312 n. 138.

[39] See, e.g., Fermat to Mersenne, p. 465, where Fermat insists on the "proof of a proposition." Throughout he refers to "paralogisms" in Descartes's argument. On pp. 466–467 Fermat's argument makes a demand of Descartes's use of *determination* that is not needed in the specific situation presented. On p. 468 Fermat is correct in saying that determination and force of motion have not been clearly distinguished but wrong in claiming that Descartes equates the two. Fermat's "imagining" (see pp. 468–474) is not at all the same as Descartes's notion of imagining, and the objection has no force. That Descartes understood Fermat's predicament appears from Descartes to Mydorge, 1 Mar. 1638, *Oeuvres,* Vol. II, esp. 17–21.

constructions to be discussed and described, not derived, "without my stopping to give you a more exact demonstration."[40]

This tone is endemic to the whole of the *Dioptrics*. Preparing to conclude the work, Descartes reminds the reader once again: "I do not add here the demonstrations of many things which belong to Geometry: for those who are somewhat versed in this science will be able to understand them adequately themselves, and I am persuaded that the others will be happier to believe me about them, than to have the trouble of reading them." To look in the *Dioptrics* for either geometry or mathematical physics in any traditional academic sense is to fail to understand Descartes's strategy: a pedagogical strategy rather than a strategy of theoretical discovery or proof.[41]

VIII. TRANSITION FROM TENNIS BALL TO LIGHT

Beginning with the situation of a tennis ball striking and penetrating water, Descartes moves the process of analogical reasoning forward a step by shifting from speed to force. Prior to this shift, he preserves an awareness that the ball, like the stick and the vat of grapes in the first discourse, provides a common-sense image whose chosen properties are only formally (descriptively) the same as the parallel properties of light. He consciously avoids inserting any sense of mechanical (causal) analogy between the situations of ball and light. After making the shift from speed to force in discussing the ball's motion, he still maintains the language of formal analogy but is able to use vocabulary and reasoning that are much closer to identity in the analogized situations. The only term that continues to require the sense of analogy rather than identity is the term *motion* as applied to the ball, for the analogous property of light is denoted by the word *action*. The ball and light remain distinct, but the descriptions of their transmissions merge perceptibly closer together.

Once the penetrating ball enters the water, its ease of passage, which is linked with its direction of passage, becomes Descartes's main concern. Just as speed and determination of the ball's motion are sharply separated earlier in the discourse, so now are resistance and direction of passage pointedly distinguished. With this distinction Descartes goes against Fermat's understanding of refraction, conceived in terms of maxima and minima of velocity and related directly to resistance. The path of motion does not derive from speed and resistance but, rather, from the directional ease of passage as determined by the surface structure of the medium. And so Descartes informs the reader that size, weight, shape, and so forth relate to resistance and not to direction of passage.

The next stage in the argument begins with a new assumption (*supposition*). Descartes wishes us to imagine that the ball is struck twice in succession, first at *A* and again at *B*, the water's surface, "by the racket *CBE*"; that is, we are now to conceive the water's surface as a second racket, imparting another stroke to the ball. Descartes carefully preserves his new position in the developing analogy by having this second strike to the ball increase the "force of its motion"

---

[40] Quoting Olscamp, trans., *Discourse on Method*, pp. 127–128 (= *Oeuvres*, Vol. VI, p. 166), 144; see also pp. 128–135.
[41] Quoting Olscamp, trans., *Discourse on Method*, p. 171. That Descartes could provide a more systematic analysis for optical questions is shown, for example, in the optical section of his *Geometry*, Bk. II.

rather than simply the motion itself. As before with the decrease in speed, the amount of increase is made up on the spot, giving a convenient value of "for example one third." This convenient fraction corresponds to the appropriate index of fraction. Thus Descartes prepares, both in language and in visual form, for the last crucial shift in the argument for refraction. Of the augmentation in the force of the ball's motion by the second racket stroke, he says: "This will have the same effect as if the ball were to meet at point *B* a body of such a nature that it could pass through its surface *CBE* one third *more easily* [emphasis added] than through the air."[42] There are two new assertions here: first, that a medium can *by its nature* either inhibit or enhance a body's passage; second, that the effect of enhancement should be understood as an easing rather than a quickening of passage. The carefully chosen phrase "more easily" (*plus facilement*) is not to be interpreted as velocity. Because each assertion relates directly to an element of the diagram and an understanding of its construction, the hypothetical artisan can more easily accept the new assertions that Descartes makes.

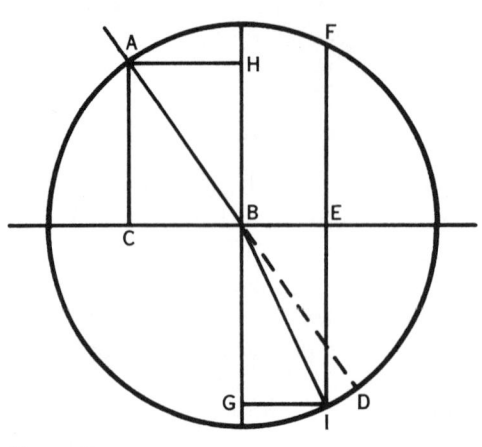

Figure 7

To complete the transition from tennis ball to light, Descartes offers his first diagram applicable directly to light and describes the passage of the ball through water in terms completely appropriate for light (see Fig. 7). Once again, the placing of reference line *FE*, determined now by the increased ease of passage (a hypothetical one third), is the only operation required to find the path of ball or light ray.[43] And this now familiar operation brings the reader to a new generalization, what is now known as the sine law of refraction. The reader, however, has learned this generalization by constructing the intersection of a radius, which represents the path of the ball's speed, or force, with a perpendicular (*FE*), which is placed in terms of the relative change in ease of penetration. Upon arriving at the generalized rule, we find that we have been led slowly to the construction of a hypothetical composite mechanism for the refraction of a *ball*. Now we know how to construct a diagram for refraction on the basis of this mechanical understanding, for each succeeding diagrammatic figure in the book provides an image and a construction for the specific mechanism, leading to a composite image as well as a composite mechanism. This composite image is not limited to the tennis ball but is also the fullest account given for light. Because it is abstract enough to describe both the motion of a ball and the action of light, and because it has all the elements—descriptive diagram and hypothesized mechanics of transmission—to be carried over to the action of light, the

---

[42] Descartes, *Oeuvres*, Vol. VI, p. 100.

[43] Only once before in the *Dioptrics* has Descartes referred directly and in detail to the passage of light rather than an analogous body: at the end of Discourse 1, where he talks of light rays coming from the candle through a cloth and being deflected at the surface. The penetration of the surface by the rays he describes, significantly, as "more or less easily"; *ibid.*, p. 93.

reader can finally make the transition from the contrived mechanics of the ball to the hypothetical and partial mechanics of light. Through the hypothesized mechanics of a ball, the student of Descartes's *Dioptrics* has learned to construct the diagram for the sine law of refraction.

IX. VOCABULARY AND THE LIMITS OF DESCARTES'S ANALOGIES

The final third of the second discourse deals directly with the refraction of light, applying and extending the mechanical understanding built up through the earlier analogies. The mechanisms are part of the hypothetical image: they do not provide the basis for an independent derivation of the refraction rule, nor do they justify the rule; they simply make the rule comprehensible. Finally, turning directly to light, Descartes concludes:

> Insofar as the action of light follows in this [i.e., with respect to force, which is Descartes's explicit base for the refraction rule], the same laws as the motion of the ball, we must say that when its rays pass obliquely from one transparent body to another which receives them more easily [*facilement*] than the first, they are deflected in such a way that they are always less inclined towards the surface of these bodies on the side of the one that receives them the more easily than on the other side, and this exactly according to the ratio of the ease with which the one as opposed to the other receives them.[44]

A. I. Sabra offers a very different understanding of the text. Whereas Descartes goes to some pains to distinguish speed from ease of passage, Sabra insists that when Descartes speaks of ease, or facility, he means velocity. "In order to have any clear idea of what . . . [Descartes] means," Sabra claims, "one is forced to operate explicitly with speeds, and thus dispose of the oppressive assertion of the instantaneous transmission of light."[45] Since Sabra has chosen to reduce the analogies in the *Dioptrics* to identities, his notion of a "clear idea" differs substantially from Descartes's. The analogy of the blind man's stick postulates and explains the passage of light without the element of time. The force of light in refraction *is analogous to* the force of the racket in contact with the tennis ball; it may be conceived as a tension or stress. Whether or not Descartes will later choose to modify his assertion of instantaneous transmission of light, this assertion is an essential part of the *Dioptrics*.[46] Consequently, the reader does better to follow the evident intent of the text than to attempt to extrapolate from it.

From the time of Fermat to the present, scholars have brought to bear types of analysis that are inappropriate to the nature of the *Dioptrics*.[47] In fact, the

---

[44] *Ibid.*, pp. 100–101.
[45] See Sabra, *Theories*, esp. pp. 109–116, quoting p. 113.
[46] Descartes does speak of action as a general term that comprehends both motion and inclination, but action and motion are not identical; see Descartes to Morin, 13 July 1639, *Oeuvres*, Vol. II, p. 204. In Descartes to Hobbes via Mersenne, 12 Jan. 1641 and 18 Feb. 1641, he speaks of the *motus* of the subtle matter by means of which *lumen* is propagated; see *Oeuvres*, Vol. III, pp. 290, 315–316. And even more to the point, in Descartes to Mersenne, 27 May 1638, he says: "As for the difficulty which you find in the communication in an instant, there is equivocation in the word, instant; . . . the word does not exclude priority of time and does not deny that each of the lower parts of the ray [of the sun] is completely dependent upon the upper in the same way as the end of a sequential motion depends wholly on the preceding parts"; *Oeuvres*, Vol. II. p. 143.
[47] For a recent example, see W. B. Joyce and Alice Joyce, "Descartes, Newton, and Snell's Law,"

work essentially is not even written for scholars. Descartes directs his explanation to the hypothetical intelligent layman, and one of his tactics for reaching that layman is avoidance of the question of the speed of light. Since light seems to be instantaneous, Descartes approaches it as an action whose laws of transmission can be stated in a way that adequately describes the motion of the ball as well. Descartes's *Dioptrics* makes much more sense when seen as an essay in the rhetoric of persuasive reason rather than an attempt at demonstration. His concepts of force and ease of passage are more acceptable on the less exacting layman's level, even though they arouse criticism at the theoretical physicist's level. Moreover, we can easily imagine Descartes's delight in offering a solution to a problem that mathematicians and philosophers have left unanswered, yet he has solved in the terms of the untutored.[48]

All that now remains for Descartes is to introduce the practical question of measuring the quantity of refraction, to remind the reader that light and moving balls are quite different in reality, and to set limits to the uses of the newly discovered law of refraction. Descartes does so succinctly.

In conducting experiments to determine an actual index of refraction for a given material, students of the *Dioptrics* cannot help but notice that light bends sharply towards the perpendicular in a transition, whereas a ball will incline in the opposite direction. The passages of ball and light are "quite the opposite" of each other, for "the nature that I [Descartes] attributed to light" is quite other than the nature of the ball. When the ball moves more easily, it is more successful in displacing the particles of the transparent body. When the light transits more easily, it makes no such displacement and therefore acts more effectively when there is more "resistance" or rigidity.[49] Thus Descartes finally casts aside the analogy between light and ball, pointing out that the two forms of transmission show opposite mechanisms when fully analyzed in mechanical rather than formal terms. Here we see again why Descartes compared the natural transmission of light with the thoroughly artificial transmission of a ball. The analogy was one of form, image, and description—a hypothetical geometrical construction. The causal mechanisms of *natural* transmission for light and ball are quite different. Descartes gives a complete geometry of motion and of light but not a complete physics of motion and of light. There is no intention and no need, given Descartes's rhetorical educational program, to provide a thorough and independent analysis of the mechanics of transmission of light across different media.[50]

Journal of the Optical Society of America, 1976, 66:1–8; the article's main purpose is to reinstate light as an example of point-particle dynamics for modern physicists, and it reduces Descartes's analogies to identities.

[48] This was his claim at the beginning of the *Dioptrics,* Discourse 1. In addition, certain convenient bits of information were not available in 1637, such as the understanding of elasticity later reached by Huygens and the observation of a measurable speed for light—much less the observation of measured variations in the speed of light.

[49] Descartes, *Oeuvres,* Vol. VI, pp. 102–103. Alan E. Shapiro, "Kinematic Optics: A Study of the Wave Theory of Light in the Seventeenth Century," *Archive for History of Exact Sciences,* 1973, *11*:155–156, notes a contradiction between this passage from *Dioptrics* and a related one from *Le monde.* But the latter passage is a limited observational statement, while the passage under discussion is a theoretical mechanical statement.

[50] Thus I would revise Mahoney's claim that Descartes's law of refraction is physical, not geometrical, and does not depend on diagrams; *Mathematical Career,* p. 380. Mahoney translates the

The discussion at the end of the discourse on refraction limits the usefulness of the law of refraction. Generally we are told to anticipate the reciprocity of angles of refraction across two different media. But Descartes makes this observation:

> There could easily be found other bodies, principally in the sky, where the refractions, proceeding from other causes, are not thus reciprocal. And there can also be found certain cases in which the rays must be curved, even though they pass through only a single transparent body, just as the motion of a ball often curves, since it is deflected in one direction by its weight, and in another by the action with which it has been impelled, or for various other reasons.[51]

While providing no further insight into what he means by this set of qualifications, Descartes does make clear that his exposition has laid out all that the idealized artisan should need to know. Other sorts of refraction or incurving of light do not occur in situations of concern to the artisan. Curved surfaces of water or glass, for example, will follow the rules already given for the transmission of light across rectilinear surfaces. The reader may proceed, assured that regular surfaces, such as spherical ones, will require only the knowledge of refraction already gained.[52] It is concerning these regular curves—in the eye and in artificial lenses—that Descartes means to instruct his "untutored" readers for the remainder of his treatise.

The rhetorical strategy in Descartes's presentation of refraction permits the identification of more than one level of discourse. Neither a proof or derivation nor an account of his path to discovery, the *Dioptrics* describes refraction in a way that admits acceptance or rejection but not disproof. Pedagogically it attempts both to convey in a straightforward manner a certain kind of understanding of a refraction diagram and to avoid discussing physical principles in full theoretical dress. This duality draws partial justification from the Cartesian claim that knowledge is the product of clear thinking rather than academic training. But Descartes can also hide behind his rhetoric of clear and untutored analogies to evade debate over theoretical principles. The rhetorical method in *Dioptrics* 1 and 2 neatly serves its author's two aims: to show the superiority of a hypothetical artisan's approach over that of the traditional sciences and to frustrate his opponent's attempts to engage him in controversy regarding the theoretical foundations of the law of refraction. Within its self-imposed limits—limits of vocabulary and rhetorical approach—Descartes's account of refraction is virtually unassailable.

---

account to show the *prior* necessity of a mechanics for constructing the account. Agreed; but what is present in *Dioptrics* is not a mechanics but a persuasive account formulated to *obviate* the need for an explicit and complete rational mechanics.

[51] Descartes, *Oeuvres,* Vol. VI, p. 104.
[52] *Ibid.,* pp. 104–105.

Early Seventeenth-Century Atomism

Theory, Epistemology, and the Insufficiency of Experiment

By Christoph Meinel

DURING THE SCIENTIFIC REVOLUTION two relatively independent developments joined forces: the mechanization of the world picture, to use Anneliese Maier's famous term, and the recognition of the crucial role to be played by observation and experiment in the establishment of a scientifically valid theory. The attempt to describe natural phenomena by means of particles and motion was appealing to the new scientific age. Within a few decades corpuscular theories of matter evolved from the obliquity of a controversial fancy into a widely accepted rationale. Compared, however, to the rise of astronomy and mechanics, this success remained ambiguous. Seventeenth-century atomism did not necessarily provide fertile soil for an understanding of material properties and processes. Its empirical background was weak, and not one of its alleged proofs would be accepted by today's scientific standards. In Galileo's inclined plane and his law of falling bodies, or in Newton's theory of colors and his *experimentum crucis* with the prism, theory and experiment, observation and conclusion, were connected in a way that is still convincing. In atomism, however, there was no experimental proof possible, although most corpuscular theories of the seventeenth century explicitly claimed to be based upon experience. But it was not until the nineteenth century that experimental results made the atomic theory at least plausible.

The questionable relationship between seventeenth-century atomism and its empirical background has been obscured to some extent by later historians. When the standard histories of atomism were written, the atomic hypothesis itself was still very much under debate. Twentieth-century historians of science, on the other hand, have all too easily taken the atom for granted. With few exceptions they dealt with these issues in terms of mere intellectual history and neglected the empirical aspects or underestimated their importance. In 1968 Hans Kangro reminded us that the empirical difficulties involved in early modern corpuscularianism are well worth being studied by historians of science.[1]

An earlier version of parts of this study was read at the VIth Joint International Conference on the History and Philosophy of Science in Ghent, Belgium, on 26 August 1986.

[1] Hans Kangro, "Erklärungswert und Schwierigkeiten der Atomhypothese und ihrer Anwendung auf chemische Probleme in der ersten Hälfte des 17. Jahrhunderts," *Technikgeschichte*, 1968, *35*:14–

In this article I give a historical typology and evaluation of the arguments presented in support of the corpuscular hypothesis during the first half of the seventeenth century. I intend to focus on authors who considered themselves empiricists; in their systems the clearest departure from merely bookish reasoning should be expected. Nevertheless, the empirical approach was embedded in a whole network of ontological, epistemological, and mathematical arguments. Such arguments created patterns of thought and habits of perception that were instrumental in the acceptance of the corpuscular view.

I. PARTICLES: PRESUPPOSITION OR PROOF?

Early in the seventeenth century, the assumption of the existence of atoms was by no means a scientific hypothesis that could be proposed without extensive justification of its empirical validity, philosophical soundness, and religious acceptability. One does not even need to go as far as Pietro Redondi in his recent *Galileo eretico* to see that atomism was a most dangerous topic indeed.[2] Its experimental confirmation would have been extremely momentous, not only for the theory of matter.

In 1624 Jean Bitaud and Antoine de Villon, two otherwise unknown Parisian scholars, announced a public tribunal directed against the doctrines of Aristotle and Paracelsus. They were assisted by Etienne de Clave, a physician and skilled chemist, who was scheduled to prove the truth of the assertions by public experiments. The theses the three authors prepared for this event were aimed at disproving the Peripatetic assumptions about matter and form, and the Paracelsian *tria prima*. They culminated in the fourteenth thesis in a clear commitment to atomism: "Omnia . . . esse in omnibus, et omnia componi ex Atomis seu indivisibilibus. Quod utrumque, quia rationi, verae philosophiae, et corporum anatomiae conforme est, mordicùs defendimus, et intrepidi sustinemus." The reaction of the authorities was surprisingly vigorous. Not only was the assembly that had gathered for the event dissolved, one of the organizers arrested, and the theses torn up, but it was also forbidden to propagate anything of this nature under penalty of death.[3] It remains unclear whether the authorities intervened because

---

36. For nineteenth-century histories see, e.g., Friedrich Albert Lange, *Geschichte des Materialismus und Kritik seiner Bedeutung in der Gegenwart*, ed. Alfred Schmidt (1866; 2nd ed., 1872; Frankfurt: Suhrkamp, 1974); Kurd Lasswitz, *Geschichte der Atomistik vom Mittelalter bis Newton*, 2 vols. (Hamburg/Leipzig, 1890); and Richard Ehrenfeld, *Grundriß einer Entwicklungsgeschichte der chemischen Atomistik* (Heidelberg: Winter, 1906). For exceptions to the rule for twentieth-century histories see Ernst Bloch, "Die antike Atomistik in der neueren Geschichte der Chemie," *Isis*, 1913, *1*:377–415; J. R. Partington, "The Origins of the Atomic Theory," *Annals of Science*, 1939, *4*:245–282; and R. Hooykaas, "The Experimental Origin of Chemical Atomic and Molecular Theory before Boyle," *Chymia*, 1949, *2*:65–80. For standard twentieth-century histories see Hélène Metzger, *Les doctrines chimiques en France du début du XVIIe à la fin du XVIIIe siècle* (1923; Paris: Blanchard, 1969); G. B. Stones, "The Atomic View of Matter in the XVth, XVIth, and XVIIth Centuries," *Isis*, 1928, *10*:445–465; Marie Boas, "The Establishment of the Mechanical Philosophy," *Osiris*, 1952, *10*:412–541; and Stephen Toulmin and June Goodfield, *The Architecture of Matter* (1962; Chicago/London: Univ. Chicago Press, 1982), pp. 137–200. For a purely semantic and philosophical treatment see Ugo Baldini, "Il corpuscolarismo italiano del Seicento: Problemi di metodo e prospettive di ricerca," in *Ricerche sull'atomismo del Seicento* (Pubblicazioni del Centro di Studi del Pensiero Filosofico del Cinquecento e del Seicento in Relazione ai Problemi della Scienza, I, 9) (Florence: Nuova Italia, 1977), pp. 1–76.

[2] Pietro Redondi, *Galileo eretico* (Microstorie, 7) (Turin: Einaudi, 1983).

[3] *Positiones publicae contra dogmata Aristotelica, Paracelsica, et Cabalistica*, thesis 14. The "omnia in omnibus" refers to the three authors' peculiar doctrine of five elements. The quotation is

an atomic theory of matter seemed to question the transubstantiation of the Eucharist. We are more concerned with the assertion, proposed by the authors, that the atomic doctrine was not only reasonable but also in accordance with chemical analysis (*corporum anatomiae conforme*) and that there were experiments or observations that would immediately decide between the competing hypotheses about the nature of matter and thus convince everybody: "Experientijs atque iteratis rationibus ita sua dicta comprobet, omniumque oculis tam apertè subjicitat, ut omnes adstantes verissima haec omnia uno simul ore fatentur."[4]

When Robert Boyle, only a generation later, began to establish his corpuscular philosophy, a mechanical theory of matter based exclusively upon the two principles "matter" and "motion," he did not dwell on the trifling task of proving the existence of atoms or corpuscles first. Instead, he presupposed them and developed his hypothesis to direct and explain the experimental operations based on it. Although Boyle admitted that there was little systematic connection between empirical facts and the corpuscular hypothesis, the operational, if not ontological, status of the corpuscles was beyond any doubt.[5] In the *Sceptical Chymist*, published in 1661, he devoted much effort to criticizing and refuting experimentally the doctrines of elements or principles proposed by the Peripatetics and Paracelsians, yet his own corpuscular alternatives were never exposed to the touchstone of the experiment.

II. THE RISE OF ATOMISM

The knowledge of classical atomism had been passed down to the Renaissance humanists through different lines of tradition. Above all, there was Aristotle's consistent refutation of it, which became an integral part of every scholar's training in philosophy. Second, the writings of Greek medicine incorporated important relics of atomism. Besides these, the doctrine of *minima naturalia*, as put forth in the Averroist tradition of commentators and more fully developed in the sixteenth century, provided a concept of small, qualitatively different parts of matter that related more closely to experience than did the atoms of the ancients.[6] The *minima*, however, were not mechanical particles and could not sim-

---

taken from an early reprint in the anonymously edited *Auctarium epitomes physicae . . . Danielis Sennerti* (Hamburg, 1635), pp. 86–91, a collection of short texts, mainly from Sennert's *Epitome scientiae naturalis*. It is clear that the Hamburg edition was commissioned by Joachim Jungius, who was also responsible for the addition of the *Positiones*. The main sources for the incident are Jean Baptiste Morin, *Réfutation des thèses erronées* (Paris, 1624); and Marin Mersenne, *La vérité des sciences contre les s[c]eptiques ou Pyrrhoniens* (Paris, 1625), pp. 78–83; see also Lynn Thorndike, *A History of Magic and Experimental Science*, Vol. VII: *The Seventeenth Century* (New York: Columbia Univ. Press, 1958), pp. 186–187.

[4] *Positiones publicae* (cit. n. 3), p. 91. In the only surviving original broadsheet (Bibliothèque Nationale, Paris, MS Dupuy 630, fol. 72) this passage is wanting, but a manuscript copy (Biblioteca Apostolica Vaticana, MS Reg. lat. 952, fols. 47–48v) confirms the quoted version.

[5] See Thomas S. Kuhn, "Robert Boyle and Structural Chemistry in the Seventeenth Century," *Isis*, 1952, *43*:12–36; and William Arthur Drumin, "The Corpuscular Philosophy of Robert Boyle: Its Establishment and Verification" (Ph.D. diss., Columbia Univ., 1973), esp. pp. 120–149.

[6] Alfred Stückelberger, *Vestigia Democritea: Die Rezeption der Lehre von den Atomen in der antiken Naturwissenschaft und Medizin* (Schweizerische Beiträge zur Altertumswissenschaft, 17) (Basel: Reinhardt, 1984); Stückelberger, "Empirische Ansätze in der antiken Atomphysik," *Archiv für Kulturgeschichte*, 1974, *56*:124–140; and Stückelberger, *Antike Atomphysik: Texte zur antiken Atomlehre und ihrer Wiederaufnahme in der Neuzeit* (Munich: Heimeran, 1979), pp. 33–38. Still the best account of the medieval doctrine of *minima naturalia* with regard to atomism is A. G. M. van Melsen, *From Atomos to Atom: The History of the Concept Atom* (Pittsburgh: Duquesne Univ.

ply be translated into corpuscular terms. Therefore, additional factors are necessary to account for the sudden rise and reluctant acceptance of atomism in the first half of the seventeenth century, just as the attractiveness of Aristotelianism was fading away. When in 1417 the lost *De rerum natura* by Lucretius, with its wealth of immediately convincing imagery, was rediscovered, it provided the proper impulse at the very best moment. The *editio princeps* appeared in 1473, only three years after the first Latin translation of Diogenes Laertius's biographical history of philosophy, the last two books of which dealt with Leucippus, Democritus, and Epicurus. Still, it was some time before its influence became evident in the natural sciences. The Italian physician Girolamo Fracastoro, in his *De sympathia et antipathia rerum* of 1545, was probably the first of the humanists to use the ancient atomic theory in explaining physical and chemical phenomena. In the following decades Hero's *Pneumatica,* another influential source for Greek atomism, went through many translations into Latin and the vernacular.[7] At the turn of the seventeenth century the debate between supporters and adversaries of the atomic doctrine had stimulated enough interest to encourage deliberate experimentation.

By that time internal problems within the Scholastic framework had weakened the Aristotelian position. One of these problems was the assumption of a *creatio ex nihilo* of substantial forms. On the basis of the eternity and uncreatedness of Being, Aristotle had assumed that in many alterations, especially when elements were transmuted or new compounds formed, the form of the product appeared out of nothing, *ex nihilo,* whereas the forms of the original bodies disappeared into nothing, *in nihilum*. This issue had been vigorously disputed by the Peripatetics ever since. In Averroes' opinion forms were merely subordinate to matter and appear from or disappear into matter, and not from or into nothing. The later Averroism, as it flourished in the School of Padua in the sixteenth century, taught that during the process of mixture the forms were only broken (*refractae*) or gradually weakened. The followers of Avicenna, on the other hand, were inclined to admit the persistence of substantial forms of the reactants in a compound, although dominated by the more powerful substantial form of the mixture. It was exactly the latter view, that a compound contained its elements *in actu,* which was favored among the Peripatetic physicians, and it was but natural that Avicenna's doctrine made people more inclined to accept unchangeable and enduring corpuscles as constituent parts of a mixture.[8] The Lucretian axiom *nihil ex nihilo,* echoing Epicurus, must have been appealing as a simple and reasonable way out of the dilemma. In addition, the axiom was also not unacceptable to an enlightened, "secular" Aristotelianism, despite the theological difficulties that arose if *nihil ex nihilo* was confronted with the biblical account of creation.[9]

Press, 1952). I use the enhanced German translation *Atom gestern und heute: Die Geschichte des Atombegriffs von der Antike bis zur Gegenwart* (Orbis Academicus, II, 10) (Freiburg/Munich: Alber, 1957).

[7] Wilhelm Schmidt, "Heron von Alexandria im 17. Jahrhundert," *Abhandlungen zur Geschichte der Mathematik,* 1898, *8*:195–214; and Marie Boas, "Hero's 'Pneumatica': A Study of Its Transmission and Influence," *Isis,* 1949, *40*:38–48.

[8] See W. Subow, "Zur Geschichte des Kampfes zwischen dem Atomismus und dem Aristotelismus im 17. Jahrhundert (Minima naturalia und Mixtio)," in *Sowjetische Beiträge zur Geschichte der Naturwissenschaft,* ed. Gerhard Harig (Berlin: Deutscher Verlag der Wissenschaften, 1960), pp. 161–191; Norma E. Emerton, *The Scientific Reinterpretation of Form* (Cornell History of Science Series) (Ithaca/London: Cornell Univ. Press, 1984), pp. 76–105.

[9] Lucretius, *De rerum natura* 1.150 ("nullam rem e nihilo gigni"), 1.215 ("natura neque ad nihilum

Influenced by nominalist thought, many natural philosophers turned toward the empirically accessible particulars of nature. In David Gorlaeus's posthumously published *Exercitationes philosophicae* of 1620 universals had no existence; only individual things, defined by their intrinsic properties, were real: that is, essence and existence, essence and properties, quantity and body, were the same.[10] Only in thought could the attributes of a body, such as number, quantity, and physicochemical properties, be distinguished from the body itself. The only reality was the reality of the physical particulars, the atoms: *nihil reale esse in corporibus praeter atomos*. Occam's razor, that the entities are not to be multiplied beyond necessity, was quoted by Gorlaeus again and again. A decade later Joachim Jungius referred to it as the *hypothesis hypotheseon*.[11]

The experience of practical men, separated from the mainstream of learning by educational and social barriers, had become more influential since the Renaissance. By the very nature of their crafts they treated matter in a nonphilosophical, purposeful way. For obvious reasons metallurgists, assayers, chemists, and apothecaries were more concerned with the properties of the products than with the theory of the processes. Seventeenth-century chemistry was a rational and pragmatic subject, devoted to medical, pharmaceutical, and metallurgical purposes.[12] If we judge it on the basis of Jean Beguin's *Tyrocinium chymicum* of 1610, the most influential chemical manual of the time, the field had divorced itself from the old dream of gold and longevity and was explicitly atheoretical and concerned primarily with substances, their essentials, and their classification as distinct species. It should also be remembered that the available theories of matter rested upon a rather limited acquaintance with metals and minerals. Alchemists and practical men, on the other hand, knew a great deal about these things, and they knew how to handle and study them experimentally. As far as the experimental approach is concerned, the alchemical heritage did much to determine the questions of early modern theory of matter. Within this context of empirical chemistry the corpuscular hypothesis gained momentum and made converts. If we examine the web of arguments presented in support of the atomic theory of matter during the first half of the seventeenth century, the chemical arguments appear particularly powerful.

III. EPISTEMOLOGICAL ARGUMENTS

Many of the arguments adduced in support of atomism were epistemological. Of these, one important group was concerned primarily with the relationship be-

---

interemat res"), et passim; Epicurus *apud* Diogenes Laertius 10.38–39. On secular Aristotelianism see John Henry, "Thomas Harriot and Atomism: A Reappraisal," *History of Science*, 1982, 20:267–296, on p. 272.

[10] David Gorlaeus, *Exercitationes philosophicae quibus universa fere discutitur philosophia theoretica* (Leiden, 1620), pp. 39–40, 84–85, 95–100. See also Lasswitz, *Geschichte der Atomistik* (cit. n. 1), Vol. I, pp. 455–463; and esp. Tullio Gregory, "Studi sull'atomismo del Seicento, II: David van Goorle e Daniel Sennert," *Giornale critico della Filosofia Italiana*, 1966, 45:44–63, on pp. 46–51.

[11] Gorlaeus, *Exercitationes philosophicae* (cit. n. 10), pp. 39, 95–96 (on distinguishing attributes; see also Gregory, "Studi sull'atomismo, II," cit. n. 10, pp. 46–48); and p. 247 (on reality; see also Gorlaeus, *Idea physicae* [Utrecht, 1651], pp. 24–25); and Joachim Jungius, *Praelectiones physicae*, ed. Christoph Meinel (Veröffentlichung der Joachim-Jungius-Gesellschaft der Wissenschaften, 45) (Göttingen: Vandenhoeck & Ruprecht, 1982), p. 96, lines 1–10.

[12] Marie Boas, *Robert Boyle and Seventeenth-Century Chemistry* (Cambridge: Cambridge Univ. Press, 1958), pp. 48–74.

tween the structures of the external world and the corresponding structures and abilities of human perception and cognition. In regard to that relationship, one solution, rigid mechanical atomism, had little to offer the more empirically minded naturalists. Since it located the observed qualities within the sensations and the mind of the observer, how could one ever be able to know about reality through experiment? A solution at the other extreme was proposed by Claude Gillermet de Bérigard, a Frenchman who taught in Pisa and Padua, knew Galileo personally, and is likely to have witnessed the condemnation of the atomistic *Positiones publicae* at Paris in 1624. In his *Circulus Pisanus* of 1643, a rather traditional dialogue, Bérigard suggested a theory of matter that might be termed a "qualitative atomism."[13] Every possible quality was, so to speak, incorporated into atoms, each atom being the corporeal hypostasis of only one quality. Consequently, there were as many different kinds of atoms as there were different qualities in nature, and only their juxtaposition and interference in macroscopic aggregates added up to the qualities we see, feel, smell, or taste. The remarkable point in Bérigard's theory is that his quality-atoms were unchangeable, so that qualities became the basic entities in nature, and the study of qualities, as it could be performed in the laboratory, would eventually lead to the basic level of reality. Thus Bérigard was able to avoid the epistemological break between the sensuous qualities and the properties of the atoms which had been such a disturbing feature of Greek atomism.[14]

Usually, however, the solution to this problem of how the primary qualities of the corpuscles produce the sensation of secondary qualities in the observer was sought somewhere between the two extremes. Thus the notion of element—and this means of course the four Aristotelian elements—was to some extent amalgamated with that of atoms. The identification of corpuscles and elements is already present in Sebastian Basso's *Philosophia naturalis adversus Aristotelem* of 1621. The author admitted, however, that it would be impossible to decide whether the particles of fire, air, water, and earth were in fact the ultimate atoms, and so he was probably the first to introduce a clear concept of secondary, tertiary, quaternary, and so on, aggregates which, in chemical reactions, behave as if they were stable particles.[15] In a similar way Daniel Sennert, a very influential yet little-studied figure, imagined the *minima naturae* or *atomi* to be smallest units of the four elements, which in turn compose the *prima mixta* as the real, experimentally treatable units of matter. The closest amalgamation of the concepts of atom, element, and pure substance that can be found before the nineteenth century, however, was reached by Joachim Jungius in 1632.[16] Here

---

[13] Claudius Berigardus, *Circulus Pisanus de veteri et peripatetica philosophia* (1643; Padua, 1661), pp. 418–425.

[14] Jürgen Mau, "Studien zur erkenntnistheoretischen Grundlage der Atomlehre im Altertum," *Wissenschaftliche Zeitschrift der Humboldt-Universität zu Berlin* (Gesellschafts- und sprachwissenschaftliche Reihe, 2, 3), 1952/53, pp. 1–20.

[15] Reijer Hooijkaas, *Het Begrip Element in zijn historisch-wijsbegeerige Ontwikkeling* (Utrecht: Schotanus & Jens, 1933), pp. 136–143, 183–190; see also Henricus Hermanus Kubbinga, "Le développement historique du concept de 'molécule' dans les sciences de la nature jusqu'à la fin du XVIIIe siècle" (diss., Ecole des Hautes Etudes en Sciences Sociales, Paris, 1983), pp. 60–73.

[16] Rembert Ramsauer, "Die Atomistik des Daniel Sennert als Ansatz zu einer deutschartigschauenden Naturforschung und Theorie der Materie im 17. Jahrhundert" (Ph.D. diss., Univ. Kiel, 1935), is the only full-length, if ideologically biased, study of Sennert's corpuscular theory. For Sennert's achievements see J. R. Partington, *A History of Chemistry,* Vol. II (London: Macmillan, 1961), pp. 271–276; for his "chemical" theory see Gregory, "Studi sull'atomismo, II" (cit. n. 10), pp.

the gap between perceivable and experimentally accessible qualities of macroscopic bodies and those of the ultimate constituents of matter had almost disappeared.

Another epistemological assumption underlying atomism was that knowledge requires the existence of some basic entities in reality upon which, as upon irreducible units or axioms, both cognition and the plurality of nature could be built. It was no less a person than Giordano Bruno who incorporated this idea—originally formulated by Nicholas of Cusa—into his notion of *minimum*. For Bruno, the existence of a smallest, indivisible unit, such as the point in geometry, the atom in physics, and the monad in metaphysics, was the matrix of reality, the measure and prerequisite of cognition.[17] However, Bruno's physical atoms had no sensuous properties, they were all of the same kind, and only in the senses did they appear endowed with specificity. But still they were the units out of which nature builds her fabric and into which bodies dissolve again.[18] Since the same principle of synthesis and analysis was valid in art and in nature, it should be not only natural but indeed necessary to proceed from the simple to the complex, once the point of departure had been found.

In 1621 Sebastian Basso, one of the most influential authors among the early corpuscularians, proposed the argument that the instauration of learning had to begin with the most basic entities of reality, namely, the physical principles or atoms, from which level all future conclusions would depend. Their exact determination would be a prerequisite for any solid science, since these principles were as important in the natural sciences as characters in typography or building materials in a construction. Consequently, they had to be preexistent, incorruptible, and finite in number.[19] Jungius's thoughts were quite similar. In his opinion a distinct science of nature required above all a finite number of principles, just as Euclidean geometry relies upon a small number of basic entities such as the point, the line, and the angle. Jungius's attempt to rebuild the system of physical knowledge belongs to the widespread quest for making both philosophy and natural science as axiomatically structured as geometry. In contrast to most of his contemporaries, Jungius insisted that only the evidence of sensuous experience and an inductive methodology would lead to the identification of these ultimate units of reality. These *hypostatical principles,* as he termed them, were not me-

---

51–63. On Jungius see Christoph Meinel, "Der Begriff des chemischen Elementes bei Joachim Jungius," *Sudhoffs Archiv,* 1982, 66:313–338, on p. 336.

[17] On Nicholas see Werner Schulze, *Zahl, Proportion, Analogie: Eine Untersuchung zur Metaphysik und Wissenschaftshaltung des Nikolaus von Kues* (Buchreihe der Cusanus-Gesellschaft, 7) (Münster: Aschendorff, 1978); and Fritz Nagel, *Nicolaus Cusanus und die Entstehung der exakten Wissenschaften* (Buchreihe der Cusanus-Gesellschaft, 9) (Münster: Aschendorff, 1984). For Bruno's theory of matter see Paul-Henri Michel, "L'atomisme de Giordano Bruno," in *La science au seizième siècle* (Histoire de la pensée, 2) (Paris: Hermann, 1960), pp. 249–264; and Michel, "Les notions de continu et de discontinu dans les systèmes physiques de Bruno et de Galilée," in *L'aventure de l'esprit* (Mélanges Alexandre Koyré, 2) (Paris: Hermann, 1964), pp. 346–359.

[18] Iordanus Bruno, *De triplici minimo et mensura* (1591), in *Iordani Bruni Nolani Opera latine conscripta,* ed. F. Tocco and H. Vitelli, Vol. I, Pt. 3 (Florence, 1889; Stuttgart/Bad Cannstadt: Frommann Holzboog, 1962), pp. 139–140.

[19] Sebastianus Basso, *Philosophiae naturalis adversus Aristotelem libri XII, in quibus abstrusa veterum physiologia restauratur et Aristotelis errores solidis rationibus refelluntur* (1621; Amsterdam, 1649), here on p. 8. On Basso and his influence see Tullio Gregory, "Studi sull'atomismo del Seicento, I: Sebastiano Basso," *G. Critico Fil. Ital.,* 1964, 43:38–65, esp. pp. 49–50; and Giancarlo Zanier, "Il macrocosmo corpuscolaristico di Sebastiano Basso," in *Richerche sull'atomismo* (cit. n. 1), pp. 77–118.

chanical atoms in the classical or in the Boylean sense, but real, chemically distinct parts that could be separated by chemical analysis.[20]

IV. ARGUMENTS BASED ON DIVISIBILITY

The second type of argument for or against atomism was based on mathematical or geometrical grounds. The ancient refutation of atomism had been based upon reasoning of this kind, since logical contradictions result if one assumes that division of continuous bodies leads to indivisible bodies, or, vice versa, that geometrical points add up to form an extended line. It was the concept of the atom as the limit of divisibility that dominated philosophical discussion from Aristotle to the Renaissance, and the arguments need not be repeated here. Their impact was still felt in the seventeenth century, for instance, in Galileo's theory of matter or in Thomas Harriot's unpublished notes on Zeno's paradoxes, in which Harriot inferred the atoms on the basis of mathematical progressions, which he considered as analogues to the structures of the corporeal world.[21] In general, however, it was more the concept of a physical atom as a constituent part that became the prevailing idea at this time. Thus the ἄτομος was replaced by the concept of principle (ἀρχή) or element (στοιχεῖον)—although "element" was intended in a formal, not in any chemical, sense as a binary relation in the form "x is an element of y" that goes back to the Aristotelian definition. For Giordano Bruno the *minimum* was clearly a relational notion that referred to the process of composing and decomposing, an idea that allowed him to distinguish mathematical and physical *minima,* at the price, however, of somewhat bizarre mathematics.[22] Yet for those who professed themselves empiricists the mathematical arguments had little to do with their scientific questions. They reluctantly dismissed the quarrel about the difference between atoms and limits in favor of a more pragmatic concept of little particles that looked as if it would be useful in the laboratory. This change of attitude is best illustrated by Sennert's dismissal

---

[20] Hans Kangro, *Joachim Jungius' Experimente und Gedanken zur Begründung der Chemie als Wissenschaft: Ein Beitrag zur Geistesgeschichte des 17. Jahrhunderts* (Boethius, 7) (Wiesbaden: F. Steiner, 1968), pp. 194–211; and Meinel, "Begriff des chemischen Elementes" (cit. n. 16). See also Christoph Meinel, *In physicis futurum saeculum respicio: Joachim Jungius und die Naturwissenschaftliche Revolution des 17. Jahrhunderts* (Veröffentlichung der Joachim-Jungius-Gesellschaft der Wissenschaften, 53) (Göttingen: Vandenhoeck & Ruprecht, 1984); and, on the quest for axiomatic structure, Hermann Schüling, *Die Geschichte der axiomatischen Methode im 16. und 17. Jahrhundert* (Studien und Materialien zur Geschichte der Philosophie, 13) (Hildesheim/New York: Olms, 1969), pp. 72–75.

[21] For Galileo's theory of matter see William R. Shea, "Galileo's Atomic Hypothesis," *Ambix,* 1970, *17*:13–27; A. Mark Smith, "Galileo's Theory of Indivisibles: Revolution or Compromise?" *Journal of the History of Ideas,* 1976, *37*:571–588; and Homer E. Le Grand, "Galileo's Matter Theory," in *New Perspectives on Galileo,* ed. Robert E. Butts and Joseph C. Pitt (The University of Western Ontario Series in Philosophy of Science, 14) (Dordrecht/Boston: Reidel, 1978), pp. 197–208. On Harriot see Henry, "Thomas Harriot" (cit. n. 9), pp. 269–271. On ancient refutations of atomism see S. Luria, "Die Infinitesimaltheorie der antiken Atomisten," *Quellen und Studien zur Geschichte der Mathematik, Astronomie und Physik,* Ser. B, 1932, *2*:106–185; and Jürgen Mau, *Zum Problem des Infinitesimalen bei den antiken Atomisten* (Deutsche Akademie der Wissenschaften zu Berlin, Institut für hellenistisch-römische Philosophie, Veröffentlichung 4) (Berlin: Akademie-Verlag, 1954), pp. 19–47.

[22] For Aristotle see Aristotle, *Physics* 1.2 (185a4); for Bruno see Lasswitz, *Geschichte der Atomistik* (cit. n. 1), Vol. I, p. 377. In *Metaphysics* 6.3 (1014b5) Aristotle used the term *element* (στοιχεῖον) metaphorically to denote any small, simple, and indivisible unit. In the seventeenth century this inconsistency was seen as a justification for the attempt to harmonize the Aristotelian and Democritean views of matter.

of mathematical arguments in matter theory. In the concluding paragraph of his chapter on atoms in the *Hypomnemata physica* he frankly declared that in physics the divisibility of the continuum was not a relevant question. The only relevant question was whether nature in generation and resolution subsists in some kind of small bodies.[23]

V. ARGUMENTS BASED ON EXPERIENCE AND EXPERIMENTS

The empirical arguments presented in favor of atomism during the first half of the seventeenth century fall roughly into six groups:

1. Extrapolations from the visible to beyond the limits of sense perception.
2. Attempts to use the microscope to extend the reach of the eyes to the intrinsic textures of matter.
3. Transport processes of material character, such as evaporation, abrasion, or growth.
4. Arguments taking the physical problems related to condensation and rarefaction, including the question of the vacuum, as their point of departure.
5. Observations related to noncorporeal species such as light, magnetism, sound, or heat.
6. Arguments derived from phenomena involving qualitative alterations of chemical substances.

Extrapolations

Arguments based upon extrapolations from macroscopic bodies were the more traditional ones, already put forward in classical antiquity; most of them in fact are in Lucretius's *De rerum natura*. They all start from the trivial experience that macroscopic bodies are distinct and go on to extrapolate from this distinctiveness to an underlying material reality. The most frequently quoted example was the one of insects that are so tiny that their third part would be below the limits of visibility. How small then, Lucretius asked, must their internal organs be, how small their heart, their eyes, their feet? And each of these is in turn composed of atoms. Although Lucretius's consideration was aimed merely at making plausible the unbelievably small size of atoms, the observation was frequently quoted in order to prove their very existence. It occurs in Bruno, Basso, Sennert, Gassendi, and Magnenus, to mention only a few.[24] On a similar level lies Lucretius's comparison of the atomic size with the size of the tiny motes dancing back and forth if a ray of sunlight falls in a dark room. Again the passage, which goes back

---

[23] Daniel Sennertus, *Hypomnemata physica* (1636), in *Danielis Sennerti Opera omnia*, 4 vols. (Lyons, 1656), Vol. I, pp. 100–172, Hypomnema 3.1.1: "De atomis et mistione," p. 119. See also Sennertus, *De consensu et dissensu Galenicorum et Peripateticorum cum Chymicis* (1619), in *Opera*, Vol. I, pp. 177–284, here Ch. 12, p. 230.

[24] Lucretius, *De rerum natura* 4.116–122; cf. Bruno, *De triplici minimo* (cit. n. 18), 1.9.12–20, p. 169; Basso, *Philosophia naturalis* (cit. n. 19), p. 15; Sennertus, *Hypomnema* 3.1.1, p. 119; Petrus Gassendi, *Animadversiones in decimum librum Diogenis Laertii, qui est de vita, moribus, placitisque Epicuri*, 2 vols. (Lyons, 1649), Vol. I, p. 221, also in Gassendi, *Syntagma philosophicum* 2.1.3, in *Petri Gassendi Opera omnia in sex tomos divisa*, 6 vols. (Lyons, 1658), Vol. I, p. 269; and Iohannes Chrysostomus Magnenus, *Democritus reviviscens sive vita et philosophia Democriti* (Leiden, 1648), p. 206.

to Democritus, was referred to by Basso, Gassendi, and many others in the seventeenth century.[25] Evidently most of these references, even if they did not always mention their sources, go back to classical authorities, not to fresh observation. The empirical facts referred to were *topoi* of the scholarly literature. Their aim was to appeal to the reader's erudition and imagination, rather than to his critical or experimental abilities. They belong to a literary tradition of figurative rhetoric, aimed at creating astonishment and, by means of astonishment, assent and persuasion. Their frequent occurrence reflects once more the overwhelming influence of Lucretius's *De rerum natura* upon sixteenth- and seventeenth-century minds, not only because of its scientific merits, but also because of its poetic and imaginative qualities.[26]

Authors only scarcely expanded on these examples; instead they simply repeated or quoted them either in action or in writing. Bérigard presents us with a carefully designed experimental verification of Lucretius's motes in the sunbeam. His aim was to exclude the possibility that the phenomenon might be caused by major particles such as normal dust. For this purpose he sealed a glass vessel and kept it quiet for a very long time to make sure that all dust had settled down. The minute reflecting particles he was nevertheless able to see inside the glass were therefore judged to be the atoms themselves. Under the heading "Atomi quibus experimentis asseri possunt" Bérigard described his experiment:

> In vase vitreo purissimus aer inclusus, nec a vento aut alio quod sciatur impulsus, tamen ad Solem ita expositus ut radij non totum collustrent, sed per foramen clausae fenestrae, ut saepe fit, medium pertranseant, oberrare videntur et ultro citroque concursare multa corpuscula, non tantùm in aere externo, ubi pulverem volaticum suspicaberis, sed etiam in eo qui multis annis vitro concluditur, et cuius pulvis, si quis est, iampridem fundum petere debuit. Atqui omnino nisi volitarent tenuia multa corpuscula, radio per duo foramina conclavis obscuri transeunte, nullus fieret luminis repercussus et si quis ibi conclusus oculos quàm maximè intenderet in eum radium, nihil tamen intueretur: At verò propius obtutum defigenti semper aliquid conspicitur, quod aliud esse non potest, quàm atomi, ad quas lumen impingens minima ex parte ad nos deflectitur, ferè enim totum inter atomorum vacuitates recta procedit.[27]

Among those who made the first, if cautious, steps toward a quantitative determination of atomic dimensions by means of experiments was Daniel Sennert. In a series of experiments designed to prove the existence and demonstrate the size of the atoms by chemical *resolutio* Sennert described, first of all, a distillation in which he made a stream of alcohol vapor pass through a sheet of writing paper that had been folded four times. From the density of the paper and its invisibly small pores one might imagine how small the atoms really were if they passed through it so freely. This was certainly an impressive observation, but again it was also a tacit reference to Lucretius, who said the same about filtration of

---

[25] Lucretius, *De rerum natura* 4.114–131; and Democritus, in Aristotle, *De anima* 1.2 (404a3); cf. Basso, *Philosophia naturalis* (cit. n. 19), p. 13; Gassendi, *Animadversiones* (cit. n. 24), Vol. I, p. 222; and Gassendi, *Syntagma philosophicum* (cit. n. 24) 1.3.7, p. 277.

[26] As historians of science (qua science) we tend to minimize the impact of merely fictional works upon scientific matters. Even superficial evaluation of the sources quoted by early modern adherents to the corpuscular theory shows, however, that their effect can be considerable. For the general impact of Lucretius's text see George Depue Hadzsits, *Lucretius and His Influence* (Our Debt to Greece and Rome, 12) (New York: Longman, 1935); and esp. Alfred Stückelberger, "Lucretius reviviscens: Von der antiken zur neuzeitlichen Atomphysik," *Arch. Kulturgesch.*, 1972, 54:1–25.

[27] Berigardus, *Circulus Pisanus* (cit. n. 13), pp. 421–422.

solutions through paper.[28] Sennert went on to give examples of various distillations, comparing the enormous volume of vapor with myriads of atoms in it to the small droplets into which they condense:

> Dum etiam spiritus vitrioli, vel alij destillantur, corpusculis eiusmodi parvis, nondum tamen planè minimis, vas recipiens saepe per biduum vel triduum continuò plenum est, et singulis momentis aliquot myriades eiusmodi corpusculorum praesentes sunt, et sibi succedunt. Exigua tamen liquoris quantitas ex iis coëuntibus et condensatis provenit: ita ut, quae eodem momento praesentia sunt atoma corpuscula, quorum tamen aliquot myriades sunt, vix unam guttam constituant.[29]

In a similar vein is Sennert's comparison between the smoky wick of an extinguished candle, which was not even the size of a pea, and the huge volume of air that was filled with its smoke: "Et quanta inter corpus compactum et in atomos resolutum sit differentia, vel candela extincta docet. Si quis enim flammam è candela accensa flatu dissipet, ellychnium fumigans, quod vix pisi magnitudinem aequat, tantam continuò atomorum copiam emittit, ut magnum aëris spatium eâ repleatur."[30]

The language in which these observations were described abounded with quantitative statements about the duration of the experiment (*biduum vel triduum*), the number of corpuscles (*aliquot myriades*), the amount of the product (*vix unam guttam*), or the size of the candlewick (*vix pisi magnitudinem*). There is not, however, the vaguest idea of a quantitative methodology behind these indications. The language of the laboratory displays its figurative and rhetorical power, aimed at the imagination of the reader and his eventual persuasion. In tribute to the new scientific age, arguments needed support from the rhetoric of the experiment. But to do justice to Sennert, we have to admit that, in this case, even the most scrupulous quantitative experimenter would not have arrived at any result.

A few decades later Johannes Chrysostomus Magnenus, professor of medicine at Pavia, suggested exact figures for the size of an atom on the basis of impressive quantitative calculations. Taking Archimedes' *Sand-Reckoner* as a model, Magnenus estimated the number of particles in a grain of incense from the volume of air filled with its scent:

> Adverti non semel granum thuris combustum fumo ita dispergi, ut locum repleverit septingentis, et amplius millionibus se majorem. Ille enim locus grana hujuscemodi facile cepisset
>
> | | |
> |---|---:|
> | secundum altitudinem | 720 |
> | secundum latitudinem | 900 |
> | in longitudine | 1.200 |
> | in superficie | 648.000 |
> | in area | 777.600.000. |
>
> Cum ergo nulla aeris sensibilis portio esset, quae odoros non haberet halitus granumque thuris aequaret cicerem, qui sine igne in partes sensibiles saltem mille dividi posset, sequitur particulas odoras sensibiles fuisse istius num[eri] 777.600.000.000. Atqui singulae illae particulae mixtae erant, nullamque fuisse probabile est, cui unus

---

[28] Lucretius, *De rerum natura* 2.391–397; and Sennertus, *Hypomnema* 3.1.1 (cit. n. 23), p. 118. The marginal heading is "Chymicae operationes atomos probant."
[29] Sennertus, *Hypomnema* 3.1.1 (cit. n. 23), p. 118.
[30] *Ibid*. The comparison also occurs in Gassendi, *Animadversiones* (cit. n. 24), Vol. I, p. 222.

Figure 1. Daniel Sennert (1572–1637). Courtesy of the E. F. Smith Memorial Collection, University of Pennsylvania Libraries, and the Beckman Center for the History of Chemistry.

ad minimum elementalium atomorum millio inesset, unde, secundum hanc regulam, fuissent in hoc thuris grano, pisi magnitudinem non superante, atomi elementales ad minimum 777.600.000.000.000.000. Ex quibus patet, quantae sit parvitatis atomus una, conjicique potest, quantus sit atomorum numerus in toto universo.[31]

But what does it prove that the number of atoms computed in this way comes surprisingly close to modern figures, as a recent historiographer did not fail to underline?[32] Were these meditations on incense atoms really scientific calculations, or merely the outcome of a boring sermon Magnenus had to listen to in his parish church? Yet he was not the only atomist to be fond of playing with

---

[31] Magnenus, *Democritus reviviscens* (cit. n. 24), pp. 206–207; see also Stückelberger, *Antike Atomphysik* (cit. n. 6), pp. 282–285.

[32] Stückelberger, "Lucretius reviviscens" (cit. n. 26), p. 19: "Zum erstenmal ist hier in der Geschichte der Atomphysik der Schritt vom qualitativen zum quantitativen Denken gemacht worden, mit einer übrigens durchaus brauchbaren Methode—später sind allgemeine Atomgewichte durch Gase errechnet worden—, die auch zu einer erstaunlich zutreffenden Größenordnung führt: er errechnet eine Anzahl von ca. $7,7 \cdot 10^{19}$ Atomen in einem Weihrauchkorn"; cf. Stückelberger, *Antike Atomphysik* (cit. n. 6), p. 60.

numbers. In 1654 Walter Charleton published exactly the same calculation—probably a mere translation from Magnenus, though the author did not disclose his source. At least Charleton should have heeded the warnings of his hero Gassendi: referring to Archimedes' attempts to compute the number of sand grains that would fit into a poppyseed, the French philosopher had already pinpointed the methodological problems involved in transferring this kind of geometric reasoning to physical matters. His conclusion was that one must not apply to physics that which the mathematicians have demonstrated abstractly.[33]

In contrast to Gassendi, the author of *Democritus reviviscens* displayed a philosophically inconsistent attitude. Although he tried to imitate the Euclidean method of presentation, his work abounded with purely dialectical reasoning and syllogistic conclusions. After the detailed calculations quoted above, he freely admitted that, apart from conjectures, one could never know anything about the absolute or even relative sizes of atoms.[34] But since Magnenus's calculations have recently been called the beginning of the quantitative methodology in atomic physics, it might be worthwhile examining more closely the attitude of this allegedly scientific mind toward experience and experiment.

There is no doubt that Magnenus favored the empirical and mechanical spirit of his age. Among the axioms of his natural philosophy is the often-repeated principle that, in physical matters, one has to philosophize on the basis of experience and judge by the senses: "In iis, quae sub sensum cadunt, posita experientia oportet philosophari, sensibilia enim per sensus judicanda sunt." He even went so far as to claim that the results of precise experimentation should be taken as self-evident presuppositions in scientific reasoning: "Experientias accurate factas tanquam principia per se nota admittere."[35] However, this empiricism was embedded in a great number of purely abstract and speculative arguments. The same inconsistency applies to the six "proofs" Magnenus gave for the existence of atoms. Five of them stem from, and were defended on, philosophical grounds alone; the sixth at least was the "experience of the chemists," for which the author referred, if somewhat vaguely, to Sennert's *Hypomnema* 3.2.[36]

The only real experiment Magnenus presented on some five pages was taken from Jacques Gaffarel's *Curiositez inouyes sur la sculpture talismanique des Persans, horoscope des patriarches et lecture des estoilles* of 1629, a widely read and controversial compendium of natural magic. From this dubious source Mag-

---

[33] Walter Charleton, *Physiologia Epicuro-Gassendo-Charltoniana: Or a Fabrick of Science Natural upon the Hypothesis of Atoms* (London, 1654), 2.4, p. 114; Archimedes, *Arenarius* 2.4; and Gassendi, *Syntagma philosophicum* (cit. n. 24), 1.3.5, p. 265. For Gassendi's atomism see Bernard Rochot, *Les travaux de Gassendi sur Epicure et sur l'atomisme, 1619–1658* (Paris: Vrin, 1944); and Lillian U. Pancheri, "The Atomism of Pierre Gassendi: Ontology for the New Physics" (Ph.D. diss., Tulane Univ., 1972).

[34] Magnenus, *Democritus reviviscens* (cit. n. 24), p. 208: "At in his nihil aliud quam conjicere possumus, quis enim novit an atomi igneae majores, minoresve sunt aqueis, et terreis? quis dignoscit aerarum parvitatem, caelorum profunditates omnes, et extimam illam Beatorum sedem?"

[35] *Ibid.*, pp. 36 (the 11th of Magnenus's principles of corpuscular philosophy), 163.

[36] *Ibid.*, pp. 176–182. It is not clear whether Magnenus had ever read Sennert's *Hypomnema* 3.2, for in it Sennert developed his doctrine of mixture from the elements solely by the concurrence of their opposing qualities under the lead of a more noble form, supported by experimental evidence that the elements, if divided into minute particles, no longer oppose each other even if their qualities are contrary. Magnenus's first five principles are as follows. 1. Natura omnem refugit infinitatem. 2. Elementa ex minimis, ergo ex indivisibilibus; at elementa sunt materia prima; ergo materia prima ex indivisibilibus, ergo ex atomis. 3. Tot creationes in materia prima quot individua. 4. Unio non sine partibus; ergo partes non sine unione. 5. Impossibilitas partium infinitarum in continuo.

nenus took the then widely discussed story of a Polish physician who kept a collection of sealed glass vessels that contained finely ground flowers of various kinds. But when a candle was put underneath such a vessel containing the ashes of a rose, the corpuscles coalesced under the influence of the heat to form a perfect rose. When the candle was removed, the appearance disintegrated again into its atoms.[37] This strange experiment, which goes back to a report given by Joseph du Chesne in 1604, was originally presented as an example of palingenesis or resuscitation of living bodies from their ashes by virtue of the formative or seminal power inherent in their salts. Curiously enough, this Paracelsian paradigm acquired some fame among the proofs of the reality of the atoms later on. Etienne de Clave, the skilled chemist and physician who had intended to prove the truth of the scandalous anti-Aristotelian theses presented in 1624, was reported to have performed daily this experiment of reproducing an herb or a flower from its ashes.[38] The phenomenon was discussed even among the most respectable scientists of the time, but this is not the place to follow up the strange vicissitudes of the palingenesis experiment, the tradition of which can be traced down to Romantic *Naturphilosophie* and nineteenth-century debates over chemical vitalism.[39]

Microscopical Perception

It was more than merely a physical analogy to mathematical extrapolations when Daniel Sennert, after an exposition of the structure of clouds and smoke, concluded that from their small drops and particles one could infer the existence of ultimate particles. Atmospheric phenomena seemed especially well suited to support this approach, for according to the theory Aristotle put forward in his *Meteorologica,* they were not perfect mixtures but somewhat incomplete aggregates of the elements. Consequently Sennert stated that whoever is close to a cloud, for example, when hiking in the mountains, will be able to confirm that clouds are not continuous bodies but accumulations of atoms. Though this sounds like an empirical statement, it was almost certainly an allusion to Lucretius's theory of the formation of clouds.

---

[37] Magnenus, *Democritus reviviscens,* pp. 183–184. I have used the Latin translation of Jacobus Gaffarellus, *Curiositez inouyez, hoc est: Curiositates inauditae de figuris Persarum talismanicis, horoscopo Patriarcharum et characteribus coelestibus* (Hamburg/Amsterdam, 1678), pp. 97–99; see also Thorndike, *History of Magic* (cit. n. 3), Vol. VII, pp. 304–309. Martin Fichman's entry on Magnenus in the *Dictionary of Scientific Biography,* ed. Charles Coulston Gillispie, 16 vols. (New York: Scribners, 1970–1980), Vol. IX, pp. 14–15, is entirely misleading when it states that "much of the experimental evidence was drawn from Daniel Sennert's *Hypomnemata Physica* and Jacques Gaffarel's *Curiositez inouyez,*" without revealing the true nature of this "experimental" evidence.

[38] See (on Etienne de Clave) *Correspondance du P. Marin Mersenne,* ed. Marie Tannery and Cornelis de Waard, Vol. I (Paris: Presses Universitaires, 1945), p. 326; see also Iosephus Quercetanus, *Ad veritatem Hermeticae medicinae ex Hippocratis veterumque decretis ac therapeusi . . . responsio* (Paris, 1604), pp. 292–294. Du Chesne admitted to having performed the experiment unsuccessfully several times until, eventually, on a cold winter day he succeeded when a solution of salts extracted from stinging nettles froze into a block of ice displaying thousands of nettlelike structures (p. 296).

[39] See the contemporary survey by Gregorius Michaelis, *Notae in Jacobi Gaffarelli Curiositates* (Hamburg, 1676), pp. 249–253. Recent studies include Jacques Marx, "Alchimie et Palingénésie," *Isis,* 1971, *62*:274–289; Allen G. Debus, "A Further Note on Palingenesis," *Isis,* 1973, *64*:226–230; Joachim Telle, "Chymische Pflanzen in der deutschen Literatur," *Medizinhistorisches Journal,* 1973, *8*:1–37; and François Secret, "Palingenesis, Alchemy and Metempsychosis in Renaissance Medicine," *Ambix,* 1979, *26*:81–92.

> Meteora certè pleraque tantùm sunt Elementarium atomorum varia congeries. Exhalationes enim et vapores, quod vulgò creditur, non corpora continua sunt, sed congeries infinitarum atomorum: id quod ex vaporibus ex aqua, quae ad ignem calefit, ascendentibus manifestum est. Hi enim etsi procul corpus continuum videantur; tamen qui prope est, aut qui in montibus aere nebuloso ambulat, vel visu discernere potest, vapores tales non esse continua corpora sed atomorum congeriem. Nubes nihil aliud sunt, nisi infinita atomorum multitudo.[40]

Seventeenth-century science was fond of the small: it discovered worlds in a drop of water, and it developed the apparatuses to open up perspectives hitherto unseen. There was a widespread enthusiasm for the magnifying glass and for the microscope, which had just been invented. Microscopy became a preoccupation with the Baconians.[41] The new instruments made it possible to come closer to the details, closer to reality, and—so it was assumed—closer to truth. What had been the exclusive domain of speculation or extrapolation for centuries was now at least potentially observable.[42] The possibilities of optical ingenuity seemed almost unlimited. In a chapter on the size of the atoms Gassendi meditated about the degree of refinement to which the borderline between man's and nature's subtlety might be extended by means of the microscope (*engyscopium*):

> Sanè enim quae nostro visui apparent esse minima, ipsi naturae maxima sunt; ac dici potest, ubi nostra industria, subtilitásque desinit, inde incipere industriam, subtilitatémque naturae. Quippe, ut videas artifices qui annuli pallâ concludant tot illas horologij parteis, quas nisi turris capacitate rudiores fabri non valeant; ita Natura plureis parteis in milij grano distinguere, quàm homo possit in Caucaso, imò in toto globo telluris. Videri id incredibilius poterat maioribus nostris, ante inventionem Engyscopij; nunc verò, quî possit, cùm videamus granum detritissimi pulveris piso amplius repraesentari, et cum distinctissimis quidem facieculis, angulísque, de quibus ne venire quidem in mentem suspicio potuisset: adeò ut cùm diameter corpusculi Engyscopio visi sit propemodum centupla ad diametrum citra ipsum visam, dicere liceat ipsum saltem ex decies centenis millibus partium esse conflatum. Saltem, inquam, nam cogita et Engyscopium perfectius, quàm hactenus viderimus, et acutissimum quemque visum consistere semper infra naturae industriam; et agnosces denique rem abire propè in immensum.[43]

What had been a merely potential aptitude of the instrument to Gassendi, carefully restricted by the *saltem* of his epistemological reservation, was presented as an empirical fact by the English physician Henry Power only a few years later. In his *In Commendation of ye Microscope,* written about 1661, the new optical means became a new, artificial eye to help the blindness of the aged world,

---

[40] Sennertus, *Hypomnema* (cit. n. 23), 3.1.1, p. 118; cf. Lucretius, *De rerum natura* 6.451–523, esp. 468–469: "Patere res ipsa et sensus, montis cum ascendimus altos."

[41] Charles Webster, *The Great Instauration: Science, Medicine, and Reform, 1626–1660* (London: Duckworth, 1975), pp. 165, 170.

[42] See Francis Bacon's often-quoted statement, "quale perspicillum si vidisset Democritus, exiluisset forte, et modum videndi atomum (quem ille invisibilem omnino affirmavit) inventum fuisse putasset"; Francis Bacon, *Novum Organum* 2.39 (1620), in *The Works of Francis Bacon,* ed. James Spedding, Robert Leslie Ellis, and Douglas Denon Heath, 14 vols., Vol. I (London, 1857), p. 307. For the ambiguous relationship between the new optical instruments and the question of truth see Hans Blumenberg, "Das Fernrohr und die Ohnmacht der Wahrheit," in Galileo Galilei, *Sidereus Nuncius: Nachricht von neuen Sternen,* ed. Hans Blumenberg (Frankfurt: Insel, 1965), pp. 7–75.

[43] Gassendi, *Animadversiones* (cit. n. 24), Vol. I, p. 207; verbatim also in Gassendi, *Syntagma philosophicum* (cit. n. 24), 1.3.6, pp. 268–269.

> By whose augmenting power wee now see more
> then all the world Has euer donn Before.
> Thy Atomes (Braue Democritus) are now
> made to appeare in bulk & figure too.
> When Archimide by his Arithmatick,
> numbred the sands, had hee But knowne this trick.
> Hee might haue seene each corn a massy stone,
> & counted them distinctly one by one.[44]

In his *Experimental Philosophy*, the first English book on microscopy, antedating by less than a year Robert Hooke's famous *Micrographia: or, Some Physiological Descriptions of Minute Bodies made by Magnifying Glasses* (London, 1665), Power, writing in 1661, claimed enthusiastically that "our Modern Enginge (the Microscope)" enabled men to

> see what the illustrious wits of the Atomical and Corpuscularian Philosophers durst but imagine, even the very Atoms and their reputed Indivisibles and least realities of Matter. . . . indeed, if the Dioptricks further prevail . . . we might hope, ere long, to see the Magnetical Effluviums of the Loadstone, the Solary Atoms of light . . . , the springy particles of Air, the constant and tumultuary motion of the Atoms of all fluid Bodies, and those infinite, insensible Corpuscles which dayly produce those prodigious (though common) effects amongst us: And though these hopes be vastly hyperbolical, yet who can tel how far Mechanical Industry may prevail; for the process of Art is indefinite, and who can set a *non-ultra* to her endeavours?[45]

To Henry Power this was certainly more than the selling rhetoric and pretentious phraseology so common in prefaces. Among the many microscopical observations described in his *Experimental Philosophy*, mainly of insects and plants, there were also a few of chemicals. For instance, Power studied traces of running mercury and found that the "atoms of Quick-silver . . . seemed like a globular Looking-glass." From the heterogeneity and particulate structure of a "cosmetical mercury precipitate" he was viewing through his lenses, he inferred that the metal atoms remain unaltered when a compound is formed and retain invisibly their true nature: "You may most plainly and distinctly see all the globular Atoms of current and quick [mercury]; besprinkled all amongst those Powders, like so many little Stars in the Firmament: which shews that those Chymical Preparations, are not near so purely exalted and prepared, as they are presumed to be; nor the Mercury any way transmuted, but meerly by an Atomical Division rendred insensible."[46] Yet as a serious scientist Power had to admit that he had not succeeded in seeing any corporeal effluvia by means of his optical device, not even those of camphor or the transpiration of human skin, although a Dr. Highmore, perhaps with better eyes and a more powerful microscope, had claimed to have seen the magnetic effluvium "in the form of a Mist to flow from

---

[44] Quoted from Thomas Cowles, "Dr. Henry Power's Poem on the Microscope," *Isis*, 1934, *21*:71–80, on p. 71, lines 9–16.
[45] Henry Power, *Experimental Philosophy, In Three Books: Containing New Experiments, Microscopical, Mercurial, Magnetical. With Some Deductions, and Probable Hypotheses, raised from them, in Avouchement and Illustration of the now famous Atomical Hypothesis* (London, 1664), Part I, sigs. b2r, c2v–[c3r]. See Charles Webster, "Henry Power's Experimental Philosophy," *Ambix*, 1967, *14*:150–178.
[46] Power, *Experimental Philosophy*, Observation 34, p. 43; Observation 35, p. 44.

the Load-Stone." Power, the meticulous observer, was entirely conscious that such an observation, could it be proved to be true, would be the *experimentum crucis* for matter theory: "This Experiment indeed would be an incomparable Eviction of the Corporeity of Magnetical Effluviums, and sensibly decide the Controversie 'twixt the Peripatetick and Atomical Philosophers."[47]

However stimulating these instruments were for the study of biology, their meaning for matter theory was ambiguous. Jungius used the magnifying glass to study textile fabric and apparently homogeneous substances such as polished surfaces. He observed that they were in fact always heterogeneous if viewed through a microscope (*anchiscopium*). In 1633, commenting on Sennert's *Epitome scientiae naturalis* of 1618, in which Sennert had shown that arguments from geometry about divisibility and continuity must not be applied to the physical sciences, Jungius remarked that until then no physical body had ever been proved to be entirely homogeneous. For no surface could be so smooth that one could not think of a more powerful microscope that would reveal its true discontinuity. Consequently, Jungius categorically stated that continuity was foreign to the realm of sensuous experience. On the other hand he had to admit that if there were no truly continuous parts in the end, infinite progression and divisibility would result. This was the vicious circle of every observational approach to the atoms.[48] Methodological difficulties and a contradictory epistemology arose if one attempted to model the real after the visible.

Material Transport

The next type of argument is on a similar level. Different kinds of transport phenomena were considered in which material substances appear or disappear unnoticed. The standard examples are well known: Smell is an efflux of corporeal particles. Clothes that are hung near the seashore become wet and dry again in the sunshine, but no vapors can be observed. A ring on the finger gets worn out, and so do tools and road surfaces, and even a stone is hollowed by constant drops of water. Diminution and growth and the almost imperceptible loss of moisture by bodies, such as the drying of bread or the slow evaporation of liquids, are material processes, although the flux of material cannot be observed.[49] In all these cases, quantitative change of material substances was recorded, and from this the existence of invisible parts of matter could be inferred. This was little more than an application of the theory of effluxion or $ἀπορροαί$ proposed by Empedocles, Democritus, and Asclepiades of Prusa and eventually expounded in Lucretius's poem.[50] In the seventeenth century these examples were repeated again and again, and similar ones were added. It was entirely in accord with the

---

[47] *Ibid.*, Observation 51, "Of Aromatical, Electrical, and Magnetical Effluxions," p. 51. The English physician and atomist Nathaniel Highmore was a friend of William Harvey's. The observation referred to seems not to come from his published writings.

[48] Joachim Jungius, *Exercitatio VI de continuo*, 1633, Staats- und Universitätsbibliothek Hamburg, Jungius-MSS, Wo. 24 (1), between fols. 24 and 25. The epistemological antinomy involved in the application of microscopy to the question of atomism is developed further in Christoph Meinel, "'Das letzte Blatt im Buch der Natur': Die Wirklichkeit der Atome und die Antinomie der Anschauung in den Korpuskulartheorien der frühen Neuzeit," *Studia Leibnitiana*, 1988, *20* (in press).

[49] Lucretius, *De rerum natura* 1.298–299, 305–310, 311–321, 322–328. The water that abrades the stone in finite portions is already an Aristotelian example; see Aristotle, *Physics* 8.3 (253b15–23).

[50] Stückelberger, "Empirische Ansätze" (cit. n. 6), passim.

traditional line of argumentation when Daniel Sennert, in 1619, discussed the growth of chalk and stalactites from clear mineral waters, phenomena which, in his opinion, pointed to the corpuscular structure of matter. Again the observer's surprise over nature's mysteries is used as the point of departure for a rhetorical stratagem: "Cùm tamen aqua quae decurrit limpidissima sit, ut quis mirari possit, quomodo ex tam perspicua et clara aqua corpus tam crassum fieri queat. Proculdubio in talibus aquis mineralis et lapidea materia in minimas particulas resoluta fuit, quae postea suo concurso et συνκρίσει saxeum et durum corpus constituunt."[51]

It is difficult to believe that this kind of argumentation was taken as a scientific, empirical demonstration. For what did it prove except that there is material transport that cannot be seen? Perhaps three aspects should be given particular consideration: the phenomena dealt with suggested that the ultimate constituents of matter were potentially observable by extended experimental effort, deductible by analogy, and provable by virtue of their actions. There is little doubt, however, that these phenomena were not adequate for definitively deciding the question of matter, and they were certainly not understood in this manner by contemporaries. Yet the frequent occurrence and repetition of these observations, the persuasive idea that truth should be visible or could be thought of in a pictorial way, infiltrated scholarly discourse and the very language of science. Atomism was an enticingly pictorial image of reality. The wealth of appealing and immediately convincing images offered by Lucretius's poem supplied the scheme according to which material change was assumed to occur in nature. Consider Bérigard's experiment: the Lucretian motes in a sunbeam, which might have been simple dust particles, could of course not be taken seriously by the more sophisticated contemporaries of Galileo. Had not the microscope shown that even the smallest items were in reality composed of much smaller ones? Hence Bérigard made that very careful and reasonable experiment with sealed glass flasks to exclude all possible dust and turbulence in the air, and he actually saw the atoms. To be precise, he saw something "quod aliud esse non potest, quàm atomi," just as Sennert perceived them in the droplets of fog and clouds or Henry Power in the mercury preparations under his microscope.[52] Why then should they proceed any further, why devise more sophisticated experiments, why bother about the quantitative side, and why seek proofs, when the atoms were more than evident? The question of atomism had become rather a matter of plain evidence than of proof or confirmation.

Condensation and Rarefaction

Unlike the instances of analogical extrapolation from sense perception, the problems that arose from condensation and rarefaction and, above all, from the question of the vacuum belong to the field of physical experimentation proper, in which one would expect more convincing departures from the traditional ways of reasoning. In ancient atomism, with its hard and impenetrable atoms, change required motion, and motion required a void space to move into. The void was

---

[51] Sennertus, *De consensu et dissensu* (cit. n. 23), p. 231. See also Sennertus, *Hypomnema* 3.1.1 (cit. n. 23), p. 117.
[52] Berigardus, *Circulus Pisanus* (cit. n. 13), p. 422.

where the atoms were not; it formed the gaps between the particles. The nature of this void, however, was a matter of endless controversy until the eighteenth century. Yet as far as the experimental side was concerned, the question had not been advanced beyond the ancient state of affairs.[53] Basically two alternatives were discussed: first, the continuous or three-dimensionally extended void (*vacuum separatum*), and, second, the more widely received idea of a discontinuous, interspersed void between the particles of matter (*vacuum disseminatum*). The latter was less difficult to accept since it did not imply the existence of a space whose dimensions were not defined by the surface of bodies.

There was one classic experiment that could be interpreted as evidence for the discontinuous structure of matter and the existence of microvacua: a vessel filled with loose ashes was reported to hold as much water as an empty vessel because the tiny ash particles are received within the pores or vacua of the water. Aristotle ascribed this observation to those who believed in the void, but he duly rejected it on the ground that two bodies cannot occupy the same space simultaneously. The observation was indeed puzzling and kept the medieval commentators busy.[54] Most of them favored the explanation given by Averroes, who, while admitting that he had never viewed the phenomenon, denied the existence of vacua. Instead, he assumed a partial corruption of water by the ashes to be responsible for the shrinkage in volume.

Francis Bacon was presumably the first to disprove the phenomenon in question experimentally. In his posthumously published *Sylva sylvarum* he used his results to ridicule the old quarrel:

> It is strange how the ancients took up experiments upon credit, and yet did build great matters upon them. The observation of some of the best of them, delivered confidently, is, that a vessel filled with ashes will receive the like quantity of water that it would have done if it had been empty. But this is utterly untrue; for the water will not go in by a fifth part. And I suppose that that fifth part is the difference of the lying close or open of the ashes; as we see that ashes alone, if they be hard pressed, will lie in less room; and so the ashes with air between lie looser, and with water closer. For I have not yet found certainly, that the water itself, by mixture of ashes or dust, will shrink or draw into less room.[55]

This was a clear departure from the traditional way of reasoning. In reality, however, the matter was not as simple as one might believe from Bacon's straightforward refutation. In the second part of Gassendi's *Syntagma philosophicum,* published posthumously in 1658, the ash experiment was referred to as the traditional proof for the existence of an interspersed void (*inane interspersum*) but rejected on both experimental and philosophical grounds ("experimentum explorando falsum deprehenditur, uti et principiis naturae repugnat"). Instead, the French philosopher proposed another and more convincing experiment in support of corpuscles and *spatiola inania* in a solution. He saturated water

---

[53] The best account is Edward Grant, *Much Ado about Nothing: Theories of Space and Vacuum from the Middle Ages to the Scientific Revolution* (Cambridge/London/New York: Cambridge Univ. Press, 1981). For the experimental side see Pierre Duhem, *Le système du monde,* Vol. III (Paris: Hermann, 1958), pp. 121–168; and Charles B. Schmitt, "Experimental Evidence for and against a Void: The Sixteenth-Century Arguments," *Isis,* 1967, 58:352–356.

[54] Aristotle, *Physics* 4.6 (213b21–22, 214b7–8); see Grant, *Much Ado about Nothing* (cit. n. 53), pp. 71–72.

with ordinary salt and found, to his great surprise, that this solution was still as capable of dissolving alum as pure water would have been: "Experiundi gratiâ Alumen conieci in Aquam per complures dies sale impraegnatum; ac tum, non sine quodam stupore succedere coniecturam vidi: scilicet alumen perinde, ac si aqua sale caruisset, exsolutum fuit; neque id modò, sed et consequenter alios praeterea saleis exsolvit, et, ut paucis dicam, ostendit quàm varia, insensibilia licet, loculamenta contineret."[56]

From this Gassendi concluded that there must be various differently shaped microvacua or *loculamenta* in the water, each kind of which receives exactly one kind of corpuscle, for example, a cubic space a cubic corpuscle such as salt, and an octahedral space an octahedral one such as alum.[57] Although Gassendi's account was somewhat vague regarding the decrease or increase of volume during this process, he must have assumed that the volume of the solution remains unaltered. Otherwise the entire argument would have been pointless. But was it not absurd to assume that the quantity of matter remains constant when another quantity is added? This is exactly what Jean-Baptiste Morin, a somewhat obscure and dubious French antiatomist and anti-Copernican, thought when he read Gassendi's report. He repeated the experiment more carefully in a glass flask with a graduated neck and found that when salt, alum, and sugar were added to water, the volume of the resulting solution was greater than that of pure water.[58] From textual evidence alone, it is difficult to judge who was correct, Gassendi or Morin. At least, the controversy could not be as easily settled by a single experiment as Bacon had assumed. In fact, either observation, the constant or the increased volume, may have been correct: for there are indeed certain salts that do not increase the volume of water when they are dissolved, and water-free alum is one of them.

Apart from explaining dissolution, the assumption of discontinuous matter and interparticulate vacua seemed especially helpful in explaining the coherence of bodies. The standard experiment was the drawing apart of two perfectly flat surfaces from direct contact, first described by Lucretius to show that during this action a void must result, since the air cannot fill the entire opened space instantaneously.[59] Originally supposed to prove the existence of a vacuum, the experiment soon acquired a crucial position among the proofs for its nonexistence. It was in this latter sense that Galileo referred to it, for in his opinion the coherence of surfaces was a perfect illustration of nature's abhorrence of a vacuum. He availed himself of the occasion to describe, through the mouth of his spokesman Salviati, a hydrostatic experiment, based upon Hero's *Pneumatica* and designed to measure the breaking force of a water column, which would give him a quantitative value for what he called "la resistenza del vacuo." It was exactly this

[55] Francis Bacon, *Sylva sylvarum; or, A Natural History* 1.34 (1627), in *Works of Bacon* (cit. n. 43), Vol. II (London, 1859), p. 354.

[56] Gassendi, *Syntagma philosophicum* (cit. n. 24), 1.2.3, p. 195; see also p. 196.

[57] The same experiment and the same conclusions reappear several times in the seventeenth century. Charleton, *Physiologia Epicuro-Gassendo-Charltoniana* (cit. n. 33), p. 31, probably took it verbatim from Gassendi. A more critical statement was that of Bérigard, who objected that there was no reason why a cubic particle, e.g., should not be received within a bigger octahedral space; see Berigardus, *Circulus Pisanus* (cit. n. 13), p. 422.

[58] Johannes Baptista Morinus, *De atomis et vacuo contra Petri Gassendi Philosophiam Epicuream* (Paris, 1650), pp. 19–20.

[59] Lucretius, *De rerum natura* 1.385–397.

"resistance of the vacuum" that he believed to be responsible for the strength and rupture tension of solid bodies. The melting of a metal, for instance, could then be explained by an influx of fire particles into the originally void spaces, so that the *vacuola* disappear and the hard metal loses cohesion. But as Galileo did not believe in an extended vacuum, he had to assume that the size of these *vacuola* is almost zero, whereas their number is beyond limit. Consequently, he assumed that solid bodies consist of an infinite number of atoms that have no extension at all (*atomi non quanti*), and an equally infinite number of dimensionless spaces between them.[60] Strangely enough this implied that liquids were entirely continuous, because they lacked internal cohesion—a conclusion that some of Galileo's contemporaries arrived at for similar reasons. Still, the physical explanation of coherence remained a major difficulty within the corpuscular framework. Why do metals melt if heated, but not if finely crushed with a hammer? Why is silver liquefied by acid, but not if ground with a file?[61]

The problem of coherence and the interspersed void became even more acute in condensation and rarefaction.[62] According to the Aristotelian doctrine a given amount of matter could assume, at different times, contrary qualities; and since dense and rare are contrary, the same amount of matter could occupy different volumes at different times. The standard example was the change from water to air or the change in volume that occurred if a liquid was heated or cooled.[63] Aristotelian matter was capable of stretching and contracting over a wide range of volumes without losing its continuity. There was indeed no corpuscular explanation of a similar simplicity available during the seventeenth century. Few authors would have admitted that the interspersed vacua could be blown up to a size that would account for the observed change of volume during evaporation. Otherwise, they would have had to admit a continuous vacuum. Even after the Torricellian vacuum had been experimentally demonstrated in 1643, it was by no means unanimously considered to be entirely void. Thus its impact on the corpuscular theory was doubtful. The first atomist to discuss the Torricellian experiment in great depth was presumably Gassendi in his *Animadversiones in decimum librum Diogenis Laertii,* completed in 1646, but even his account of the extended void remained vague about its meaning for the theory of matter.[64] Besides, Gassendi believed it was solely an artificial phenomenon with no equivalent in nature.

Instead of admitting the void, most authors held that there was some kind of ether or spirit between the impenetrable particles of gross matter. This medium filled the spaces between the vapor atoms and glued the corpuscles of solid bodies together. This was also the explanation favored by Gorlaeus, who identi-

---

[60] Galileo Galilei, *Discorsi e dimostrazioni matematiche* (1638), in *Le Opere di Galileo Galilei,* edizione nazionale, 20 vols. (Florence: Barbèra, 1964–1966), Vol. VIII (1965), pp. 59, 62–63, 66–67. To illustrate this last assumption about atoms and spaces Galileo used the paradox known as Aristotle's wheel; see Israel E. Drabkin, "Aristotle's Wheel," *Osiris,* 1950, 9:162–198. See also the works on Galileo's theory of matter cited in n. 21 above.

[61] Galileo, *Discorsi* (cit. n. 60), Vol. VIII, p. 85. See also E. C. Millington, "Theories of Cohesion in the Seventeenth Century," *Ann. Sci.,* 1941/47 (publ. 1945), 5:253–269.

[62] Grant, *Much Ado about Nothing* (cit. n. 53), pp. 70–74.

[63] Aristotle, *Physics* 4.9 (217a20–b19); and Aristotle, *De generatione et corruptione* 1.5 (321a10–29).

[64] Gassendi, *Animadversiones* (cit. n. 24), Vol. I, pp. 424–444; and Gassendi, *Syntagma philosophicum* (cit. n. 24), 1.2.5, pp. 203–216.

fied it with the air, which he believed to be capable of neither expansion nor compression. Basso imagined a *spiritus* or a very thin corporeal substance to go between the particles of expanding air in order to prevent the formation of a vacuum.[65] It should not be overlooked, however, that the reintroduction of an active spirit or ether into atomism undermined the advantages and theoretical consistency of the corpuscular theory.[66] If a mechanical explanation of condensation and rarefaction was sought, the ether was certainly not a convincing solution. Other alternatives were equally weak. Magnenus, who strongly rejected the vacuum and denied any real condensation or rarefaction, offered a strange blend of atomism and Aristotelian stretchability of matter. He endowed his atoms with the ability to expand almost indefinitely in two dimensions while the third dimension shrinks accordingly, so that the surface remains constant, or "isoperimetric" as he termed it. The idea was that an extremely expanded atom would dilute and augment, so to speak, the space it occupied before. In this way expansion could be explained without admitting the inflation of interparticulate vacua. From these examples it is clear that, in contrast to classical atomism, the vacuum, be it dispersed or extended, was not a *conditio sine qua non* for a corpuscular theory in the seventeenth century. Plenist corpuscularians such as Bérigard, Basso, or Descartes witness that it was entirely acceptable to assume corpuscles without admitting the void.[67]

Atoms of Light

One last type of physical argument should be mentioned at least briefly: the evidence of a particulate structure of matter derived from nonmaterial phenomena such as sound, heat, magnetism, electricity, or light. Most of the early attempts to substantialize or materialize the immaterial would lead us too far away from the main purpose of this study. They originate from ancient atomism and were taken up again by many seventeenth-century authors. Explicit formulations, however, such as Basso's atoms of heat and cold, Gorlaeus's atoms of time, or the quality-atoms imagined by Bérigard, remained exceptional.[68] Only light was frequently regarded as a substance that consisted of tiny particles, thin

---

[65] Cornelis de Waard, *L'expérience barométrique, ses antécédents et ses explications: Etude historique* (Thouars: Imprimerie Nouvelle, 1936), pp. 29–32, 85–90; Gorlaeus, *Exercitationes philosophicae* (cit. n. 10), pp. 248–249; and Basso, *Philosophia naturalis* (cit. n. 19), p. 300. As to the nature of this subtle matter, Basso pointed to some kind of Stoic spirit or ether (p. 306).

[66] Lasswitz's judgment is especially harsh in that regard: "Die intramolekularen Zwischenräume können daher wieder nur durch etwas ausgefüllt werden, das ebenfalls ein Körper ist und doch eigentlich kein Körper sein soll. Es ist ein durch Abstraktion von allen empirischen Qualitäten gewonnenes Destillat, ein verblaßter, unklarer Körper, heiße er nun Luft, Äther oder Spiritus, durch dessen Einschaltung alle jene Schwierigkeiten wieder eingeführt werden, die der Begriff des Korpuskels ablösen sollte. Gibt es überhaupt kontinuierliche, qualitative Substanzen, so hat die Unveränderlichkeit der Korpuskeln ihren systematischen Wert vollständig verloren." Lasswitz, *Geschichte der Atomistik* (cit. n. 1), Vol. I, p. 517.

[67] Magnenus, *Democritus reviviscens* (cit. n. 24), pp. 415–428, 254, 262. For Descartes's spongelike concept of rarefaction see Renatus Descartes, *Principia philosophiae* (1644), 2.6–7, in *Oeuvres de Descartes,* ed. Charles Adam and Paul Tannery, 13 vols. (Paris: Vrin), Vol. VIII (1982), Pt. 1, pp. 43–44. Descartes's theory of matter occupies a somewhat peculiar position. See also Boas, "Mechanical Philosophy" (cit. n. 1), pp. 442–460; and Francesco Trevisani, "La teoria corpuscolare in Cartesio dal 'Traité du Monde' ai 'Principia,' " in *Ricerche sull'atomismo* (cit. n. 1), pp. 179–223.

[68] For Basso see Basso, *Philosophia naturalis* (cit. n. 19), pp. 413–421; and Zanier, "Macrocosmo corpuscolaristico" (cit. n. 19), pp. 99–109. On Gorlaeus see Lasswitz, *Geschichte der Atomistik* (cit. n. 1), Vol. I, pp. 458–459. For Bérigard see Berigardus, *Circulus Pisanus* (cit. n. 13), pp. 418–425.

enough to go through the pores of even hard and dense bodies. Lucretius had imagined atoms of light in order to explain the almost instantaneous transmission of light and images through space. Such atoms were extensively used by Bérigard and questioned by Gorlaeus, reflecting, however, more of a literary tradition than an empirical one. This is equally true for Gassendi, whose theory of light and vision was little more than a compromise between the ancient doctrine of corpuscular effluvia, the Lucretian *simulacra,* and Scholastic assumptions about the propagation of immaterial species through a medium.[69] In that regard the Cartesian theory of light is remarkable, since it made contact with experience and optical experimentation at various points, even if derived from an a priori conception of matter.[70] In purely optical treatises, on the other hand, little reference to matter theory is to be found. After all, optics was then a branch of mathematics or physiology, the methods and results of which were not normally considered to solve physical puzzles.

It is, therefore, on the borderline between physics and optics that we meet with an early example of how optical observations interacted with matter theory. In a series of letters exchanged between 1606 and 1608, Thomas Harriot and Johannes Kepler discussed the phenomenon of partial reflection and partial transmission of light in diaphanous bodies. The question was, how could an apparently continuous body transmit and reflect at the same time? Harriot imagined that the continuity was only in our senses, whereas in reality some corporeal parts at the surface resist the rays and therefore reflect, while other rays penetrate into the vacua between them, are reflected within the body, and leave it, scattering in many directions. The most striking example of this kind was probably gold, an entirely homogeneous, dense, and opaque body that reflects light like a mirror. But a thin gold foil reflects and is translucent at the same time. For Harriot, it seemed absurd to assume that a single homogeneous substance can be endowed simultaneously with two opposing qualities, transparency and opacity. Kepler, however, did not want to follow Harriot *ad atomos et vacua* and suggested keeping optics and matter theory distinct and accepting exactly this contradictory assumption.[71]

We cannot go into the many difficulties that arose from argumentation based upon the corpuscular nature of light. It seems to have played but a minor role in seventeenth-century atomism, except for the conventional explanation of transparency and diaphaneity by assuming that atoms of light pass through the pores of matter. But it must be remembered that, in the tradition of Newton's *Opticks,* the corpuscular nature of light became a most powerful argument in eighteenth-century theories of matter.[72]

---

[69] Lucretius, *De rerum natura* 4.176–215; Berigardus, *Circulus Pisanus,* pp. 426–430; Gorlaeus, *Exercitationes philosophicae* (cit. n. 10), p. 109; Gassendi, *Animadversiones* (cit. n. 24), Vol. I, pp. 236–260; and Gassendi, *Syntagma philosophicum* (cit. n. 24), 1.4.11–12, pp. 421–441. For the details of Gassendi's theory see Rochot, *Travaux de Gassendi* (cit. n. 33), pp. 100–102; and Wolfgang Detel, *Scientia rerum natura occultarum: Methodologische Studien zur Physik Pierre Gassendis* (Quellen und Studien zur Philosophie, 14) (Berlin/New York: de Gruyter, 1978), pp. 197–204.

[70] A. I. Sabra, *Theories of Light from Descartes to Newton* (London: Oldbourne, 1967), pp. 17–135.

[71] Harriot to Kepler, 2 Dec. 1606, in Johannes Kepler, *Gesammelte Werke,* ed. Max Caspar (Munich: Beck), Vol. XV (1951), pp. 365–368, on pp. 367–368; and Kepler to Harriot, 2 Aug. 1607, *ibid.,* Vol. XVI (1954), pp. 31–32, on p. 32. See also Robert Hugh Kargon, *Atomism in England from Hariot to Newton* (Oxford: Clarendon, 1966), pp. 26–27. For a critical reassessment of Harriot's atomism see Henry, "Thomas Harriot" (cit. n. 9).

[72] For the role of optical arguments in Newtonian matter theory see Arnold Thackray, " 'Matter in

Chemical Change

The "chemical" arguments proposed in support of the atomic view of matter fall basically into two categories: arguments that refer to processes during which a new *mixtum* or compound is generated, and proofs based upon the recovery of constituents from a mixture. In either case an explanation was required of how distinct particles interact and of how, from this interaction, new qualities emerge that were not originally present in the reactants. The emergence of a new form during substantial alterations, the so-called *eductio formae,* was indeed the great theme of early seventeenth-century Peripatetic theory of matter.[73] The traditional alternative explanations were descent of the form from the *forma caeli* or from a *dator formae,* and eduction from the potentiality of matter. Divorced from these learned speculations, the common attitude of metallurgists and practical alchemists was to ignore such abstract questions while naively taking the original reactants as the true constituents of a compound. Was it not reasonable to assume, for example, that the ingredients required to make up a complicated medicine were actually present in this preparation, with all their respective virtues? This pragmatic solution was favored by many iatrochemical authors. They were trained in the laboratory and little troubled by philosophical scruples. Jean Beguin's *Tyrocinium chymicum* of 1610 is a well-known example of this kind of approach.[74] It was not apt, however, to satisfy the needs of a natural philosopher. Consider a sweet and ripe fruit, said Giovanni Nardi, the editor of a Florentine edition of Lucretius's *De rerum natura,* attacking Sennert's corpuscular interpretation of substantial change: where were the ripeness and sweetness beforehand? The same little part (*eadem portiuncula*) that was previously astringent and bitter is now soft and sweet, but no significant change of weight has occurred. How then could this maturation be explained without admitting a new substantial form?[75]

Similar difficulties arose when decompositions such as chemical analyses were considered. The products that resulted from dissolution or *diacrisis* were commonly believed to be "parts" of the original compound. The terms *part* or *constituent,* sometimes *element,* implied the relation between a whole and its parts. Though this seems fairly reasonable, it led to incredible difficulties if applied to chemical processes. For whether something was regarded as composition or decomposition was often a matter of chance. As long as quantitative alterations, especially the decrease or increase of weight, were not systematically taken into account, there was no means of distinguishing between the two alternatives. Consequently, the dissolution of wood by fire would result in its "parts" of smoke and ash, the dissolution of milk in its "parts" of whey, butter, and cheese.

It is easy to understand that here purely mechanical action, and especially

---

a Nut-Shell': Newton's Opticks and Eighteenth-Century Chemistry," *Ambix,* 1968, *15*:29–53; and Henry John Steffens, *The Development of Newtonian Optics in England* (New York: Science History Publications, 1977), pp. 27–54.

[73] Emerton, *Scientific Reinterpretation* (cit. n. 8); and Gregory, "Studi sull'atomismo, II" (cit. n. 10), pp. 55–62.

[74] R. Hooykaas, "The Discrimination between 'Natural' and 'Artificial' Substances and the Development of Corpuscular Theory," *Archives Internationales d'Histoire des Sciences,* 1948, *1*:640–651, on p. 643; and Hooykaas, "Experimental Origin" (cit. n. 1), p. 77.

[75] Titus Lucretius Carus, *De rerum natura libri VI unà cum paraphrastica explanatione et animadversionibus Joannis Nardii* (Florence, 1647), p. 37.

local motion, was not a sufficient explanation for the requirements of a chemical philosophy of matter, though some kind of motion was generally accepted as a prerequisite for the formation of a compound. But as Magnenus stated it, any undirected local motion of particles would account for disintegration only, not for concord and unition.[76] This then was the point of entry for additional hypotheses that did not originally belong to or were even contradictory to the principles of atomism. Assumptions of this kind were the corporeal ether, the active spirit, Neoplatonic concepts of sympathy and antipathy, or the teleology of noble or directing forms that act upon moving particles.

It is significant that Sennert explicitly admitted that, given the dimness of human cognition, there was no way of proving the mechanism by which the unity of parts was brought about and the form of the new compound generated. Practical chemist as he was, he preferred to leave these questions to others and went on to argue that at least one thing was certain, namely, that every mixture could be resolved into those parts out of which it was originally constituted. In other words, the substantial identity of the constituent parts must persist unchanged; otherwise there would be a generation of new constituents during the process of resolution and decay: "Hoc certum est, mistum quodlibet in ea, è quibus primò constitutum est, resolvi posse: et proinde formas elementorum non aboleri. Aliàs enim in resolutione et putredine fieret nova elementorum generatio."[77]

This was a most important step in the "chemical" argumentation in favor of corpuscles. For now it was no longer necessary to deal exclusively with the process of substantial change and the emergence of new qualities. Instead the question of the corpuscles' existence was reduced to a test of identity in a cyclic process that could easily be performed by the chemical means of the time. If then the identity of the original reactants and the products of decomposition could be demonstrated experimentally, the persistency of this material carrier would have been proved, no matter how many alterations had meanwhile occurred. This was clearly a departure from the former preoccupation with the quiddity of processes, and a shift from the ontological level of atomism to something that might be called a "black box theory" of chemical change.

Reduction to the Pristine State. There is ample evidence that the ground for this new perspective had been laid by the pragmatism and atheoretical attitude of metallurgists and iatrochemists. Imagine a goldsmith who during an alloying or reducing operation did not make the concept of substantial identity the very basis of his trade. However, such people were barely literate, and we do not know their theoretical suppositions about the connection between the material they were working on and its properties. Yet on the fringes of the learned tradition were a few noteworthy exceptions, among them Angelus Sala, an Italian who spent most of his life in Germany as a court physician, pharmacist, and adviser

---

[76] Magnenus, *Democritus reviviscens* (cit. n. 24), p. 197. Cf. the often-quoted definition, "mistio est motus corporum minimorum ad mutuum contactum, ut fiat unio," by Julius Caesar Scaliger, *Exotericarum exercitationum liber XV. de subtilitate ad Hieronymum Cardanum* (1557; Frankfurt, 1665), exerc. 101.1, p. 331.

[77] Sennertus, *De consensu et dissensu* (cit. n. 23), Ch. 12, p. 230; see also Gregory, "Studi sull'atomismo, II" (cit. n. 10), pp. 53–55. The second sentence, quoted here, however, points once more to the relational character of the notion *elementum*: something that is generated anew cannot, by definition, be the product of a re-solution. For us this conclusion is of course a tautology.

on commercial subjects. In his *Anatomia vitrioli* of 1609, which had to be translated from the vernacular since the author did not even know Latin, he described the formation and decomposition of copper vitriol pragmatically and to some extent also quantitatively. In doing so Sala distinguished, as usual, between transmutations, which occurred without major material additives and implied destruction of the old and emergence of a new substantial form, and changes that resulted from a mere juxtaposition or separation of finely divided particles. These latter processes were either *coniunctiones,* such as the alloying of gold and silver to yield electron or the apposition of particles of copper, acid, and water to yield vitriol, or *reductiones,* by means of which these little particles were reassembled into their former coherent state: "Reductio autem est operatio quaedam, per quam recolligimus et in unam massam coadunamus rem quampiam quae in minutissimas particulas dispersa et dilatata erat . . . et interim tamen *per Reductionem in pristinum suum statum et essentiam* revocatur et reducitur."[78]

The standard example was a solution of gold in *aqua regia* (a mixture of hydrochloric and nitric acids) and its precipitation using metallic silver. From Sala's account it is clear that he considered processes of this kind as mere divisions and rearrangements of metal atoms in which the metals retain their substantial identity, although "hidden" because of their dispersion into single atoms. As a practical chemist Sala did not bother with the nature of this "hiding" of qualities. Instead he accepted the reduction to its pristine metallic state as sufficient evidence for the persistence of "gold" throughout this process of dissolution and recovery. Even the delicate question whether gold in the form of *aurum potabile* retained its medical virtues and was therefore "real" gold—a much discussed topic at the time since potable gold was believed to be an almost universal medicine—was dismissed by Sala with the words that, instead of listening to the testimony of hundreds of authorities, one need only pay attention to the craftsmen: their expertise showed that gold could be recovered in its pristine form from such solutions without any damage to its qualities: "Possemus de hoc ipso vel sexcenta auctorum testimonia adducere, nisi magis adtenderemus ad id, quod agant isti artifices, quam quid loquantur. . . . si in manus experti artificis incidat, â spiritibus salium et sulphuris adpactis, liberari possit, et deinde pristinae suae formae, sine ulla qualitatum suarum laesione, restitui."[79] The naive corpuscularianism behind this conception of chemical change implied, however, that Sala regarded vitriol and all metal salts as mere aggregates of corpuscles, at the price of abandoning the idea of the homogeneity and substantiality of such compounds. At least from the chemical point of view this was a bit too simple, even for his contemporaries, and did not explain why properties of the atoms, such as their metallic character, disappear in the dispersed state and reappear when they are precipitated.

Hence Daniel Sennert, the learned professor of medicine, preferred to maintain the teleology of noble forms in order to account for the specific properties of the mixture. Expanding on Sala's corpuscular approach, Sennert conceived a

---

[78] Angelus Sala, *Anatomia vitrioli* (3rd. ed., Leiden, 1617), pp. 96–97 (emphasis added). On Sala see Hooykaas, "Discrimination" (cit. n. 74), pp. 646–648; and Hooykaas, "Experimental Origin" (cit. n. 1), p. 78.

[79] Angelus Sala, *Chrysologia seu examen auri chymicum* (Hamburg, 1622), 3.6.5, sig. K3r (on the "hiding"); and 2.1, sig [F6v–7r] (quotation).

more sophisticated argumentation, based on a great number of empirical observations of what he classified as reductions to the pristine state. The first types of examples referred to were simple distillations and sublimations of such substances as alcohol, sulfuric acid, and sulfur. Behind this was the assumption that sublimations in particular were merely mechanical operations—the chemists' pestle, as Sennert called them—by means of which bodies were mashed into their atoms.[80] Another and chemically more sophisticated argument for the atomic theory was taken from the reduction of different mercury compounds to running *mercurium vivum,* which retained its substantial form through all chemical and physical alterations.

> Mercurius praecipitatus si cum oleo tartari seu sale tartari per deliquium soluto teratur, sal, quod Mercurio adhaeret, unitur sali tartari, et Mercurium deserit, unde ille vivificatur. Ita si Mercurius sublimatus calci vivae misceatur, et Retortae indatur, sal vitrioli, et communis qui sublimatio inest, calci vivae adhaeret, atque ita argentum vivum in pristinam naturam redit et vivificatur; quomodo etiam cinnabaris in argentum vivum reducitur. . . . Et omnino quàm multas formas externas corpora naturalia cum aliis mista, salvâ manente formâ substantiali, induere possint, vel unus Mercurius docet, qui tot formas externas induit, ut meritò πολύμορφος dicatur. Mutatur in aquam limpidam, in liquorem butyro similem, sublimatur, praecipitatur, redigitur in pulverem, in vitrum, plumbi, auri, argenti, ut etiam laminari possit, figuram, et nescio quas alias formas induit; quas tamen omnes deponit, et pristinam ac nativam formam induit, si id quod ei admiscetur, ab eo separetur.[81]

To prove that there were indeed real atoms of mercury involved, and not just new substantial forms generated, Sennert mentioned an often-quoted observation from the nightmares of Paracelsian medication: ointments and fumigations with mercury, "in quibus argentum vivum in atomos redactum totum corpus penetrat," had the effect that a coin put in the patient's mouth became amalgamated —not to mention the findings of an autopsy.[82] In all these cases the reduction to the pristine state, *reductio in pristinum statum,* was the decisive criterion.

The same is true for another type of experiment presented by Sennert in order to prove that it was not the substantial form of a mixture that preserved the identity (*forma essentialis speciei*) of its compounds, but in fact the atoms themselves. He fused gold and silver together to obtain an entirely homogeneous alloy. Then he poured aqua fortis, or nitric acid, on it. The silver was dissolved, whereas the gold settled to the bottom as a finely divided sediment. He separated the two phases and precipitated the silver from the solution to yield another equally fine powder. Eventually, he melted each powder and obtained gold in the first, silver in the latter case.

> Etsi vero atomi illae sint minimae corpuscula: tamen in iis formae essentiales specierum integrae manent, ut modò dixi, et ipsa experientia testatur. Si enim simul aurum et argentum fundantur, atomi auri et argenti ita per minima misceantur, ut nullo sensu hae ab illis discerni queant. Interim utraeque suas formas integras servant. Quod vel ex eo patet, quòd, si massae isti aqua fortis affundatur, argentum solvitur, et

---

[80] Sennertus, *Hypomnema* 3.1.1 (cit. n. 23), p. 118 (reductions); and Sennertus, *De consensu et dissensu* (cit. n. 23), Ch. 19, p. 274 (pestle). A similar view was held by Berigardus, *Circulus Pisanus* (cit. n. 13), p. 422. At that time there was no unequivocal criterion to distinguish between simple and complex phenomena in chemistry.
[81] Sennertus, *Hypomnema* 3.1.1, p. 118.
[82] *Ibid.,* pp. 118–119.

in liquorem abit, aurum verò formâ pulveris remanet. Argentum solutum si praecipitetur, formâ pulveris subtilissimi subsidet. Uterque pulvis, si seorsim fundatur, in pristinum aurum et argentum abit.[83]

All these experiments were of course not "invented" by Sennert, and it is even irrelevant whether he actually performed them. But he was the first to connect them systematically in order to use the *reductio in pristinum statum* as an argument in favor of atomism. It is interesting to see that Sennert, after having arrived at this conclusion in the *Hypomnemata physica* of 1636, had to revise his former ideas concerning the transmutation of metals.

Transmutation of Metals. Old and fairly well-documented evidence for metallic transmutation in fact existed, evidence that was apt to be used in defense of both the Aristotelian doctrine of substantial alteration and the alchemists' quest for gold. The facts were so plain that, as Sennert remarked, it would be a waste of time to argue about them. Several springs and rivers, especially in Smolník in Slovakia and near Goslar in the Harz Mountains, had the peculiar property that a piece of iron, on being immersed for some time, became true copper, first at the surface and later throughout. The same phenomenon could equally well be produced artificially by dipping an iron bar into a glass of water that contained blue vitriol. This process, called cementation, resulted in the transmutation of a vile metal into a nobler one. It had been known for a long time and described by authors whose credibility was beyond any doubt, such as Agricola, Libavius, and Cesalpino. Cementation was even known commercially for producing fine metallic copper—and is, by the way, still used today for the exploitation of low-grade ores and the recycling of scrap copper.[84]

There were, however, authors who doubted these observations and categorically denied the possibility of transmutation, among them the Lorraine physician Nicolas Guibert, writing in 1603. It is amusing to read how Sennert, in the 1629 edition of his *De chymicorum cum Aristotelicis et Galenicis consensu ac dissensu,* ridiculed Guibert's objections, and not because he was at that time still an Aristotelian, but because transmutation had been proved experimentally. It was the practical experience of metallurgists and assayers, accumulated over centuries, that Sennert introduced to support the view, contrary to Guibert's, that the copper produced by cementation was true copper and even of greater purity than that usually obtained from its ores. "Quasi per totum Imperium Romanum, fide publicâ, omniumque artificium et Docimastarum consensu non esset notissimum, cuprum illud genuinum esse, imò eo, quod multis in locis è terra effoditur, praestantius; et Guiberto, nescio quas ratiunculas in contrarium afferenti, plus

---

[83] *Ibid.,* p. 119. In another context the same process was also discussed by Scaliger in 1557 and Bodin in 1605; see Hooykaas, "Discrimination" (cit. n. 74), pp. 643–646.

[84] Sennertus, *De consensu et dissensu* (cit. n. 23), Ch. 2, p. 182; Georgius Agricola, *De natura eorum quae effluunt ex terra,* ed. Georgius Fabricius (Wittenberg, 1612), 2.10, p. 235; Agricola, *De natura fossilium,* ed. Georgius Fabricius (Wittenberg, 1612), 2.2, p. 701; Andreas Libavius, *Defensio et declaratio perspicua alchemiae transmutatoriae opposita Nicolai Guiperti* (Oberursel, 1604), pp. 216–233; and Andreas Caesalpinus, *De Metallicis libri III* (Nuremberg, 1602), 1.6, p. 17. For the commercial application see G. E. Löhneyss, *Bericht vom Bergkwerck* (Zellerfeld, 1617), IX, Bergordnung 5, p. 332, with reference to the Rammelsberg near Goslar. The pros and cons of transmutation appear at the beginning of Sennert's work under the heading *De veritate et dignitate chymiae,* the locus classicus for the defense of alchemy!

quàm tot artificum censurae et non unius saeculi experientiae, fidei habendum, et umbratilis ad pulpitum speculatio experientiae tot artificum praeferenda sit. Homine imperito nihil est ineptius!"[85]

But when the posthumous edition of Sennert's collected works was published, the editor added to this paragraph a note that he had found among Sennert's manuscript remains. In it Sennert admitted that the alleged transmutation was presumably a mere separation of copper from its vitriolic solution by means of iron ("videtur nimirum probabile, ferrum in aes non mutari, sed aes ex aquis vitriolatis saltem separari, ferri beneficio"), and he added that the reaction was equivalent to the one by which silver is precipitated from aqua fortis by means of copper. He correctly considered each process as an exchange of substances that retained their chemical identity and were, therefore, not transmuted. This view was also supported by the observation that from a given amount of natural or artificial vitriolic water only a certain amount of copper could be obtained. To explain this exchange Sennert suggested a mechanism whereby the "salt" that was "united" with copper in vitriolic water tried to "dissolve" the iron that had been added. In doing so it released the copper atoms with which it had been "united" until then, and these atoms, abandoned by the "salt," sank to the bottom, where they could be collected: "Nimirum dum cuprum in aqua salsa solutum est, si iniiciatur ferrum, tum sal illud, quod est in aqua, etiam ferrum solvere conatur, et aggreditur, atque ita atomos cupri, cum quibus se univit, deserit, quae à sale derelictae, ad fundum descendant, quod praecipitari dicitur."[86]

From his account it is evident that Sennert's change of mind was due to his conversion to atomism. Indeed he referred to the appropriate passage in the *Hypomnemata physica*. Sennert even went a step further and conceived an experiment by which one could prove that the process of cementation was truly quantitative: one ounce of copper was dissolved in sulfuric acid to obtain an acid, blue solution of copper vitriol. Then two ounces of iron filings were added. The vessel was kept in a warm place until all of the iron had disappeared, the blue solution turned colorless, and a red precipitate settled at the bottom. This precipitate was washed, dried, and reduced, to yield exactly one ounce of copper—"quantum scilicet solutum fuit"—the same amount that had been dissolved before.[87]

Again the method of *reductio in pristinum statum* was applied to demonstrate the existence of atoms. Sennert was thus able to avoid the difficulties involved in explaining what happened to the properties of the metal when it "united" with

---

[85] Sennertus, *De consensu et dissensu* (cit. n. 23), Ch. 2, p. 182. For Guibert see Partington, *History of Chemistry* (cit. n. 16), Vol. II, p. 268.

[86] Sennertus, *De consensu et dissensu*, Ch. 2, p. 182. The addition seems to appear first in the edition Frankfurt/Wittenberg 1653; in the corresponding paragraph of the last edition he supervised— Daniel Sennertus, *De chymicorum cum Aristotelicis and Galenicis consensu ac dissensu* (Wittenberg, 1629), pp. 10–11—Sennert still displayed his belief in transmutations.

[87] Sennertus, *De consensu et dissensu*, Ch. 2, p. 182. From Sennert's account it is not entirely clear whether he in fact performed this experiment, for he went on to say that one evaporates the remaining liquid to dryness and finds exactly two ounces of vitriol, corresponding to the two ounces of iron added. The correct value, however, would be almost 10 ounces $FeSO_4 \cdot 7\ H_2O$, 6 ounces $FeSO_4 \cdot H_2O$, or 5½ ounces $FeSO_4$, depending on the degree of heat applied in drying the residue. Since we have no manuscripts, it is difficult to judge what Sennert actually did or wrote. In my opinion, the most likely explanation would be that he was so exclusively concerned with the *reductio in pristinum statum* that only in the first part of the reaction did the quantitative side seem relevant to him, whereas the rest may have been added somewhat negligently in order to complete the balance of substances. It is unreasonable to assume either that the famous iatrochemist was unaware that iron metal cannot be reobtained by this process, or that he simply mixed up iron and green vitriol.

Figure 2. Joachim Jungius (1587–1657). Courtesy of the Smith Collection and the Beckman Center.

the "salt" in the solution, or whether the particles of copper, salt, and water form an essentially uniform mixture (as required by chemical experience) or a mere juxtaposition of parts (as imagined by Sala).

Sennert was presumably not aware that in 1630 Joachim Jungius had reached a similar conclusion.[88] Jungius's approach, however, was different. He neither took the *reductio ad pristinum statum* as the decisive criterion nor concerned himself with the quantitative aspect of cementation. He correctly observed that it was an exchange of material and by no means a transmutation, because, during the process, the color of the solution gradually changed from blue to greenish—the color of the iron vitriol—and once all the dissolved copper had been consumed, no further iron was "transmuted" ("ubi tantum aeris, quantum in se continuit, rursus exspuit, ferrum amplius ita transmutari nequit"), since with green

---

[88] The chronology is not entirely sure, but it is reasonable to assume that the posthumously added paragraph from *De consensu et dissensu chymicorum* was written after the *Hypomnemata* of 1636. Since Jungius's ideas of 1630 were not published until 1661, and Sennert died in 1637, it is unlikely that there was any mutual influence in this area.

vitriol alone, the reaction would not take place. However, Jungius's point of view was different from Sennert's, as was his experimental technique. Whereas Sennert used an acid solution that slowly dissolved the iron, Jungius obviously referred to a neutral solution in which the process was not accompanied by simultaneous dissipation of the metal by the acid. Hence the iron objects Jungius dipped into the solution did not disappear but retained their original form. Taken out after a while, they were covered with a layer of copper, so that one could carefully pull the iron out of the copper sheath.

> Bacilla ferrea in aquis caeruleo vitriolo praegnantibus ita aere quasi vestiri, ut ferrum ex eo tanquam è vaginâ educi possit; verum etiam clavos et hujusmodi alia ferramenta in puteis aquâ hujusmodi plenis temporis diuturnitate tandem aerea inveniri. Nulla tamen hic transmutatio intervenit, sed permutatio potius. . . . Quodsi id sensim fiat et longo temporis spatio atomis aeris in locum atomorum ferri subeuntibus aliquando potest, ut ferramenta eandem figuram servantia aerea tandem inveniantur.[89]

The explanation suggested by Jungius was a corpuscular one: the copper atoms exchanged places with the iron atoms. Were it not through an exchange of tiny atoms, one atom taking exactly the space its counterpart had just occupied, it could not be understood why the external form of the metal was completely retained during the process. For the gross form of an object, such as the form of a nail, was of course entirely fortuitous and could therefore exert no formative power whatsoever. Jungius's argument came more from a topological or crystallographic point of view than from a chemical one. The only quantitative statement we meet in his account—"spiritus sulfuris . . . aes . . . à se dimittit et tantumdem ferri vicissim complectitur et quasi deglutit"—seems to suggest a one-to-one exchange of atomic positions.[90]

Other scholars of Jungius have interpreted this same passage entirely differently, namely, as a quantitative statement. But this has led them into severe difficulties, which the interpretation offered here may avoid.[91] In addition there is another argument in favor of the mere spatial or topological sense of Jungius's *tantumdem*.[92] The original manuscript, written around 1629/30, reads only *tantumdem ferri rursus deglutit*, whereas the additional *vicissim complectitur* came in when Jungius first dictated the text shortly afterward. *Vicissim complectitur*, however, seems an indication that he wanted to put more stress on the spatial

[89] Jungius, *Praelectiones physicae* (cit. n. 11), p. 234, lines 20–21, 3–17 (longer quotation).
[90] *Ibid.*, p. 234, lines 7–13. The passage reads: "Nam spiritus sulfuris, qui hospitium habet in aquâ istâ, quia vel promptius ferrum ut imperfectius metallum quam aes corrodere et, ut ita loquar, perdomare potest, vel quia majore sympathia erga illud afficitur, aes qod hactenus insedit, cum quo hactenus in mistum sive, ut Chymici loquuntur, magisterium coaluit, à se dimittit et tantumdem ferri vicissim complectitur et quasi deglutit."
[91] Emil Wohlwill, *Joachim Jungius und die Erneuerung atomistischer Lehren im 17. Jahrhundert: Ein Beitrag zur Geschichte der Naturwissenschaft in Hamburg* (Abhandlungen aus dem Gebiete der Naturwissenschaften X, 2) (Hamburg, 1887), p. 58; Kangro, "Erklärungswert" (cit. n. 1), pp. 32–33; and Walter Pagel, "Chemistry at the Cross-Roads: The Ideas of Joachim Jungius," *Ambix*, 1969, *16*:100–108, on p. 103. Kangro, *Joachim Jungius* (cit. n. 20), pp. 85–86, tried to save Jungius's quantitative methodology by assuming that he was thinking in terms of atomic volumes.
[92] Jungius's famous disciple Bernhard Varenius interpreted it spatially in *Geographia generalis* (1650), 2nd ed., ed. Isaac Newton (Cambridge, 1681), 1.17.11, p. 199: "Cupreae aquae particulae in ablatarum ferrearum locum reponuntur sive ibi haerent, dum allabuntur cum fluente aqua." However, the idea was not developed any further and seems unrelated to similar seventeenth-century ideas about the formation of crystals; see also Emerton, *Scientific Reinterpretation* (cit. n. 8), pp. 126–153.

aspect, the one-to-one exchange of atomic positions that were surrounded by neighboring positions—a process that he must have considered as something rather unusual and unlike other transmutations. In fact, in his first draft of the text he called it *metamorphosis* (a change by *translocatio partium*), a term that does not appear elsewhere in this work.[93]

The Range of Chemistry. Taken together, the chemical evidence presented by various authors to support the corpuscular view of matter supplied good empirical reasons for regarding natural bodies as divisible into much smaller ones that somehow retained their specific properties and could frequently be recombined to yield the original body. It is not surprising that most experiments proposed in that context referred to metals, metal salts, their aqueous solutions, and mineral waters. There are both chemical and historical reasons for this choice. Metals and salts were the best-known classes of substances, easy to obtain, fairly easy to treat by well-established laboratory techniques, and simple enough to fit into a crude theoretical framework. Among them the noble metals gold, silver, copper, and mercury occupied the most prominent place since their compounds could most easily be reduced to the respective metal and thereby identified. On the other hand there was a great interest in mineral waters, stemming from late sixteenth-century Paracelsianism. Within the iatrochemical tradition the analysis of natural mineral waters became a prominent branch of medical chemistry and was in fact the driving force behind the development of modern chemical analysis. For this *via humida* new experimental techniques were required that had not been available to the traditional *via sicca,* the "dry" analysis by means of fire.[94] The outstanding example of this new orientation is Robert Boyle's *Sceptical Chymist* of 1661, with its harsh attack on the use of fire in chemical analysis.

Yet the range and depth of experimental proofs for the atoms as such remained limited. The old experiments were more frequently referred to in writing than repeated or extended in the laboratory. During the first half of the seventeenth century attempts to widen the experimental basis were exceptional and hardly convincing, not even for contemporaries. Most of these attempts appear as rather tentative excursions from the better-established field of inorganic chemistry into the vegetable and animal kingdoms, mounted more or less for the sake of comprehensiveness and in order to display the universality of the corpuscular hypothesis. Sennert's treatment is symptomatic in this regard. After a lengthy presentation of chemical proofs in favor of the atoms he went on to argue that not only inanimate bodies were composed of atoms but also "some" animate bodies: "imò dantur atomi non solùm corporum inanimatorum, sed et animatorum quorundam"; atoms that were apt to carry, retain, and propagate the *anima,* the substantial form of a living body. Sennert was thinking of the seeds of plants and

---

[93] In a later revision of the manuscript Jungius replaced *metamorphosis* by *transmutatio;* see Jungius, *Praelectiones physicae* (cit. n. 11), p. 234, lines 24–26. *Transmutatio* was Jungius's general term for substantial alterations—be they by *syncrisis,* by *diacrisis,* or by rearrangement of particles; see *ibid.,* p. 70, line 24–p. 71, line 7.

[94] Allen G. Debus, "Fire Analysis and the Elements in the Sixteenth and the Seventeenth Centuries," *Ann. Sci.,* 1967, *23*:127–147. See also Gernot Rath, "Die Mineralquellenanalyse im 17. Jahrhundert," *Sudhoffs Arch.,* 1957, *41*:1–9; Allen G. Debus, "Solution Analyses prior to Robert Boyle," *Chymia,* 1962, *8*:41–61; and Noel G. Coley, " 'Cures without Care': 'Chemical Physicians' and Mineral Waters in Seventeenth-Century English Medicine," *Medical History,* 1979, *23*:191–214.

the sperm of animals, but also of spontaneous generation.[95] However, the indications of corpuscular structure in animate matter that Sennert presents are brief, conventional, and disappointing: the growth of tartar crystals in an entirely clear wine; the presence of chalk crystals in arthritic joints; the transmission of diarrhea to a suckling child whose mother had drunk milk from goats fed on laxative herbs; and finally the inhomogeneity of milk, blood, and bones. Sennert even includes the rather misplaced standard reference to the *acari,* which Aristotle had taught as being the smallest existing animals.[96]

In this context it would be worthwhile to compare the experimental approach of the early seventeenth-century "chemical" atomists with that of Robert Boyle. In Boyle's experiments "organic" material occupied a central place. In the "historical part" of his *Considerations and Experiments,* where the experimental background for his corpuscular philosophy was most fully presented, the first four "observations" were in fact experiments with animal and vegetable bodies: the development of a chicken in an egg, the growth of plants in sealed glass containers, the engrafting of a scion onto another plant, and the decay of a rotten cheese. All of these examples were aimed at demonstrating that qualitative change could occur in closed systems without any material additive and must therefore be explained by a change of texture within the corpuscular structure of matter. The corpuscles themselves, however, were taken for granted. The same applies to Boyle's subsequent ten "experiments," most of which were carried out with inorganic material. Their aim was to show how, by means of mechanical operations, the internal texture of a body could be altered and, by this means, a change of qualities and even true transmutations effected. Only the first experiment bears some resemblance to traditional "proofs" for the atoms. White camphor was dissolved in sulfuric acid to yield a reddish solution that lacked the characteristic smell of camphor. On addition of an excess of water, however, "the camphire that lay concealed in the pores of the menstruum, will immediately disclose itself, and immerse in its own nature and pristine form."[97] Yet Boyle's interpretation differed notably from the conventional one based upon the *reductio in pristinum statum.* In his opinion this cyclic process was meant to be a proof neither for the existence of imperceptible camphor corpuscles nor for their reception into the vacua of the liquid, which was taken for granted. For Boyle the experiment was an illustration of how mechanical action, such as dissolution, could alter heaviness, color, transparency, odor, fixity, and volatility of bodies because their intrinsic texture was altered. The properties of a body were seen as mechanical responses to outside objects and not as innate qualities whatsoever.

---

[95] Sennertus, *Hypomnema* 3.1.1. (cit. n. 23), p. 119. All of this is dealt with in great length in the subsequent *Hypomnemata,* IV ("De generatione viventium") and V ("De spontanea viventium generatione") (1636), in Sennert, *Opera* (cit. n. 23), Vol. I, pp. 123–172.

[96] Aristotle, *Historia animalium* 5.32 (557b8); and Sennertus, *Hypomnema* 3.1.1, p. 119.

[97] Robert Boyle, *Considerations and Experiments touching the Origin of Qualities and Forms* (1666), in *The Works of the Honourable Robert Boyle,* ed. Thomas Birch, 6 vols. (London, 1772), Vol. III, pp. 66–112, esp. pp. 76–112; quoting p. 76. Boyle's rendering of the camphor experiment is, by the way, quite correct. At room temperature camphor is easily soluble in sulfuric acid and can be precipitated again by means of an excess of water. The colors mentioned by Boyle, however, were evidently due to an impurity; see Thomas Stewart Patterson, E. F. M. Dunn, C. Buchanan, and J. D. Loudon, "The Rotation-Dispersion of Camphor, Camphoroxime, iso-Nitrosocamphor, and Oxymethylene Camphor," *Journal of the Chemical Society* (London), 1932, *134*:1715–1747, on pp. 1719–1721.

The difference between Boyle's mechanical corpuscular philosophy and the earlier "chemical" atomism is evident.

VI. CONCLUSION

It is difficult to believe that empirical arguments of the six types analyzed here were enough to convince those who did not already share the atomic view of matter. But since atomism was still a controversial issue, and its adherents maintained that it was rooted in experience, why then did they not devote much more effort to widening and strengthening the empirical basis? After all, there was little truly conclusive evidence in favor of the atoms that could not have been easily dismissed from a Peripatetic or Neoplatonic point of view. Writing in the 1660s even Boyle had to concede that "the intelligible [i.e., corpuscular] philosophy, . . . seems hitherto not to have so much as employed, much less produced, any store of experiments."[98]

As a matter of fact, the mere extrapolations from the visible to an underlying invisible reality were epistemologically questionable, to say the least. The phenomena of distillation, evaporation, growth of crystals, and so on, were appropriate for showing that something material was transferred from one place to another, but they did not prove its corpuscular nature. In addition, in almost all of these cases the experiences referred to were little more than variations on classical themes. Even when true experiments were carried out, they were often merely practical performances from a common repertory of literary paradigms. The corpuscular theories of light, magnetism, and electricity echoed the ἀπορροαί and ἀποφορά of Greek medicine without new experimental support. The corpuscular interpretation of rarefaction and condensation remained questionable, since the entire problem of how the atoms interact and cohere was open. The introduction of a material ether disposed of these difficulties while creating new ones by the strange hybridization of particulate and continuous matter. As far as experimental support for the corpuscular theory was concerned, the chemists and iatrochemists offered indeed the most convincing, yet still inconclusive arguments, based, to some extent, upon new facts, new techniques, and new methodologies.

By its very nature the chemical approach was pragmatic, realistic, and eclectic. The majority of chemists worked on real matter and real properties in a purposeful way. After all, they wanted to sell a product or to cure a patient. They simply could not afford to rely too closely upon a rigid theory. Needless to say, contrary to a stubborn historiographical myth, Boyle's clockwork universe "proved a sterile and occasionally adverse intellectual climate for an understanding of the processes underlying chemical change."[99] Hence it was not the mechanical philosophy that was to succeed in chemistry, but a noncommittal, substance-oriented notion of the corpuscle as something closer to an elementary particle or a small amount of substance, corresponding to something real in the chemists' vessels and furnaces and endowed with sensible properties. Jungius's

---

[98] Robert Boyle, "A Physico-Chymical Essay, containing an Experiment, with some Considerations touching the differing parts and redintegration of Salt-Petre," in *Certain Physiological Essays* (1661), in *Works*, Vol. I, pp. 359–376, on p. 375.

[99] Kuhn, "Robert Boyle" (cit. n. 5), p. 15; see also Boas, "Mechanical Philosophy" (cit. n. 1), pp. 494–499.

Figure 3. Robert Boyle (1627–1691). Courtesy of the Smith Collection and the Beckman Center.

maxim "si sensilia principia sufficiunt, quid opus est insensilia insuper sensilium rerum principia adsciscere?" epitomized a widely shared attitude.[100]

As in Baconian science, truth was not an ontological category but a social one, confirmed by utility; similarly, the undetermined atoms of the chemists were merely useful means for practical and explanatory ends, at best compatible with the experimental results, though not derived from them. Empiricism and realism were to prevail over the philosophically consistent atomism of the late seventeenth century. By the standards of Boyle's mechanical philosophy and John Locke's insistence on the epistemological status of the corpuscles, there was nothing of its kind in the eighteenth century. The new interest focused on elements and affinity, not on atoms and motion.[101] Authors of chemical textbooks relegated the corpuscular theory of matter to the introductory chapters of their works and had little recourse to it when they discussed the properties of sub-

---

[100] Jungius, *Praelectiones physicae* (cit. n. 11), p. 100, lines 11–13.
[101] Peter Alexander, *Ideas, Qualities and Corpuscles: Locke and Boyle on the External World* (Cambridge/London/New York: Cambridge Univ. Press, 1985); Boas, "Mechanical Philosophy," pp. 505–520; and Arnold Thackray, *Atoms and Powers: An Essay on Newtonian Matter-Theory and the Development of Chemistry* (Harvard Monographs in the History of Science) (Cambridge, Mass.: Harvard Univ. Press, 1970).

stances and the operations of chemistry. The particles were taken for granted, and their ontological and epistemological status did not even become a matter of debate. This noncommittal character enabled the resulting notion of corpuscle to assume whatever requirements future research would find convenient. The requirement of decisive proof or falsification by means of experiments and, what is more, the very question of the truth-value of the corpuscular view of matter were dismissed in favor of a merely operational link between theory and reality.[102]

It is not yet entirely clear by what exact mechanism the corpuscular theory, despite the obvious lack of experimental support, was able to win so many adherents among those who considered themselves empirical scientists after only a few decades of vigorous pros and cons. In any case, it would be mistaken to describe the steep rise of atomism as "a triumph of patient experimental research over metaphysical speculation,"[103] unless we admit that science proceeds by inferring correct theories from inadequate experiments. The acceptance of corpuscularianism cannot be reduced to a single cause, and least of all to the experimental progress of science alone. The arguments and rhetorical stratagems in defense of atomism operated, as we have seen, on many different levels simultaneously. They came from epistemological, mathematical, and empirical points of view, not to mention the theological and metaphysical ones. Their stratification, interdependence, and respective momentum need further study. The aim of this study was but to evaluate the more empirical grounds. They were rooted in the common heritage of ancient natural philosophy, but they also incorporated new experiences from the crafts tradition. Among them three lines of argumentation were especially powerful: (1) the new visual approach to reality, enabled by the recently invented microscope and based upon the bold hope that truth might be made visible by extended technical effort; (2) the readiness of practicing chemists and metallurgists to take material objects as a reality that needed no further ontological determination; and (3) the persuasive appeal of the pictorial scheme supplied by Lucretius's poetic imagery, which offered an immediately convincing way of picturing material processes on the basis of everyday experience within the visible world.

[102] Robert Boyle, *About the Exellency and Grounds of the Mechanical Hypothesis* (1674), in *Works* (cit. n. 97), Vol. IV, p. 77.
[103] Hooykaas, "Experimental Origin" (cit. n. 1), p. 79.

Robert Boyle and Structural Chemistry in the Seventeenth Century

By Thomas S. Kuhn

STUDIES of the impact of atomism upon seventeenth-century chemistry have usually emphasized the inevitable clash between any particulate theory of matter and the Aristotelian dictum that "if 'combination' has taken place, the compound *must* be uniform in texture throughout — . . . [Therefore] so long as the constituents are preserved in small particles, we must not speak of them as 'combined.'"[1] Such studies, by isolating the atomistic principle that macroscopic bodies are aggregates of stable microscopic corpuscles, have accentuated the benefit accruing to the theoretical chemist from a "corpuscular" metaphysic. Of course a philosophy which holds that simple particles do not perish in their compounds would create, it is said, an atmosphere favorable to the development of modern chemical theory.[2] But the complex of beliefs described as chemical atomism has normally included commitment to more than a particulate theory of matter, and although the simplification which treats all atomisms as mere particulate theories has illuminated portions of the history of chemistry,[3] it has also been misleading. Atomism has not invariably provided a fertile soil for the growth of chemical theory. More specifically, the form of atomism developed by

---

[1] Aristotle, *De Generatione et Corruptione* (Oxford Translation), 328ᵃ 8–12.

[2] Compare the more rhapsodic judgment of Ida Freund, who, in evaluating the role of atomism in chemistry prior to Dalton, states: "All previous advance in the establishment of clearer conceptions concerning chemical combination and chemical change must be considered as due to the use of a corpuscular theory of matter." *The Study of Chemical Composition* (Cambridge: at the University Press, 1904), p. 284. This evaluation of atomism recurs implicitly in many of the quotations below.

[3] Pierre Duhem exploits this simplification particularly effectively in *Le Mixte* (Paris: C. Naud, 1902), pp. 1–45. R. Hooykaas, The Experimental Origin of Chemical Atomic and Molecular Theory before Boyle, *Chymia*, 2, 65–80, 1949, consistently develops a similar viewpoint. By contrast L. Mabilleau distorts the perspective by occasionally inverting it; in his *Histoire de la philosophie atomistique* (Paris: Félix Alcan, 1895), pp. 384–396, he seems to imply that because alchemical writers sometimes indicated that elements endure materially in compounds, they had adopted the entire metaphysical doctrine of the classical atomists.

philosophers and applied to physics in the seventeenth century [4] embraced concepts inconsistent with the development of such fundamental chemical notions as element and compound. These impediments to chemistry are manifest in the chemical theory of the "corpuscular philosopher" Robert Boyle.[5]

Discussions of Boyle's chemical theories have customarily implied the opposite opinion that his atomism was conducive to a "modern" belief in the endurance of elements in their compounds and to the recognition of analysis and synthesis as fundamental tools of the working chemist. Masson, for example, diverges from a prevalent evaluation only in his enthusiasm:

> Boyle had, and habitually used, a *corpuscular* notion of matter; and to a very large extent it was derived from and supported by his own experiments. . . . He made a guess [i.e., atomism] as wide and free from arbitrary assumptions as it could consistently be, found it an inspiration as a working hypothesis, and exploited it with experiments as far as he could; and his interpretations of results are clearest to us when they are expressed in terms of it. . . . [With the aid of this hypothetico-experimental technique] Boyle dragged chemistry out of a welter of sophisms and charlatanry, endowed it with life and ideals and material, and set it forward upon the true path. Chemists can nowadays use terms as brief as "analysis," "synthesis," "chemical element," each of which covers great groups of phenomena; it was Boyle who had to hew out, to array, describe, and to assess these great groups, in order for there to be known anything to bestow names upon. . . .[6]

Similar judgments can be found in many of the early histories of chemistry which treat Boyle's work in any detail.[7] Contemporary histories [8] note the century separating

---

[4] Excellent discussions of various aspects of the development and conceptual application of atomism during the seventeenth century will be found in: K. Lasswitz, *Geschichte der Atomistik vom Mittelalter bis Newton* (Hamburg und Leipzig: Verlag von Leopold Voss, 1890), still the best history of atomism for the period it covers; F. A. Lange, *History of Materialism* (translated from the 3rd German ed. of 1873, London: Kegan Paul, Trench, Trubner & Co., Ltd., 1925); E. Meyerson, *Identity and Reality* (translated from 3rd French ed. of 1926, London: George Allen & Unwin Ltd., 1930), particularly chapters 2, 4, & 10; E. A. Burtt, *The Metaphysical Foundations of Modern Physical Science* (Revised ed., New York: Harcourt Brace & Co., 1932); M. Boas, The Establishment of the Mechanical Philosophy, to appear in *Osiris*, *10*, 1952. Dr Boas's discussion is particularly useful, and her monograph includes a convenient and comprehensive bibliography. I am much indebted to her for permission to examine the manuscript before publication.

[5] For good general discussions of Boyle's "corpuscular philosophy" see M. Boas, *op. cit.*, and, Boyle as a Theoretical Scientist, *Isis*, *41*, 261-268, 1950. Useful supplements are provided by: E. Bloch, Die antike Atomistik in der neueren Geschichte der Chemie, *Isis*, *1*, 377-415, 1913; C. T. Harrison, Bacon, Hobbes, Boyle, and the Ancient Atomists, *Harvard Studies and Notes in Philology and Literature*, *15*, 191-218, 1933; P. P. Wiener, The Experimental Philosophy of Robert Boyle, *The Philosophical Review*, *41*, 594-609, 1932; J. C. Gregory, *A Short History of Atomism from Democritus to Bohr* (London: A. C. Black Ltd., 1931). In this essay only a few special aspects of the "corpuscular philosophy" are considered.

[6] I. Masson, *Three Centuries of Chemistry* (London: Ernest Benn Ltd., 1925), pp. 70, 77.

[7] H. Kopp, for example, begins an analysis of Boyle's corpuscular theory with the words: "If one wishes to date the existence of chemistry . . . from the time at which its task was clearly represented as the accumulation of knowledge about how the heterogeneity of materials results from differences in their composition, and what the demonstrable constituents of differing bodies are, then one must salute Boyle as the first man conscious of such a task for chemistry and as the one who ushered in the period in which chemistry was pursued as a branch of natural science." *Beiträge zur Geschichte der Chemie* (Braunschweig: Friedrich Vieweg und Sohn, 1875), drittes stück, p. 165. Or compare E. von Meyer, *A History of Chemistry* (3rd English ed. translated from 3rd German ed., London: Macmillan and Co., Ltd., 1906), p. 111 f. Meyer relates Boyle's corpuscular theory of matter to his clarification of the notion of element and his unique ability "to draw a sharp distinction between mixtures and chemical compounds. . . . No one before him had grasped so clearly and treated so successfully the main problem of chemistry,— the investigation of the composition of substances." Again, Duhem (*op. cit.*, p. 17) views Boyle's application of the corpuscular theory of matter as the first source of "the notion of a *simple substance* such as that provided by Lavoisier and his contemporaries."

[8] For examples see the analyses of Boyle's work in: J. R. Partington, *A Short History of Chemistry* (2nd ed., London: Macmillan and Co., Ltd., 1948); and E. J. Holmyard, *Makers of Chemistry* (Oxford: at the Clarendon Press, 1931).

the apparent modernity of Boyle from Lavoisier's chemical revolution as well as the numerous equivocations in Boyle's hasty and voluminous writings, and so their evaluations are, properly, less categoric. It is said that: Boyle was not so clear and dogmatic as could have been wished for his time, or that: he himself was unable to evolve experimental methods of deciding whether or not a given substance is to be considered an element. But it is still felt that the corpuscular and experimental philosophies, as Boyle combined them, are the source of the first clear statement of the problems and concepts of modern theoretical and experimental inorganic chemistry; "Boyle apparently did not know how to use in his experimentation his own ideas . . . , but others made unconscious use of them until, more than a century after . . . [their] appearance . . . , Lavoisier brought them clearly to the surface of thinking, exploited them experimentally, and obtained their almost immediate and entirely general acceptance." [9] Boyle is no longer "the father of chemistry," but he might have been.

One particularly interesting extension of this common thesis was provided by the late Ernst Bloch.[10] He believed that Boyle's research was immediately followed by a simultaneous decline in the caliber of chemical experiment and theory, and he credited the inception and duration of this relapse to the influence of Newton:

In the same country where mechanical chemistry had achieved its greatest success through the work of Boyle and Mayow, it met an all-powerful scientific opponent. Newton, who was at heart sympathetic to mechanistic teachings, considered it impossible to explain gravity as a result of [mechanical] impacts. And since his unparalleled success in the mathematical explanation of gravitational forces held the center of his attention and of that of his contemporaries, . . . his work caused a reaction against the entire mechanical method. This so strongly affected chemistry that in England it was immediately directed to the new path. . . . In France, where the Cartesian school flourished, three decades were required before the new [occult] view prevailed in chemistry; then, however, chemistry was deprived of the mechanical method — and not to its benefit." [11]

The apparent decline of eighteenth-century chemical thought after Boyle had been noted by other historians,[12] and the necessity of explaining Boyle's failure to effect a lasting modification of chemical theory may be responsible for the current overestimate of the role of the phlogiston theory in determining the course of chemical conceptualization during the century separating Boyle and Lavoisier. Boyle's work, it is said, evoked little response, because his successors became immediately engrossed in the phlogiston theory of combustion; [13] theoretical chemistry lagged until Lavoisier, who, in creating the chemical revolution, independently recreated many of Boyle's fundamental chemical concepts. A second, more acute, analysis of eighteenth-century chemical thought is provided in a sketch by Meyerson [14] and in the penetrating studies

---

[9] T. L. Davis, The First Edition of the Sceptical Chymist, Isis, 8, p. 71, 1926. Davis subsequently modified his evaluation of the role of the definition of an element in Boyle's work, but this alteration does not affect the opinion quoted above (see note 59 and accompanying discussion below).

[10] Op. cit. Bloch, however, carefully notes that Boyle's chemical corpuscular theory was much more than a particulate theory of matter.

[11] Ibid., p. 402.

[12] Kopp, for example, notes (op. cit., p. 182): "If one consults the chemical textbooks which were best known and most influential at the end of the seventeenth century to discover the prevalent doctrines about chemical composition and the ultimate constituents of chemical compounds, one finds no trace of the opinions promulgated by Boyle; instead, doctrines whose errors had been seen by Boyle and even by van Helmont were retained."

[13] I concur in L. T. Davis's brief evaluation of the role of the phlogiston theory (Boyle's conception of Element compared with that of Lavoisier, Isis, 16, 82, 1931): "the phlogiston doctrine had for a long time but little effect upon the thinking of chemists. Boerhaave ignored it, and phlogiston became of importance only when it became an object of attack." Davis's significant remark is indirectly confirmed by J. R. Partington and D. McKie in their four definitive articles on the phlogiston theory (Historical Studies on the Phlogiston Theory, Annals of Science, 2–4, 1937–1939). As their citations show, the widespread discussion, development and application of the theory occurs almost entirely after 1770. The view that the phlogiston theory had a negligible role in the apparent deterioration of eighteenth century chemistry is directly opposed to a prevalent appraisal exemplified in Holmyard, op. cit., p. 143, and in H. Butterfield, The Origins of Modern Science (London: G. Bell and Sons Ltd, 1949), chapter XI.

[14] Meyerson, op. cit., chapter X.

by his student Hélène Metzger.[15] They portray the chemical revolution as proceeding not from a sudden break, like Boyle's, with the concepts of alchemy and scholasticism, but through an almost continuous extension and elaboration of peripatetic and iatrochemical concepts. Yet on either analysis Boyle's theoretical and programmatic suggestions lie outside the major tradition of seventeenth- and eighteenth-century chemistry. And so Boyle himself is often portrayed as an isolated "precursor" of Lavoisier, as a man who failed because the "time was not ripe."

But there is an alternative analysis of the impact upon chemical thought of the corpuscular philosophy as it was understood and developed by Boyle and a few of his contemporaries; and there is an associated alternative evaluation of Boyle's chemical concepts. On this view, to be delineated below, the same atomistic, mechanical metaphysic which led Boyle and his contemporaries to the first clear descriptions of the "clock-work universe"[16] and which provided so many fruitful new problems and new concepts to seventeenth century physics, proved a sterile and occasionally adverse intellectual climate for an understanding of the processes underlying chemical change. Boyle was not an isolated "precursor," but a man who brought to its most developed form a type of chemical conceptualization consonant with a major tendency of the scientific thought of his day. His failure to exert an important influence upon the future course of chemical theory was due, not to an inability to "fit" his scientific contributions to "the times," but to specific shortcomings of his chemical doctrines themselves, shortcomings which markedly differentiate his opinions from the later doctrines of Lavoisier. Boyle's views about elements and compounds are by no means so modern nor, for the seventeenth century, so advanced as they have sometimes been made to appear.

This essay will document the viewpoint sketched above. An examination of the important doctrinal differences between Boyle's corpuscular theory and the particulate theories of earlier seventeenth century chemists permits the discovery of the source of his divergence within the dynamical atomistic tradition of seventeenth century physical science. The effects of this "mechanical philosophy" upon Boyle's views of transmutation, the elements, and chemical instruments can then be examined successively. And the conclusion will indicate that Boyle's opinions on these controversial issues were one important cause of the rejection by many of his contemporaries and successors of major tenets of his theoretical chemistry.[17]

The application to chemistry of particulate theories of matter did not originate with Boyle. As Mabilleau and Hooykaas[18] have shown, numerous mediaeval alchemists as well as many renaissance chemists and physicians believed that the four Aristotelian elements (or the two or three alchemical principles)[19] endured as small (usually

---

[15] H. Metzger, *Les Doctrines chimiques en France du début du XVII* à la fin du XVIII* Siècle*, Tome I (Paris: Presses Universitaires de France, 1923); *Newton, Stahl, Boerhaave et la doctrine chimique* (Paris: Félix Alcan, 1930); *La Philosophie de la matière chez Lavoisier* (Paris: Hermann et C¹ᵉ, 1935).

[16] For the seventeenth century "clock-work universe" see particularly the works of Boas and Lange cited in notes 4 & 5.

[17] Boyle's works (excepting *The Sceptical Chymist*) will be cited and quoted from the five-volume folio edition of *The Works of the Honorable Robert Boyle*, edited by Thomas Birch (London: A. Millar, 1744). The source of quotations will be indicated parenthetically in the text: (II 375) refers to vol. 2, p. 375. For the convenience of readers *The Sceptical Chymist* will be cited in the more available Everyman's Library edition (London: J. M. Dent & Sons Ltd., 1911). Quotations from this edition will be indexed parenthetically: as (187) for Boyle's definition of an element on p. 187.

[18] *Op. cit.*, and see also R. Hooykaas, The discrimination between "natural" and "artificial" substances and the development of corpuscular theory, *Archives Internationales d'Histoire des Sciences*, 4, 640–651, 1948.

[19] The words "element" and "principle" will normally be used interchangeably below. This accords with the usage of Boyle, who finds the same arguments applicable to both and so normally distinguishes between them only historically (29).

indivisible) corpuscles in their compounds without modification of their essence or form. By the middle of the seventeenth century, through the writings of such men as Sennert, Basso, Magnen, Etienne de Clave, and Jung, particulate theories were widely known and vehemently defended.[20]

These particulate theories are not "modern." The chemical elements they employ are always very like those of the peripatetics and iatro-chemists, so it is difficult for a contemporary reader to discover the source of their authors' frequently violent rejections of older chemical concepts. But these writers do employ just those "modern" notions of element, compound, and corpuscular combination normally ascribed to the subsequent writings of Boyle. For Etienne de Clave "the elements are simple bodies of which all compounds are originally constituted and into which these compounds are, or can be, ultimately resolved." [21] Lasswitz [22] discovers in Sennert, as Wohlwill [23] had found in Jung, the same conceptual basis for the distinction between the atom (*minima*) of an element and the molecule (*prima mixta*) of a compound that is so frequently exploited by Boyle.[24] These are theories which justify the high appraisal of atomism sketched at the beginning of this essay: a compound is no longer required to be uniform "to the eye of a Lynx;" [25] elements endure unmodified in their compounds; and no element can be transmuted into another.

But this particulate view of matter is not Boyle's corpuscular theory. On the contrary, Boyle repeatedly criticizes these theories (though he applauds the experimental skills of their authors) without distinguishing them from those of the "chymists" or "peripateticks." [26] Explicitly he attacks the use by these early atomists of the Aristotelian or iatro-chemical elements and their continuing reliance upon "occult" qualities. But implicitly Boyle differentiates his constructive theory of chemical composition from those of his atomistic predecessors by a further novelty which he never ceases to emphasize. While particulate chemistry has a continuing precedent

---

[20] Lasswitz, *Atomistik*; Boas, Mechanical Philosophy; and G. B. Stones, The Atomistic view of matter in the XV, XVI, & XVII centuries, *Isis, 10*, 444-464, 1928; provide the most complete account of the doctrines of these chemist-philosophers. For Jung see especially, E. Wohlwill, *Joachim Jungius und die Erneuerung Atomistischer Lehren im 17. Jahrhundert* (Hamburg: L. Friedrischen & Co., 1887). Wohlwill's translation of Jung's disputations of 1642 has been separately reprinted with an introduction by A. Meyer: *Joachim Jungius, Zwei Disputationen über die Prinzipien (Teile) der Naturkörper* (Hamburg: 1928).

[21] *Nouvelles lumière philosophique des vrais principes et elemens de nature, & qualité d'iceux* (Paris: chez Olivier de Varennes, 1641), p. 39: "les elemens sont corps simples, qui entrent premierement en la composition des mixtes, & ausquels ces mixtes se resoluent, ou se peuuent resoudre finalement." And compare the similar definitions offered on pp. 27, 40, & 260. De Clave devotes three chapters (pp. 274-302) to proving that these elements actually exist permanently in their compounds, and he also insists that one can only discover the number and nature of these permanent elements by experience: "aucun n'a iamais bien cognu (& ne peut cognoistre) la composition des choses, que par leur resolution, qui est une reduction des corps composez en leurs principes" (p. 39). He recognizes that the oils obtained by distilling various natural products are distinct, but he states that such differences are due to impurities, and he describes a variety of laboratory distillation and fermentation processes which will effect the purification (pp. 46 ff.). For the similar concepts of Jung, see the quotation cited in note 52, below.

[22] *Op. cit.*, vol. 1, p. 449.

[23] *Op. cit.*

[24] See the discussion of Boyle's use of these concepts, below.

[25] Aristotle, *De Generatione*, 328^a, 16.

[26] It can plausibly be argued that Boyle's criticisms of the "chymists" are directed exclusively at those of his predecessors who employed a particulate theory. In one of the few passages in which he explicitly distinguishes the views of the "peripateticks" from those of the "chymists," the "peripateticks" are described as adherents of a continuous (Aristotelian) theory of matter, and the "chymists" are treated as proponents of the particulate theory derived from "the antient philosophers that preceded Aristotle." (79 ff.) Shaw, in a note to his translation of Boerhaave's *Elementa Chemiae*, remarks that: "the chemists . . . speak of *elements*, as the very primary corpuscles whereof mixt bodies are composed: a way of conceiving, which subjects them to infinite difficulties, and is the foundation of a good part of the objections made against them by Mr. Boyle." *A New Method of Chemistry* (Third Edition Corrected, London: T. & T. Longman, 1753), vol. 1, p. 158. Whether or not Boyle meant the particulate theorists when he spoke of the "chymists," he criticized the former extensively. His treatment of Sennert in *The Sceptical Chymist* is particularly revealing (147, 164, 166, 167).

since Galen, Boyle's theory derives from the radically different, "new" or "mechanical philosophy," which views inorganic phenomena as the manifestations of a dynamical atomism and the universe as a "*self-moving engine.*" (II 474) [27]

In consequence, as will be shown, Boyle derives the qualitative characteristics of natural substances, not from the permanent characteristics of the ultimate component corpuscles as had his predecessors, but from the manner in which the corpuscles (ultimate or not) are arranged and moved relative to each other in their compounds. Sennert and the earlier atomists correlated sensory or laboratory properties with the nature (or form) of the ultimate components; Boyle correlates them with the motions of the corpuscles and with the structure in which the relatively neutral components are arranged.[28]

That then, which I chiefly aim at, is to make it probable to you by experiments, (which I think hath not yet been done,) that almost all sorts of qualities, most of which have been by the schools either left unexplicated, or generally referred to I know not what incomprehensible substantial forms, may be produced mechanically; I mean by such corporeal agents, as do not appear either to work otherwise than by virtue of the motion, size, figure, and contrivance of their own parts, (which attributes I call the mechanical affections of matter, because to them men willingly refer the various operations of mechanical engines:) (II 459).

The novelty here is a novelty of means, not of ends. Boyle's ultimate objective remains that of his "peripatetick" and "chymical" predecessors. In spite of his reintroduction of the Epicurean distinction between primary and secondary qualities, he is as concerned as his opponents to explain the causes of qualities and to trace their

---

[27] Lasswitz and Wohlwill remark on the manner in which Sennert and Jung emphasize their participation in an ancient tradition; Boyle, on the contrary, emphasizes the novelty of the "corpuscular" opinion. These seventeenth-century authors are thus noting an important distinction which has too often escaped modern historians. Particulate theories of the structure of matter can be distinguished significantly, in their historical and philosophical development, from the cosmological dynamical atomisms of the ancient and modern Epicureans.

The unraveling of these two traditions is beyond the scope of the present essay. But it is worth remarking, tentatively and in the most general terms, that the particulate theories of matter (and the frequently associated theories of the vacuum) embodied in the medical writings of Galen and the engineering works of Hero of Alexandria were never entirely lost from view and exerted a marked influence upon European thought for at least a century before the reemergence of the developed metaphysical systems of the Epicureans. The particulate theories associated with neo-Platonic and Pythagorean philosophies are, perhaps, a third tradition which, in the sixteenth century, partially elided with the Galenic and Heronic theories. (Compare: C. de Waard, *L'Expérience barometrique* (Thouars: Dr J. Gamon, 1936), and M. Boas, *Hero's Pneumatica — A Study of its Transmission and Influence*, Isis, *40*, 38–48, 1949.)

The particulate theories are also philosophically distinct from the dynamical atomisms, for the second promoted (though they did not necessitate, logically or historically) a belief in an infinite universe, the breakdown of the role of absolute position in physics, and the recognition of a distinction between primary and secondary qualities. The particulate theories were compatible with a more nearly peripatetic metaphysic. A study of the emergence, after 1620, of true dynamical atomisms (Gassendi, Descartes, Boyle, Newton) from (or alongside of) the preexisting particulate theories of such disparate figures as Telesio, Bruno, Galileo, Sennert, Basso, etc. is a great desideratum.

[28] Jung is, in some respects, an exception to this generalization about the early seventeenth-century particulate chemists. In the first of the *Zwei Disputationen* he recognizes the possibility of a reaction, *metasynkrisis*, in which new qualities are produced by rearrangements alone. But he reserves this reaction for application to cases which defy explanation in terms of analysis or synthesis (*diakrisis* or *synkrisis*), and he is not certain that any such reactions will be discovered (§§ 73, 78).

Jung's conception of *metasynkrisis* is, despite Bloch (*op. cit.*, p. 400), quite distinct from Boyle's conception of change of quality by corpuscular rearrangement. Boyle sees rearrangement as a fundamental component of all chemical process, and he devotes much of his attention to the search for chemical and mechanical changes in which "there appears not to intervene in the patient or subject of the change, any thing but a mechanical alteration of the mechanical structure or constitution." This sort of alteration, Boyle feels, provides the best evidence for the corpuscular hypothesis, since: "if, by a bare mechanical change of the internal disposition and structure of a body, a permanent quality, confessed to flow from its substantial form, or inward principle, be abolished, and, perhaps, also immediately succeeded by a new quality mechanically producible; ... such a phaenomenon will not a little favor that hypothesis, which teaches, that these qualities depend upon certain contextures, and other mechanical affections of the small parts of the bodies, that are endowed with them." (III 567)

evolution within chemical reactions. As Meyerson has said: "During the seventeenth century the prestige of peripateticism, as a philosophical and scientific doctrine, diminishes little by little. But the theories of qualities which had sprung from them continue to dominate chemistry in a still more absolute form precisely because they are free from the purely logical equipment of Aristotle's doctrine." [29] Boyle's chemical researches can scarcely be understood without a realization of the surprising extent to which he exemplifies this important generalization. But though his aim is identical with that of the older chemical theorists, his conceptual tools are new to chemistry. Boyle will employ only the "mechanical philosophy." He will reduce all qualities to "matter and motion" (II 461); and his new chemistry will take its unique form from this reduction.

Boyle's insistence upon explanation in terms of a dynamical atomism places him in a major seventeenth century scientific tradition — a tradition deriving proximately from Bacon, Gassendi, and Descartes [30] and ultimately from the metaphysical writings of the Democritean and Epicurean philosophers. The admirable studies of Boyle's philosophy by Dr Boas make superfluous the elaboration of this thesis. But the extent of Boyle's debt to and freedom within an explicit tradition must still be considered in order to discover whether the seventeenth century's "climate of opinion" rather than Boyle's private aberration produced his new chemistry.

On the question of originality, Boyle himself is relatively clear and accurate. He claims it at only two points: as the originator of the phrase "corpuscular philosophy," [31] and as the first to apply systematic experimentation to the elucidation, elaboration, and verification of the dynamical atomism for which he had supplied a name.[32] To these might be added his novel but unfounded ontological reduction of the number of primary qualities to two: "matter and motion." [33] Boyle, who invented the title "corpuscular philosophy," claims no originality in the description of its doctrine. On the contrary, he represents himself as the spokesman for the corpuscularians as a group, and he conceives his mechanical-corpuscular philosophy as a least common denominator of ancient and modern Epicurean and of Cartesian opinion. In the introduction to *The Origin of Forms and Qualities*, the treatise which supplies the most systematic account of his scientific metaphysic, Boyle describes himself as one "who here write[s] rather for the Corpuscularians in general, than any party of them" (II 455), and he sees it as his task "not only to devise hypotheses and experiments, but to examine and improve those, that are already found out." (II 452)

The conception of his task as that of a publicist and experimentalist for the previously developed "mechanical philosophy" can be traced to the beginning of Boyle's public career. Even before the publication of *The Sceptical Chymist*, Boyle had proclaimed,

---

[29] *Identity and Reality*, p. 331.

[30] Because Descartes denied the existence of the void, he is frequently denied the title of atomist. But the historical and ideological reasons for describing Descartes' "particulate plenum" as atomistic are overriding. For bibliography and an outline of the case, see Boas, Mechanical Philosophy, chapter V. Bacon, too, is frequently denied the title, but see Boas, Boyle as a theoretical scientist, and Harrison, Bacon, Hobbes, Boyle.

[31] *Origin of Forms and Qualities* (II 454): "that philosophy, which I find I have been much imitated in calling Corpuscularian."

[32] See the last quotation in the text above. Boas has pointed out (Boyle as a Theoretical Scientist) that this objective is implicit in much of Bacon's programmatic writing.

[33] The complete reduction follows, for Boyle, from an elaboration of the following (II 461): "local motion seems to be indeed the principal amongst second causes, and the grand agent of all that happens in nature: for though bulk, figure, rest, situation, and texture do concur to the phaenomena of nature, yet in comparison of motion they seem to be, in many cases, effects, and in many others, little better than conditions, or requisites."

... notwithstanding these things, wherein the Atomists and Cartesians differed, they might be thought to agree in the main, and their hypotheses might by a person of a reconciling disposition be looked on as, upon the matter, one philosophy. Which because it explicates things by corpuscles, or minute bodies, may (not very unfitly) be called corpuscular; ...
[By this consideration] ... I was invited to try, whether, without pretending to determine the abovementioned controverted points, I could, by the help of the corpuscular philosophy, in the sense newly given of that appellation, associated with chymical experiments, explicate some particular subjects more intelligibly, than they are wont to be accounted for, either by schools or the chymists. (I 228)

Boyle takes the corpuscular philosophy as given. "I do not," he said at the beginning of his career, "expect to see any principles proposed more comprehensive and intelligible than the corpuscularian or mechanical." (I 374) [34] It is only the detail of specific corpuscular mechanisms that he expects to elicit from his experiments. His scepticism and distrust of philosophical system enables him to refuse lengthy dialectic about such metaphysical points as the infinite divisibility of the atom, and the existence of a Cartesian *materia subtilis*; his eclecticism allows him to diverge from both Descartes and Gassendi in developing the corpuscular mechanisms for heat, light, etc.; but neither his eclecticism nor his scepticism extends to doubts that some corpuscular mechanism underlies each inorganic phenomenon he investigates, and his conviction is prior to experiment. Boyle saw no contradiction in coupling the sentence: "I presume it will easily be taken notice of, that in the following history I have declined the asserting of any particular hypothesis, concerning the adequate cause of cold," with "I shall represent in the first place, that the account, upon which we are wont to judge a body to be cold, seems to be, that we feel its particles less vehemently agitated than those of our fingers." (II 239 f.)

◆

Boyle's faith in the corpuscular principles of the "mechanical philosophy" is the major source of his new emphasis in chemistry upon structure, configuration, and motion, as well as a cause of his rejection of explanations in terms of inherent characteristics of the ultimate corpuscles. For example, the analogy to the machine is a primary determinant of the emphasis upon structure or configuration. Boyle and many of his contemporaries regard the universe as an engine like "the formerly mentioned clock of Strasburg, ... [in which] the various motions of the wheels and other parts concur to exhibit the phaenomena designed by the artificer in the engine," (I 446) and they transfer this metaphysical conception from the macrocosm to the inorganic microcosm. Boyle believes that "we are not to look upon the bodies we are conversant with, as so many lumps of matter, that differ only in bulk and shape, ... but rather as bodies of peculiar and differing internal textures, as well as external figures: on the account of which structures many of them must be considered as a kind of engines." (IV 252) But the properties (qualities) of an engine

---

[34] Exaggerated estimates of Boyle's originality in the development of the corpuscular philosophy are sometimes based upon Boyle's often quoted remark (I 194) that he refrained from "seriously and orderly reading over" the works of Descartes, Gassendi, and Bacon, in order that he "might not be prepossessed with any theory." But in the same work (*Physiological Essays*, I 227 f.) he points out that having "purposely refrained from acquainting myself thoroughly with the intire system of either the Atomical, or the Cartesian, or any other whether new or revived philosophy" does not make him less fit to serve as spokesman for the corpuscularians, for "having divers years before read the lives of the Atomical, among other philosophers, in *Diogenes Laertius*; and having sometimes occasionally heard mention made of divers Epicurean and Cartesian notions, and having hence framed to my self some general, though but imperfect, idea of the way of philosophizing my friends esteemed; I thought I might, without a more particular explicit inquiry into it, say something to illustrate some notions of it. ..." Which is as categoric as could be wished. The absurdity of supposing Boyle unacquainted with the work of his predecessors has also been emphasized by Boas (*op. cit.*) and Gregory (*Short History of Atomism*), who point out that if Boyle only skimmed the major works of earlier authorities, his extensive citations of them display an enviable ability to absorb the gist of an argument while only "transiently consulting" (I 194) the text.

are dependent upon its structure, rather than upon the matter from which it is built; "in a watch, the number, the figure, and coaption of the wheels and other parts is requisite to the shewing the hour, and doing the other things, that may be, performed." (II 461) Qualities derive from the totality of parts in appropriate mutual relation.

Other facets of the mechanical philosophy reinforce the new emphasis on configuration. Since antiquity dynamical atomisms had been thought to imply the infinity of the universe and, in consequence, the impossibility of the physical efficacy of absolute position. The dynamical atomist therefore had to derive physical efficacy from relative position, that is, from configuration. Descartes, for example, in a typical passage dealing with the proper concept of translation notes:

a translation is made from the vicinity of one contiguous body to the vicinity of another and not from one place to another; because, as explained above, the signification of position varies, and depends upon our understanding. But when we understand motion to be translation from the vicinity of contiguous bodies, . . . we cannot attribute to a mobile more than one motion at a time.[35]

And Boyle is participating in the same conceptual aspect of atomism when he writes:

every distinct portion of matter, whether it be a corpuscle or a primary concretion, or a body of the first, or of any other order of mixts, is to be considered, not as if it were placed in vacuo, nor as if it had relation only to the neighboring bodies, but as being placed in the universe constituted as it is, amongst an innumerable company of other bodies. . . . (III 77)[36]

Finally this emphasis upon configuration is confirmed in the application of the mechanical philosophy to chemistry by the necessity of accounting for a greater variety of qualities than can easily be deduced from those mechanical affections, size and shape, which might inhere permanently in the corpuscles:

the multiplicity of qualities, that are sometimes to be met with in the same natural bodies, needs not make men reject the opinion we have been proposing, by persuading them, that so many differing attributes, as may be sometimes found in one and the same natural body, cannot proceed from the bare texture and other mechanical affections of its matter. For we must consider each body, not barely as it is in itself, an intire and distinct portion of matter, but as it is a part of the universe, and consequently placed among a great number and variety of other bodies, upon which it may act and by which it may be acted on in many ways. . . . (II 464)

The factors enumerated above as the source of Boyle's novel emphasis upon structure or configuration could be reiterated as determinants of the emphasis upon motion as a source of chemical qualities. In classical atomisms motion of the corpuscles was the cause of the perception and of the alteration of qualities, and in the mechanical corpuscular theories of the seventeenth century motion was a cause of quality itself as well as of perception and change of quality.[37] By Boyle's time this emphasis had been further reinforced by the contemporary concern with problems of celestial and terrestrial dynamics and by the search for laws of motion.[38]

---

[35] R. Descartes, *Principia Philosophiae* (Ultima Editio, Amsterdam: Daniel Elzevir, 1664), Pars Secunda, XXVIII.

[36] Compare Boyle's remarks to the same effect at II 464 f., 467, and at III 75, 82 ff. The insistence upon considering bodies in the universe "*constituted as it is*" is also typical of Descartes, after 1630, though not of the atomists generally. This emphasis is the source of one of Descartes' major quarrels with Galileo. (See, A. Koyré, *La Loi de la chute des corps* (Paris: Hermann et Cⁱᵉ, 1939), pp. 45-54 *passim*.

[37] Compare Descartes' statement (*op. cit.*, XXIII): "All the properties which we clearly perceive in . . . [matter], are reducible to the single attribute that it is divisible and movable by parts; and thus it is capable of all the affections which we perceive to follow from the motion of its parts," with Boyle's remark quoted in note 33, above, and with the quotation at the end of this section.

[38] Dynamics is almost the only one of the sciences pursued in the seventeenth century to which Boyle did not direct a special treatise. His interest in and knowledge of contemporary work in the field (excepting that of Hooke and Newton, to which, surprisingly, I find no reference) is, however, indicated by frequent discussions of results due to Galileo and Mersenne (e.g., III 178, 189), by his occasional references to the effects of "uniform" or "diform" or "elliptical" motions (III 78), and by a variety of references in his correspondence with Oldenburg (e.g., V 253, 312, etc.). A similar remark might be addressed to Boyle's knowledge of astronomy.

From these facets of the contemporary scientific metaphysic proceeds Boyle's constant concern to correlate qualities with corpuscular motions invisible to the senses. This concern culminated in the special tract on *The Great Effects of Even Languid and Unheeded Motion*, in which sub-sensory motions are shown to cause a number of special physical effects and in which Boyle asserts that:

whilst a whole body, or the superficies, that includes it, retains its figure, dimensions and distance from other stable bodies, that are near it, the corpuscles, that compose it, may have various and brisk motions and endeavors among themselves. (IV 266 f.) [39]

Enduring qualities of stable bodies may be attributed to enduring sub-sensory motions within the bodies, and occasionally such motions are considered the primary source of the secondary qualities, as in:

local motion hath, of all other affections of matter, the greatest interest in the altering and modifying of it; since it is not only the grand agent or efficient among second causes, but is also often times one of the principal things that constitutes the form of bodies. (II 471)

Such motions are for Boyle the necessary and sufficient causes of such physical phenomena as heat, and electrical and magnetic attractions; they also play, though less directly, a role in his chemistry.

The new emphasis upon structure and motion in Boyle's analysis of the qualities has marked effects on his specifically chemical doctrines; in particular it is a major determinant of his views on transmutation, the existence of chemical elements, and the role of instruments in the natural and artificial production of chemical species. These opinions, now to be examined in more detail, set Boyle apart from the major chemical tradition of his day, and they may plausibly account for the marked neglect of his theoretical chemical doctrines by his contemporaries and successors.

Professor Sarton and other contemporary scholars [40] concerned with seventeenth century chemistry have pointed out that the corpuscular philosophy provides a theoretical basis for a belief in the possibility of transmuting base metals to gold, a remark normally intended to explain or excuse this apparently archaic residue in Boyle's [41] (or Newton's) thought. The belief in the possibility of obtaining gold from base metals was widespread in this period (though many important chemists denied the possibility of such "multiplications"); [42] and such transmutations were readily reconciled with most of the mystical or rational chemical theories of the day, for the

---

[39] Boyle's conviction that sub-sensory motions endure and produce qualities first appears at a much earlier date. See the discourse about "Of Absolute Rest in Bodies," appended to the second (1669) edition of the *Physiological Essays*.

[40] G. Sarton, Boyle and Bayle, The Sceptical Chemist and the Sceptical Historian, *Chymia*, 3, 155–189, 1950; see particularly pp. 160 ff. R. J. Forbes, Was Newton an Alchemist?, *Chymia*, 2, 27–36, 1949. D. McKie, Newton and Chemistry, *Endeavour*, 1, no. 4, 1942. L. T. More, *Life and Works of the Honorable Robert Boyle* (New York: Oxford University Press, 1944), p. 221. More's evaluation of Boyle's views on transmutation most nearly coincides with the opinion developed below.

[41] Boyle's appropriately qualified belief in the possibility of obtaining gold by transmutation is expressed throughout his works. Typical examples of the evidence he employs to support his conviction are: the letter to Glanville (V 244) in which he states himself persuaded "upon good grounds, that, though most of these stories [of the transmutation of base metals to gold] be untrue, yet they are not all so," and in which he retails an account of a successful transmutation; his recurrent references to a *menstruum* he commands which will debase gold (e.g., 215); and the occasional explicit but fragmentary accounts of experiments which may eventuate in transmutation, particularly: An experimental discourse of quicksilver growing hot with gold, *Phil. Trans.*, 11, 515–533, 1675/6; and the tract on *The Degradation of Gold*. (IV 13–19)

[42] Compare Metzger, *Les Doctrines*, chapter 2. "Multiplication" is a transmutation in which a base metal is "seeded" with gold to generate a much larger quantity of the precious metal.

metals were normally considered to be closely related modifications of one or two of the elements sulphur and mercury. Boyle, on this analysis of his opinion, simply adopted the major chemical tradition of his day in the absence of adequate experimental evidence to the contrary.

But to make Boyle, in his views on transmutation, a child of his times is to miss that vital and indispensable novelty of corpuscular chemistry which Boyle continually proclaimed. Committed to deriving the secondary qualities of bodies from the relative positions and motions of their qualitatively neutral corpuscles, Boyle was bound to the conclusion that by sufficient rearrangement of positions and motions one could obtain, *not simply gold from lead, but anything from almost anything.*

since bodies, having but one common matter, can be differenced but by accidents, which seem all of them to be the effects and consequents of local motion, I see not why it should be absurd to think, that (at least among inanimate bodies) by the intervention of some very small addition or subtraction of matter, (which yet in most cases will scarce be needed,) and of an orderly series of alterations, disposing by degrees the matter to be transmuted, almost of any thing, may at length be made any thing: as, though out of a wedge of gold one cannot immediately make a ring, yet by either wire-drawing that wedge by degrees, or by melting it, and casting a little of it into a mould, that thing may be easily effected. And so though water cannot immediately be transmuted into oil and much less into fire; yet if you nourish certain plants with water alone, (as I have done,) till they have assimilated a great quantity of water into their own nature, you may, by committing this transmuted water . . . to distillation in convenient glasses, obtain, besides other things, a true oil, and a black combustible coal, (and consequently fire;) both of which may be so copious, as to leave no just cause to suspect, that they could be any thing near afforded by any little spirituous parts, which may be presumed to have been communicated by that part of the vegetable, that is first put into the water, to that far greater part of it, which was committed to distillation. (II 474)

Such a conviction is nearly unique among seventeenth century chemists;[43] indeed van Helmont appears to be the only major chemist who would agree with Boyle.[44] None of the earlier particulate theorists believed that one of the elements could be transmuted to another, and even those alchemists and iatro-chemists who did not hold

---

[43] The most important figures (excepting the Helmontians) who concur in some form of Boyle's doctrine that anything can be made of anything are the early Cartesians — particularly Descartes himself and Rohault. They produced no developed chemical theory; most of their remarks on chemical subjects are integral portions of their development of the Cartesian cosmogony, the mechanical evolution of the universe. They supposed that the various chemical species had arisen gradually through the mechanical fabrication of base matter, and they occasionally implied that transmutation by refabrication had virtually unlimited potentiality. (See Hartsoeker's criticism of the "Cartesian System," cited in note 88, below; and Metzger, *ibid.*, pp. 128 ff.)

[44] The partial but significant coincidence of van Helmont's, Boyle's, and Descartes' opinions about matter, transmutation, and the elements provides a riddle which could profitably be explored further. Van Helmont was anything but a dynamical atomist. Although many of his experiments were well known during the second half of the seventeenth century (Descartes apparently did not read van Helmont at all), his mystical, metaphysical, and methodological views were so totally alien to the members of the "mechanical" tradition that the direct influence upon them of van Helmont's *theoretical* writings was probably negligible. Except in a few of the experiments they employ as documentation (particularly that of the willow tree), the two traditions appear independent, at least after the middle of the century. Perhaps their parallel views about the malleability of base matter arose from a common concern to explain the creation. Van Helmont claimed the Book of Genesis as a primary source, and the dynamical atomisms of the seventeenth century were shaped by the necessity of avoiding the infinite temporal regress and the associated atheism of the original Epicurean theories. But this conjecture and the riddle from which it springs are beyond the scope of this essay.

On the parallelism of van Helmont and the mechanists see: Metzger, *ibid.*, p. 262 n.; Metzger's version of the views of K. Sprengler is not entirely accurate, so see his *Histoire de la Médecine* (translated from the German by A. J. L. Jourdan, Paris: Deterville and Th. Desoer, 1815), vol. 5, p. 51; also see D. McKie, Newton and Alchemy. On van Helmont's theories of matter and chemical change see: Metzger, ibid., pp. 165 ff.; J. R. Partington, Joan Baptista van Helmont, *Annals of Science*, 1, 359-384, 1936; and W. Pagel, *The Religious and Philosophical Aspects of van Helmont's Science and Medicine* (Baltimore: The Johns Hopkins Press, 1944). On Descartes' ignorance of van Helmont see: P. N. de Mévergnies, *Jean-Baptiste van Helmont, philosophe par le feu* (Paris: E. Droz, 1935), p. 21.

particulate theories of matter seem to have held that transmutations could occur only between closely related chemical species (e.g., within the metalline kingdom).[45]

Boyle is quite aware that his opinion is radical, but he considers this novelty one of the primary arguments against his contemporaries and in favor of the corpuscular philosophy. For example, after describing, in the *Origin of Forms and Qualities*, the transformation of water into a white powder, Boyle remarks to Pyrophilus:

And if you do acquiesce in what hath been already done, you will, I presume, think it no mean confirmation of the corpuscularian principles and hypotheses. For if, *contrary to the opinion, that is so much in request* among the generality of modern physicians and other learned men, . . . *the elements themselves are transmuted into one another*, and those simple and primitive bodies, which nature is presumed to have intended to be the stable and permanent ingredients of the bodies she compounds here below, may be artificially destroyed, and (without the intervention of a seminal and plastick power) generated or produced: if, I say, this may be done, and that by such slight means, why may we not think, that the changes and metamorphoses, that happen in other bodies, which are acknowledged by the moderns to be far more liable to alterations, may proceed from the local motion of the minute or insensible parts of matter, and the changes of texture, that may be consequent, thereunto? (II 521, italics added)

Such use of the *fact of transmutation* as an argument against the "chymists" recurs frequently in Boyle's writings, at least until 1682. It figures first in the *Physiological Essays*;[46] it provides a major thesis to *The Sceptical Chymist* (particularly "The Second Part," pp. 63 ff.), in which for reasons to be examined below Boyle multiplies examples of partial and complete transmutations; and it is repeated incessantly in the *Forms and Qualities* and in the *Mechanical Qualities*.[47] Further, the polemical utility of the transmutation reaction leads Boyle to focus much of his attention in the laboratory and in his writings upon reactions which he can persuade himself take place "without addition" or "without additaments."[48] Among these are the transmutation of water by plants and by heat; the generation of chicks, through a closed shell, from the white of an egg; and those purely chemical reactions, particularly the formation of metallic oxides, which Boyle can consider true transmutations.

The emphasis upon transmutation in Boyle's writing accords with a definite historical pattern. As Aristotle had established in the *De Generatione*, transmutation, that is the alteration without use of major additament of an apparently elementary substance, provides the best evidence for the dependence of qualities upon "forms," that is, upon modifiable, rather than upon enduring characteristics of base matter. Aristotle's classic example was the conversion of water to air by fire. And Boyle in correlating quality with configuration and motion had chosen again to attach it to

---

[45] Metzger, *ibid.*, points out that the most prominent argument employed in the seventeenth century to justify the continued pursuit of the transmutation of base metals was that the metals formed a family whose most developed member was gold. The alchemists therefore endeavored to complete quickly in the laboratory a slow process of natural maturation which had been accidentally interrupted. There was little thought in this period of making gold from non-metallic starting materials.

[46] "But that some of the principal of the hermetick opinions may be more handsomely accommodated by the notions of the Phoenician [corpuscular] hypotheses, than by the common philosophy of elements and substantial forms, . . . may appear from hence, that whereas the schools generally declare the transmutation of one species into another, . . . to be against nature, and physically impossible; the corpuscular doctrine . . . seems much more favorable to the chymical doctrine of the possibility of working wonderful changes, and even transmutations." (I 229 f.)

[47] "And as for those Spagyrists, that admit, as most of them are granted to do, that all kinds of metals may be turned into gold, by a very small proportion of what they call the philosophers elixir, one may, I think, shew them, from their own concessions, that divers qualities may be changed, even in such constant bodies as metals, without the addition of any considerable proportion of the simple ingredients, to which they are wont to ascribe those qualities; provided the agent, (as an efficient rather than as a material cause,) be able to make a great change in the mechanical affections of the parts whereof the metal it acts upon is made up." (III 600)

[48] Significant phrases of this sort recur throughout Boyle's writings. For typical contexts see above, or *The Sceptical Chymist*, pp. 176 ff.

a "form" [49] (though not an occult one), and so he had returned to the study of transmutations, the original source of evidence for the opinion.

By the seventeenth century this was an appreciable degeneration in chemical conceptualization. Most of the working "peripateticks" and "chymists" whom Boyle criticized had departed from the original metaphysical Aristotelian position, though many retained the vocabulary. The important "forms" had been "substantialized"; in most reactions they were attached permanently to matter. The mystical chemists of the day required at least a material seed as the vehicle for the chemical qualities, and the growing school of particulate chemistry was grounding "forms" permanently in eternally immutable material particles. They had ceased to expect many major qualitative changes without correspondingly large increments (or decrements) of matter, and they sought to explain change without recourse to transmutation. Jung "taught that if the evidence for the intra-conversion of the elements vanishes, so too does the Aristotelian proof of the existence of primary matter, i.e., of matter free from all substantial and accidental forms," [50] and he accordingly reserved *metasynkrisis* for the as yet undocumented transmutation reaction. And Sennert was considered to have made a major contribution to the new chemical tradition when he pointed out (though he was not the first) that the action of fire on water produced water-vapor rather than air, and so eliminated a reaction which had constituted primary evidence for the peripatetic doctrine.[51]

Boyle, though he frequently employs against the "chymists" as a group many arguments developed by the more recent particulate school, is in fundamental disagreement with their fundamental tenet. For example, Jung, some of whose work was known to Boyle, defended the opinions that:

It is not true, that all sublunary bodies (i.e. the four elements) can be transformed one into the other.

All bodies are not compounded of the usual four elements.

We agree with Sennert that there are other Principles of compound bodies besides the (4) elements and that the analysis of compounds does not always [immediately] produce the primary elements.

The axiom which maintains that every natural body is compounded of whatever substances it can be [immediately] resolved into is false;

But it is entirely true that every body is compounded of those substances into which it is ultimately analysed.[52]

Boyle employs the second, third, and fourth of these theses repeatedly, particularly in *The Sceptical Chymist*. But since he is in qualified disagreement with the first thesis (see above) and in unqualified disagreement with the last (see below), the significance of the intervening arguments is, when they occur in Boyle's writings, almost purely dialectic. Boyle is again widening the range of possible transmutations, which his particulate predecessors had narrowed.

◈

Of course Boyle does not believe that every reaction is a transmutation. He employs in his discussion of chemical reactions two sorts of invisible corpuscles: the "very solid" *minima naturalia* which though "mentally, and by divine omnipotence divisible, . . . nature doth scarce ever actually divide," and the *prima mixta* which are formed as primary concretions from the *minima* and which "remain intire in great variety of

---

[49] Boyle's decision to correlate the qualities with modifiable characteristics of matter is an implicit consequence of his atomism, but Bacon, from whom Boyle borrowed so liberally, made the same correlation entirely explicit. The Baconian "Forms" are precisely specified as the invariable correlates of individual qualities, and an important reason for the study of "Forms" is, for Bacon, that it will enable man to superinduce in matter the qualities to which the individual "Forms" correspond. (See the Novum Organum, Book II, Aphorisms II–V.)

[50] Quoted by Wohlwill from a student of Jung's; *Joachim Jungius*, p. 29.

[51] Lasswitz, *Atomistik*, vol. 1, p. 445.

[52] Wohlwill, *op. cit.*, p. 27. The quotation is again taken from a student of Jung's, and it is employed here because of the convenience of the formulation. The same views are expressed repeatedly in the *Zwei Disputationen*.

sensible bodies, and under various forms or disguises." (II 470 f.) The *minima* are never directly exemplified in nature, but the *prima mixta* play the role of the elementary atoms or molecules of various naturally occurring bodies (gold, silver, mercury, sulphur, etc.). In many reactions change of quality is associated either with a rearrangement of the *prima mixta* (mercury to mercury oxide and vice versa) or with the secondary union of the *prima mixta* of two relatively elementary substances (synthesis and analysis of the mercury sulphides).

So Boyle does recognize reactions in which one or more substances are preserved in spite of a drastic change of qualities; he uses the concepts of compound and of analysis and synthesis employed by his predecessors. But for him these reactions are distinguishable, though only with difficulty, from the reactions in which the *prima mixta* are themselves recompounded,[53] and this second sort of reaction is the more basic. Boyle normally treats the *prima mixta* as stable and indivisible only with respect to the particular class of reactions or laboratory manipulations with which he is then concerned; with respect to other reactions or manipulations these same *prima mixta* may be considered as themselves compounded of the (compound) corpuscles of other natural bodies.[54] And since even the most stable and elementary natural bodies (like gold) are compounded of *prima mixta*, which are only relatively stable, rather than of the *minima*,[55] there is scarcely a conceivable limit to the number of bodies which can be generated from an apparently elementary and homogeneous starting material:

I consider that . . . there are few bodies whose minute parts stick so close together, to what cause soever their combination be ascribed, but that it is possible to meet with some other body, whose small parts may get between them, and so disjoyn them; or may be fitted to cohere more strongly with some of them, than those some do with the rest; or at least may be combined so closely with them, as that neither the fire, nor the other usual instruments of chymical anatomies will separate them. These things being premised, I will not peremptorily deny, but that there may be some clusters of particles . . . [which are not broken or distorted in any known compounds or by any known agents. But in the more usual case, above,] two thus or continue unraised *per modum unius*, (as they speak) or as one entire corpuscle. As in a corpuscle of sal-armoniac, whether it be a natural or factitious thing, or whether it be perfectly similar, or compounded of differing parts, I look upon the entire corpuscle, as a volatile portion of matter; and so I do on a corpuscle of sulphur, though experience shews, when it is kindled, that it has great store of acid salt in it, but which is not extricated by bare sublimation: and so colcothar of vitriol falls under our consideration, as a fixed body without enquiring what cupreous or other mineral, and not totally fixed parts, may be united with the earthly ones; since the fires, we expose it to, do not separate them." (III 610)

---

[53] For example, in the *Particular Qualities* Boyle introduces the notion of compounding as a tenth distinct means by which the corpuscular philosophy can account for change of quality: "And as those conventions of the simple corpuscles, that are so fitted to adhere to, or be complicated with one another, constitute those durable and uneasily dissolluble clusters of particles, that may be called the primary concretions or elements of things: so these themselves may be mingled with one another, and so constitute compounded bodies; and even those resulting bodies may by being mingled with other compounds, prove the ingredients of decompounded bodies; and so afford a way, whereby nature varies matter, which we may call (10) mixture, or composition; *not that the name is so proper as to the primary concretion of corpuscles; but because it belongs to a multitude of associations, and seems to differ from texture*, with which it hath so much affinity, as perhaps to be reducible to it, in this, that always in mixtures, but not still in textures, there is required a heterogeneity of the component parts." (III 77, italics added)

[54] ". . . in the following notes about volatility and fixedness, when I speak of the corpuscles, or minute parts of a body, . . . [I mean] such corpuscles, whether of a simple, compounded, or decompounded nature, as have the particles they consist of so firmly united, that they will not be totally disjoined, or dissipated, by that degree of fire or heat, wherein the matter is said to be volatile, or to be fixed. But these combined particles will, in their aggregate, either ascend,

[55] Just what qualities Boyle imagines would be exhibited by an aggregation of the *minima naturalia* not further grouped into *prima mixta* is difficult to determine, for he never takes any natural substance to be of this sort. But a hint may be provided by an otherwise obscure passage in *The Sceptical Chymist* (180) in which Boyle, while discussing the incompleteness of the resolution achieved by fire, states: "if in chymical resolutions the separated substances were pure and simple bodies, and of a perfect elementary nature; no one would be indued with more specifick vertues, than another; and their qualities would differ as little as do those of water." This is Aristotle's base matter again, illustrated with the same example used by Aristotle in describing ultimate homogeneity (*De Generatione*, 328^a, 12).

combining corpuscles losing that shape, or size, or motion, or other accident, upon whose account they were endowed with such a determinate quality or nature, each of them really ceases to be a corpuscle of the same denomination it was before; and from the coalition of these there may emerge a new body, as really one, as either of the corpuscles was before they were mingled, or, if you please, confounded. (87 f.) [56]

But the best evidence of the mutability of the *prima mixta* is obtained by a comparison of Boyle's descriptions of various chemical reactions. It is the prevalence of such mutations that permits Boyle, in the *Fluidity and Firmness* (as elsewhere), to treat elementary water as composed of its own "watery" corpuscles, and to maintain simultaneously that under the action of plants this same water is transformed without additaments to four or more other substances each of which is an aggregate of its own *prima mixta*. The same remark applies to the corpuscles of sulphur, which enter into the sulphides of mercury as undistorted entities, but which are also fabricated from the *prima mixta* of antimony (45 ff., Boyle's elementary "antimony" is actually antimony sulphide).

Though Boyle, to permit the corpuscular philosophy to explain a greater variety of qualitative changes, uses concepts very like the analysis and synthesis of substances composed of relatively stable corpuscles, these are not such basic mechanisms of reaction in his chemistry as they were in that of Jung or Sennert or de Clave. And so the presence of these mechanisms does not qualify his belief in the thesis that anything can be made of anything; they simply make certain transformations more difficult. Boyle does *not* believe that everything can be made *immediately* of anything.

~§

The chemical theory described above is incompatible with belief in the existence of enduring elements. It is therefore surprising to find Boyle so often regarded as a man seen at his best in his demolition of the Aristotelian "elements" and the Paracelsan "principles," in place of which he substituted that definition of an element which is now universally adopted.[57] Such evaluations of Boyle's famous definition of a chemical element [58] are not so frequent now as they have been. They were particularly cogently criticised by the late T. L. Davis in an article which concludes "that 'The Sceptical Chymist' was not written for the purpose of setting forth a conception of element — and that the conception which it does set forth was not original with Boyle." [59]

Davis's conclusion can be criticized only for its reserve. Boyle intends to, and

---

[56] Compare: "the corpuscles, into which a body is dissipated by the fire, may by the operation of the same fire have their figures so altered, or may be by associations with one another brought into little masses of such a size and shape, as not to be fit to make sensible impressions on the tongue." (209) Other references to the breaking of *prima mixta* will be found in *Forms and Qualities* (e.g., II 471 ff.) and elsewhere in the works, but for the most part this process is described implicitly.

[57] Boas (Mechanical Philosophy) is, so far as I know, the only historian to point out explicitly that Boyle did not believe in the existence of elements.

[58] This may be an opportune moment to correct an error found in certain recent literature dealing with Boyle. L. T. More states (*Life and Works*, p. 244, text and note) and Sarton (Boyle and Bayle, p. 166) concurs that Boyle's frequently quoted definition occurs first in the second edition (1679/80) of *The Sceptical Chymist*. More's error is apparently due to his confusing "The Sixth Part" of the first edition, subtitled "A Paradoxical Appendix . . . ," with the extra appendix added in the second edition. The definition, as normally quoted, occurs in "The Sixth Part" of the 1661 edition, p. 350. I am indebted to Professor A. B. Lamb of Harvard for permitting me to compare the text usually quoted with his copy of the first edition.

[59] T. L. Davis, Boyle's Conception of Element, p. 82. Unfortunately Davis tries to show that Boyle's definition is not only the one most prevalent in the seventeenth century but also the one employed in antiquity by Aristotle and Thales, a thesis which is not supported by the Aristotelian texts. But, in spite of the counter argument which L. T. More (*ibid.*, p. 253) seems to base upon a partial quotation, Boyle's definition is equivalent to the one Davis quotes from Digby. The criticism which can perhaps be levelled at Digby's definition — its lack of explicit relation to experiment — cannot be employed against the previously quoted explicit and implicit definitions provided by de Clave and by Jung.

should, be taken entirely literally when he introduces his now famous definition with the phrase: "I now mean by elements, *as those chymists that speak plainest do by their principles* . . ." (187, italics added. It is surprising to find the italicized phrase replaced by ". . ." when the definition is quoted.) And, as Davis maintains, *The Sceptical Chymist* was not composed as a prelude to the reiteration of an old definition which Boyle did not know how to apply to the practice of chemistry. But this does not imply (as Davis does) that the book had nothing novel to say about the prevalent notion of an element. On the contrary, one of Boyle's objects in composing the work was to cast doubt upon the possibility of discovering any natural substance or group of substances which could fulfill the definition.

Until almost the end of *The Sceptical Chymist*, this thesis remains implicit. In the first five chapters Boyle restricts himself to demolishing explicitly the claim to the title "element" of a variety of natural substances which his opponents consider elementary. But Boyle's readiness to challenge the propriety of calling each and any *specific* substance an element and his continuing emphasis upon transmutations, particularly upon productions *de novo* of apparently elementary bodies (e.g. 64, 160), implies destructive doubt of the conceptual foundations from which the definition of element derives. The cumulative impact of such sentences, experimentally documented, as: "a portion of matter may, without being compounded with new ingredients, by having the texture and motion of its small parts changed, be easily, by the means of the fire, endowed with new qualities, more differing from them it had before, than are those which suffice to discriminate the chymists principles from one another" (145 f.), must surely discredit any conceptual scheme which assumes that chemistry will ever discover elementary and immutable substances.

In any case, this was the intended effect of the argument of *The Sceptical Chymist* upon the personages of the dialogue itself. At the end of the book, after Carneades (who speaks for Boyle throughout the dialogue) has concluded his attack upon the explicit doctrines of his predecessors, his "chymical" opponent, Eleutherius, declares for himself and for friends "who had likewise taken notice of the same thing": "I halfe expected, Carneades, that after you had so freely declared your doubting, whether there be any determinate number of elements, you would have proceeded to question whether there be any elements at all." And Carneades replies that: "it is not absurd to doubt, whether, for ought has been proved, there be a necessity to admit any elements, or hypostatical principles, at all." (186) The definition of the element (as used by those "who speak plainest") is then provided "to prevent mistakes," and Carneades proceeds to argue that chemists can profitably dispense with the concept of element altogether.[60]

At the conclusion of the first edition of *The Sceptical Chymist* Boyle leaves the issue of the existence of elements in doubt. But the equivocation need not reflect Boyle's personal doubts, for, as he declares throughout the book, his purpose is not to proclaim his "own opinion on the subject in question . . . , but only to shew you that neither . . . [the peripatetick nor the chymical] doctrines hath been satisfactorily proved

---

[60] The present interpretation of Boyle's intent in "The Sixth Part" of *The Sceptical Chymist* is not entirely consistent with the sentence Boyle employs in closing his definition: "now whether there be any one such body to be constantly met with in *all, and each, of* those that are said to be elemented bodies, is the thing I now question." (187, italics added) The difficulty is presented by the italicized words and would be removed if they were omitted, as they are by Shaw, in his abridgment. (*The Philosophical Works of the Honourable Robert Boyle Esq., Abridged, Methodized, . . . by Peter Shaw,* M.D. (2nd ed., London: W. Innys and R. Manby, 1738), vol. 3, pp. 336 f.)

Apparently the difficulty is due to Boyle's cavalier habits of composition, for the present form is inconsistent with the rest of the book. If the italicized phrase is included, the issue raised is one that has been repeatedly discussed earlier and answered categorically in the negative. Also the present form does not provide a "paradox" and is not equivalent to the truly paradoxical issue Boyle raised immediately before the definition (see above).

by the arguments commonly alledged on its behalfe." (16) In any case, Boyle's opinion is made successively clearer, if never completely unequivocal, in his later works.

In *The Origin of Forms and Qualities* (1663) he repeated his earlier criticisms of the "chymists," elaborated the mechanisms of corpuscular chemistry without recourse to "elementary bodies," and provided the qualified proclamation that anything can eventually be made from anything. Seventeen years later he delineated his views even more precisely by attaching to the second edition of *The Sceptical Chymist* a new appendix titled "Experiments & Notes about the Producibleness of Chemical Principles." In this tract Boyle undertook to show that many of the substances best qualified for the title elements could, in fact, be produced by transmutation from a variety of other elementary starting materials. And he considered this an important demonstration because: "if the bodies they call principles be produced *de novo* how will it be demonstrable, that nature was obliged to take those principles made ready to her hand, when she was to compound a mixt body?" (I 376) Lest doubt of his opinion remain, Boyle announced that he had intended to follow this appendix with another called "*Doubts whether there be any Elements, or material Principles of mixt bodies, one or more, in the sense vulgarly received.*" (I 373) [61]

Boyle may not have felt sure that no enduring elements could ever be discovered. He did feel that no such substances had yet been discovered and that the search for them would not engender chemical progress. Accordingly much of his effort as an experimentalist and tract writer was directed to demonstrating that any substance eligible for the title of element (because obtained by resolution and not further resolvable) could be transmuted or corrupted to some other apparently elementary body or bodies. This aim of his research may plausibly provide a rational motivation for his continuing efforts to "debase" gold or to obtain it by transmutation. Gold did satisfy the prevailing definition of an element, and Boyle was unable to alter it sufficiently to inhibit its immediate recovery. So Boyle could and did explain all the standard reactions of gold as simple transpositions or combinations of its *prima mixta* without distortion or division. Gold had the characteristics of an element; it was a continuing challenge to Boyle's view that anything can be made of anything. Boyle never relinquished the experimental effort to make gold conform to his scheme.[62]

This conjectural reconstruction of Boyle's motive in striving to transmute gold receives occasional confirmation from the manner in which the hints about his "*menstruum*" and his experiments on "degradation" enter the chemical writings. For example, in "The Sixth Part" of *The Sceptical Chymist*, after Carneades has argued against the existence of any elements, Eleutherius attempts to use gold as a counter example:

I must make bold to try whether you can as luckily answer your own arguments, as those of your antagonists, I mean (pursues . . . [Eleutherius]) that part of your concessions, wherein you cannot but remember, that you supplyed your adversaries with an example to prove that there may be elementary bodies, by taking notice that gold may be an ingredient in a multitude of differing mixtures, and yet retain its nature, notwithstanding all that the chymists by their fires and corrosive waters are able to do to destroy it.

---

[61] Boyle also appended to the *Icy Noctiluca* (1681/2) a brief tract called: "A Chymical Paradox . . . Making it Probable, That chymical principles are transmutable; so that out of one of them others may be produced." This tract was originally intended for the second edition of *The Sceptical Chymist*, and it opens with the words: "I adventured many years ago, in the *Sceptical Chymist*, to lay down some reasons of questioning, whether the fire be the true and proper instrument of analysing mixed bodies, and do but dissociate their principles or ingredients, without altering or compounding them anew. But I shall now present you a discovery, that will perhaps make you think the vulgar opinion of chymists to be less fit to be doubted of, than rejected." (IV 90)

[62] Bacon had again preceded Boyle in selecting the production of gold to exemplify the crowning achievement of the scientist who understood the nature of latent corpuscular mechanism. (*Novum Organum*, Book II, Aphorisms IV & V. See also note 49, above.)

I sufficiently intimated to you at that time (replies Carneades) that I proposed this example, chiefly to shew you how nature may be conceived to have made elements, not to prove that she actually has made any; ... But ... to answer more directly to the objection drawn from gold, I must tell you, that though I know very well that divers of the more sober chymists have complained of the vulgar chymists, as of mountebanks or cheats, for pretending so vainly, as hitherto they have done, to destroy gold; yet I know a certain menstruum (which our friend [Boyle] has made, and intends shortly to communicate to the ingenious) of so piercing and powerful a quality, that if notwithstanding much care, and some skill, I did not much deceive myself, I have with it really destroyed even refined gold, and brought it into a metalline body of another colour and nature, as I found by tryals purposely made. (215 f.)

The *menstruum*, to which Boyle alludes throughout his chemical writings, was never perfected to his complete satisfaction; until the end of his life his references to it remain vague and fragmentary. But his purpose throughout may well have been the one so explicitly stated above: "to answer more directly to the objection drawn from gold" to the thesis that there are no elements in nature. Surely this conjectural reconstruction of his motive is more creditable to Boyle than the more usual apologia for his excessive concern to discover the "elixir" of the alchemists.

Whether creditable or not, Boyle's continuing emphasis upon metalline and non-metalline transmutations is symptomatic of an analysis of chemical change which hindsight proves unsuitable to the promotion of rapid progress in seventeenth-century chemistry. And the disadvantages, isolable in retrospect, of the corpuscular philosophy as a metaphysical source of direction in the choice, evaluation, and analysis of experiments are not exhausted by the discussion of Boyle's views on transmutation and the elements.

For example, reactions displaying the transmutation of apparently homogeneous starting materials through the action of some external agent like fire are of particular interest to Boyle both because they document the fundamental tenet of the corpuscular philosophy and because they are the types of reaction which most clearly display the specific micro-mechanical correlates of individual qualities. Committed to the quest for the corpuscular structures and motions underlying the multiplicity of particular qualities, Boyle was constrained to emphasize those reactions in which a maximal change of quality is produced by an *efficient cause*, an agent, without the major intervention of a *material cause*, an additament. (The distinction is Boyle's; see note 47, above.) These are, for a corpuscular philosopher, the simplest reactions — the ones from which the specific nature of a structural alteration can with most probability be inferred. They are, almost of necessity, emphasized at the expense of the apparently more complex reversible analyses and syntheses which had been so closely correlated with the belief in enduring elements.[63]

The high pragmatic value of the reaction which displays large qualitative change under the action of an external agent or instrument is nowhere made entirely explicit in Boyle's writings. But it can be inferred with great plausibility from an examination of almost any sample of his chemical writings. And it has several facets besides the emphasis upon transmutation already noted.

The evaluation is, for example, implicit in Boyle's recurrent use of mechanical examples (which he does not always regard as metaphorical) to document chemical generalizations for which adequate chemical evidence is lacking. To show that "there may be many changes, as to quality, produced in a body without visibly adding, or

---

[63] Hooykaas, "natural" and "artificial" substances, indicates the historical and philosophical interdependence of the belief in enduring elements and of the study of the reversible reaction. See also his paper, Atomic and Molecular Theory before Boyle.

taking away any ingredient, barely by altering the texture, or the motion of the minute parts it consists of:" (III 75) Boyle cites the increased ductility produced in silver by hammering, the magnetization of a needle by the magnetic effluvia of a loadstone, and the production of red fumes from spirit of nitre in a sealed vial by the action of the focused rays of the sun. All three phenomena are considered analogous: change of quality is produced by an efficient agent of rearrangement, that is, by the hammer, the particles of effluvia, and the particles of sunlight. This is the model reaction.

In another more revealing passage Boyle endeavors to show: "That a body by a mechanical change of texture may acquire or lose a fitness to be wrought upon by . . . unheeded agents . . ." The effect of fire in making tartar deliquescent provides his first example, but he notes an important shortcoming in the experiment and replaces it with a more perfect exemplar:

But in regard that to make the change the greater, part of the tartar must be driven away by the fire, I shall rather make use of an[other] example . . . ; for having taken a loadstone, and . . . heated it and cooled it, though it had lost so little by the fire, that the eye took no notice of its being changed either as to shape or bulk, yet the operation of the fire, . . . did so diversly alter the disposition of it in reference to the magnetic effluvia of the earth, that I could presently and at pleasure change and realter the poles of the stone. . . . (III 86 f.)

In both the physical and chemical alteration (which Boyle does not, of course, see as distinct species) the fire is the efficient cause of rearrangement. But the second example is, for Boyle, the better demonstration, because, in the absence of material alteration, additament or decrement, the effect of the instrument is least obscured. Still, Boyle does not believe that loss of tartar in the first experiment destroys the analogy. On the contrary, it is simply an accidental effect of the fire which he does not know how to eliminate; the two alterations are, in essence, the same.

In the experiment with tartar, fire assumes its typical role in Boyle's chemistry, the role of an instrument.[64] Boyle's first criticism of the "chymists" had been directed at their assumption that "nothing can be obtained from a body by fire which was not pre-existent in it," (24) and he develops this criticism at length in his other chemical writings. As a destructive corrective to his opponents' naivete, Boyle's objection is unexceptionable. But his constructive alternative makes fire an instrument which, appropriately applied, can fabricate novel substances virtually independent of the starting material. For example:

. . . [probably] salts may be produced *de novo*, . . . [when] by the action of the fire or other fit agents, small portions of matter may be so broken into minute parts, and these fragments may be so shaped and connected, as, . . . to compose a body capable of being dissolved in water, and of affecting the organs of taste [as a salt does]. (I 377)

The conception of fire as an instrument which, in the hands of a skillful chemist, fabricates a new substance not pre-existent in the starting material provides a directive hypothesis, not a dogma, for Boyle's analysis of experiments. He is aware that in many distillations fire simply separates the components of a mixture, and he continues to employ fire for purifications as well as for transmutations. But where the outcome of an experiment is equivocal, Boyle's analysis places fire in the role of creative agent. That fire probably fabricates, from the base matter in organic materials, the earth, oil, spirit, and phlegm, obtained in the distillation of plants is a major thesis of *The Sceptical Chymist* and its various appendices. When an earthy residue is repeatedly recovered in the distillation of water, Boyle, though noting other possibilities, prefers to suppose that fire has transmuted the water. (II 519 ff.) When Boyle hears of an oil which appears difficult to purify since it continues to yield a pitchy residue after several distillations, he writes:

---

[64] Metzger, *Les Doctrines*, pp. 259 f., provides an excellent analysis of the instrumental role of fire in Boyle's theory.

But it was more congruous to my hypothesis [italics added], to conjecture, that the *caput mortuum* he complained of, was not, (at least after the first or second distillation,) a more gross or saeculent part of the oil, separated from the more pure, but a new compound produced, as other concretes also might be, by the operation of the fire. (IV 90)

Since the fractional distillation of organic oils is a slow process, Boyle is able to gratify his tendency to see transmutation where others see separation. And it is reactions analysable in this manner which receive most attention in his chemical writings.

Boyle devotes considerable attention to the study of this major instrument, fire. He notes, for example, that fire may also enter compounds materially, and he studies in detail the penetration of glass by fire particles which enter the pores of metallic calces. But in these famous experiments (which are conducted to elucidate the nature of the instrument, fire, not of the process, calcination), Boyle does not seem to view the material entry of fire as relevant to the calcination process itself. He describes experiments in which fire enters metals without calcining them (III 341 ff.) as well as experiments in which fire enters metals and rearranges them. (III 350 ff.) He seems to regard the fire particles in metallic calces as unavoidable corporeal impurities not relevant to the alteration of color and texture. Where experiment permits it, he will argue that fire played no *material* role in a reaction. (See, for example, the long discussion at IV 93.) Fire remains, first, an instrument.

But fire is not Boyle's only chemical instrument. In the *Fluidity and Firmness* water is frequently treated as an instrument which rearranges salts in the processes of solution and crystallization. "The First Part" of *The Sceptical Chymist* (particularly 47 ff.) cites numerous reactions in which separations and rearrangements not producible by fire or water alone are promoted by *menstrua* containing alkalisate bodies which act as "openers" of the transmuted body but which do not, at least in all cases, participate materially in the reaction. (51) Boyle thinks it possible that many additaments, useful in chemical operations with fire "do not, as ingredients, enter the composition of the obtained body, but only diversify the action of the fire upon the concrete"; (54) and he utilizes this mechanism in his analysis of the famous experiment with nitre (I 230 ff.) to which such frequent reference is made in his chemical writings. When bits of carbon dropped upon hot saltpeter cause a flame, Boyle views the carbon as an agent (almost a catalyst) which enhances the activity of the fire and permits a separation impossible for fire alone. Only the matter of the saltpeter is conceived to enter the reaction materially; the carbon, like the fire, is viewed as an instrumental agent.

In these analyses of chemical reactions Boyle displays his novel conception of the chemist as an artificer who fabricates in the microscopic realm as the mechanic does in the macroscopic. To Boyle the chemist is a micro-mechanic whose function is to produce "an orderly series of alterations" by operations analogous to "wire-drawing" or to "melting . . . and casting." (see above) His archetype of the chemist is the "skilful artist," almost the alchemical "adept," who desires to achieve, with the aid of appropriate instruments, "the greatest and most difficult changes, . . . rationally to be attempted among durable and inanimate bodies." [65] So Boyle views chemistry as an art of manipulative fabrication.

---

[65] ". . . the same body, meerly by different ways of ordering it, may be easily enough brought to afford, either acid, or inflammable, or volatile, commonly called urinous spirits, *as the skillful artist pleases*." (I 388, italics added) "I pretended hereby to devise a way of turning an acid salt into an alkali, which seems to be one of *the greatest and most difficult changes, that is rationally to be attempted among durable and inanimate bodies*." (II 510, italics added)

These passages also illustrate Boyle's emphasis upon striking qualitative changes, an aspect of his thought previously noted in his pursuit of transmutation. The present delineation of Boyle's conception of chemistry again requires the explicit description of an attitude which is only implicit in the writings. But passages which, like the above, emphasize alteration due to manipulation rather than to combination can be retrieved at will in Boyle's chemical writings.

The earlier particulate chemists like Jung and de Clave had promulgated a radically different conception of chemistry:

> Spagyrie is so named because it teaches the means of extraction and separation, . . . in order finally to reach ultimate simple purity [of the elements] . . : for chemistry or the Spagyrical philosophy having reached that last purification and reduction of mixed and compounded bodies, clearly sees their ingredients or, if you prefer, their essential parts; and in consequence [it sees] the quantity of each element, and their primary and secondary qualities: and finally the mixtures with other [elements] from which mixed and compounded bodies result.[66]

This quotation presents, in embryo, the concept of chemistry whose birth we celebrate as the chemical revolution. It is the same conception which Lavoisier stated more maturely in the words:

> The object of chemistry, in the experimental examination of the various natural substances, is to decompose them so that it may *separately examine the different substances which compose* them . . . Chemistry proceeds toward its goal and its perfection by dividing, subdividing, and resubdividing again; we do not know where this successful progression will cease.[67]

Boyle opposed this cenceptual definition of chemistry in the laboratory and in his chemical tracts from the publication of the *Physiological Essays* in 1661 to the publication of *A Chymical Paradox* in 1681/2.

⁂

Boyle's structural theory of chemical qualities and his associated instrumental theory of chemical change exerted little apparent influence upon the subsequent development of chemical concepts.[68] After 1670 almost all chemists (including the Stahliens) employed a particulate theory of matter, and during the last third of the seventeenth century many chemists rejected, or attempted to reject, the occult qualities used in the chemical theories of their predecessors.[69] But the theories of matter of these later corpuscularians resemble the static atomisms of Jung, de Clave, and Sennert far more closely than they do the dynamical atomism of Boyle. These chemists correlated quality with the enduring characteristics (particularly shape) of the indestructible atoms of the elements; they denied the possibility of mutual transmutation of the various elements; and, during the eighteenth century, they gradually extended the list of elementary substances.[70] In the late seventeenth and early eighteenth century most

---

[66] De Clave, *Nouvelle Lumière Philosophique*, translated from the fourth printed side in the unpaginated preface. The defense of a similar concept constitutes a major thesis of Jung's first *Disputation*.

[67] A. L. Lavoisier, *Traité Élémentaire de Chimie* (2nd ed., Paris: chez Cuchet, 1793), p. 193 f. Metzger, *La Matière chez Lavoisier*, shows that Lavoisier, too, attempted to trace some of the qualities through chemical reactions, which would make even closer the parallelism of the two definitions of chemistry, above.

[68] Boerhaave may have been directly influenced by some of Boyle's conceptions. At least he occasionally employs parallel notions. He considers fire, *menstrua*, etc. as "instruments" and he defines his science in the words: "*Chemistry is an art which teaches the manner of performing certain physical Operations, whereby bodies cognizable to the senses, or capable of being render'd coognizable, and of being contain'd in vessels, are so changed, by means of proper instruments, as to produce certain determined effects; and at the same time discover the causes thereof; for the service of various arts . . .*" (*New Method of Chemistry*, vol. 1, p. 65) In Boerhaave's chemistry these conceptions do not have the same significance that they have in Boyle's, for Boerhaave is quite convinced of the small range of possible transmutations.

I am indebted to Professor H. Guerlac of Cornell University for calling my attention to Boerhaave's instrumentalism. The responsibility for suggesting a parallelism to an implicit aspect of Boyle's thought is, however, my own.

[69] For the development of particulate theories during the eighteenth century see Metzger, *Newton, Stahl, Boerhaave*. The Stahlien particulate theories are considered in detail on pp. 117-129.

[70] The expansion of the list of elements is occasionally explicit, as in Stahl's tripartite division of the category, "earth" (*ibid.*, pp. 130 ff.), but more frequently the extension is implicit in the restriction of the range of possible transmutations. Boerhaave, for example, does not supply a new list of elements and makes very little use of the old one, except in so far as he employs names like "salt" and "earth," etc. as classificatory categories. He apparently takes no final position about the elements. But when he proclaims that the particles of the different metals do not have their "internal nature" changed by *menstrua* (*New Method of Chemistry*, vol. 1, p. 494), that earth cannot be obtained from metals (*ibid.*, p. 484), and that "the different metals . . . are resolvable into different

chemists participated in the current anti-peripatetic and pro-experimental emphases of the "new philosophy," but they did not continue Boyle's attempt to embrace chemical concepts within the dynamical corpuscular theory of the "clock-work universe." [71]

Gassendi and Descartes, who provided so much of the conceptual basis of the corpuscular philosophy, produced no developed chemical theory.[72] Newton is the major seventeenth century scientist who pursued Boyle's goal: the application of the new dynamical atomism to the elucidation of experimental chemistry. And Newton's results show important parallels to Boyle's.[73] Both believed that all natural substances were mechanical aggregates of small corpuscles (Boyle's *prima mixta*) and that these corpuscles of natural substances were in turn compounded of smaller, absolutely elementary corpuscles (Boyle's *minima*).[74] Throughout his scientific career Newton, like Boyle, believed in and investigated natural and artificial transmutation,[75] and he frequently employed illustrative examples provided by Boyle. Newton wrote that nature "seems delighted with transmutations," and he documented the view by citing such examples as the formation of the mercury oxides, the "changing of bodies into light, and light into bodies," and the change "of water, by frequent distillations . . . into fix'd earth, as Mr. *Boyle* has tried." [76] Indeed the emphasis upon transmutation in Newton's fragmentary chemical writings exceeds even Boyle's.

Newton did not explicitly maintain that anything can be made of anything. On the contrary, at the end of his scientific career he wrote to Locke that "there is one argument against . . . [the transmutation of base metals to gold], which I could never find an answer to." [77] But though this reservation commends Newton's skepticism, it does not really distinguish his theoretical chemistry from Boyle's. Newton, like Boyle, denied the specificity of the elementary corpuscles: he believed that with sufficiently powerful instruments of dissolution and fermentation "everything can be reduced to water." [78] And this was equivalent to Boyle's denial of the existence of chemical elements.

---

elements, both in respect of nature, and number" (*ibid.*, p. 103), he is leading the way toward a chemistry in which each metal (or its calx) will be considered a distinct element.

[71] The occasional exceptions who do pursue Boyle's goal are early eighteenth century figures like Hartsoeker and Privat de Molière. But the effect of their effort is to alter the original dynamical atomism, for by their time the elements were almost universally considered eternally distinct.

[72] See note 43, above.

[73] Useful studies of Newton's atomism are included in the works cited in note 4, above; in A. Koyré, The Significance of the Newtonian Synthesis, *Archives Internationales d'Histoire des Sciences*, 29, 291–311, 1950; and in F. Cajori, Ce Que Newton doit à Descartes, *L'Enseignement Mathématique*, 25, 7–11, 1926. The effects of atomism upon Newton's chemistry have been studied by Forbes, Was Newton an Alchemist?; D. McKie, Some Notes on Newton's Chemical Philosophy, *Philosophical Magazine & Journal of Science*, 33, 847–870, 1942; and S. I. Vavilov, Newton and the Atomic Theory, in *The Royal Society: Newton Tercentenary Celebrations* (Cambridge: at the University Press, 1947), pp. 43–55. Newton's published remarks on chemical subjects are largely contained in his *Opticks* (2nd and later editions) and in his De Natura Acidorum. For convenience these will be cited from vol. 4, of S. Horsley, *Isaaci Newtoni opera quae exstant omnia* (London: J. Nichols, 1782).

[74] Newton, in the 31st Query to the *Opticks* (*ibid.*, p. 260), asserts the eternal indivisibility of the *minima*, which differentiates his view from Boyle's.

[75] For evidence of Newton's early concern with transmutation see the letter to Ashton, printed as an appendix to vol. 1. of Sir D. Brewster's *Memoirs of the Life, Writings, and Discoveries of Sir Isaac Newton* (Edinburgh: 1855). For the continuation of the concern, see below.

[76] 30th Query to the *Opticks*, *ibid.*, p. 241.

[77] P. King, *The Life of John Locke* (London: 1830), vol. 1, p. 413.

[78] De Natura Acidorum, *op. cit.*, p. 398. Possibly this remark of Newton's indicates that he thought the *minima* were atoms of water. It is probable that Newton intended an exception to be made for gold in the uniformity of possible degradations, for he also wrote; "Mercury and *Aqua regia* can penetrate the pores which lie between the particles of the ultimate composition, but not others. If a *menstruum* could penetrate the other pores, or if the particles of gold of the first and second composition could be separated, gold would become fluid and malleable. If gold could be fermented, it could be transformed into any other body." (*ibid.*, p. 400) Newton may have believed that the "fermentation" of gold was a theoretical but not a practical possibility, but see T. S. Kuhn, Newton's "31st Query" and the Degradation of Gold, *Isis*, 42, 296–298, 1951.

Newton's implicit repudiation of the chemical elements was universally ignored by his contemporaries and successors. But Boyle's parallel views were often noted and rejected. Boyle's contemporary, John Mayow, for example, wrote:

I do not think we ought to agree with recent philosophers, who believe that fire can be produced by the subtle particles of any kind of matter if they are thrown into violent agitation. In fact, while the Peripatetics formerly assigned a distinct quality for almost every natural operation and multiplied *entia* unnecessarily, the Neoterics on the other hand maintain that all natural effects result from the same matter, its form and its state of motion or of rest alone being changed, *and that consequently any thing whatever may be obtained from any thing*. But in truth this *new philosophy* seems to depart too far from the doctrine of the ancients, and I have thought it better to take an intermediate path. It would certainly be a reasonable supposition that certain particles of matter which are unlike in no other respect than in the form and extremely solid and compact contexture of their parts, differ so much that *by no natural power can they be changed one into another, and that the* Elements *consist of primary, and in this way peculiar, particles*.[79]

And Mayow proceeds to list five elements — mercury (his "nitro-aerial spirit," sulphur, salt, water, and earth — from which he believes all natural bodies can be compounded. Reading "spirit" for "mercury" and "oil" for "sulphur," Mayow's list is identical with the roster of elements provided thirty-three years earlier by de Clave in the *Nouvelle Lumière Philosophique*.[80] Mayow is advancing within the older tradition. His conception of the elements differs from de Clave's primarily in its independence of occult qualities.

Mayow's attitude is like that of the more famous and influential continental chemist, Nicholas Lemery. Almost the only contentious remarks in Lemery's sober and systematic *Cours de Chimie* are reserved for those "doubting Scepticks (who make it their business to doubt of everything)." [81] And Lemery specifies this group as composed of the

modern Philosophers [who] would perswade us, that it is altogether uncertain, whether the substances which are separated from bodies, and are called *Chymical Principles*, do effectually exist and are naturally residing in the body before: these do tell us that *fire* by rarifying the matter in time of distillation is capable of bestowing upon it such an alteration as is quite different from what it had before, and so of forming the *Salt, Oil,* and other things which are drawn from it.[82]

This rejection of Boyle's views is of peculiar interest because it occurs as an integral part of Lemery's explicit delineation of the major chemical tradition linking de Clave (and his faction) with Lavoisier. Lemery believes that "*Chymistry* is an Art that teaches how to separate the different substances which are found in Mixt bodies," and that the "*Principle*[s which are the end products of this separation] . . . must not be understood in too nice a sense: for the substances which are so called, are only *Principles* in respect to us, and as we can advance no farther in the division of bodies." [83] These opinions he expresses in a preface which necessarily includes the

---

[79] J. Mayow, *Medico-Physical Works*, Old Ashmolean Reprints V (Oxford: 1926), p. 17. The italics, excepting "entia," are added. The reference is clearly to Boyle and probably to van Helmont.

[80] *Op. cit.*, pp. 40 ff.

[81] N. Lemery, *A Course of Chemistry* (translated from the 5th French ed. by W. Harris, M.D., 2nd ed. enlarged, London: Walter Kettilby, 1686), p. 7.

[82] *Ibid.*, p. 6. Lemery's condemnation may be compared with Boyle's statement that: "it may without absurdity be doubted whether or no the differing substances obtainable from a concrete dissipated by the fire were so existent in it in that forme (at least as to their minute parts) wherein we find them when the analysis is over, that the fire did only disjoyne and extricate the corpuscles of one principle from those of the other wherewith before they were blended." (64)

[83] *Ibid.*, pp. 2, 5. Lemery also points out (p. 9): "The *five Principles* are easily found in *Animals* and *Vegetables*, but not so easily in *Minerals*. Nay there are some *Minerals*, out of which you cannot possibly draw so much as *two*, nor make any separation at all (as Gold and Silver) whatsoever they talk, who search with so much pains for the *Salts, Sulphurs* and *Mercuries* of these metals." This was a major

explicit repudiation of Boyle's contemporaneous doctrines. Lemery is a far less philosophical chemist than Boyle, and he relates his views more closely to experiment. He rejects occult qualities and believes that quality depends upon shape. But he makes little use of explicit corpuscular explanation, and his list of elements is the same as de Clave's. His tremendous prestige [84] may provide an index of the continuing prevalence of a chemical tradition which could not embrace Boyle's opinions.

Finally, the extent of the domination of chemical thought by the old but evolving theory of the chemical elements may be gauged by the alteration it finally necessitated in the original dynamical atomism from which Boyle and Newton had drawn so heavily. Newton himself had denied the chemical specificity of the ultimate atoms which composed the mechanical universe. But the eighteenth century Newtonian chemists neglected this aspect of his thought and emphasized instead the attractive and repulsive forces which he had hypothesized to govern the interactions and accretions of the *minima*. In spite of Newton's explicit disclaimer,[85] these forces were, in the eighteenth century, usually taken to be inherent in the material corpuscles whose behavior they governed. They became "occult" forces which determined the characteristic properties of the various chemical substances. And so they constituted, contrary to Newton's intention, a primary source of that specific differentiation of elementary particles which Newton and Boyle had attempted to eliminate from chemical theory.[86]

The original Cartesian mechanical theory underwent a similar transformation. By the beginning of the eighteenth century as strict and literal a "mechanist" as Nicolas Hartsoeker could combine a sectarian adherence to Cartesian physical concepts [87] with a brusque rejection of their original chemical correlates:

... according to the Cartesian System, all bodies of the visible world can be converted into any other imaginable body, and in consequence fire can change to air, air to water, and water into any other terrestrial body; or [water can] return to air, or fire, &c., than which nothing is more absurd.[88]

For the older Cartesian opinion that chemical species continually evolve from base matter Hartsoeker substituted the view that:

Water never changes to air or salt, nor air or salt to any other substance; but all substances endure eternally, and they do not have a different nature today than they had at the most distant time or than they will have in all the centuries to come ... Gold will always remain gold ... and the same for mercury ... and for all the other metals.[89]

Such eighteenth century alterations in the conceptual structure of seventeenth century dynamical atomism resulted in an integration of dynamical corpuscular theories with the older particulate chemical theories of matter. Robert Boyle's chemical

---

argument in Boyle's critique of the "chymists," but it had been pointed out previously by Jung (see above). It is not clear what proportion of the "chymists" Boyle attacked held the view he attributed to them. A more detailed recapitulation of Lemery's views occurs in Metzger, *Les Doctrines*.

[84] Lemery's *Cours de Chemie* first appeared in 1675. There were ten subsequent French editions, and the book was repeatedly translated into English, Latin, German, Italian, and Spanish. No work of Boyle's compares in popularity.

[85] See Newton's letters to Bentley (Horsley, *Opera*, vol. 4, pp. 437 f.) in which he declares that though he cannot now give a mechanical explanation of gravity, he refuses to regard it "as essential and inherent to matter."

[86] This development of Newtonian chemical atomism in the eighteenth century is discussed brilliantly and at length by Metzger, *Newton, Stahl, Boerhaave*. See also Bloch, *Die Antike Atomistik*.

[87] Hartsoeker, in 1706, still attempted to develop planetary astronomy without the benefit of the Newtonian gravitation theory. Gravitation he thought occult, so he felt impelled to abandon the mathemetical formulation as well.

[88] N. Hartsoeker, *Conjectures Physique* (Amsterdam: H. Desbordes, 1706), p. 124.

[89] Quoted by Metzger, *Les Doctrines*, pp. 131 f.

researches, completed before the integration occurred, illustrate the extent and the conceptual consequences of the initial divergence of the two atomistic traditions.

At the beginning of his career in chemistry Boyle pointed out to his contemporaries that:

it is sometimes conducive to the discovery of truth, to permit the understanding to make an hypothesis, . . . [so] that by the examination how far the phaenomena are, or are not, capable of being solved by that hypothesis, the understanding may, even by its own errors, be instructed. For it has been truly observed by a great philosopher, that *truth does more easily emerge out of error than confusion.* (I 194, italics added) [90]

This brilliant intuition of the methodology of scientific research was unintentionally documented by its author's contributions to the development of chemistry.

Boyle undertook to eliminate from chemistry the antique and mystical residues of mediaeval and renaissance thought, and he endeavored to reformulate the science in terms of the most progressive and fruitful scientific hypotheses of his day. Boyle's brilliant destructive criticism of naive theories of the chemical elements probably facilitated the gradual elaboration of theories more consonant with experiment during the eighteenth century. Certainly his efforts resulted in fundamental experimental discoveries and in the isolation of novel problems whose further exploration advanced chemical theory and practice.

But Boyle's constructive attempt to replace existing theories of the elements by a conceptual scheme derived from the prevalent metaphysical atomism of the seventeenth century was a failure. The conviction that chemical qualities could be derived from the mechanical structure of the "clock-work universe" promoted the opinions that there were no chemical elements, that any substance could be transmuted to any other, and that the object of the chemist was to fabricate novel substances by micro-mechanical operations upon the neutral corpuscles of base matter. This view of chemistry and the chemist was rejected by many of Boyle's contemporaries and successors, because it conflicted with the still prevalent conception of chemistry as an art of separation and combination whose ultimate objectives were the isolation of elements and the determination of composition. A retrospective glance at the history of seventeenth and early eighteenth century chemistry suggests that the true progenitors of Lavoisier's chemical revolution were necessarily among Boyle's opponents.

---

[90] Boyle's "great philosopher" is, of course, Francis Bacon: "citius emergit veritas ex errore quam ex confusione." (*Novum Organum*, Book II, Aphorism XX.)

Newton's Alchemy and His Theory of Matter

By B. J. T. Dobbs

INTRODUCTION

WHEN ISAAC NEWTON DIED INTESTATE in 1727, he inadvertently—or perhaps intentionally—provided employment for generations of scholars yet unborn. Laughing, he said to John Conduitt not long before he died "that he had said enough for people to know his meaning,"[1] but the books and papers he left behind have not always borne out his words. A source of particular confusion has been the relationship between Newton's alchemy and his published theory of matter, and it is the purpose of this essay to clarify Newton's meaning in these areas.

The primary thrust of Newton's published theory of matter has always been reasonably well understood.[2] Despite occasional misinterpretations and distortions, everyone has agreed that Newton's thoughts on matter make up an intelligible system, one that could be used and refined by later students of philosophy and the physical sciences. Not so with his alchemical interests, which from the moment of their discovery tended to arouse emotions of disdain, bewilderment, horror, and disbelief.

Alchemy was even troublesome for Newton himself. In his copy of the 1698 edition of Lemery's *Chymistry*, Newton turned down the page in his characteristic fashion to point to the following story:

> But the saddest consideration of all is, to see a great many of them [the alchemists], who have spent all the flower of their years, in this desparate concern, in which nevertheless they pertinaciously run on, and consume all they have, at last instead of recompense for their miserable fatigues, reduced to the lowest degree of poverty. *Penotus* will serve us for an instance of this nature, among thousands of others, he died a hundred years old wanting but two, in the Hospital of *Yverdon* in *Switzerland*, and he used to say before he died, having spent his whole life in vainly searching after the Philosopher's stone, *That if he had a mortal Enemy he did not dare to encounter openly, he would advise him above all things to give himself up to the study and practice of* Alchymy.[3]

A shorter version of this paper was read before the History of Science Society, Toronto, 18 Oct. 1980. I wish to thank the colleagues, friends, and referees who have made suggestions.

[1] "A remarkable and curious conversation between Sir Isaac Newton and Mr. Conduitt," in Edmund Turnor, *Collections for the History of the Town and Soke of Grantham, Containing Authentic Memoirs of Sir Isaac Newton* . . . (London, 1806), pp. 172–173, on p. 173.

[2] The most recent full-length study is Ernan McMullin, *Newton on Matter and Activity* (Notre Dame/London: Univ. Notre Dame Press, 1978); see the bibliography for the older literature. See also I. Bernard Cohen, "Isaac Newton," *Dictionary of Scientific Biography* (New York: Scribners, 1970–1980, Vol. X, pp. 42–101).

[3] Nicholas Lemery, *A Course of Chymistry; containing An easie Method of Preparing those Chymical Medicines which are used in Physick* . . . , trans. James Keill (3rd Eng. ed. from the 8th Fr.;

When he noted that passage, Newton perhaps had Hooke, Leibniz, or Flamsteed in mind for the "mortal Enemy he did not dare to encounter openly," but he must surely also have identified himself with Penotus, for he was already at least fifty-six and had been struggling with alchemy for thirty years.[4]

Historians of recent decades have, however, chipped stubbornly away at the problem of Newton's alchemy, and the feeling that it must have a serious and coherent relationship with his theory of matter has been vindicated. We can now see that Newton used alchemy as a critical counterweight against the inadequacies of ancient and contemporary atomism, inadequacies regarding cohesion and activity, life and vegetation, and the dominion and providence of God. His final formulations on the nature of matter and the powers associated with it grew naturally out of alchemical, theological, metaphysical, and observational concerns.

IN THE BEGINNING: MATTER

Sometime during his student years Newton became an eclectic corpuscularian, choosing elements of matter theory from Descartes, Gassendi (via Charleton), Boyle, Hobbes, Digby, and More and leaving us a record of his thoughts in the *Quaestiones quaedem philosophicae* of his student notebook. The nature "Of ye first mater" had him somewhat bemused. It was clearly not composed of mathematical points and parts, nor of a "simple entity before division indistinct," nor was it infinitely divisible. For a short period he concluded with Henry More in favor of *minima naturalia*, but a subsequent cancellation of that passage in the notebook presumably shows that he reopened the question.[5]

At first Newton was a plenist. By postulating a subtle aether, a medium imperceptible to the senses but capable of transmitting effects by pressure and impact, mechanical philosophers had devised a convention that rid natural philosophy of incomprehensible occult influences acting at a distance (e.g., magnetic attraction and lunar effects). For Newton just such a mechanical aether, prevading the whole world and making it a plenum, became an unquestioned assumption. By it he explained gravity and, to a certain extent, the cohesion of particles of matter.

The question of cohesion had always plagued theories of discrete particles, atomism having been criticized even in antiquity on this point. The cohesion of living forms seems intuitively to be qualitatively different from anything that the random, mechanical motion of small particles of matter might produce. Nor does atomism explain even mechanical cohesion in inert materials very well, for it requires the elaboration of *ad hoc*, unverifiable hypotheses about the geometric

---

London, 1698), pp. 62–63 (Trinity College, Cambridge, call no. NQ.8.118; John Harrison, *The Library of Isaac Newton* [Cambridge/London/New York: Cambridge Univ. Press, 1978], item 938, p. 177).

[4]On Newton's turn to alchemy in 1668, see B. J. T. Dobbs, *The Foundations of Newton's Alchemy, or "The Hunting of the Greene Lyon"* (Cambridge/London/New York: Cambridge Univ. Press, 1975), pp. 121–125.

[5]A. Rupert Hall, "Sir Isaac Newton's Note-book, 1661–65," *Cambridge Historical Journal*, 1948, 9:239–250; Richard S. Westfall, "The Foundations of Newton's Philosophy of Nature," *British Journal for the History of Science*, 1962/63, *1*:171–182; Richard S. Westfall, *Force in Newton's Physics: The Science of Dynamics in the Seventeenth Century* (London: Macdonald; New York: American Elsevier, 1971), pp. 324–326. Newton's student notebook (Cambridge University Library [CUL] MS Add. 3996), the indispensible starting point for a study of Newton's thought in many areas, is being edited for Cambridge University Press by J. E. McGuire and Martin Tamny.

configurations of the atoms or else speculation about their quiescence under certain circumstances. In the various forms in which corpuscularianism was revived in the seventeenth century, the problems remained and variants of ancient answers were redeployed. Descartes, for example, held that an external pressure from surrounding subtle matter just balanced the internal pressure of the coarser particles that constituted the cohesive body. Thus no special explanation for cohesion was required: the parts cohered simply because they were at rest close to each other in an equilibrated system. Gassendi's atoms, on the other hand, stuck together through the interlacing of antlers or hooks and claws, much as the atoms of Lucretius had before them. Charleton found not only hooks and claws but also the pressure of neighboring atoms and the absence of disturbing atoms necessary to account for cohesion.[6] Francis Bacon introduced certain spirits or "pneumaticals" into his speculations. In a system reminiscent of the Stoics, those ancient critics of atomism, Bacon concluded that gross matter must be associated with active, shaping, material spirits, the spirits being responsible for the forms and qualities of tangible bodies, producing organized shapes, effecting digestion, assimilation, and so forth.[7] For Newton during his student years, with his mechanical aether ready at hand, a pressure mechanism seemed sufficient to explain cohesion; he rejected quiescence but affirmed that "ye close crouding of all ye matter in ye world" might account for it. Yet he also examined a geometric approach ("Whither hard bodys stick together by branchy particles foulded together"), and he even then suggested that "it may be some other power by wch matter is kept close together."[8]

It was to be a long, circuitous, even tortuous journey that carried Newton away from these rather indefinite reflections on matter in his student *Quaestiones*. Within a very short time, perhaps five years at the most, he had begun to modify his mechanical philosophy with an alchemical one, and about 1669 he prepared a short paper containing a series of alchemical propositions.

Gold, silver, iron, copper, tin, lead, mercury, and "magnesia" are all the species of the art, he said, and all of them are from one root.[9] If Newton had been concerned with the unity of matter during his brief years of strict mechanism, no evidence from his early papers has yet appeared. Investigators (myself included) have often tacitly assumed that Newton accepted the unity of matter from contemporary mechanical philosophy. Certainly talk among the mechanical philosophers

---

[6]Lancelot Law Whyte, *Essay on Atomism: From Democritus to 1960* (1961; New York/Evanston: Harper & Row, 1963); E. C. Millington, "Theories of Cohesion in the Seventeenth Century," *Annals of Science*, 1941–1947, 5:253–269.

[7]Bacon's early atomism gave way in his mature years to a position thought to have been influenced by his chemical/alchemical readings, a position much closer to Stoicism than has generally been recognized. Robert Hugh Kargon, *Atomism in England from Hariot to Newton* (Oxford: Clarendon Press, 1966), pp. 43–54; J. C. Gregory, "Chemistry and Alchemy in the Natural Philosophy of Sir Francis Bacon, 1561–1626," *Ambix*, 1938, 2:93–111; Charles W. Lemni, *The Classic Deities in Bacon: A Study in Mythological Symbolism* (1933; New York: Octagon Books; 1971), esp. pp. 74–109; A. A. Long, *Hellenistic Philosophy: Stoics, Epicureans, Sceptics* (London: Duckworth, 1974), pp. 152–156; Graham Rees: "Francis Bacon's Semi-Paracelsian Cosmology," *Ambix*, 1975, 22:81–101; Rees, "Francis Bacon's Semi-Paracelsian Cosmology and the Great Instauration," *ibid.*, pp. 161–173; Rees, "The Fate of Bacon's Cosmology in the Seventeenth Century," *Ambix*, 1977, 24:27–38; Rees, "Matter Theory: A Unifying Factor in Bacon's Natural Philosophy?" *ibid.*, pp. 110–125; Rees, "Francis Bacon on Verticity and the Bowels of the Earth," *Ambix*, 1979, 26:202–211.

[8]Westfall, *Force*, pp. 334, 402 (n. 19).

[9]Keynes MS 12A, "Propositions," King's College, Cambridge, fol. 1r. Quotations from this and Keynes MS 18, cited n.16 below, are by permission of the Provost and Fellows of King's College, Cambridge.

of the particles of one catholic and universal matter did nothing to undermine ancient doctrines of prime matter, even though among Presocratics, Aristotelians, and alchemists *materia prima* was not considered particulate.[10] Yet here we find Newton's lucid expression for the unity of matter in an alchemical context—"all species are from one root." Whether he had first absorbed the notion in a mechanical or other philosophical context or not, his early alchemical work evidently secured him in a conviction from which he never seriously wavered, and the doctrine of the unity of matter and its transmutability became a part of the published record of his views. The first edition of the *Principia* (1687) carried the most explicit statement: "Any body can be transformed into another, of whatever kind, and all the intermediate degrees of qualities can be induced in it."[11] There is a passage in all editions of the *Principia* in which Newton speculates that matter falling to earth from the tails of comets might be condensed into all types of earthly substances.[12] In his later years Newton stated the doctrine a number of times: in his small tract *On the Nature of Acids*, in the *Opticks*, and to David Gregory, who duly recorded it among his memoranda.[13] Although that most uncompromising statement from the *Principia* of 1687 disappeared in subsequent editions, and although the later *Opticks* passages demonstrate some possible ambiguities, the consensus of recent studies is that Newton maintained to the end his belief in the inertial homogeneity and transformability of matter.[14]

Although alchemy and mechanism do share the doctrine of the ultimate unity of matter, it seems impossible to find a mechanical counterpart for the active, vitalistic alchemical agent Newton introduced into his "Propositions" around 1669. There he called the agent by its code name "magnesia," a term that evoked for the alchemists all the mysterious properties of the magnet and expressed their understanding that certain substances had the capacity to draw into themselves the active vivifying celestial principle necessary for life. Newton aligned "magnesia" with the metals in being from "one root," but he added that magnesia is the only species that revivifies.[15] Newton had become preoccupied with a process of disorganization and reorganization by which developed species of matter might be

[10] Dobbs, *Foundations*, p. 46.

[11] Isaac Newton, *Philosophiae naturalis principia mathematica* (London, 1687), p. 402, Hypothesis III. Subsequent changes are given in Isaac Newton, *Philosophiae naturalis principia mathematica*, 3rd. ed. (1726) with variant readings, ed. Alexandre Koyré and I. Bernard Cohen with Anne Whitman, (2 vols.; Cambridge: Harvard Univ. Press, 1972), Vol. II, pp. 550–553; see also Alexandre Koyré, "Newton's 'Regulae Philosophandi,'" in *Newtonian Studies* (Cambridge: Harvard Univ. Press, 1965), pp. 261–272.

[12] *Sir Isaac Newton's Mathematical Principles of Natural Philosophy and His System of the World* (1729, trans. Andrew Motte), ed. Florian Cajori, 2 vols. (1934; Berkeley/Los Angeles: Univ. California Press, 1962), Vol. II, p. 542; Newton, *Principia* (Koyré and Cohen), Vol. II, p. 758.

[13] The published variants of the tract are (1) John Harris, *Lexicon Technicum or an Universal English Dictionary of Arts and Sciences*, 2 vols. (facs. of London ed. of 1704–1710; New York/London: Johnson Reprint Corp., 1966), Vol. II, sig. b3v–b4v; (2) *Isaaci Newtoni opera quae exstant omnia*, ed. Samuel Horsley (5 vols.; London, 1779–1785), Vol. IV: pp. 397–400; (3) *The Correspondence of Isaac Newton*, Vol. III, ed. H. W. Turnbull (Cambridge: Cambridge Univ. Press, 1961), pp. 205–214. Isaac Newton, *Opticks, or A Treatise of the Reflections, Refractions, Inflections & Colours of Light* (based on the 4th London ed. of 1730; New York: Dover, 1952), pp. 266–269, 394. David Gregory, *Isaac Newton and Their Circle: Extracts from David Gregory's Memoranda, 1677–1708*, ed. W. G. Hiscock (Oxford: The Editor, 1937), pp. 30–31.

[14] Dobbs, *Foundations*, pp. 199–204, 231–232; J. E. McGuire, "Transmutation and Immutability: Newton's Doctrine of Physical Qualities," *Ambix*, 1967, *14*:69–95; Arnold Thackray, *Atoms and Powers: An Essay on Newtonian Matter-Theory and the Development of Chemistry* (Cambridge: Harvard Univ. Press, 1970), esp. pp. 8–42.

[15] Keynes MS 12A, fol. 1r. Cf. Dobbs, *Foundations*, pp. 159–160.

radically reduced, revivified, and led to generate new forms. The alchemical agent responsible for these changes is vitalistic and universal in its actions; it is a "fermental virtue" or "vegetable spirit" and is eventually to become the force of fermentation of the *Opticks*.[16] In the "Propositions" it is the agent that confounds into chaos and then aggregates anew the particles of matter.

> This and only this is the vital agent diffused through all things that exist in the world.
> And it is the mercurial spirit, most subtle and wholly volatile, dispersed through all places.
> This agent has the same general method of operating in all things, namely, excited to action by a moderate heat, it is put to flight by a great one, and once an aggregate has been formed, the agent's first action is to putrefy the aggregate and confound it into chaos. Then it proceeds to generation.[17]

From what sources has Newton derived his ideas on the universal vital agent that he here busily attaches to seventeenth-century mechanism? Quite possibly from alchemical sources only, at this early stage in his development, though his vitalistic ideas were soon reinforced by other sources.

Vitalism seems to belong to the very origins of alchemy. In the early Christian centuries metals had not been well characterized as distinct species, but were thought of instead as rather like modern alloys, with variable properties, but even more as like a mix of dough, into which the introduction of a leaven might produce desired changes by a process of fermentation, or even like a material matrix of unformed matter, into which the injection of an active male sperm or seed might lead to a process of generation. By analogy alchemists referred to this critical phase of the alchemical process as fermentation or generation, and the search for the vital ferment or seed became a fundamental part of their quest.[18] Similar ideas occur in Aristotle and are commonplace in Newton's time.[19]

Inspired by his interest in a vital agent, Newton had begun to grope his way toward mending the deficiencies of ancient atomism and contemporary corpuscularianism. He had concerned himself with life and cohesion. He now sought the source of all the apparently spontaneous processes of fermentation, putrefaction, generation, and vegetation (that is, everything associated with normal life and growth, such as digestion and assimilation, vegetation being originally from the Latin *vegetare*, to animate, enliven). These processes produced the endless variety of living forms and could not be relegated to the mechanical actions of gross corpuscles, a point he had made explicit by the mid-1670s.[20] Mechanical action could never account for the process of assimilation, in which food stuffs were turned into the bodies of animals, vegetables, and minerals. Nor could it account for the sheer variety of forms in the world, all of which had somehow sprung from the common matter. As Newton was finally to say in the General Scholium to the *Principia*, "Blind metaphysical necessity [i.e., mechanical action], which is

---

[16]See Keynes MS 18, "Clavis," published in Dobbs, *Foundations*, pp. 251–255, and *loc. cit.* in nn. 39, 53–57 below.

[17]Keynes MS 12A, fol. 1v.

[18]Marcellin Pierre Eugene Berthelot, *Les origines de l'alchimie* (1885; Paris: Librairie des Sciences et des Arts, 1938), pp. 240–241.

[19]See, e.g., Aristotle, *De generatione animalium*, 1.1, 715b26; 1.16, 721a8; 2.3, 736b30; 2.4, 738b23; 1.2, 716a17. Cf. *loc. cit.* nn. 23, 57 below.

[20]See discussion of Burndy MS 16, Smithsonian Institution, Washington, D.C., below (nn. 26, 28).

certainly the same always and everywhere, could produce no variety of things." Ultimately, it was God who was responsible, God, who in his wisdom and with his dominion, his providence, and the final causes known only to himself, produced "all that diversity of natural things which we find suited to different times and places."[21] But God was the ultimate cause, and what Newton desired to locate in the natural world was the more proximate cause of the phenomena of life that was God's agent in these matters.

The most comprehensive answer to such problems in antiquity had been given by the Stoics. The Stoics postulated a continuous material medium, the tension and activity of which molded the cosmos into a living whole and the various parts of the cosmic animal into coherent bodies as well. Compounded of air and a creative fire, the Stoic *pneuma* was related to the concept of the "breath of life" that escapes from a living body at the time of death and allows the formerly coherent body in which it had resided to disintegrate into its disparate parts. Although always material, the *pneuma* becomes finer and more active as one ascends the scale of being, and the (more corporeal) air decreases as the (less corporeal) fire increases. Thus the Deity, literally omnipresent in the universe, is the hottest, most tense and creative form of the cosmic *pneuma* or aether, pure fire or nearly so. The cosmos permeated and shaped by the *pneuma* is not only living, it is rational and orderly and under the benevolent, providential care of the Deity. Though the Stoics were determinists, their deity was immanent and active in the cosmos, and one of their most telling arguments against the atomists was that the order, beauty, symmetry, and purpose to be seen in the world could never have come from random, mechanical action. Only a providential God could produce and maintain such lovely, meaningful forms. The universe, as a living body, was born when the creative fire generated the four elements; it lived out its lifespan, permeated by vital heat and breath, cycling back to final conflagration in the divine active principle, and ever regenerated itself.[22]

The original writings of the Stoics were mostly lost, but not before ideas of *pneuma* and *spiritus* came to pervade medical doctrine, alchemical theory, and indeed the general culture with form-giving spirits, souls, and vital principles. Spiritualized forms of the *pneuma* entered early Christian theology in discussions of the immanence and transcendence of God and of the Holy Ghost, just as the Stoic arguments that order and beauty demonstrate the existence of God and of providence entered Christianity as the "argument from design." The creative emanations of Stoic fire melded with the creative emanations of light in Neoplatonism. In addition to this broad spectrum of at least vaguely Stoic ideas, excellent, though not always sympathetic, summaries of philosophical Stoicism were available in Cicero, Seneca, Plutarch, Diogenes Laertius, and Sextus Empiricus.

By the seventeenth century ideas compatible with Stoicism were very widely diffused, and latter-day Stoics, Pythagoreans, Platonists, and Peripatetics all vied with each other in celebrating the occult virtues of a cosmic aether that was the

---

[21]Newton, *Principia* (Mott-Cajori), Vol. II, p. 546.

[22]See Shmuel Sambursky, *Physics of the Stoics* (1959; London: Hutchinson, 1971); Stephen Toulmin and June Goodfield, *The Architecture of Matter* (1962; New York: Harper & Row, 1966), esp. pp. 92–108; Mary B. Hesse, *Forces and Fields: The Concept of Action at a Distance in the History of Physics* (1962; Westport, Conn.: Greenwood Press, 1970), esp. pp. 74–79; and John M. Rist, ed., *The Stoics* (Berkeley/Los Angeles/London: Univ. California Press, 1978), esp. Robert M. Todd, "Monism and Immanence: The Foundations of Stoic Physics," pp. 137–160, and Michael Lapidge, "Stoic Cosmology," pp. 161–185.

vehicle of a pure, hidden creative fire.[23] Nonetheless, such a vital aether was to be found in its most developed form in philosophical Stoicism. It is possible, as Newton's concern for the processes of life and cohesion grew apace in the early 1670s, that he amplified his mechanical philosophy further by a close reading of the material available to him from classical Rome: Cicero, Diogenes Laertius, Plutarch, Seneca, and Sextus Empiricus were in his library.[24] Such reading would have affected his alchemy only in reinforcing certain critical ideas, for most of his early alchemical sources were distinctively Neoplatonic in tone, and in them the universal spirit or soul of the world already permeated the cosmos with its fermental virtue.[25] But Stoic ideas would have affected his views on the mechanical aether of his student years. We must conclude that if Newton had not read the Stoics, then he must independently have reached answers similar to theirs when confronted with similar problems, for by about 1674 the original mechanical aether of his *Quaestiones* had assumed a strongly Stoic cast.

The new vital aether is described in a long alchemical treatise left untitled by Newton but usually known by its initial phrase, "Of nature's obvious laws and processes in vegetation." The earth is "a great animal," he says, "or rather an inanimate vegetable [that] draws in aethereal breath for its daily refreshment and vital ferment and transpires again with gross exhalations." He goes on to describe this aethereal breath as a "subtle spirit," "nature's universal agent, her secret fire," and the "material soul of all matter." The similarity between this particular Newtonian aether and the Stoic *pneuma* is unmistakable: they are both material and both somehow inspire the forms of bodies and give to bodies the continuity and coherence of form that is associated with life. Newton expands upon that theme, saying that the earth, "according to the condition of all other things living, ought to have its times of beginning, youth, old age, and perishing," and that the subtle aethereal agent is the "only ferment and principle of all vegetation."[26] One may trace the vivid imagery of the earth-animal back through Stoic and Neoplatonic commentators on Plato, as one may trace the "perishing" of the earth to which Newton alludes here forward to his later convictions regarding a final cosmic conflagration.[27]

In this treatise Newton makes a sharp distinction between vegetation and mechanism. "Nature's actions," he says, "are either vegetable or purely mechanical," and as mechanical he lists, among other things, "vulgar chemistry." Vulgar

---

[23]Mary Anne Atwood, *Hermetic Philosophy and Alchemy: A Suggestive Inquiry into "The Hermetic Mystery" with a Dissertation on the more Celebrated of the Alchemical Philosophers* (New York: Julian Press, 1960), p. 78; D. P. Walker, *Spiritual and Demonic Magic from Ficino to Campanella* (1969; Notre Dame/London: Univ. Notre Dame Press, 1975); Walker, *The Ancient Theology: Studies in Christian Platonism from the Fifteenth to the Eighteenth Century* (London: Duckworth, 1972); Walker, "Medical Spirits: Four Lectures," Boston Colloquium for the Philosophy of Science, 27 Oct.–19 Nov. 1981.

[24]Harrison, *Library*, pp. 118–119, items 379–383; p. 133, item 519; p. 219, items 1330–1331; p. 236, items 1486–1490; p. 237, item 1503.

[25]Dobbs, *Foundations*, passim.

[26]Burndy MS 16, fol. 3v (cit. n. 20). I have modernized spelling and punctuation in all quotations from this MS. See B. J. T. Dobbs, "Newton Manuscripts at the Smithsonian Institution," *Isis*, 1977, 68:105–107. I have dated it about 1674 because of its great resemblance to "An Hypothesis of Light," which Newton sent to the Royal Society in 1675. Cf. Isaac Newton to Henry Oldenburg, 7 Dec. 1675, in *Correspondence*, Vol. I, pp. 362–389.

[27]*Plato's Cosmology: The Timaeus of Plato*, ed. and trans. Francis MacDonald Cornford (1937; Indianapolis/New York: Bobbs-Merrill, n.d.), p. 332 (the significance of this passage was first called to my attention by Samuel Scolnicov). David Kubrin, "Newton and the Cyclical Cosmos: Providence and the Mechanical Philosophy," *Journal of the History of Ideas*, 1967, 28:325–346.

chemistry may readily be identified with the kind of chemistry he discussed in the *Opticks*, the operations of which take place only among the "grosser" particles of matter. As he says here in 1674, "all the operations in vulgar chemistry (many of which to sense are as strange transmutations as those of nature) are but mechanical coalitions or separations of particles . . . and that without any vegetation." Newton admits that to many it may seem that all the "changes made by nature" may be done the same way, "that is by the sleighty transpositions of the grosser corpuscles, for, upon their disposition only, sensible qualities depend." But he argues that such is far from being the case. There is a "vast and fundamental" difference between vulgar chemistry and vegetation, which requires that we have recourse to some further cause.[28]

He continues with a development of themes from his earlier "Propositions" on the subtle nature of the vital agent, on its sensitivity to heat, and on its universality—emphasizing all the while the great differences between vulgar and vegetable chemistry.

> 6 There is, therefore, besides the sensible changes wrought in the textures of grosser matter, a more subtle, secret, and noble way of working in all vegetables which makes its products distinct from all others; and the immediate seat of these operations is not the whole bulk of the matter, but rather an exceeding subtle and unimaginably small portion of matter diffused through the mass, which, if it were separated, there would remain but a dead and inactive earth. And this appears in that vegetables are deprived of their vegetable virtue by any small excess of heat, the tender spirit being either put to flight or at least corrupted thereby (as may appear in an egg), whereas those operations which depend upon the texture of the grosser matter (as all those in common chemistry do) receive no damage by heats far greater. . . .
>
> 7 'Tis the office therefore of those grosser substances to be medium or vehicle in which rather than upon which those vegetable substances perform their actions.
>
> 8 Yet those grosser substances are very apt to put on various external appearances according to the present state of the invisible inhabitant, as to appear bones, flesh, wood, fruit, etc. Namely, they consisting of differing particles, watery, earthy, saline, airy, oily, spiritous, etc., those parts may be variously moved one among another according to the acting of the latent vegetable substances and be variously associated and concatenated together by their influence.[29]

In these distinctions between mechanical and vegetable chemistry Newton is working toward his famous hierarchical system of parts and pores arranged in three-dimensional netlike patterns. Let us take his statement about the "grosser corpuscles" from 1674: "upon their disposition only, sensible qualities depend." It is identical in meaning with one from the *Opticks* in which Newton describes the building up of the hierarchies of matter "until the Progression end in the biggest Particles on which the Operations in Chymistry, and the Colours of natural Bodies depend, and which by cohering compose Bodies of a sensible Magnitude."[30]

Or let us take another statement from 1674: grosser substances "consisting of differing particles, watery, earthy, saline, airy, oily, spiritous, etc., these parts may be . . . variously associated and concatenated together. . . ." These smaller units of watery, saline, and the like particles form subunits of the largest ones, and their characteristics are drawn from contemporary chemical systems of (usually) five or six chemical elements or principles derived from sixteenth- and seven-

---

[28] Burndy MS 16, fol. 5r–5v.
[29] *Ibid.*, fols. 5v–6r.
[30] Newton, *Opticks*, p. 394.

teenth-century combinations of Aristotelian matter theory (four elements: earth, air, fire, water) with Paracelsian (three principles: salt, sulfur, mercury). Similar intermediate particles appear in both the *Opticks* and in *On the Nature of Acids*.[31]

Furthermore, when Newton discusses putrefaction in 1674, he not only expands upon the view expressed in the "Propositions" a few years earlier but also adumbrates the final section of *On the Nature of Acids:*

> Nothing can be changed from what it is without putrefaction. . . .
> No putrefaction can be without alienating the thing putrefied from what it was.
> Nothing can be generated or nourished (but of putrefied matter).
>
> * * *
>
> Her [Nature's] first action is to blind and confound mixtures into a putrefied chaos. Then they are fitted for new generation or nourishment.[32]

It is putrefaction that reduces matter to its ultimate state of disorganization, where the particles of matter are all alike and hence can be remodeled in any form whatsoever. Although the later notion of pores is missing in the 1674 tract on vegetation, in other respects this version of Newton's matter theory closely resembles *On the Nature of Acids*. There, if a menstruum could adequately penetrate the pores of gold, or if gold could "ferment," it could be reduced to its most primordial particles. Then "it could be transformed into any other substance. And so of tin, or any other bodies, as common nourishment is turned into the bodies of animals and vegetables."[33]

Although Newton's early alchemical papers contain both the doctrine of the ultimate unity of matter and the rudiments of a hierarchical system of parts, they do not exhibit all aspects of his final theory of matter. Missing are the pores devoid of matter which later become an essential feature of his structured hierarchies. Missing also is any mention of forces, although the addition of forces to matter theory proved to be his most significant modification of corpuscularianism. Consideration of these further developments perforce leads us into some of Newton's theological concerns, and into a brief excursion on work preliminary to the *Principia*, as well as into more of his alchemical labor.

IN THE INTERIM: FORCES

From the beginning Newton was troubled by a theological problem. As I have argued elsewhere, Newton, like his older contemporaries Isaac Barrow and Henry More (and many others), was alarmed at the atheistic potentialities of the revived corpuscularianism of their century, particularly of Cartesianism.[34] Although the ancient atomists had not really been atheists in any precise sense, they had frequently been so labeled because their atoms in random mechanical motion received no guidance from the gods. Descartes, Gassendi, and Charleton had been at pains to allay the fear that the revived corpuscular philosophy would carry the stigma of atheism adhering to ancient atomism. They had solved the problem, they thought, by the simple expedient of having God endow the particles of matter with

---

[31] See Dobbs, *Foundations*, pp. 219–221, for a discussion of Newton's attempt to achieve significant union between chemistry and mechanical philosophy with his concept of "chymical" subunits.
[32] Burndy MS 16, fol. 5r; Newton's parentheses; my brackets.
[33] Newton, *Correspondence*, Vol. III, pp. 207, 211.
[34] Dobbs, *Foundations*, esp. pp. 100–105.

motion at the moment of creation. All that resulted then was due not to random corpuscular action but to the initial intention of the Deity.[35] Later writers, going further, had carefully instated a Christian Providence among the atoms (where the ancients of course had never had it). Only Providence could account for the obviously designed concatenations of the particles, and so, via Christianity, a fundamental Stoic critique actually came to be incorporated into seventeenth-century atomism. This development was all to the good in the eyes of most Christian philosophers: atomism now supported religion, because without the providential action of God the atoms could never have assumed the lovely forms of plants and animals so perfectly fitted to their habitats. Though present in Christianity from a very early period, this argument from design assumed unparalleled importance in the seventeenth century, and if the new astronomy had raised doubts about the focus of Providence upon such an obsure corner of the cosmos, the new atomism seemed to relieve them.[36] The difficulty came when one began to wonder how Providence operated in the law-bound universe emerging from the new science, and that difficulty was especially severe in the Cartesian system, where only matter and motion were acceptable explanations. Even though Descartes had argued that God constantly and actively supported the universe with his will, in fact it seemed to Henry More and others that Descartes's God was in danger of becoming an absentee landlord, one who had set matter in motion in the beginning but who then had no way of exercising his providential care.

Newton faced this theological difficulty squarely and directly. We have already seen him insisting that God produces "all that diversity of natural things which we find suited to different times and places," and we have seen his speculations on the vital agent that reduces matter to its primordials and then "proceeds to generation," making "all that diversity of natural things" from the common matter. It is now time to make the connection in Newton's mind explicit: the alchemical active principle—the vital spirit of which he was in hot pursuit—was no more and no less than the agent by which God exercised his providential care among the atoms. God's use of such a "creature" to effect his purposes seemed good to Newton, for it seemed to enhance God's power, as he succinctly observed regarding the creation of matter:

> If any think it possible that God may produce some intellectual creature so perfect that he could, by divine accord, in turn produce creatures of a lower order, this so far from detracting from the divine power enhances it; for that power which can bring forth creatures not only directly but through the mediation of other creatures is exceedingly, not to say infinitely, greater.[37]

---

[35]Kargon, *Atomism*, pp. 64, 67–68, 87–89; Margaret J. Osler, "Descartes and Charleton on Nature and God," *J. Hist. Ideas*, 1979, 40:445–456.

[36]Jacob Viner, *The Role of Providence in the Social Order: An Essay in Intellectual History* (Philadephia: American Philosophical Society, 1972), pp. 8–9.

[37]CUL MS Add. 4003; see Isaac Newton, *Unpublished Scientific Papers of Isaac Newton: A Selection from the Portsmouth Collection in the University Library, Cambridge*, ed. and trans. A. Rupert Hall and Marie Boas Hall (Cambridge: Cambridge Univ. Press, 1962), pp. 108, 142. In "Neoplatonism and Active Principles: Newton and the *Corpus Hermeticum*," in Robert S. Westman and J. E. McGuire, *Hermeticism and the Scientific Revolution: Papers Read at a Clark Library Seminar, March 9, 1974* (Los Angeles: Clark Memorial Library, Univ. California, 1977), p. 107, J. E. McGuire has cited a portion of this statement of Newton's to support precisely the opposite point. McGuire argues that Newton "emphatically rejects" intermediary agents between God and the world, but McGuire could have reached his conclusion only by omitting (as he does) Newton's clear statement that an intermediary "so far from detracting from the divine power enhances it. . . ." Cf. also Newton's statement, cited at n. 64, that "the supreme God doth nothing by himself which he can do by others."

So Newton had no theological qualms concerning an active universal agent that acted for God in the world to produce and guide the processes of vegetation. In the words of an early eighteenth-century alchemist whom Newton was later to study diligently, the alchemical spirit was God's "Vicegerent," and it was through alchemy that man could *"pierce through the external shell of things, to the internal working Spirit . . . and so become an Opener and Manifester of the Wonders of God in Nature."*[38] Newton understood alchemy to be one of the most, if not the most, important of his many studies, for if all went well he could demonstrate God's action in the world in an absolutely irrefutable fashion by demonstrating the operations of the nonmechanical vegetable spirit, and thus lay the specter of atheism to rest forever more. With some appreciation of the momentous issues at stake for Newton, let us return to his study of and speculations on "the internal working Spirit" in all things.

In his treatise "On nature's obvious laws and processes in vegetation," Newton used the term "vegetation" in quite a general sense, applying it to all three kingdoms of nature—mineral and animal as well as vegetable. He says, furthermore, that his "vegetable spirit is radically the same in all things,"[39] and he is greatly interested in working out the similarities and dissimilarities of its actions in the three kingdoms. At first he suggests that the vegetable spirit is none other than the aether that pervades all things, and that congealed aether, interwoven with the grosser texture of sensible matter, constitutes the material soul of all things.[40] This is the same material aether, so similar to that of the Stoics, that he has said is daily inspired by the earth-animal and which also here accounts for the action of gravity. But then Newton decides that the aether is more probably only the vehicle for "some more active spirit" entangled in it. Here we find the influence of Neoplatonism returning, for he suddenly sees what this active spirit might be: perhaps it is "the body of light." In a rapture of insight he sets out parallels and relationships between the vegetable spirit and light:

> This spirit perhaps is the body of light because both have a prodigious active principle, both are perpetual workers. 2 Because all things may be made to emit light by heat. 3 The same cause (heat) banishes also the vital principle. 4 'Tis suitable with infinite wisdom not to multiply causes without necessity. 5 No heat is so pleasant and bright as the sun's. 6 Light and heat have a mutual dependence on each other, and no generation without heat. Heat is a necessary condition to light and vegetation. [Heat excites light and light excites heat; heat excites the vegetable principle and that increases heat.] No substance so indifferently, subtly, and swiftly pervades all things as light, and no spirit searches bodies so subtly, piercingly, and quickly as the vegetable spirit.[41]

Newton never fully abandoned the moment of truth he captured in this remarkable passage. As we have just seen, in his speculations about the vegetable spirit he was seeking the source of all the God-directed processes of generation and growth in the natural world, processes that produced the endless variety of living forms and could not be relegated to the mechanical actions of gross corpuscles.

---

[38] Cleidophorus Mystagogus, *Mercury's Caducean Rod: Or, The great and wonderful Office of the Universal Mercury, or God's Vicegerent, Displayed . . .* (London, 1702), sig. A3v.

[39] Burndy MS 16, fol. 6r; cf. Keynes MS 12A, fol. 1v.

[40] Burndy MS 16, fols. 3v–4r; cf. Newton, *Correspondence*, Vol. I, pp. 363–366, and Dobbs, *Foundations*, pp. 204–206.

[41] Burndy MS 16, fol. 4r; Newton's brackets.

Newton's alchemical papers continue to show a preoccupation with the workings of this spirit, this active principle of vegetable actions.

Newton did not remain absolutely committed to the idea that the vegetable spirit was the body of light, however. In his 1674 tract on vegetation, he was, as we have seen, thinking in terms of a material agent, carried by the aether but composed of finer and more active particles. Newton's various aethers differed from the Stoic *pneuma* in that they were all particulate, the continuity of the medium in Stoic thought apparently having little appeal to him at first. In fact one and all of Newton's aethers, from the very earliest version in his student notebook to the final self-repulsive aether of the *Opticks*, were composed of fine particles of matter.[42] These particles, he realized (probably as he was composing the *Principia*), could only serve as a drag on the motion of the planets and the comets.[43] Hence during the 1680s he turned away from aethereal mechanisms, which in the early years had served not only to explain gravity but also to explain chemical reactions and cohesion, and toward theories of interparticulate forces that could account for the operations of chemistry and the problems of cohesion and capillarity.

The demise of his cosmic aether undoubtedly led Newton to reconsider his interstitial aether as well, and perhaps at this time he thought seriously about the void of ancient atomism. His copy of the *De rerum natura* of Lucretius, albeit showing signs of concentrated study, could not have been purchased before 1686.[44] However, Newton was later to pin his argument for spaces empty of matter—his pores—in the larger particles to certain experimental facts: that light can be transmitted through "pellucid" bodies without being stifled and lost, as it surely would be if it hit a truly solid particle, and also that magnetic and gravitational forces could pass through even very dense and opaque bodies "without any diminution."[45] Whatever the source of his ideas, his grosser corpuscles came to consist not just of increasingly complex parts, as in the 1670s, but of hierarchical structures of parts and pores. In the hierarchies the parts were held together not by the presence of an aether (either mechanical or active) but by interparticulate forces. Newton's pores were thus not truly void, for they were permeated by forces, and in the end all he seems to have accepted from ancient atomism was the limited divisibility of matter. The primitive particles never wear out or break, and their permanence guarantees that nature will be "lasting," with the "Changes of corporeal Things . . . placed only in the various Separations and new Associations and Motions of these permanent Particles. . . ."[46]

When the basic aether was discarded, what was then the fate of the vegetable spirit that had been entangled in the aether, and did Newton continue to suppose that the spirit might be "the body of light"?

---

[42] Newton, *Opticks*, pp. 347–354, esp. 350–353.

[43] CUL MS Add. 4003 (Newton, *Unpublished Papers*, pp. 113, 147): "For if the aether were a corporeal fluid entirely without vacuous pores, however subtle its parts are made by division, it would be as dense as any other fluid, and it would yield to the motion of bodies through it with no less sluggishness. . . ." In my opinion previous investigators placing this MS in the late 1660s or early 1670s have been seriously misdating it. It simply could not have been written from the plenist viewpoint of Newton's early manhood and must date at the earliest from the early 1680s. Cf. the recent review of Newton's changing ideas on the aether, subtle media, and vortices in I. Bernard Cohen, *The Newtonian Revolution* (Cambridge/London/New York: Cambridge Univ. Press, 1980), pp. 114–120.

[44] Harrison, *Library*, item 990, p. 183.

[45] Newton, *Opticks*, pp. 266–269.

[46] *Ibid.*, p. 400.

As to the second question, Newton continued to think that light at least contributed to the activity of matter, that it could enter into the composition of bodies and in turn be emitted by bodies. "Are not gross Bodies and Light convertible into one another," he asked in the *Opticks*, "and may not Bodies receive much of their Activity from the Particles of Light which enter their Composition?"[47] But he says "much of their Activity," not all of it, whereas in the treatise on vegetation he had been quite clear that if the vegetable spirit were separated from the mass of the body, "there would remain but a dead and inactive earth."

In any event, the animating vegetative principle had been broadly conceived by the alchemists. It effected the "acuation" or activation of matter, and the process frequently was called "illumination," but it would be a mistake to think that Newton or any other alchemist limited this alchemical spirit to the nature of visible light.[48] True, it was associated with the light of Genesis. The light was God's creature and an obvious candidate for the active agent that God pressed into service to help with the rest of creation. We find tract after alchemical tract describing the alchemical process, utilizing "illumination," as a microcosmic model of God's actions at the beginning of time. But the active principle could also be identified with the spirit of God that moved upon the face of the waters in Genesis, and indeed it was sometimes so identified, which left the whole question in a state of delightful ambiguity. Seventeenth-century spirits inhabited a gray area between the corporeal and the incorporeal, shading over into either realm as required by the exigencies of the problem to be explained.[49] Which was the vegetable spirit? Corporeal or incorporeal?

Newton's ideas seem to have partaken fully of this ambiguity as he continued to worry the alchemical literature for information on the vegetation of metals. Metals are the only part of the mineral kingdom that vegetate in his opinion, other mineral substances being formed mechanically.[50] Vegetation in metals is thus the simplest case for study, the vegetation of the animals and vegetables in the other kingdoms being obviously more complex. So in the vegetation of metals lay the most accessible key to the problem of nonmechanical action.

In a long series of papers on the regimens of the great work of alchemy (dating from the 1670s until the 1690s), Newton describes again and again the living processes he thinks are occurring at the lowest levels of the hierarchies of matter to form the metals. The final climactic manuscript in the series, written in the mid-1690s, is a draft for a formal treatise, one chapter of which is devoted to "the first agent," that without which nothing is done, without which life and activity will not ensue. The first agent is the bond between the male and female aspects of the material that enables them to unite and produce "the young king," "the more noble offspring," which is the true beginning of the work and which we may interpret as the particles at the first level of composition in the Newtonian hierarchies.[51]

As to the corporeal or incorporeal nature of this bond, Newton gives us little

[47]*Ibid.*, p. 374; McMullin, *Newton on Matter and Activity*, pp. 84–94.
[48]B. J. T. Dobbs, "Newton's Copy of *Secrets Reveal'd* and the Regimens of the Work," *Ambix*, 1979, 26:145–169.
[49]Walter Pagel, *Paracelsus: An Introduction to Philosophical Medicine in the Era of the Renaissance* (Basel/New York: Karger, 1958), *passim*; McMullin, *Newton on Matter and Activity*, pp. 54–56.
[50]Burndy MS 16, fol. 3r.
[51]Dobbs, "Newton's Copy of *Secrets Reveal'd*"; Isaac Newton, "Praxis," Babson MS 420, Babson Institute Library, Babson Park, Mass., pp. 10–12, cited by permission of the Babson Institute.

indication, and it seems likely that he was as incapable of fully resolving that question for the bonds between the particles of micromatter as he was for the gravitational bonds among celestial bodies. Hence his recourse to the concept of interparticulate forces, which skirted the metaphysical problem while still offering an explanation for cohesion and vegetation.

Newton's contemporaries, accustomed as they were to the idea of action by mechanical impact only, found these forces of Newton's unsavory, and indeed Newton himself often seemed uneasy and defensive about them. As we have already noted, he even put forward one last aethereal hypothesis for gravity in the *Opticks*. It has been argued by the Halls, however, that Newton's aethers were always presented in a tentative way, whereas he was much more forthright in statements about forces. Forces did really exist, though he had no way of explaining their actions, and they certainly were physical, though not material, and were measurable by their effects. They were, furthermore, law-bound; they were natural forces, and they acted by regular and natural means even though not mechanical in the seventeenth-century sense. Newton had Samuel Clarke explain this to Leibniz with respect to gravity, and Newton clearly thought that the interparticulate forces were strictly analogous to the force of gravity. Although the metaphysical status of the forces remained in doubt, Newton continued to argue for their usefulness in explaining the phenomena of nature.[52]

In his most general statement on the forces associated with matter Newton had the following to say:

> It seems to me . . . that these Particles have not only a *Vis inertiae*, accompanied with such passive Laws of Motion as naturally result from that Force, but also that they are moved by certain active Principles, such as is that of Gravity, and that which causes Fermentation, and the Cohesion of Bodies.[53]

These several forces or "Principles" serve a variety of functions. The passive laws are those of impact phenomena, and Newton simply means here to indicate that for the particles of micromatter they are essentially the same as for large bodies, dependent upon the mass and the impressed forces. Gravity, on the other hand, though it had originally been one of the passive principles in Newton's thinking, had finally come to reside among the active ones.[54] Like the passive laws of motion, the law of gravity applies to the small particles just as it does to large bodies. The other active principles, those that cause fermentation and cohesion, apply only to the realm of micromatter. Elsewhere Newton also argues for a longer-range repulsive "Virtue" that comes into play past the very short range of the attractive cohesive force.[55] It also acts only with respect to micromatter.

Fermentation is, as we have seen, a concept that Newton used with considerable frequency, both in his alchemical papers and in his published works. It seems to refer to any sort of process that could agitate and rearrange the particles of micromatter, and it appears in discussions of condensation, rarefaction, the emission of light or volatile substances from solid ones, putrefaction, generation, vegetation,

---

[52]A. Rupert Hall and Marie Boas Hall, "Introduction to Part III, Theory of Matter," in Newton, *Unpublished Papers*, pp. 183–213.

[53]Newton, *Opticks*, p. 401.

[54]The evolution of Newton's thought on the nature of the gravitational force (a fascinating topic and related to his alchemical studies) will be explored in B. J. T. Dobbs, *The Janus Faces of Genius: The Role of Alchemy in Newton's Thought* (forthcoming).

[55]Newton, *Opticks*, p. 395.

and the assimilation of nutriment. When working with the cohesive force, fermentation tends to draw the primordials together into the various forms of sensible matter:

> ... the particles of bodies ... are impelled towards one another and cohere.... Depending on the force and manner of the coming together and cohering of the particles, they form bodies which are hard, soft, fluid, elastic, malleable, dense, rare, volatile, fixt.... The particles of such bodies can very easily be agitated by a vibrating motion ... ; through such agitation, if it is slow and continuous, they little by little alter their relative arrangements, and by the force by which they cohere they are more strongly united, as happens in fermentation and vegetation, by which the rarer substance of water is gradually converted into the denser substances of animals, vegetables, salts and stones.[56]

In rarefaction, on the contrary, fermentation worked against the force of cohesion, agitating the particles so violently that they broke away from the control of the cohesive force and came under the influence of the force of repulsion. In processes of solution, for example, when particles rush together so violently that they grow hot, the relatively large composite particles of "air, vapours and exhalations" may be expelled, or very small particles may be given off as light. "Putrid vapours" may likewise "shine from the agitation due to putrefaction." By fermentation Newton thus seems to have meant almost any process that worked in opposition to a static state of either cohesiveness or dispersion among the particles. It *changed* particulate relationships and thus may be seen as a general term comprehending all the rearrangements of the small particles of matter.[57]

By postulating active forces of cohesion, repulsion, and fermentation at work between and among the small particles of matter, Newton attempted to formalize his long-standing concern for problems of cohesion and activity, for life and vegetation, and for the specific mode of God's providential care of the world. The force of fermentation is especially closely related to his alchemical work, for it is the force that initiates and controls changing relationships among the particles. It seems clearly to be derived from the "fermental virtue" and "vegetable spirit" of earlier years. Whether the forces acted in a corporeal or a noncorporeal way, however, Newton was not able to say, and he did not rest quite easy with his formal statement on them. Just as he inserted a final version of the aether in the *Opticks* to explain gravity, so also he appended a final suggestion for interparticulate forces to the General Scholium of the *Principia*, in the form of "a certain most subtle spirit."[58]

---

[56] CUL MS Add. 3965 (Newton, *Unpublished Papers*, pp. 302–308), the Halls' translation slightly modified.

[57] Though Newton articulated his ideas on fermentation more precisely than his contemporaries, his basic conception does not seem to have differed substantially from their very broad one. Originally from the Latin *fervere*, to boil, fermentation referred to any internal working of a substance which changed its properties as well as to the action of a leaven. Only after the time of Pasteur was the term restricted to a relatively small class of organic reactions produced by identifiable "ferments." The *OED* cites many examples from Newton's period, many of them in context with putrefaction, generation, or vegetation. Medical and physiological applications were also common. See Robert G. Frank, Jr., *Harvey and the Oxford Physiologists: Scientific Ideas and Social Interaction* (Berkeley/Los Angeles/London: Univ. California Press, 1980), pp. 165–169; Audrey B. Davis, *Circulation Physiology and Medical Chemistry in England, 1650–1680* (Lawrence, Kan.: Coronado Press, 1973); Walter Pagel, *New Light on William Harvey* (Basel/New York: Karger, 1976), pp. 155–156.

[58] Newton, *Principia* (Mott-Cajori), Vol. II, p. 547; Marie Boas Hall and A. Rupert Hall, "Newton's Electric Spirit: Four Oddities," *Isis*, 1959, 20: 473–476; CUL MS Add. 3965 (Newton, *Unpublished Papers*, pp. 355–364).

That spirit was "electrical attraction unexcited": electricity that had not been made to expand through excitation by friction. The electrical spirit was a "subtle medium" that permeated solid bodies, a "most active" medium that could emit light, and Newton had come to realize that this unexcited electrical medium might be the true physical manifestation of the vegetable spirit.[59] In a manuscript entitled "De vita & morte vegetabili," he made that identification quite explicit, giving cohesion, diversity, continuity of living form, assimilation, and vegetation all an electrical explanation:

> The particles of bodies coalesce and cohere in diverse ways by means of the electric force.
>
> * * *
>
> And [as] by the friction of an electric body its attraction is extended, so also by the action of living the electric force of living parts is extended, and by a strong attraction it happens that those parts both conserve their proper form and situation and impart their nutriment by degrees in the same manner as a magnet converts iron into a magnet and fire converts bodies into fire and leaven converts paste into leaven: but with the ceasing of vegetable life that vital attraction ceases and its absence immediately begins the action of dying which we call corruption and putrefaction.[60]

But we should note that Newton variously used the language of "force," "spirit," and "medium" in these passages and so leaves the question still shrouded in ambiguity.

IN THE END: SPIRIT

For Newton, it was theologically unacceptable to designate the forces that generated activity in nature as intrinsic components of matter. Activity—the generation of activity—was the province of divinity; to attribute to "brute matter" the capacity for initiating motion would lead to atheism. Newton was always aware of the danger inherent in attributing activity to matter, and he always insisted that his forces acted only between particles. They were not really a part of matter itself, but were manifestations of God's activity in nature, a position that subsequent philosophers soon explicitly rejected.[61]

Given Newton's insistence upon the divine generation of activity, however, it hardly mattered whether his electrical "force," "spirit," or "medium" was corporeal or incorporeal. Where Newton said "active" in his discussions of forces, we really should understand that a divine spirit is there at work either directly or indirectly, and that divine spirits, despite the ambiguity of the broad seventeenth-century usage of the term "spirit," are unequivocally incorporeal. Since a divine spirit was necessarily at work behind any active force that generated motion, then if an active force proved to be incorporeal, its operations would be direct evidence of the operation of divinity in the universe. If it proved to be corporeal, then an

---

[59]Joan L. Hawes, "Newton and the 'Electrical Attraction Unexcited,'" *Ann. Sci.,* 1968, *24*:121–130; Newton, *Unpublished Papers,* pp. 357, 362.

[60]CUL MS Add. 3970, as trans. by Joan L. Hawes in "Newton's Two Electricities," *Ann. Sci.,* 1971, *27*:95–103.

[61]McMullin, *Newton on Matter and Activity;* J. E. McGuire and P. M. Heimann, "The Rejection of Newton's Concept of Matter in the Eighteenth Century," in *The Concept of Matter in Modern Philosophy,* ed. Ernan McMullin (Notre Dame/London: Univ. Notre Dame Press, 1978), pp. 104–118; P.M. Heimann and J. E. McGuire, "Newtonian Forces and Lockean Powers: Concepts of Matter in Eighteenth-Century Thought," *Historical Studies in the Physical Sciences,* 1971, *3*:233–306.

incorporeal spirit must stand behind it. Behind the universal force of gravity, whether its operations were effected by corporeal or incorporeal means, stood the literal omnipresence of God: "He [God] is omnipresent, not *virtually* only, but also *substantially;* for virtue cannot subsist without substance." ". . . [T]hose ancients who more rightly held unimpaired the mystical philosophy as Thales and the Stoics, taught that a certain infinite spirit pervades all space *into infinity,* and contains and vivifies the entire world."[62]

Similarly, in the end it did not matter much to Newton whether the interparticulate forces operated in nature by corporeal or incorporeal means.[63] He found a satisfactory solution to that problem, at least for himself, by assuming that the forces at play in micromatter were also subsumed by a spiritual being, just as the force of gravity was. The being was Christ, who acted as God's viceroy in these matters. Taking literally the beginning passages of the Gospel of John, Newton identified Christ with the Word and argued that he was with God before his incarnation, "even in the beginning," and that he was God's active agent throughout time, speaking to Adam in paradise and appearing to the patriarchs and Moses "by the name of God": "For the father is the invisible God whom no eye hath seen nor can see." Christ wrestled with Jacob and gave the law on Mount Sinai; after his resurrection his testimony was "the spirit of prophecy":

> He [Christ] is said to have been *in the beginning with God* & that *all things were made by him* to signify that as he is now gone to prepare a place for the blessed so in the beginning he prepared & formed this place in which we live, & thenceforward governed it. For the supreme God doth nothing by himself which he can do by others.[64]

These views on the nature of Christ are related to Newton's Arianism, for he goes on to argue that "God & his son cannot be called one God upon account of their being consubstantial," but that they may and should be called one God through a "unity of Dominion, . . . the Son receiving all things from the Father, being subject to him, executing his will, . . . & so is but one God with the Father as a king & his viceroy are but one king."[65] Thus Christ is the viceroy, the spiritual being that acts as God's agent in the world, a very unorthodox Christ indeed but one whose many duties keep him engaged with the world throughout time. A part of his function is to insure God's continued relationship with his creation; Newton's God is in no danger of becoming an absentee landlord, for he always has the Christ transmitting his will into action in the world.

The duties assigned to Newton's Christ would seem to place him in charge of such active natural entities as the vegetable spirit, which, we have argued above,

---

[62]*Principia* (Mott-Cajori), Vol. II, p. 545 (cf. *David Gregory's Memoranda,* p. 30); CUL MS Add. 3965.12, as cited in J. E. McGuire and P. M. Rattansi, "Newton and the 'Pipes of Pan,'" *Notes and Records of the Royal Society of London,* 1966, *21*:108–143. Cf. Westfall, *Force,* pp. 395–398, 418–422 (nn. 180–186); Edward Grant, *Much Ado About Nothing: Theories of Space and Vacuum from the Middle Ages to the Scientific Revolution* (Cambridge/New York: Cambridge Univ. Press, 1981), pp. 240–255.

[63]For further exploration of the problem see Dobbs, *Janus Faces of Genius* (cit. n. 54).

[64]Yahuda MS 15, Jewish National and University Library, Jerusalem, as quoted in David Castillejo, *The Expanding Force in Newton's Cosmos as Shown in his Unpublished Papers* (Madrid: Ediciones de Arte y Bibliofilia, 1981), pp. 61–62; Newton's italics.

[65]Yahuda MS 15, as quoted in Castillejo, *Expanding Force,* p. 74; on Newton's Arianism see Richard S. Westfall, *Never at Rest: A Biography of Isaac Newton* (Cambridge/London/New York: Cambridge Univ. Press, 1980), pp. 311–319.

Newton saw as exercising God's providential care in shaping "[a]ll that diversity of natural things which we find suited to different times and places," God's alchemical "Vicegerent." Newton apparently thought that when organized matter first arose from chaos, Christ, as God's executive, directed the vegetative, nonmechanical processes between the most minute primordials ("in the beginning he prepared & formed this place in which we live"), then continued to direct the vegetative operations of nature ("& thenceforward governed it"). For, Newton said in the General Scholium, natural diversity "could arise from nothing but the ideas and will of a Being necessarily existing."[66] It was the Christ, united with God in a "unity of Dominion" though not of substance, that put the ideas into effect. Once the religious significance of Christ's cosmological function appears, and it becomes apparent that the most intimate operations of Providence may be studied in nonmechanical, "vegetable" chemistry, then, to return to our former point, it must have seemed irrelevant to Newton whether the motions were directly effected by corporeal or incorporeal means. The forces were active principles and thus necessarily subsumed ultimately by a divine spirit.

CONCLUSION

The relationship between Newton's alchemy and his published theory of matter has been a persistent problem in Newtonian scholarship for two and a half centuries. The last half century, however, has yielded some intensive investigations of other aspects of Newton's work which provide an intellectual matrix for the resolution of the problem. Especially when considering Newton's theological concerns, one can now understand his intense interest in the alchemical process, for he saw it as the epitome of God's providential, nonmechanical action in the world.

In addition, new evidence from Newton's alchemical papers, especially from the 1674 treatise on vegetation, in which Newton makes explicit the relationships between vegetable and mechanical chemistry, makes it possible to trace an evolutionary process in which alchemy and corpuscularianism interact to produce his published theory of matter. We have shown that his doctrine of the unity of matter and its transformability appear very early in an alchemical context; that his final notion of structured particulate hierarchies was formulated in the 1670s, when he differentiated between common chemistry and vegetable action; and that the active principles that operate between and among the small particles of matter in the *Opticks* are identical with those that so operate in the alchemical papers. Whether they be called forces, virtues, media, principles, or spirits, and whether they operate by corporeal or incorporeal means—all that is in the end only of secondary importance, for activity requires divinity, and nonmechanical action indicates the presence of the divine in the natural order. Universal gravity demonstrates the omnipresence of God the Father; vegetable actions in micromatter indicate continuing supervision of the world by God's viceroy, the Christ. But now perhaps Newton has finally said enough for us to grasp his meaning, and we may conclude with that final triumphant cry by which he summed up his life's work, both for himself and for his public: "And thus much concerning God; to discourse of whom from the appearances of things does certainly belong to Natural Philosophy."[67]

[66]*Principia* (Mott-Cajori), Vol. II, p. 546.
[67]*Ibid.*

Totius in verba

Rhetoric and Authority in the Early Royal Society

By Peter Dear

> Simplicio is confused and perplexed, and I seem to hear him say, "who would there be to settle our controversies if Aristotle were to be deposed?"
> —Sagredo, in Galileo's *Dialogue Concerning the Two Chief World Systems,* trans. Stillman Drake.

THE SCIENTIFIC REVOLUTION reached a stage of consolidation in the second half of the seventeenth century. The ferment of new ideas, and the conscious rejection of old, had resulted in the formation of an identifiable community of practitioners with a shared ideal of natural inquiry. The clearest indicators of this consolidation are the scientific societies that arose in the 1650s and 1660s, foremost among them the Académie Royale des Sciences in Paris and the Royal Society of London. These groups stand as testimony to a new attitude toward knowledge of nature.[1]

This article, concentrating on the Royal Society, examines how the cooperative investigation of nature both shaped and was made possible by the new forms of natural knowledge generally associated with the Scientific Revolution. The present study, however, does not mean to suggest that these new conceptions emerged suddenly with the formation of the Royal Society, nor that they were especially English. (Some recent literature on the connections between experimental philosophy and ideological currents in Restoration England is curiously Anglocentric.)[2] Instead, the Society is treated here as a convenient focus at the end of a process of change.

I would like to thank Michael Mahoney, Pauline Dear, Tom Broman, and John Carson for discussion and advice, and Steven Shapin for his encouragement.

[1] On scientific societies in general, see Harcourt Brown, *Scientific Organizations of Seventeenth Century France* (Baltimore: 1934); Martha Ornstein, *The Role of Scientific Societies in the Seventeenth Century* (1913; New York: Arno Press, 1975).

[2] See esp. James R. Jacob and Margaret C. Jacob, "The Anglican Origins of Modern Science: The Metaphysical Foundations of the Whig Constitution," *Isis,* 1980, 71:251–267, which claims that the connection between religious ideology and matter theory reveals the "social genesis of the conceptual revolution that culminated in the Newtonian synthesis, the hallmark of the Scientific Revolution" (p. 254). This article uses a remarkably conventional view of the Scientific Revolution, one resting on a European perspective, to justify the broader significance of a very focused analysis of the English context.

Frontispiece to Thomas Sprat's History of the Royal Society of London *(3rd ed., London, 1722), displaying the Society's motto, suggested by John Evelyn, adopted in 1662. From "Nullius addictus iurare in verba magistri," Horace, Epistles 1. 1. 14. Courtesy of the E. F. Smith Memorial Collection, University of Pennsylvania.*

The Royal Society has been the most intensively examined of the seventeenth-century scientific academies. Studies have been made of the scientific work carried out by Fellows of the Society and of the Society as the representative of particular social and ideological movements in Restoration England. In addition, work has been done on the epistemological positions promoted by many of the Fellows.[3] The present article considers the Royal Society from a microsocial perspective, as the embodiment of an ideal of cooperative research. Although the Society did little to control natural philosophical endeavor in a strong, programmatic, and corporate sense, as will be discussed below, its very existence and the desire of so many people to belong to it are themselves indicators of the Royal Society's significance *qua* society. The Royal Society is worth studying, not because it followed any enforced and exclusive code of practice, but because it can be taken as a symbol of particular conceptions of natural philosophy in this period, especially of *cooperative* natural philosophy.

For cooperative inquiry to be conducted, it was necessary to agree upon standards and criteria; conversely, the credibility of a contribution to the common endeavor had to be determined by appeal to some sort of authority. The place of experience in the "new philosophy" played a crucial part in this process, first, because the idea of experiment and observation played an important role in the Scientific Revolution generally and in the Royal Society in particular, and, second, because polemical antischolastic writings of the period opposed experience to an alleged reliance on ancient authority. I shall argue below that not only did authority take on a new guise, but that in its doing so, the concept of experience and its place in natural knowledge also changed.

The significance of the image of the "book of nature" so frequently invoked in the sixteenth and seventeenth centuries may lie precisely in this relationship between a new structure of authority and a new view of experience. Just as a written text needs to be read and at some level to be glossed, so the "book of nature" must be interpreted. In using the latter text, the early Fellows of the Royal Society employed new techniques of exegesis. The final product of their work was the book, letter, or article in which that work became a property of the whole community. Hence an examination of form will contribute perhaps at least as much as an examination of content to an understanding of what the new philosophy, as exemplified by the early Royal Society, actually was.

---

[3] For the scientific work of the Society see, e.g., Dorothy Stimson, *Scientists and Amateurs: A History of the Royal Society* (New York: Schuman, 1948). For social aspects, see Michael Hunter, *Science and Society in Restoration England* (Cambridge: Cambridge Univ. Press, 1981); Hunter, "The Social Basis and Changing Fortunes of an Early Scientific Institution: An Analysis of the Membership of the Royal Society, 1660–1685," *Notes and Records*, 1976, *31*:9–114; Hunter, *The Royal Society and its Fellows, 1660–1700: The Morphology of an Early Scientific Institution* (British Society for the History of Science Monographs, 4) (Chalfont St. Giles, Bucks.: British Society for the History of Science, 1982); James R. Jacob, *Robert Boyle and the English Revolution: A Study in Social and Intellectual Change* (New York: Burt Franklin, 1977); and Jacob, *Henry Stubbe, Radical Protestantism and the Early Enlightenment* (Cambridge: Cambridge Univ. Press, 1983), esp. Ch. 5. On epistemological positions see esp. Barbara Shapiro, "History and Natural History in Sixteenth- and Seventeenth-Century England: An Essay on the Relationship Between Humanism and Science," in *English Scientific Virtuosi in the 16th and 17th Centuries* (Los Angeles: William Andrews Clark Memorial Library, 1979), pp. 3–55; Shapiro, *Probability and Certainty in Seventeenth-Century England* (Princeton, N.J.: Princeton Univ. Press, 1983); and also, esp. on Boyle, Henry G. Van Leeuwen, *The Problem of Certainty in English Thought 1630–1690* (The Hague: Martinus Nijhoff, 1963).

I

As the frontispiece to Thomas Sprat's *History of the Royal Society* testifies, the early Fellows liked to think of themselves as following the precepts and method of Francis Bacon. There, beside their royal patron, sat the noble *Artium Instaurator,* as if to lend his seal of approval to the Fellows' activities.[4] The Society itself, therefore, had a ready-made analogy: "*Solomon's House* in the NEW ATLANTIS, was a Prophetick Scheam of the ROYAL SOCIETY," said Glanvill; indeed, Bacon's "SOCIETY of *Experimenters* in a *Romantick Model*" had inspired the virtuosi intentionally to "set themselves on work upon this *grand* Design."[5] Certainly, such claims indicate an awareness that cooperative inquiry was a novel sort of cognitive enterprise, one demanding that knowledge be handled according to new, special criteria. But those criteria did not come from Bacon's account of Solomon's House. That notional institution divided knowledge into its component parts, with personnel assigned to distinct, autonomous tasks all contributing to the whole. The structure reflected Bacon's belief that the acquisition of knowledge was somehow an automatic process, once the correct procedure was followed.[6]

Despite Bacon's ideal of a cooperative investigative institution, the Royal Society in reality failed even to act as a successful coordinator of the projects of individuals. Shared ideals among the early Fellows did not, as a rule, translate into shared projects or programs of research; half-hearted attempts at promoting the compilation of Baconian histories of trades bore little or no fruit, despite Sprat's extravagant claims.[7] The Royal Society was more of a club than a college. If it did not resemble Solomon's House, however, the Royal Society did have a collective voice. The minutes of its meetings, its *imprimatur,* and, although technically run by Oldenburg in a private capacity, the *Philosophical Transactions*—all served to underline the corporate character of the Society, and hence the notion of cooperative endeavor—of what the work overseen (at least) by the official apparatus of the Society ought to be.

As a first approximation, one can describe the "research ideal" of the society as an association of individuals with common interests who were engaged in work that was commonly perceived to be of value; cooperation in the work itself was not part of this ideal. What common values did the members share? First,

---

[4] Thomas Sprat, *History of the Royal Society* (London, 1667); facs. ed. Jackson I. Cope and Harold Jones (Saint Louis: Washington Univ. Press, 1958), frontispiece. For the view of Bacon as a convenient emblem for the early Royal Society see, e.g., Barbara Shapiro, *John Wilkins, 1614–1672: An Intellectual Biography* (Berkeley: Univ. California Press, 1969), pp. 205–206. Margery Purver, *The Royal Society: Concept and Creation* (Cambridge, Mass.: MIT Press, 1967), argues for the basic Baconianism of the early Society; but see the review by Charles Webster, "The Origins of the Royal Society," *History of Science,* 1967, 6:106–128.

[5] Joseph Glanvill, "An Adress to the Royal Society," in *Scepsis Scientifica* (London, 1665), facs. ed. in Glanvill, *The Vanity of Dogmatizing: The Three Versions,* intro. by Stephen Metcalf (Sussex: Harvester Press, 1970), p. [22]; Glanvill, quoted in Purver, *Royal Society,* p. 96.

[6] Francis Bacon, *The New Atlantis,* in *The Works of Francis Bacon,* ed. J. Spedding, R. L. Ellis, and D. D. Heath, 7 vols. (London, 1859–1874), Vol. III, pp. 125–166, esp. pp. 164–165.

See Walter E. Houghton, Jr., "The History of Trades: Its Relation to Seventeenth-Century Thought," in Philip P. Wiener and Aaron Noland, ed., *Roots of Scientific Thought: A Cultural Perspective* (New York: Basic Books, 1957), pp. 354–381. For Sprat's comments, see Sprat, *History,* pp. 155–157.

there had to be tacit agreement on the nature of knowledge. This, in turn, implied a common standard of authority, and ancient authority, the Fellows habitually claimed, no longer served. But what constituted ancient authority?

For Aristotle, who was to become the preeminent "ancient authority," phenomena were, literally, *data,* "givens." They were statements about how things behave in the world, and they were to be taken into account when discussing topics concerning nature. The immediate sources of phenomena were diverse: common opinion and the assertions of philosophers, as well as sense-perception. Given these statements, a system of syllogistic reasoning yielded, in principle, a theoretical description and explanation of them. This was "dialectical" rather than "demonstrative" reasoning, based on probable rather than certain grounds. Because "probable" meant "worthy of approbation" rather than simply "likely," the probability of a statement was intimately bound up with matters of authority: the more authoritative the source, the more probable the statement.[8]

The Middle Ages and Renaissance continued this approach toward phenomena, that is, statements about how the world behaves. Since the focus of scholastic pedagogical procedure was on disputation, emphasis was given to the proper *form* of an argument. The argument's premises were almost the least interesting part.[9] The establishment of a premise usually consisted of quoting an authority for it—above all, Aristotle. For medieval and Renaissance scholastics, as for Aristotle himself, attaching the name of an authority to a statement of experiential fact rendered it probable and hence suitable for use in argument. R. W. Southern has pointed out the almost subservient role that experience played within the framework of medieval philosophical discourse; Robert Grosseteste rebuked those who would "form their own opinions from their experiments without a foundation of doctrine." Criticizing another's premises, therefore, involved demonstrating their inconsistency with other commonly accepted statements, and the essence of such statements was their generality, as in the example "bees reproduce parthenogenetically"—all bees, at all times.[10] An "experience" was thus a statement of how things are, or how they behave, and it was taken to have been originally constructed from a large number of individual sensory impressions. As Aristotle described in the *Posterior Analytics* and Buonamici, in the sixteenth century, echoed: "Great is the power of experience which arises from the memory of things which sense time and again supplies; for indeed memory comes from repeated sensation. Many memories of the same

---

[8] See G. E. L. Owen, "Aristotle: Method, Physics and Cosmology," *Dictionary of Scientific Biography,* ed. Charles C. Gillispie, 16 vols. (New York: Scribner's, 1970–1980) (hereafter *DSB*), Vol. I, pp. 250–258, esp. 252–254; Stephen Gaukroger, *Explanatory Structures: A Study of Concepts of Explanation in Early Physics and Philosophy* (Atlantic Highlands, N.J.: Humanities Press, 1978), pp. 91–92. On premodern concepts of probability, see Ian Hacking, *The Emergence of Probability* (Cambridge: Cambridge Univ. Press, 1975), Chs. 1–5.

[9] On scholasticism as a pedagogical "method," and the role of disputation, see George Makdisi, "The Scholastic Method in Medieval Education: An Inquiry into its Origins in Law and Theology," *Speculum,* 1974, *49*:640–661, esp. pp. 642–648.

[10] The example comes from Aristotle, *De animalibus historiae,* 9.42. See R. W. Southern, commentary on medieval science, in *Scientific Change,* ed. Alistair C. Crombie (New York: Basic Books, 1963), pp. 301–306, on p. 305, and for a comparison of Grosseteste's concept of "experimentum" to "experience" in the Aristotelian sense, see Bruce S. Eastwood, "Medieval Empiricism: The Case of Grosseteste's Optics," *Speculum,* 1968, *43*:306–321, esp. p. 321. See also James McEvoy, *The Philosophy of Robert Grosseteste* (Oxford: Clarendon Press, 1983), esp. pp. 206–211.

thing grant the means of one experience."[11] But one did not need to have acquired such experiences directly in order to use them in argumentation, provided that they could be drawn from the statements of a weighty authority, and thereby rendered probable.

Jean Buridan, for example, when discussing whether or not the earth moves, remarked that "the last appearance that Aristotle notes is more demonstrative [than the previous] to the matter in hand. This is that an arrow projected from a bow directly upward falls again in the same spot of the earth from which it was projected. This would not be so if the earth were moved with such velocity." The "experience" here is a general statement about how things habitually behave; furthermore, like a purely rational argument, it is ascribed to Aristotle. Buridan does not regard the point as conclusive, and he marshals counterarguments to the interpretation of the experience as proof of the earth's stability, but he does not question the truth of the experience itself.[12]

It was this subordination of experiential statements to the structure of argument, without subjecting them to much investigation, which prompted the seventeenth-century charges of slavish adherence to ancient authority, rather than any widespread and genuinely uncritical acceptance of Aristotelian texts. We can see this in the famous passage concerning the motion of bodies on a moving earth in Galileo's *Dialogue Concerning the Two Chief World Systems*. When Simplicio, defending the claim that a ball dropped from the mast of a moving ship falls aft of the mast's foot, is asked, "Have you ever made this experiment of the ship?" he is made to reply: "I have never made it, but I certainly believe that the authorities who adduced it had carefully observed it. Besides, the cause of the difference [between the effects on a moving and a stationary ship] is so exactly known that there is no room for doubt." The subsumption of experience by argumentative structure is here quite evident; Galileo's spokesman, Salviati, uses it as the basis of his criticisms.[13]

The "ancient authority" against which the seventeenth-century proponents of new approaches to natural philosophy argued was located in authoritative *texts*. Those texts laid down the accepted framework of discourse and interpretation, and this framework, supported by the position of the commentary as the standard philosophical genre, tended to come from Aristotle (or, in the case of medicine, Galen). Because the texts determined the character and function of statements of general experience, the authoritativeness of such statements derived from the place they held in the text, and the place the text held in natural philosophical inquiry, rather than from a blind faith in the assertions of Aristotle. The way in which an authoritative text could inform the structure of natural philosophical discourse even when the genre of commentary was not chosen can be seen in an example from Albertus Magnus. Albertus, discussing trees, says that his procedure will be to "consider first the whole diversity of the parts":

[11] Quoted in Charles B. Schmitt, "Experience and Experiment: A Comparison of Zabarella's View with Galileo's in *De motu*," *Studies in the Renaissance*, 1969, *16*:80–138, on pp. 90–91, deriving from Aristotle, *Posterior Analytics*, 2.19.

[12] Jean Buridan, *Questions on the Four Books on the Heavens and the World of Aristotle*, in Marshall Clagett, *The Science of Mechanics in the Middle Ages* (Madison: Univ. Wisconsin Press, 1961), p. 596.

[13] Galileo Galilei, *Dialogue Concerning the Two Chief World Systems*, trans. Stillman Drake (Berkeley: Univ. California Press, 1967), p. 144.

"First, however, we shall only cite these differences and afterward assign the causes of all the differences. If we did not follow Aristotle, however, but others, we would surely proceed otherwise. We say, therefore, with Aristotle, that certain plants called trees have gum—as the pine tree, resin and almond gum, myrrh, frankincense, and gum arabic." In effect, Aristotle was an "authority" because he was the author of texts used habitually as loci for the discussion of particular subjects. If one wished, for example, to write about the nature of the heavens, the appropriate procedure was to compose a commentary on *De caelo*. Within that framework, "experience" had a particular meaning and role.[14]

In rejecting ancient authority, therefore, the "moderns" were not rejecting the ancients themselves, but the role their writings played in intellectual inquiry. Again, Galileo remarked that "it is the followers of Aristotle who have crowned him with authority, not he who has usurped or appropriated it to himself." The blind adherence of the schoolmen (so the argument ran) was at fault; there was nothing especially authoritative about The Philosopher. John Wilkins, later in the century, put it this way: "And though we do very much honour *Aristotle* for his profound judgment and universall learning, yet are we so farre from being tyed up to his opinions, that persons of all conditions amongst us take liberty to discent from him, and to declare against him, according as any contrary evidence doth engage them, being ready to follow the Banner of truth by whomsoever it shall be lifted up."[15]

Whether or not one questions the sincerity of such pronouncements (and there seems little reason to do so), it remains clear that Aristotle's image had changed. He was no longer The Philosopher; he was, like Socrates, a man, and he had been shackled by the state of learning in his own time. Consequently, although he had ceased to be an Authority (one would not wish any more to write commentaries on his works in order to elucidate nature), Aristotle was not himself to be reviled, for he certainly had been learned, as Henry Oldenburg, the Royal Society's secretary, acknowledged: "We say heartily, Read *Aristotle,* read him

---

[14] Albertus Magnus, *On Plants,* 1.2, trans. in *A Source Book in Medieval Science,* ed. Edward Grant (Cambridge, Mass.: Harvard Univ. Press, 1974), p. 692 (the text Albertus used was not, in fact, by Aristotle, although it was widely believed to be at the time). It was not just the assimilation of ancient, particularly Aristotelian, texts in the thirteenth century that promoted the commentary as a major scholarly form; the dominance of various forms of "Aristotelianisms" well into the Renaissance meant the continuing dominance of commentaries on Aristotle. See, e.g., Charles B. Schmitt, *A Critical Survey and Bibliography of Studies on Renaissance Aristotelianism 1958–1969* (Padova: Editrice Antenore, 1971); Schmitt: "Toward a Reassessment of Renaissance Aristotelianism," *History of Science,* 1973, *11*:159–193; and Schmitt, *Aristotle in the Renaissance* (Cambridge, Mass.: Harvard Univ. Press, 1983). For a short discussion of medieval commentary see Neal W. Gilbert, *Renaissance Concepts of Method* (New York: Columbia Univ. Press, 1960), pp. 27–31. For a description of various genres of commentary, see C. H. Lohr, "Renaissance Latin Aristotle Commentaries: Authors A–B," *Studies in the Renaissance,* 1974, *21*:228–289, on pp. 228–233. Edward Grant, in "Aristotelianism and the Longevity of the Medieval World View," *History of Science,* 1978, *16*:93–106, attributes the survival of the Aristotelian world view well into the seventeenth century to the status of the commentary as the prime vehicle of natural philosophical discussion. See also Ian Maclean, "The Interpretation of Natural Signs: Cardano's *De subtilitate* versus Scaliger's *Exercitationes,*" in *Occult and Scientific Mentalities in the Renaissance,* ed. Brian Vickers (Cambridge: Cambridge Univ. Press, 1984), pp. 231–252, esp. pp. 239–240. A curious echo of this genre of natural-philosophical writing is Samuel Clarke's edition of Jacques Rohault's textbook, *Physica* (London, 1697), wherein Rohault's Cartesian text is glossed with lengthy dissenting Newtonian notes by Clarke.

[15] Galileo, *Dialogue,* p. 111; John Wilkins, *Vindiciae Academiarum* (Oxford, 1654), facs. in Allen G. Debus, *Science and Education in the Seventeenth Century* (London: Macdonald, 1970), p. 196.

in his own Stile; read him entirely and fully; not feeding onely on his Ulcers and Excrescensies. . . . chuse his best Vertues, examine and weigh all his Mathematical Illustrations, descend to his particulars." And after this indulgence, Oldenburg urges, the seeker after truth should "hasten to our *Christian* Philosophers, and they will forth-with acquaint you with the true Works and wonderful Contrivances of the Supreme Author."[16] The "Supreme Author," he suggests here, has replaced the Authority, Aristotle. The natural philosopher's proper task is to write a commentary on nature, not on the works of the Stagirite. Oldenburg's recommendation that Aristotle should be read "in his own Stile," that is, in classical Greek, serves to emphasize Aristotle's new status as a historical figure and his corresponding irrelevance to natural philosophy.

Although the people we have just quoted agreed that the authority of the ancients should be rejected, they did not agree on what should replace it. Bacon, too, had claimed that, despite his innovations, the "honour of the ancient authors remains untouched, since the comparison I challenge is not of wits or faculties, but of ways and methods";[17] nonetheless, he and Galileo would scarcely have seen eye-to-eye on most substantive questions concerning knowledge of the natural world. Similar philosophical differences existed between members of the early Royal Society. William Petty, for example, believed that the world was made up of atoms; furthermore, as he declared to his fellow academicians, there were reasons for postulating the existence of two types of atom: "Male and Female created He them." After dabbling in Cartesianism, Henry More, also a Fellow of the Royal Society, subscribed to the idea of an "hylarchick spirit" which acted as an active principle in nature. Henry Power saw himself as a Cartesian, while to Newton, of course, such a position was anathema.[18]

Such deep-rooted disagreements formed no basis for uniting the membership of a philosophical society, and collective distrust of older patterns of authority scarcely constituted an adequate basis. Yet the Royal Society prospered. A new source of authority had in fact come into being, a means of facilitating a cooperative enterprise dedicated to the expansion of natural knowledge. This new authority overrode individual differences in ontological commitment.

II

Thomas Sprat, in a passage verifiable by means of a brief perusal of the *Philosophical Transactions*, emphasized the diversity of those contributing to the Royal Society's work: "We find many Noble Rarities to be every day given in, not onely by the hands of Learned and profess'd Philosophers; but from the

---

[16] *Philosophical Transactions*, 1666, 2:413.

[17] Francis Bacon, *The New Organon*, Aphorism 32, trans. James Spedding (Indianapolis: Bobbs-Merrill, 1960), p. 46.

[18] See Robert Kargon, "William Petty's Mechanical Philosophy," *Isis*, 1965, 56:63–66, on p. 65; William H. Austin, "Henry More," *DSB*, Vol. IX, pp. 509–510 (for a strong reaction against More's ideas on spiritual agencies by another Fellow of the Royal Society, see Robert Hooke, *Lampas*, in *Early Science in Oxford*, ed. R. T. Gunther, 14 vols., Oxford, 1923–1945, Vol. VIII, pp. 154–208, on pp. 182–195); and, Marie Boas Hall's introduction to Henry Power, *Experimental Philosophy* (London, 1664), facs. ed. (New York: Johnson Reprint, 1966); and Charles Webster, "The Discovery of Boyle's Law and the Concept of the Elasticity of the Air," *Archive for the History of the Exact Sciences*, 1965, 2:441–502, esp. pp. 460 ff. As for Newton, all of *Principia*, Bk II, is openly anti-Cartesian.

Shops of *Mechanicks;* from the Voyages of *Merchants;* from the Ploughs of *Husbandmen;* from the Sports, the Fishponds, the Parks, the Gardens of *Gentlemen.*" He continued: "If we only requir'd perfect *Philosophers,* to manage this employment, it were another case. For then I grant it were improbable, that threescore, or an hundred such should meet in one time." All that is required are "plain, diligent and laborious observers."[19] All could, in principle, participate in the Society's activities. But in order to do this, they needed to conform to certain standards so that they could each assume the mantle of a new kind of authority.

When a Fellow of the Royal Society made a contribution to knowledge, he did so by reporting an experience. That experience differed in important respects from the definition informing scholastic practice; rather than being a generalized statement about how some aspect of the world *behaves,* it was instead a report of how, in one instance, the world had *behaved.* Consider, for example, Robert Boyle on an air-pump experiment:

> But when after this, the feathers being placed as before, we repeated the experiment by carefully pumping out the air, neither I nor any of the bystanders could perceive anything of turning in the descent of the feathers; and yet for further security we let them fall twice more in the unexhausted receiver, and found them to turn in falling as before; whereas when we did a third time let them fall in the well exhausted receiver, they fell after the same manner as they had done formerly.[20]

An "experience" was now, it appears, an event of which the observer was a part. Boyle's report conveys the impression of an actual, discrete event and of the observer's central role in it, not only by his careful recounting of the facts, but also by his use of the first person, active voice. This form of presentation is overwhelmingly the rule in the writings of the Fellows, regardless of the subject matter. Henry Power, discussing his microscopical examinations of the "sycomore locust," said: "I could not only see its eyes, which are red, goggled and prominent; but also I could see them perfectly latticed." Christopher Merrett describing his work on horticulture, reported: "*Anno* 1665, I made the following Experiment with 3 *May-Cherry Trees.*"[21]

Accounts of other people's experiences were cast in the same way as those reported at first hand: the veracity of the report clearly depended on the original experience of a specified person on a particular occasion. No hearsay evidence could be admissible, as the following excerpt from the Royal Society's *Philosophical Transactions* reports:

> There was very lately produced a Paper, containing some observations, made by Mr. *Hook,* about the Planet Mars; in the Face whereof he affirmed to have discov-

---

[19] Sprat, *History,* p. 72.

[20] Quoted in Marie Boas Hall, *Robert Boyle on Natural Philosophy* (Bloomington: Indiana Univ. Press, 1966), p. 349. Steven Shapin, "Pump and Circumstance: Robert Boyle's Literary Technology," *Social Studies of Science,* 1984, *14*:481–519, discusses the function of Boyle's use of detailed accounts as a form of "virtual witnessing" designed to establish "matters of fact." The importance of the latter as a form of social boundary marker is discussed in Simon Schaffer, "Making Certain" (essay review of Shapiro, *Probability and Certainty,* cit. n. 4), *Social Studies of Science,* 1984, *14*:137–152. See also Steven Shapin and Simon Schaffer, *Leviathan and the Air-Pump: Hobbes, Boyle and the Experimental Life* (Princeton, N.J.: Princeton Univ. Press, forthcoming), Ch. 4).

[21] Power, *Experimental Philosophy,* p. 32; report from Christopher Merrett, *Phil. Trans.,* 1666, *2*:455.

ered, in the late months of *February* and *March,* that there are several *Maculae* or *Spotted parts,* changing their place, and not returning to the same *Position,* till the next ensuing night near about the same time. . . . This short intimation of it, is intended onely to invite others, that have opportunity, timely to make Observations, (either to confirm, or rectify) before *Mars* gets out of sight.[22]

The actuality of a discrete event was the central point to be established in any contribution to the cooperative philosophy of the Royal Society, and the tokens of good faith employed by a virtuoso, that is, the circumstantial details, often ran to excruciating length. Robert Boyle is notorious in this respect.

We took a slender and very curiously blown cylinder of glass, of nearly three foot in length, and whose bore had in diameter a quarter of an inch, wanting a hair's breadth: this pipe, being hermetically sealed at one end was, at the other, filled with quicksilver, care being taken in the filling, that as few bubbles as was possible should be left in the mercury. Then the tube being stopt with the finger and inverted, was opened, according to the manner of the experiment, into a somewhat long and slender cylindrical box (instead of which we are now wont to use a glass of the same form) half filled with quicksilver: and so, the liquid metal being suffered to subside, and a piece of paper being pasted on level with its upper surface, the box and tube and all were by strings carefully let down into the receiver.[23]

Much of this last passage is written in the passive voice. This practice was not uncommon in those sections of a report dealing with the setting-up of an experiment; the experimenter would abandon it when the "experience" itself came to be described (as we saw in the previous quotation from Boyle). Frequently, a recipe-like format appeared, the reader being provided with a set of instructions for experimental procedure. Again, the writer would switch to a description of what had happened, once the preparations for the experience were out of the way.[24] The procedure could always be repeated; the event could never be. When a contribution to the *Philosophical Transactions* by Richard Lower breached the usual protocol, the editorial hand of Henry Oldenburg apparently intervened. The piece, concerning a way of "*making a* Dog *draw his Breath exactly like a* Wind-broken Horse," explained the technique involved by means of instructions. However, the outcome was improperly presented: "You will plainly see him draw his breath exactly like a *Wind-broken Horse*." The absence of a discrete experience as the culmination of the report appears to have deprived it of proper credentials; accordingly, the deficiency is compensated for by these words, which preface the article: "This Experiment was made before the *Royal Society,* Octob. 17. 1667. after it had been tried by the Author in private some while before."[25] An experience was linked to a particular time and place.

---

[22] *Phil. Trans.,* 1666, *1*:198. I am not aware of Hooke's apparent observation of the axial rotation of Mars having been noted previously; Cassini had first reported it earlier that same year (1666); see René Taton, "Cassini I," *DSB,* Vol. III, pp. 100–104, on p. 101.

[23] Quoted in Hall, *Robert Boyle,* pp. 329–330.

[24] See, e.g., "An extract of a letter lately written by an observing friend of the *Publisher,* concerning the vertue of *Antimony,*" *Phil. Trans.,* 1668, *3*:774. The experience proceeds thus: "I tried that a Boare, to whom I had given an ounce of crude Antimony at a time, puting him into the Sty, would be fat a fortnight before another, having no Antimony, upon the like feeding . . ." The "recipe" section begins: "The manner of using it, is this. Take one drachme of crude Antimony powder'd for one Horse, and when you given him his Oats in a morning . . . "

[25] *Phil. Trans.,* 1667, *2*:544–546, on pp. 545, 544.

Examples can be chosen at random to show how the presentation of experiments or observations (both "experiences") served to place the reporter at the center of an event. The described experience was *discrete;* it was a single, historical occurrence, not a generalized statement. These things, we seem to be told, had happened by the action of or in the presence of a particular person, at a particular time and place. The specificity and consequent verisimilitude of the presentation lent the described experience an authority functionally equivalent to, but different from, that deriving from the use of an authoritative ancient text. With the rejection of ancient authority, a new way of supporting one's statements was needed; no longer were detailed references to Aristotle acceptable as premises for assertions or as frames within which to construct arguments. The new approach to knowledge entailed the rooting of claims about the natural world in discrete events. Boyle criticized those who "neither themselves ever took the pains to make trial of [experiments], nor received from any credible persons that professed themselves to have tried them";[26] for knowledge about the world was to be provided by a definite occurrence happening to a particular person. The resulting style of presentation allowed no clear distinction to be made between a "natural historical" (in the modern sense) and an "experimental" report; each was, in the same way, given as an experience defined in space and time by an actor, the observer. The credentials that established the actuality of the event were provided by surrounding the description by a wealth of circumstantial detail. This detail generally included information regarding time, place, and participants, together with additional extraneous remarks about the experience, all serving to add verisimilitude. (John Beale, in one contribution, was careful to note that a mackerel left to rot so as to exhibit phosphorescence had been prepared by his cook.[27]) Frequent use of the active voice served to strengthen the impression of an actual event and the observer's central role within it. All of this constituted the report of an experience, and the literary form itself represented a credential formerly supplied by references to an appropriate ancient, or by a different literary form—the commentary. The ancients themselves suffered a change of image consonant with the new authority. The microscope, said Henry Power, had revealed a whole new world of living creatures, "so that were *Aristotle* now alive, he might write a new History of Animals."[28]

That the new conception of authority had become a virtual *sine qua non* of the experimental philosophy is evident in the paper sent by Isaac Newton to the Royal Society in 1672. Newton had taken pains to cast the piece in the correct mold, so that it should carry the proper weight. What is interesting is that the scenario he describes, the experience he records, is a fabrication.

> To perform my late promise to you, I shall without further ceremony acquaint you, that in the beginning of the Year 1666 (at which time I applied my self to the grinding of Optick glasses of other figures than Spherical,) I procured me a Triangular glass-Prisme, to try therewith the celebrated *Phaenomena* of *Colours*. And in order thereto having darkened my chamber, and made a small hole in my window-shuts, to let in

---

[26] Quoted in Hall, *Robert Boyle*, p. 131.
[27] "An Experiment To examine, what *Figure*, and *Celerity* of *Motion* begetteth, or encreaseth *Light* and *Flame*," *Phil. Trans.*, 1666, 1:226–228.
[28] Power, *Experimental Philosophy*, preface (unpaginated), pp. [7]–[8].

a convenient quantity of the Suns light, I placed my Prisme at his entrance, that it might be thereby refracted to the opposite wall.[29]

This famous account, it has been amply shown, was a spurious description of the origin of Newton's ideas on light. Newton's research was conducted over a period of some years, and his presentation to the Royal Society is an idealized version representing a whole series of investigations.[30] He gave the Society just the sort of thing which it required—an event, in which he was the central participant. When delivering his optical ideas to those undergraduates (if any) who attended his *lectiones opticae* at Cambridge in 1670, however, Newton adopted a completely different strategy, one designed to meet the demands of a different forum, a different approach to knowledge. Instead of an event, Newton provided a demonstration, and he gave his conclusion at the beginning, as a postulate to be proved, rather than at the end, as a deduction from a discrete series of experiments. After having stated baldly that "Rays in respect to the Quantity of Refraction differ from one another," Newton presented a geometrically defined exposition of that assertion: "For a farther Illustration of this, let EFG (in *Fig.* 1) be any refracting surface, suppose of Glass, and let any Line OF be drawn meeting it in F." Following this precise, formal enunciation, Newton proceeded with the proof: "Our Opinion in this Matter being thus briefly explained, that you may not think we have declared to you Fables instead of Truth, we shall immediately produce the Reasons and Experiments on which these things are founded." Then came the prism experiment—but not in the form of an historical narration: "Let (in *Fig.* 2) F be any Hole in the Wall or Shut of a Chamber Window, through which the solar Rays OF may be transmitted, all other Avenues being every where carefully closed, lest the Light may enter elsewhere. . . . Then let be applied to that Hole the triangular glass Prism AaBbCc." The difference between Newton's presentation in the Cambridge lectures and his presentation in the letter to the Royal Society demonstrates that the Society (as did Cambridge) embodied a particular approach to natural philosophy. Cooperative experimental research required that Newton construct a discrete event as the foundation for his claims—not as a cynical attempt to gain credibility, but as a matter of propriety.[31]

In at least one instance, then, the event described did not actually take place. But since part of the purpose of the rhetorical form we have examined was to establish that the event did take place, a further factor, that is, the identity and

---

[29] Isaac Newton, "New Theory about *Light* and *Colors*," *Phil. Trans.*, 1672, 6:3075–3087; on pp. 3075–3076.

[30] See, e.g., Richard S. Westfall, "The Development of Newton's Theory of Color," *Isis*, 1962, 53:339–358, esp. pp. 351–352; Westfall, *Never at Rest: A Biography of Isaac Newton* (Cambridge: Cambridge Univ. Press, 1980), pp. 156–174, 211–222, esp. pp. 156–164; Thomas S. Kuhn, "Newton's Optical Papers," in *Isaac Newton's Papers and Letters on Natural Philosophy*, ed. I. Bernard Cohen (Cambridge, Mass.: Harvard Univ. Press, 1958), pp. 27–45, esp. pp. 33–34, n. 11: "The implication of Newton's account of 1672 is wrong in that Newton did not proceed so directly or so immediately from the first prism experiment to the final version of the theory as the first paper would imply."

[31] Isaac Newton, *Optical Lectures read in the Publick Schools of the University of Cambridge* (London, 1728), pp. 5, 7; on the strategy see Kuhn, "Newton's Optical Papers," p. 32, n. 10, p. 35. The formal quality of Newton's lectures is consonant with their being originally delivered in Latin, unlike his letter in English to the Royal Society. *Optical Lectures*, a translation of the original "lectiones opticae," has recently been edited by Alan Shapiro in *The Optical Papers of Isaac Newton*, Vol. I: *The Optical Lectures, 1670–72* (Cambridge: Cambridge Univ. Press, 1984).

status of those involved, clearly held some importance. We have seen Sprat boasting of the wide diversity of people who contributed to the Royal Society's project, but it should not be imagined that all were therefore equal. The humble could play their part, but not in quite the same way as the mighty. Gentility certainly added weight in its own right: "To the truth of these Relations [concerning two elderly people who suddenly sprouted new teeth], not onely the said *Joseph Shute* and *Maria Stert* [the recipients of the teeth], have put the one his name, the other her Mark, the third and seventh of *January*, 1666. but also Sir *William Strode*, and Mr. Colepresse have subscribed the same, as believing the Relation to be true." Nobility sometimes served to reduce the required amount of circumstantial description needed for accreditation. "This story Sir R. *Moray* affirmed to have received from the *Earl* of *Weymes*, Brother in Law to the Lord *Sinclair*, as it was written to him from *Scotland*."[32] And Boyle's credibility was probably not unconnected to the fact that he was "doubly Honourable (both for his parts and parentage)"[33]—though in his case details did not thereby suffer. The Society was very conscious of its social position; as Sprat's listing shows, among the Fellows were included a great number of churchmen and aristocrats who had no real connection with experimental philosophy.[34] The Royal Society as a group thereby gained not only some small measure of additional financial patronage, but also social prestige, which could itself be turned to evidential advantage. Hooke, in order to lend force to a purported observation, remarked that he had "found by a most certain Experiment, which I exhibited before divers illustrious Persons of the *Royal Society*, that the Refraction of Water was greater than that of Ice, though some considerable Authors have affirm'd the contrary."[35]

At the other end of the scale were the humble gardeners and merchants: "Here follows a Relation . . . which is about the new *Whale-fishing* in the *West-Indies* about the *Bermudas*, as it was delivered by an understanding & hardy Sea-man, who affirmed to have been at the killing work himself." Gentlemen were trustworthy just because they *were* gentlemen; lowly folk possessed a different asset. Boyle passed on a paper concerning ambergris with the following comment: "And probably you will be invited to look on this account, though not as compleat, yet as very sincere, and on that score Credible, if you consider, that this was not written by a Philosopher to broach a *Paradox*, or serve an *Hypothesis*, but by a Merchant or Factor for his Superiors, to give them an account of a matter of fact."[36] Everyman had one advantage his betters could so easily lack—he was unlikely to be prejudiced by hypotheses. The importance of this trait shows from judgments made of those who might be so prejudiced. A review in *Philosophical Transactions* of a book by Hevelius not only stressed Hevelius's attempts to show what "care and diligence" he had employed in his observations of a comet, but also noted that these qualities showed that he had

---

[32] *Phil. Trans.*, 1666, *1*:381, 45.
[33] Power, *Experimental Philosophy*, preface, p. [18].
[34] See Sprat, *History*, pp. 431–433, for a list of members, and Hunter, *Fellows of the Royal Society*, for a full analysis.
[35] Robert Hooke, *Micrographia* (London, 1665), p. 220; on patronage see Shapiro, *John Wilkins*, pp. 194–195.
[36] *Phil. Trans.*, 1665, *1*:11; *ibid.*, 1673, 7–8:6113.

not been "prepossessed by any *Hypothesis*."[37] Boyle's merchant was *a priori* less suspect on this score, and that lent authority to his assertions.

III

The Royal Society's empiricism was rooted in the authority of the individual reporter as the actor in a well-defined, particular experience. The strict limitation that this created on claims to knowledge was essential to the successful functioning of the Society; the various "theoretical" positions held by its members were far too diverse for a consensus to be possible except in a very general way. Sprat expressed it in terms of avoiding error: the Society's work ought to be entrusted "not to a *Company* all of *one mind*"; the "heat of invention" should not lead the experimental philosopher to "swallow a deceit too soon."[38] Whatever Sprat's own reasons for stressing the point, this also happened to be the common view of the Fellows. Henry Power approvingly quoted Boyle on the matter:

> When a writer, saith he, acquaints me onely with his own thoughts or conjectures, without inriching his discourse with any real Experiment or Observation, if he be mistaken in his Ratiotination, I am in some danger of erring with him, and at least am like to lose my time, without receiving any valuable compensation for so great a loss: But if a Writer endevours, by delivering new and real Observations or Experiments, to credit his Opinions, the case is much otherwayes; for, let his Opinions be never so false (his Experiments being true) I am not obliged to believe the former, and am left at my liberty to benefit my self by the latter.[39]

One problem with hypotheses, of course, was that they could not be grounded in the bedrock of the new authority. The speculations of one person no longer found acceptable justification in ancient precedent, nor could they by way of compensation assume the status conferred by the device of participant and discrete experience. Hypotheses remained a matter of choice, at best of heuristic, and could play no active part in the furtherance of the experimental philosophy that was the basis of the Royal Society's cooperative scheme.[40] Outside the limits of the Society's activities, however, the individual natural philosopher could quite properly utilize hypotheses. Hooke's well-known disclaimer at the beginning of his *Micrographia* makes just this point. Praising the Society for "avoiding *Dogmatizing,* and the espousal of any *Hypothesis* not sufficiently grounded and confirm'd by *Experiments,*" he apologizes for any "*Expressions,* which may seem more *positive* then YOUR Prescriptions will permit"; if he has gone too far, " 'tis fit that I should declare, that it was not done by YOUR Directions." Theorizing was not in keeping with the Society's function, but it was not ruled out *tout court*.[41]

---

[37] *Ibid.*, 1666, *1*:303.
[38] Sprat, *History*, p. 73.
[39] Power, *Experimental Philosophy*, preface, pp. [18]-[19].
[40] For a good treatment of hypothesis in the early Royal Society, see Marie Boas Hall, "Science in the Early Royal Society," in Maurice Crosland, ed., *The Emergence of Science in Western Europe* (New York: Science History Publications, 1976), pp. 57-77; also Shapiro, *Probability and Certainty*, p. 44-61.
[41] Hooke, "To the Royal Society," preface to *Micrographia*. Barbara Shapiro, in her examinations of this question, does not make a clear distinction between the propriety of hypothesis for individual

If a Fellow wished some of his theoretically informed research to be added to the Royal Society's work, then, he had to elevate its status above that of a conceit and provide it with all the appropriate trappings of authority. In the third number of the *Philosophical Transactions* is a fairly lengthy report concerning "*Some Observations and Experiments upon May-Dew*" communicated by "That ingenious and inquisitive Gentleman, Master *Thomas Henshaw*." K. Theodore Hoppen has pointed out that May dew had important alchemical significance, thereby explaining Master Henshaw's great concern with such a surprising substance. None of this, however, is even hinted at in the *Philosophical Transactions* report. Instead, one reads, in the usual style: "He observed, that at the beginning, within twenty-four hours, a slimy film floated on the top of the water"; "He found the glass almost filled with an innumerable Company of small Flyes"; and so forth.[42] The theoretical justification for the experiments could not properly be presented; what took its place was a correctly accredited experience.

John Wallis's theory of the tides, read to the Society in 1666 and printed in the *Philosophical Transactions*, presented similar problems. Precisely because it was a theory (being a modification of Galileo's, adjusted so as to include and account for monthly periodicity), it conformed rather uncomfortably to the format of the Royal Society.[43] However, one consequence of Wallis's model was that the highest tides should occur not at the equinoxes, as was popularly held to be the case, but during February and November. This circumstance provided him with an opportunity not only to seek confirmation for his theory, but also to justify presenting it by reporting experiences. After giving the testimony of the people of Romney Marsh, which he had heard before he had ever considered questioning the usual opinion, Wallis proceeded to bolster his claim in the correct fashion:

> And since that time, I have my self very frequently observed (both at *London* and elsewhere, as I have had occasion) that in those months of *February* and *November*, (especially *November*) the Tides have run much higher, than those at other times: Though I confess, I have not been so diligent to set down those Observations, as I should have done. Yet this I do particularly very well remember, that in *November* 1660 (the same year that his Majesty returned) having occasion to go by Coach from the *Strand* to *Westminster*, I found the Water so high in the middle of King-street, that it came up, not onely to the Boots, but into the body of the Coach; and the

---

investigation and its suspect character within the common forum of debate. E.g.: "The turning point in England seems to have come somewhere about 1650, for from the 1650s to Newton's hostile pronouncements at the end of the century, a substantial portion of English scientists found theory, conjecture, and hypothesis extremely useful" ("History and Natural History," p. 32), a position upheld also in *Probability and Certainty*, esp. Ch. 2. Aside from asking who in England before 1650 (apart from Bacon) represented a strongly contrasting empiricism, one should bear firmly in mind the strongly circumscribed place of hypothesis within the auspices of the Royal Society. This social rather than epistemological requirement sprang from the Fellows' recognition of its importance in that particular forum rather than from any rigid enforcement by a corporate apparatus.

[42] *Phil. Trans.*, 1665, *1*:33–36, quoting pp. 33, 34, 35. On Henshaw, see Stephen Pasmore, "Thomas Henshaw, F.R.S. (1618–1700)," *Notes and Records*, 1982, *36*:177–188; on Maydew see K. Theodore Hoppen, "The Nature of the Early Royal Society," *British Journal for the History of Science*, 1976, *9*:246.

[43] See Thomas Birch, *The History of the Royal Society of London* (London, 1756–1757), Vol. I, pp. 88–89; *Phil. Trans.*, 1666, *1*:263–289, esp. 271–275. For a brief secondary account, see E. J. Aiton, "Galileo's Theory of the Tides," *Annals of Science*, 1954, *10*:44–57, on pp. 50–54.

> *Pallace-yard* (all save a little place near the *West-End*) overflow'd; as likewise the Market-place; and many other places; and their Cellars generally filled up with Water.[44]

Rather like Newton, though in a less thoroughgoing way, Wallis tried to increase the acceptability of an hypothesis by adhering as much as possible to the proper evidential form.[45]

IV

The style of science espoused by the Fellows of the Royal Society was more important than the substance of that science. The form of their research reports, as well as prefatory and programmatic statements, demonstrates an ethic of investigation suitable to the ideal of cooperative research. This form also indicates that a fundamental change in concepts of experience and authority in natural philosophy had occurred in the seventeenth century, a change that underlay contemporary charges of scholastic vassalage to ancient authority. Ancient authority was rooted in the use of the writings of, above all, Aristotle as the proper vehicles of philosophical discussion, and experience appeared in the form of a generalized statement of fact, its role dictated by the overall structure of discourse. By contrast, when a Fellow of the Royal Society utilized experience, he provided a report of a discrete event, independent of the setting of an ancient text or conceptualization.[46] In such reports, the operator or observer (the two were equivalent) was central to the episode recounted—and episodes they were. Located, explicitly or implicitly, at a precise point in space and time, the observer's reported experience of a singular phenomenon constituted his authority. Newton resorted to fabricating an event as a means of conferring this authority on his optical theories.

This article has touched only briefly on mathematical sciences. There are, in fact, very few mathematical contributions to be found in the *Philosophical Transactions,* and the indications are that they fitted only uneasily into the Royal Society's work. Perhaps the most famous of this period are those of John Wallis and Christopher Wren on collision. Both papers are in Latin, like most other mathematical pieces, whereas English contributors on other topics usually wrote in their native language. The authority accruing to this material seems to have lain somewhat in doubt, since mathematical forms of argument and demonstration could not be fitted readily into the usual form of presentation. Accordingly, Wren's paper is prefaced by remarks, in English, calculated to make it more palatable. The theory is "verified by many Experiments, made by Himself and

---

[44] *Phil. Trans.,* 1666, *1*:275–276.

[45] See esp. Zev Bechler, "Newton's 1672 Optical Controversies: A Study in the Grammar of Scientific Dissent," in *The Interaction Between Science and Philosophy,* ed. Yehuda Elkana (Atlantic Highlands, N.J.: Humanities Press, 1974), pp. 115–142; Newton presents his hypothesis as if it were not a hypothesis at all.

[46] Certain Fellows, in particular John Wilkins, advocated the development of a "plain style" of language in which to couch natural philosophy, and Wilkins, as is well known, attempted to develop a philosophical language. The welcome accorded the latter by the Royal Society was, however, less than resounding, and Hans Aarsleff suggests that this was due to its overly rationalist, essentialist implications. Aarsleff, "John Wilkins," *DSB,* Vol. XIV, pp. 361–381, which contains a wealth of relevant references, and M. M. Slaughter, *Universal Languages and Scientific Taxonomy in the Seventeenth Century* (Cambridge: Cambridge Univ. Press, 1982).

that other excellent Mathematician *M. Rook* before the Royal Society, as is attested by many Worthy Members of that Illustrious Body."[47] The departures in these mathematical papers from the usual format show that the Royal Society could not comfortably accommodate together the disparate traditions and criteria of experimental and mathematical researches.

Simon Schaffer's recent stress on the importance of the "matter of fact" both to Restoration science and to the ideological position of the Royal Society stems from his claim that the matter of fact posits "certain forms of social organization."[48] His remarks relate to the specificities of the English context, and are extremely valuable in that regard. My purpose in the present article, however, is to suggest themes constitutive of the Scientific Revolution as a whole, looking beyond the particular situation of the early Royal Society while using it as a case example. Whatever the concerns underlying the natural philosophy of the early Fellows, and the ideological concomitants of the Society's existence, the Royal Society was recognized throughout Europe as a body devoted to a form of natural inquiry, that had adherents in all parts of the continent. The Society's empiricism, especially that of Boyle, was closely paralleled by the approach in the 1650s of the Florentine Accademia del Cimento, a group operating in a very different social context from that of the English philosophers. The activities of the Royal Society were also consonant with those of the Académie des Sciences in this period, despite the widespread notion that the Académie was concerned preeminently with mathematical sciences.[49] The continental prestige of the Royal Society sprang from roots independent of its own domestic programs and concerns (although the latter are essential to a full understanding of science in Restoration England), and this article has been concerned with features of the natural philosophy of the Society which placed it in the wider context of all of Europe. The early Fellows certainly used the "matter of fact" to promote their own particular ideological ends, but they did not invent it; they deployed it. What Schaffer calls the "matter of fact" was, we have seen, constituted by a literary form which simultaneously redefined the role of experience in natural philosophy and provided a new basis of authority. That the Royal Society could itself exploit this redefinition of authority is suggested by Schaffer's remarks, but to understand its genesis requires a broader perspective.

The older forms of experience and authority were paradigmatically embodied in the genre of commentary. The new forms were embodied in a research report that provided details of discrete events. The change from one form to the other can be traced in the relative fortunes of these literary forms in natural philosophy. During the course of the seventeenth century, within the scholastic tra-

---

[47] *Phil. Trans.*, 1669, *3*:864 (Wallis), 867 (Wren), quoting p. 867. Perhaps this discomfort with mathematical material is connected with the distinction drawn by Thomas Kuhn in his "Mathematical versus Experimental Traditions in the History of the Physical Sciences," *Journal of Interdisciplinary History*, 1976, *7*:1–31, reprinted in Kuhn, *The Essential Tension* (Chicago: Chicago Univ. Press, 1977), pp. 31–65.

[48] Schaffer, "Making Certain," p. 140.

[49] On the Accademia del Cimento, see W. E. Knowles Middleton, *The Experimenters: A Study of the Accademia del Cimento* (Baltimore: Johns Hopkins Univ. Press, 1971); and Richard Waller, *Essayes of Natural Experiments* (London, 1684; facs. ed., New York: Johnson Reprint, 1964), a translation of the Accademia's *Saggi* into English under the auspices of the Royal Society. On the early Académie des Sciences, see Roger Hahn, *The Anatomy of a Scientific Institution: The Paris Academy of Sciences, 1666–1803* (Berkeley: Univ. California Press, 1971), Ch. 1.

dition itself, commentaries on Aristotle's natural philosophical works increasingly gave way to textbooks and compendia organized along thematic lines.[50] Although the substance of these works remained for the most part rooted in Aristotle and his commentators, the freedom of structure implied a loosening of the older forms of authority and, one may conjecture, encouraged the development of alternative techniques of justification. Mersenne, in many ways a transitional figure, moved from an apologetic defense of Aristotelian authority in the early 1620s to a position strikingly similar to that seen among the early Fellows of the Royal Society. In a letter to Jean Rey in 1632 he attempted to convince his correspondent that heavy bodies do not fall at a faster rate than light ones: "Truly, I am astonished at what you distrust of my experiment [*expérience*] of the equal speed of an iron bullet and a wooden bullet: for if several persons of quality who have seen and made the experiment with me were simply to swear [*signer*] solemnly to you, they would witness it to you authentically."[51] Authenticity here devolves from the witnessing of an event, not from the authorship of an authoritative text.

Because the changes in literary form and the changes in the effective meanings of authority and experience are firmly integrated, it is difficult to assign logical priority. They may not be separable at all, but together may be seen as constituting a new structure of natural philosophical discourse and practice. The reasons for the appearance of this new structure are hard to discern, but the social uses of this structure in the English context may point the way to a broader understanding of the Scientific Revolution. The importance of skeptical arguments in the sixteenth and seventeenth centuries arose in large measure, as Richard Popkin and others have shown, from their value in religious disputes. From Catholic attacks on Protestant theology in the sixteenth century to the Latitudinarians' probabilistic approaches to knowledge in Restoration England, the ability to cast general doubt on one's opponents' claims to knowledge served as a powerful weapon.[52] But in order to maintain a front against cognitive anarchy, criteria of certainty, albeit reduced, were required. In the France of the 1620s one finds the "mitigated skepticism" of Mersenne and Gassendi; in the England of the 1660s and 1670s one finds a kind of probabilism which allowed "moral certainty" to properly accredited experiential facts.[53] The authority in which that "moral certainty" resided for the Royal Society's Fellows was conferred by a postscholastic literary form which also defined a new meaning for "experience." The Society's cooperative venture was itself based on the ability of this new literary form to define the level at which cooperation could occur.

---

[50] See Patricia Reif, "The Textbook Tradition in Natural Philosophy 1600–1650," *Journal of the History of Ideas*, 1969, *30*:17–32. John Heilbron notes the change among Jesuit writers in the seventeenth century from writing commentaries to writing treatises in natural philosophy; see Heilbron, *Electricity in the 17th and 18th Centuries: A Study of Early Modern Physics* (Berkeley/Los Angeles: Univ. California Press, 1979), pp. 108–110.

[51] Mersenne, *Correspondance du P. Marin Mersenne, religieux minime*, Vol. III, ed. Cornelis de Waard (Paris: Centre de la Recherche Scientifique, 1969), pp. 274–275. On Mersenne see esp. Robert Lenoble, *Mersenne, ou la naissance du mécanisme* (Paris: J. Vrin, 1943).

[52] See esp. Richard H. Popkin, *The History of Scepticism from Erasmus to Spinoza* (Berkeley/Los Angeles: Univ. California Press, 1979), Ch. 1, on p. 14; see also Schaffer, "Making Certain."

[53] See Popkin, *History*, Ch. 7; for the humanist origins of this doctrine see Peter Dear, "Marin Mersenne and the Probabilistic Roots of 'Mitigated Scepticism,'" *Journal of the History of Philosophy*, 1984, *22*:173–205; and on the 1660s, Shapin, "Pump and Circumstance."

The House of Experiment in Seventeenth-Century England

By Steven Shapin

> That which is not able to be performed in a private house will much less be brought to pass in a commonwealth and kingdom.
> —William Harrison, *The Description of England* (1587)

MY SUBJECT is the place of experiment. I want to know where experimental science was done. In what physical and social settings? Who was in attendance at the scenes in which experimental knowledge was produced and evaluated? How were they arrayed in physical and social space? What were the conditions of access to these places, and how were transactions across their thresholds managed?

The historical materials with which I am going to deal are of special interest. Seventeenth-century England witnessed the rise and institutionalization of a program devoted to systematic experimentation, accompanied by a literature explicitly describing and defending practical aspects of that program. Nevertheless, aspects of the historiography manifest in this paper may prove of more general interest. Historians of science and ideas have not, in the main, been much concerned with the siting of knowledge production.[1] This essay offers reasons for systematically studying the venues of knowledge. I want to display the network of connections between the physical and social setting of inquiry and the position of its products on the map of knowledge. I shall try to demonstrate how the siting of knowledge-making practices contributed toward a practical solution of epistemological problems. The physical and the symbolic siting of experimental work was a way of bounding and disciplining the community of practitioners, it was a

---

For criticisms of earlier versions of this paper, I thank Peter Galison, J. V. Golinski, Owen Hannaway, Adi Ophir, Trevor Pinch, Simon Schaffer, and members of seminars at the universities of Bath, Harvard, Illinois, London (University College), Melbourne, Oxford, Pennsylvania, and the Hebrew University of Jerusalem. For permission to quote from the Boyle Papers, I thank the Council of the Royal Society of London.

[1] An outstanding exception is Owen Hannaway, "Laboratory Design and the Aim of Science: Andreas Libavius versus Tycho Brahe," *Isis*, 1986, 77:585–610. See also Peter Galison, "Bubble Chambers and the Experimental Workplace," in *Observation, Experiment, and Hypothesis in Modern Physical Science*, ed. Peter Achinstein and Owen Hannaway (Cambridge, Mass.: MIT Press, 1985), pp. 309–373; Larry Owens, "Pure and Sound Government: Laboratories, Playing Fields, and Gymnasia in the Nineteenth-Century Search for Order," *Isis*, 1985, 76:182–194; and, although not concerned with knowledge-making processes, Sophie Forgan, "Context, Image and Function: A Preliminary Enquiry into the Architecture of Scientific Societies," *The British Journal for the History of Science*, 1986, 19:89–113.

way of policing experimental discourse, and it was a way of publicly warranting that the knowledge produced in such places was reliable and authentic. That is to say, the place of experiment counted as a partial answer to the fundamental question, Why ought one to give one's assent to experimental knowledge claims?

I start by introducing some connections between empiricist processes of knowledge making and the spatial distribution of participants, pointing to the ineradicable problem of trust that is generated when some people have direct sensory access to a phenomenon and others do not. I then mobilize some information about where experimental work was in fact performed in mid to late seventeenth-century England, focusing upon sites associated with the work of the early Royal Society and two of its leading fellows, Robert Boyle and Robert Hooke. The question of access to these sites is then considered: who could go in and how was the regulation of entry implicated in the evaluation of experimental knowledge? The public display of the moral basis of experimental practices depended upon the form of social relations obtaining within these sites as much as it did upon who was allowed within. Indeed, these considerations were closely related, and I discuss how the condition of gentlemen and the deportment expected of them in certain places bore upon experimental social relations and, in particular, upon the problems attending the assessment of experimental testimony. The paper concludes by analyzing how the stages of experimental knowledge making mapped onto physical and symbolic patterns of movements within the rooms of a house, particularly the circulation between private and public places.

ON THE THRESHOLD OF EXPERIMENT

The domestic threshold marks the boundary between private and public space. Few distinctions in social life are more fundamental than that between private and public.[2] The same applies to the social activities we use to make and evaluate knowledge. On either side of the threshold the conditions of our knowledge are different. While we stand outside, we cannot see what goes on within, nor can we have any knowledge of internal affairs but what is related to us by those with rights of access or by testimony still more indirect. What we cannot see, we must take on trust or, trust being withheld, continue to suspect. Social life as a whole and the social procedures used to make knowledge are spatially organized.[3] The threshold is a social marker: it is put in place and maintained by social decision

---

[2] See, e.g., Erving Goffman, *The Presentation of Self in Everyday Life* (London: Allen Lane, Penguin Press, 1969), Ch. 3; Richard Sennett, *The Fall of Public Man* (Cambridge: Cambridge Univ. Press, 1974), esp. Ch. 5; Shirley Ardener, "Ground Rules and Social Maps for Women: An Introduction," in *Women and Space: Ground Rules and Social Maps*, ed. Ardener (London: Croom Helm, 1981), Ch. 1; Clark E. Cunningham, "Order in the Atoni House," in *Right & Left: Essays on Dual Symbolic Classification*, ed. Rodney Needham (Chicago: Univ. Chicago Press, 1973), pp. 204–238.

[3] Anthony Giddens, *The Constitution of Society: Outline of the Theory of Structuration* (Cambridge: Polity Press, 1984), Ch. 3 (esp. his discussion of the work of the Swedish geographer Torsten Hägerstrand); Bill Hillier and Julienne Hanson, *The Social Logic of Space* (Cambridge: Cambridge Univ. Press, 1984), esp. pp. ix–xi, 4–5, 8–9, 19. For Foucauldian perspectives see Michel Foucault, "Questions on Geography," in Foucault, *Power/Knowledge: Selected Interviews and Other Writings, 1972–1977*, ed. Colin Gordon (Brighton: Harvester Press, 1980), pp. 63–77; Foucault, *Discipline and Punish: The Birth of the Prison*, trans. Alan Sheridan (New York: Vintage, 1979); and Adi Ophir, "The City and the Space of Discourse: Plato's Republic—Textual Acts and Their Political Significance" (Ph.D. diss., Boston Univ., 1984).

and convention. Yet once in place it acts as a constraint upon social relations. The threshold acts as a constraint upon the distribution of knowledge, its content, quality, conditions of possession, and justification, even as it forms a resource for stipulating that the knowledge in question really is the thing it is said to be.

Within empiricist schemes of knowledge the ultimate warrant for a claim to knowledge is an act of witnessing. The simplest knowledge-producing scene one can imagine in an empiricist scheme would not, strictly speaking, be a social scene at all. It would consist of an individual, perceived as free and competent, confronting natural reality outside the social system. Although such a scene might plausibly be said to be the paradigm case of knowledge production in seventeenth-century empiricist writings, it was not, in fact, recommended by the writers. Three sorts of problems were recognized to attend the privacy of solitary individual observation.[4] First, the transformation of mere belief into proper knowledge was considered to consist of the transit from the perceptions and cognitions of the individual to the culture of the collective. Empiricist writers therefore looked for the means by which such a successful transit might be managed. The second problem was connected with the view that the perceptions of postlapsarian man were corrupt and were subject to biases deriving from interest. Although these factors could not be eliminated, their consequences might be mitigated by ensuring that both witnessing and the consideration of knowledge claims took place in a social setting. Third, there were often contingent practical problems attending the circumstances of observation, which meant that social relations of some kind had to be established for the phenomena in question to be dealt with. Certain observations, particularly in the natural history sciences but also in experimental science, could, for instance, be made only by geographically privileged persons. In such cases there was no practical way by which a witnessing public could be brought to the phenomena or the phenomena brought to the public. Testimony was therefore crucial: the act of receiving testimony constituted a rudimentary social scene, and the evaluation of testimony might occur in an elaborately constructed social scene.

English empiricists did not think that testimony could be dispensed with, but they worked strenuously to manage and discipline it. Most empiricist writers recognized that the bulk of knowledge would have to be derived from what one was told by those who had witnessed the thing in question, or by those who had been told by those who had been told, and so on. If, however, trust was to be a basis for reliable knowledge, the practical question emerged: Whom was one to trust? John Locke, among others, advised practitioners to factor the creditworthiness of the source by the credibility of the matter claimed by that source.[5] One might accept the report of an implausible phenomenon from a creditworthy

[4] For a survey of the evaluation of evidence in this setting, see Barbara J. Shapiro, *Probability and Certainty in Seventeenth-Century England: A Study of the Relationships between Natural Science, Religion, History, Law, and Literature* (Princeton, N.J.: Princeton Univ. Press, 1983), esp. Ch. 2; for treatment of experimental practice in these connections, see Steven Shapin and Simon Schaffer, *Leviathan and the Air-Pump: Hobbes, Boyle, and the Experimental Life* (Princeton, N.J.: Princeton Univ. Press, 1985), esp. Ch. 2.

[5] John Locke, *Essay Concerning Human Understanding,* in Locke, *Works,* 10 vols. (London, 1823), Vol. III, on pp. 97–100; and John Dunn, "The Concept of 'Trust' in the Politics of John Locke," in *Philosophy in History: Essays on the Historiography of Philosophy,* ed. Richard Rorty, J. B. Schneewind, and Quentin Skinner (Cambridge: Cambridge Univ. Press, 1984), pp. 279–301.

source and reject plausible claims from sources lacking that creditworthiness. Credibility as an attribute ascribed to people was not, therefore, independent of theories of what the world was like. One might calibrate persons' credibility by what it was they claimed to have witnessed, just as one might use their accepted credibility to gauge what existed in the world.[6] Nevertheless, credibility had other sources: certain kinds of people were independently known to be more trustworthy sources than others. Roughly speaking, the distribution of credibility followed the contours of English society, and that it did was so evident that scarcely any commentator felt obliged to specify the grounds of this creditworthiness. In such a setting one simply knew what sorts of people were credible, just as one simply knew whose reports were suspect.[7] Indeed, in certain instances Robert Boyle recommended that one ought to credit the testimony of things rather than the testimony of certain types of persons. Discussing one of his hydrostatical experiments of the 1660s, Boyle argued that "the pressure of the water in our . . . experiment having manifest effects upon inanimate bodies, which are not capable of giving us partial informations, will have much more weight with unprejudiced persons, than the suspicious, and sometimes disagreeing accounts of ignorant divers, whom prejudicate opinions may much sway, and whose very sensations, as those of other vulgar men, may be influenced by predispositions, and so many other circumstances, that they may easily give occasion to mistakes."[8] When in 1667 the Royal Society wished to experiment on the transfusion of animal blood into a human being, they hit upon an ingenious solution to the problem of testimony posed by such an experiment. The subject, Arthur Coga, was indigent and possibly mad (so it was expedient to use him), but he was also a Cambridge graduate (so his testimony of how he felt on receipt of sheep's blood might be credited).[9]

EXPERIMENTAL SITES

One of the considerations that recommended the program of systematic artificial experimentation launched in the middle of the seventeenth century by Boyle and his associates was that experimental phenomena could be arranged and produced at specified times and places. Such phenomena were disciplined, and disciplined

---

[6] This is an observational version of what Harry Collins has called "the experimenter's regress": H. M. Collins, *Changing Order: Replication and Induction in Scientific Practice* (London: Sage, 1985), Ch. 4.

[7] Even in legal writings centrally concerned with the evaluation of testimony, the need to spell out the grounds of persons' differential credibility was apparently rarely felt; see, e.g., Shapiro, *Probability and Certainty*, pp. 179–188 (cit. n. 4); Julian Martin, " 'Knowledge Is Power': Francis Bacon, the State and the Reform of Natural Philosophy" (Ph.D. diss., Univ. Cambridge, 1988), esp. Ch. 3; cf. Peter Dear, "*Totius in verba*: Rhetoric and Authority in the Early Royal Society," *Isis*, 1985, 76:145–161, on pp. 153–157.

[8] Robert Boyle, "An Hydrostatical Discourse, occasioned by the Objections of the Learned Dr. Henry More" (1672), in Boyle, *Works*, ed. Thomas Birch, 6 vols. (London, 1772) (hereafter **Boyle, Works**), Vol. III, pp. 596–628, on p. 626. In other circumstances Boyle elected to credit the testimony of divers: see Boyle, "Of the Temperature of the Submarine Regions," *ibid.*, pp. 342–349, on p. 342; and Shapin and Schaffer, *Leviathan and the Air-Pump* (cit. n. 4), pp. 217–218.

[9] For the Coga episode, see *The Correspondence of Henry Oldenburg*, ed. A. Rupert Hall and Marie Boas Hall, 13 vols. (Madison: Univ. Wisconsin Press; London: Mansell/Taylor & Francis, 1965–1986), Vol. III, pp. 611, 616–617; Vol. IV, pp. xx–xxi, 6, 59, 77; "An Account of the Experiment of *Transfusion*, practised upon a *Man* in London," *Philosophical Transactions*, 9 Dec. 1667, No. 30, pp. 557–559; Henry Stubbe, *Legends No Histories* (London, 1670), p. 179.

witnessing might be mobilized around them. What sorts of places were available for this program? What conditions and opportunities did they provide? Put simply, the task resolves into the search for the actual sites of seventeenth-century English experiment. Where and what was the laboratory?

Two preliminary cautions are necessary. The first is a warning against verbal anachronism. The word *laboratory* (or *elaboratory*) was not in common English usage at the middle of the seventeenth century. For example, despite his extensive description of ideal experimental sites, I cannot find the word used by Francis Bacon, in *The New Atlantis* or elsewhere. As Owen Hannaway has shown, there is some evidence of medieval Latin usage (*laboratorium*), but the word did not acquire anything of its modern sense until the late sixteenth century. It seems that the word was transmitted into English usage in the late sixteenth century, carrying with it alchemical and chemical resonances.[10] Among scores of English usages I have registered through the 1680s, I have not encountered one in which the space pointed to was one without a furnace, used as a nonportable source of heat for chemical or pharmaceutical operations. The word did become increasingly common during the course of the seventeenth century, although even by the early eighteenth century it was not used routinely to refer to just any place dedicated to experimental investigation.[11] On the founding of the Royal Society there were a number of plans for purpose-built experimental sites, none of which materialized, even though the new Oxford Ashmolean Museum (1683) did contain a chemical laboratory in its basement.[12] By the end of the century there still did not exist any purpose-designed and purpose-built structure dedicated to those non-heat-dependent sciences (such as pneumatics and hydrostatics) that were paradigmatic of the experimental program. The new experimental science was carried on in existing spaces, used just as they were or modified for the purpose.

Second, the status of spaces designated as laboratories and of experimental venues generally in seventeenth-century England was intensely contested. Were they private or public, and what status ought they to have? In the rhetoric of

---

[10] See Hannaway, "Laboratory Design" (cit. n. 1), pp. 585–586; and (on alchemical usage) Shapin and Schaffer, *Leviathan and the Air-Pump*, p. 57 and note. For reference to the "laboratory" as an intensely private space, see Gabriel Plattes, "Caveat for Alchymists," in Samuel Hartlib, comp., *Chymical, Medicinal, and Chyrurgical Addresses* (London, 1655), p. 87.

[11] I shall use the term more loosely, although it should be clear from the context what sort of place is being referred to. Like Hannaway ("Laboratory Design," p. 585), I accept that an intensive investigation of scientific sites would be obliged to take in such places as the anatomical theater, the astronomical observatory, the curiosity cabinet, and the botanic garden.

[12] For futile planning by the Royal Society in the late 1660s to construct experimental facilities in the grounds of Arundel House, see Michael Hunter, "A 'College' for the Royal Society: The Abortive Plan of 1667–1668," *Notes and Records of the Royal Society*, 1983–1984, *38*:159–186. Wren's plan (p. 173) called for "a fair Elaboratory" in the basement of the proposed house. Before the society was founded there were several proposals for experimental colleges which included laboratories: see [William Petty], *The Advice of W.P. to Mr. Samuel Hartlib . . .* (London, 1648; rpt. in *The Harleian Miscellany*, Vol. VI [London, 1745]), pp. 1–13 (pp. 5, 7 for the "Chymical Laboratory"); John Evelyn to Robert Boyle, 3 Sept. 1659, in *The Diary and Correspondence of John Evelyn*, ed. W. Bray, 3 vols. (London, 1852), Vol. III, pp. 116–120. A year after the society's foundation a plan emerged for a "Philosophical Colledge," again with "great Laboratories for Chymical Operations": Abraham Cowley, *A Proposition for the Advancement of Experimental Philosophy* (London, 1661), p. 25. For the Ashmolean laboratory, see R. F. Ovenell, *The Ashmolean Museum 1683–1894* (Oxford: Clarendon Press, 1986), pp. 16–17, 22; and Edward Lhuyd to John Aubrey, 12 Feb. 1686, quoted in S. Mendyk, "Robert Plot: Britain's 'Genial Father of County Natural Histories,'" *Notes Rec. Roy. Soc.*, 1985, *39*:159–177, on p. 174 n. 28 (for confusion about what space was designated by "ye Labradory").

English experimental philosophers, what was wrong with existing forms of practice was their privacy. Neither the individual philosopher in his study nor the solitary alchemist in his "dark and smokey" laboratory was a fit actor in a proper setting to produce objective knowledge (see Figure 1).[13] In contrast, spaces appropriate to the new experimental program were to be public and easy of access. This was the condition for the production of reliable knowledge within.[14] In stipulating that experiment was to take place in public spaces, experimental philosophers were describing the nature of the physical and social setting in which genuine knowledge might be made.

The performance and the consideration of experimental work in mid to late seventeenth-century England took place in a variety of venues. These sites ranged from the apothecary's and instrument maker's shop, to the coffeehouse, the royal palace, the rooms of college fellows, and associated collegiate and university structures. But by far the most significant venues were the private residences of gentlemen or, at any rate, sites where places of scientific work were coextensive with places of residence, whether owned or rented. The overwhelming majority of experimental trials, displays, and discussions that we know about occurred within private residences. Instances could be enumerated ad libitum: the laboratory equipped for Francis Mercury van Helmont at Anne Conway's Ragley House in Warwickshire; the role of Towneley House in Lancashire in the career of English pneumatics; Clodius's laboratory in the kitchen of his father-in-law Samuel Hartlib's house in Charing Cross; Kenelm Digby's house and laboratory in Covent Garden after the Restoration; the Hartlibian laboratory worked by Thomas Henshaw and Thomas Vaughan in their rooms at Kensington; William Petty's lodgings at Buckley Hall in Oxford, where the Experimental Philosophy Club originated in 1649; Thomas Willis's house, Beam Hall, where the club met during the early 1660s.[15]

In the following sections of this paper I shall try to display the conditions and

---

[13] Robert Boyle, "The Sceptical Chymist," in Boyle, *Works*, Vol. I, pp. 458–586, on p. 461, intended to draw "the chymists' doctrine out of their dark and smokey laboratories" and bring "it into the open light." The contemporary Dutch-Flemish pictorial genre of "the alchemist in his laboratory" generally depicted alchemical workplaces in this way, without the *necessary* implication of criticism: see C. R. Hill, "The Iconography of the Laboratory," *Ambix*, 1975, *22*:102–110; Jane P. Davidson, *David Teniers the Younger* (London: Thames & Hudson, 1980), pp. 38–43; and Davidson, " 'I Am the Poison Dripping Dragon': Iguanas and Their Significance in the Alchemical and Occult Paintings of David Teniers the Younger," *Ambix*, 1987, *34*:62–80 (who diverges from Hill in her acceptance that Teniers's many paintings of the laboratory genre are probably accurate and informed representations of such sites).

[14] Thus accusations that the Royal Society's meeting places were *not* public might count as particularly devastating: see, e.g., Thomas Hobbes, "Dialogus physicus de natura aeris" (1661), in Hobbes, *Latin Works*, ed. Sir William Molesworth, 5 vols. (London, 1839–1845), Vol. IV, pp. 233–296, on p. 240; Shapin and Schaffer, *Leviathan and the Air-Pump* (cit. n. 4), pp. 112–115, 350; and Stubbe, *Legends No Histories* (cit. n. 9), "Preface," sig. *3.

[15] Among secondary sources that are relatively rich in material relating to these sites, see Robert G. Frank, Jr., *Harvey and the Oxford Physiologists: A Study of Scientific Ideas* (Berkeley/Los Angeles: Univ. California Press, 1980), Ch. 3; Charles Webster, *The Great Instauration: Science, Medicine and Reform 1626–1660* (London: Duckworth, 1975), esp. pp. 47–63, 89–98, 130–157; R. T. Gunther, *Early Science in Oxford*, 15 vols. (Oxford: privately printed, 1923–1967), Vol. I, pp. 7–51; Betty Jo Dobbs, "Studies in the Natural Philosophy of Sir Kenelm Digby," Parts I–III, *Ambix*, 1971, *18*:1–20; 1973, *20*:144–163; 1974, *21*:1–28; Ronald Sterne Wilkinson, "The Hartlib Papers and Seventeenth-Century Chemistry, Part II," *Ambix*, 1970, *17*:85–110; Lesley Murdin, *Under Newton's Shadow: Astronomical Practices in the Seventeenth Century* (Bristol: Adam Hilger, 1985); and Michael Hunter, *Science and Society in Restoration England* (Cambridge: Cambridge Univ. Press, 1981).

Figure 1. The Alchemist in His Laboratory; *painting from about 1649 by David Teniers the Younger (1610–1690). This is typical of very many such paintings by Teniers and other Dutch and Flemish artists of the mid to late seventeenth century. Note particularly the presence of only one "master" in the scene, the fact that assistants are either clearly subservient or backgrounded, and the prominence of texts. Courtesy of the Philadelphia Museum of Art, John G. Johnson Collection.*

opportunities presented by the siting of experiment in the private house. In particular, I shall point to the role of conditions regulating access to such venues and to conventions governing social relations within them. I will argue that these conditions and conventions counted toward practical solutions of the questions of how one produced experimental knowledge, how one evaluated experimental claims, and how one mobilized and made visible the morally adequate grounds for assenting to such claims. To this end I shall concentrate on three of the most important sites in the career of experiment in mid to late seventeenth-century England: the various residences and laboratories of Robert Boyle, the meeting places of the Royal Society of London, and the quarters occupied by Robert Hooke.

Robert Boyle. Boyle had laboratories at each of the three major residences he successively inhabited during his mature life. From about 1645 to about 1655 he was mainly in residence at the manor house of Stalbridge in Dorset, an estate acquired by his father, the first Earl of Cork, in 1636 and inherited by his youngest son on the earl's death in 1643. By early in 1647 Boyle was organizing a chemical laboratory at Stalbridge, perhaps with the advice of the Hartlibian circle, whose London laboratories he frequently visited.[16] Late in 1655 or early in

---

[16] Robert Boyle to Lady Ranelagh, 6 Mar. 1647, in Boyle, *Works,* Vol. I, p. xxxvi, and Vol. VI, pp.

1656 he removed to Oxford, where his sister Katherine, Lady Ranelagh, had searched out rooms for him in the house of the apothecary John Crosse, Deep Hall in the High Street. He was apparently able to use Crosse's chemical facilities, and his own rooms contained a pneumatic laboratory, where, assisted by Hooke, the first version of the air pump was constructed in 1658–1659.[17] During his Oxford period Boyle also had access to a retreat at Stanton St. John, a village several miles to the northeast, where he made meteorological observations but apparently did not have a laboratory of any kind.

Boyle was away from Oxford for extended periods, staying sometimes at a house in Chelsea, sometimes with Katherine in London, and sometimes with another sister, Mary Rich, Countess of Warwick, at Leese (or Leighs) Priory in Essex.[18] In 1666 he had Oldenburg look over possible lodgings in Newington, north of London, but there is no evidence that he ever occupied these. And he periodically stayed at Beaconsfield in Buckinghamshire, possibly at the home of the poet Edmund Waller. But Oxford remained his primary residence and experimental workplace until he moved into quarters with Katherine at her house in Pall Mall in 1668. This was a house (actually two houses knocked into one) assigned to Lady Ranelagh by the Earl of Warwick in 1664. It stood on the south side of Pall Mall, probably on the site now occupied by the Royal Automobile Club. Although luxury building in this area was proceeding apace in the Restoration, at the time Boyle moved in Pall Mall still retained a rather quiet and semi-rural atmosphere. During the 1670s Boyle's neighbors included Henry Oldenburg, Dr. Thomas Sydenham, and Nell Gwyn.

Boyle's laboratory in Katherine's house was probably either in the basement or attached to the back, and there is some evidence to suggest that one could obtain access to the laboratory from the street without passing through the rest of the house.[19] The unmarried Boyle seems to have dined regularly with his sister, who was a major social and cultural figure in her own right, living "on the publickest scene," and who entertained his guests at the family table.[20] He remained there until his death in 1691, which closely followed Katherine's.

39–40; G. Agricola to Boyle, 6 Apr. 1668, *ibid.,* Vol. VI, pp. 650–651; and James Randall Jacob, "Robert Boyle, Young Theodicean" (Ph.D. diss., Cornell Univ., 1969), pp. 129–138. For the Stalbridge house, see R. E. W. Maddison, "Studies in the Life of Robert Boyle, F.R.S., Part VI: The Stalbridge Period, 1645–1655, and the Invisible College," *Notes Rec. Roy. Soc.,* 1963, *18*:104–124; Maddison, *The Life of the Honourable Robert Boyle F.R.S.* (London: Taylor & Francis, 1969), Ch. 2; Nicholas Canny, *The Upstart Earl: A Study of the Social and Mental World of Richard Boyle, First Earl of Cork 1566–1643* (Cambridge: Cambridge Univ. Press, 1982), pp. 68, 73, 98–99.

[17] Lady Ranelagh to Boyle, 12 Oct. [1655], in Boyle, *Works,* Vol. VI, pp. 523–524; Maddison, *Life of Boyle,* Ch. 3.

[18] For Chelsea, see *The Diary of John Evelyn,* ed. E. S. de Beer (London: Oxford Univ. Press, 1959), pp. 410, 417; Henry Oldenburg to Robert Boyle, 10 Sept. 1666, in Oldenburg, *Correspondence* (cit. n. 9), Vol. III, pp. 226, 227n.; and Maddison, *Life of Boyle* (cit. n. 16), p. 94. For Leese, see Boyle to Oldenburg, 13 June 1666, in Oldenburg, *Correspondence,* Vol. III, p. 160; Mary Rich, *Memoir of Lady Warwick: Also Her Diary, from A.D. 1666 to 1672* (London: Religious Tract Society, 1847?), esp. pp. 51, 161–163, 242–243; and Maddison, *Life of Boyle,* pp. 74, 132, 142.

[19] Lady Ranelagh to Boyle, 13 Nov. [1666] (as given, but more likely to be 1667), in Boyle, *Works,* Vol. VI, pp. 530–531 (where Katherine offers her "back-house" to be converted into a laboratory); Thomas Birch, "Life of Boyle," *ibid.,* Vol. I, pp. vi–clxxi, on pp. cxlv, cxxix; John Aubrey, *Brief Lives,* ed. Oliver Lawson Dick (Harmondsworth: Penguin, 1972), p. 198; and Maddison, *Life of Boyle,* pp. 128–129, 133–137, 177–178.

[20] Gilbert Burnet, *Select Sermons . . . and a Sermon at the Funeral of the Honourable Robert Boyle* (Glasgow, 1742), pp. 204–205; Maddison, *Life of Boyle,* pp. 134–135; and Webster, *Great Instauration,* pp. 61–63.

The Royal Society. After its founding in 1660, the weekly meetings of the Royal Society were held in Gresham College in Bishopsgate Street, originally in the rooms of the professor of geometry, afterward in rooms specially set aside for its use. The Great Fire of London in September 1666 made Gresham College unavailable, and temporary hospitality was extended by Henry Howard, later sixth Duke of Norfolk, at Arundel House, his residence in the Strand. The society met there for seven years, from 1667 to 1674, until Gresham become available again (see Figures 2 and 3). Gresham continued to be its home until 1710, when the society for the first time became the owner of its premises, purchasing the former home of a physician, Crane Court in Fleet Street.[21]

During the 1660s and 1670s the society was continually searching for alternative accommodation and making plans, all of which proved abortive, for purpose-built quarters of its own. In the event, for the first half century of its existence the public business of the Royal Society was transacted largely within places of private residence. Arundel House was unambiguously such a place, and Gresham College, built in the late sixteenth century as the residence of the great merchant banker Sir Thomas Gresham and transformed into a place of public instruction in 1598, had by the 1660s changed its character. When the Royal Society met there, it was a place where some professors lived and taught; where other sinecurist professors lived and did not teach; and where still others, who were not professors, lived in quarters hired out to them. According to its modern historian, Gresham College had by the mid 1670s "declined from a seat of learning into a lodging house." The significance of Arundel House in seventeenth-century English culture and social life cannot be overestimated. Until his death in 1646 it was the residence of Thomas Howard, second Earl of Arundel, who (despite the Catholicism he abandoned in 1616) as Earl Marshal was the head of the English nobility and the "custodian of honour." Arundel was one of the greatest collectors and patron of the arts of his age, and the house that contained his collections was made into a visible symbol of how a cultivated English gentleman ought to live. Indeed, his patronage of the educationalist Henry Peacham resulted in the production of an influential vade mecum for the guidance of English gentlemen. Arundel and his circle set themselves the task of modeling and exemplifying the code of English gentility, drawing liberally upon Italianate patterns. His grandson Henry Howard continued the great Arundel's proclivities, and it was through the encouragement of his friend John Evelyn that the society was offered space in the gallery of Arundel House and, ultimately, became one of the beneficiaries of the celebrated Arundel Collection of books, manuscripts, and objets d'art.[22]

---

[21] D. C. Martin, "Former Homes of the Royal Society," *Notes Rec. Roy. Soc.,* 1967, *22*:12–19; I. R. Adamson, "The Royal Society and Gresham College 1660–1711," *Notes Rec. Roy. Soc.,* 1978–1979, *33*:1–21; and Charles Richard Weld, *A History of the Royal Society,* 2 vols. (London, 1848), Vol. I, pp. 80–85, 192–198. For a satirical account of the Royal Society at Gresham, and esp. of the Repository and "the *Elaboratory-keepers* Apartment," see [Edward Ward], *The London-Spy,* 4th ed. (London, 1709), pp. 59–60.

[22] Hunter, "A 'College' for the Royal Society" (cit. n. 12); and Adamson, "The Royal Society and Gresham College," pp. 5–6. For Arundel House see David Howarth, *Lord Arundel and His Circle* (New Haven, Conn.: Yale Univ. Press, 1985); and Graham Perry, *The Golden Age Restor'd: The Culture of the Stuart Court, 1603–42* (Manchester: Manchester Univ. Press, 1981), Ch. 5. For Peacham see Henry Peacham, *The Complete Gentleman . . . ,* ed. Virgil B. Heltzel (1622, 1634, 1661; Ithaca, N.Y.: Cornell Univ. Press, 1962), pp. ix–xx.

Robert Hooke. On the founding of the Royal Society Robert Hooke was still serving Boyle as his technical assistant, lodging with Boyle in Oxford and when in London staying at least occasionally with Lady Ranelagh. When in November 1662 he was appointed by the society to the position of curator of experiments, Boyle was thanked "for dispensing with him for their use." By the next year Hooke was made a fellow (with charges waived) and was being paid by the society to lodge in Gresham College four days a week.[23] In 1664 he was elected professor of geometry at Gresham, with its associated lodgings, and there he remained, even during the society's absence at Arundel House, until his death in 1703. His quarters apparently opened behind the college "reading hall" and contained an extensive pneumatical, mechanical, and optical workshop, supplemented in 1674 by a small astronomical observatory constructed in a turret over his lodgings.[24]

The conditions in which Hooke lived and worked were markedly different from those of his patron Robert Boyle. Margaret 'Espinasse has vividly described his personal life at Gresham, where he "lived like a rather Bohemian scientific fellow of a college." His niece Grace was sharing his quarters from 1672 (when she was eleven years old) and was evidently sharing his bed sometime afterward. Hooke was also having sexual relations with his housekeeper Nell Young and, on her departure, with her successors. To what extent Hooke's domestic circumstances were known to his associates among the fellowship is unclear, though it is possible that there was some connection between those circumstances and the relative privacy of his rooms. It was Hooke who visited his high-minded patron Boyle; Boyle almost never visited Hooke. Hooke's relations with his various technicians were, in a different way, also very intimate. He took several of them into his lodgings, where they were treated in a manner intermediate between sons and apprentices (three of them becoming fellows of the Royal Society and one succeeding him as curator). Although his rooms were rarely frequented by gentlemen fellows on other than scientific and technical matters, and although his table was not a major venue for their discourses, Hooke lived on a public stage. He circulated through the taverns and the coffeehouses of the City of London and was a fixture at the tables of others. Hooke's place of residence, probably the most important site for experimental trials of Restoration England, was in practice a private place, while he himself lived an intensely public life. It is questionable indeed whether Hooke's quarters constituted a "home" in seventeenth-century gentlemanly usage. It was a place fit for Hooke to live and to work; it was not a place fit for the reception and entertainment of gentlemen.[25]

[23] Thomas Birch, *The History of the Royal Society of London*, 4 vols. (London, 1756–1757), Vol. I, pp. 123–124, 250; and Oldenburg to Boyle, 10 June 1663, in Boyle, *Works*, Vol. VI, p. 147.
[24] Margaret 'Espinasse, *Robert Hooke* (London: Heinemann, 1956), pp. 4–5; John Ward, *Lives of the Professors of Gresham College* (London, 1740), pp. 91, 178; and Adamson, "The Royal Society and Gresham College" (cit. n. 21), p. 4. The contents of Hooke's rooms at his death in 1703 are detailed in Public Record Office (London) MS PROB 5/1324. I thank Dr. Michael Hunter for a transcript of this inventory, which he will shortly publish.
[25] 'Espinasse, *Robert Hooke*, pp. 106–107, 113–127, 131–138, 141–147. The major source for Hooke's domestic life and activities is his diary for periods from the early 1670s: *The Diary of Robert Hooke, M.A., M.D., F.R.S., 1672–1680*, ed. Henry W. Robinson and Walter Adams (London: Taylor & Francis, 1935); and *The Diary of Robert Hooke, Nov. 1688 to March 1690; Dec. 1692 to Aug. 1693*, in Gunther, *Early Science in Oxford* (cit. n. 15), Vol. X, pp. 69–265. I have surveyed the diurnal

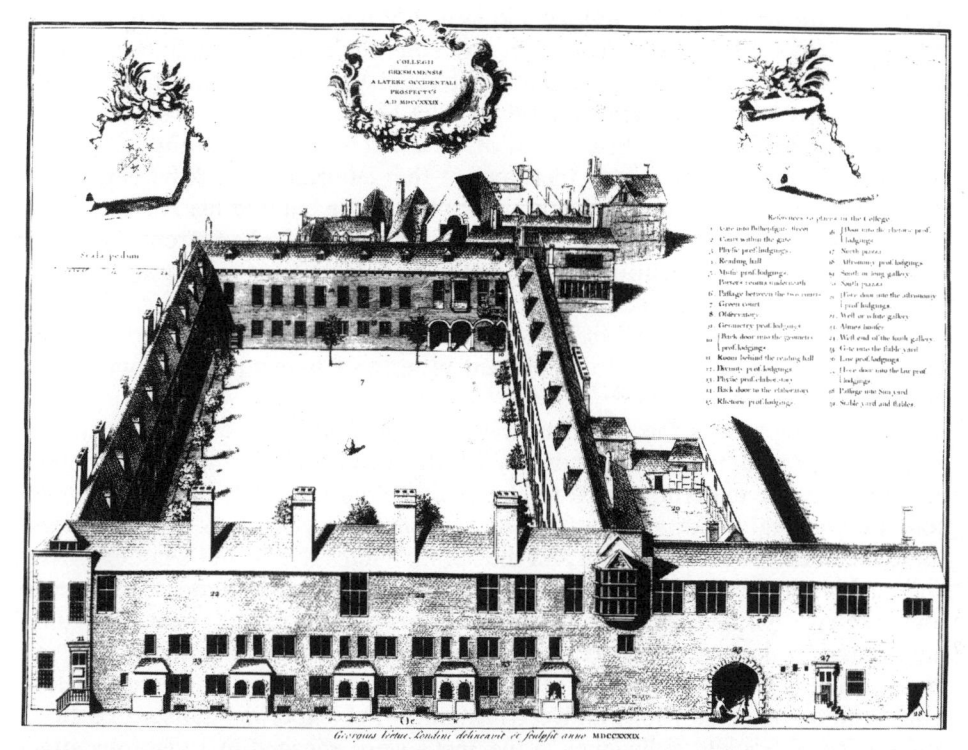

Figure 2. Gresham College in Bishopsgate Street, London; engraving from about 1739 by George Vertue. Gresham College was the meeting place of the Royal Society during 1660–1666 and 1674–1710. The lodgings of Robert Hooke, as professor of geometry, are at the far right corner of the quadrangle. Note also the turret above his rooms, used for astronomical observations. The Royal Society originally met in the rooms of the professor of astronomy at the near left corner of the quadrangle, later in various other rooms. The West Gallery where the Royal Society kept its Repository is in the foreground on the top floor. The college reading hall, where professors lectured, is the large pitched-roof structure behind Hooke's rooms. From John Ward, Lives of the Professors of Gresham College (London, 1740); courtesy of Special Collections, University of Edinburgh Library.

ACCESS

The threshold of the experimental laboratory was constructed out of stone and social convention. Conditions of access to the experimental laboratory would flow from decisions about what kind of place it was. In the middle of the century those decisions had not yet been made and institutionalized. Meanwhile there were a variety of stipulations about the functional and social status of spaces given over to experiment, and a variety of sentiments about access to them.

To the young Robert Boyle the threshold of his Stalbridge laboratory constituted the boundary between sacred and secular space. He told his sister Katherine that "*Vulcan* has so transported and bewitched me, that as the delights I taste

---

patterns of Hooke's life and tried to relate them to his place in the experimental community: Steven Shapin, "Who Was Robert Hooke?" in *Robert Hooke: New Studies,* ed. Michael Hunter and Simon Schaffer (Woodbridge, Suffolk: Boydell & Brewer, forthcoming). An unreflectively Freudian account of Hooke's sex life is found in Lawrence Stone, *The Family, Sex and Marriage in England 1500–1800* (London: Weidenfeld & Nicolson, 1977), pp. 561–563.

in it make me fancy my laboratory a kind of *Elysium,* so as if the threshold of it possessed the quality the poets ascribe to *Lethe,* their fictions made men taste of before their entrance into those seats of bliss, I there forget my standish [inkstand] and my books, and almost all things."[26] The experimenter was to consider himself "honor'd with the Priesthood of so noble a Temple" as the "Commonwealth of Nature." And it was therefore fit that laboratory work be performed, like divine service, on Sundays. (In mature life Boyle entered his Pall Mall laboratory directly after his morning devotions, although he had apparently given up the practice of experimenting on the Sabbath.)[27] In the 1640s he told his Hartlibian friends of his purposeful "retreat to this solitude" and of "my confinement to this melancholy solitude" in Dorset. But it was said to be a wished-for and a virtuous solitude, and Boyle complained bitterly of interruptions from visitors and their trivial discourses.[28]

Transactions across the experimental threshold had to be carefully managed. Solitude appeared both as a mundanely practical consideration and as a symbolic condition for the experimentalist to claim authenticity. Models of space in which solitude was legitimate and out of which valued knowledge emerged did exist: these included the monastic cell and the hermit's hut. The hermit's hut expressed and enabled individual confrontation with the divine; the solitude of the laboratory likewise defined the circumstances in which the new "priest of nature" might produce knowledge as certain and as morally valuable as that of the religious isolate. Here was a model of space perceived to be insulated from distraction, temptation, distortion, and convention.[29] Yet experimentalists like Boyle and his Royal Society colleagues in the 1660s were engaged in a vigorous attack on the privacy of existing forms of intellectual practice. The legitimacy of experimental knowledge, it was argued, depended upon a public presence at some crucial stage or stages of knowledge making. If experimental knowledge did indeed have to occupy private space during part of its career, then its realization as authentic knowledge involved its transit to and through a public space.

This transit was particularly difficult for a man in Boyle's position to accomplish and make visible as legitimate. He presented himself as an intensely private man, one who cared little for the distractions and rewards of ordinary social life. This presentation of self was successful. Bishop Burnet, who preached Boyle's funeral sermon, described him as a paragon: "He neglected his person, despised

---

[26] Boyle to Lady Ranelagh, 31 Aug. 1649, in Boyle, *Works,* Vol. VI, pp. 49–50, and Vol. I, p. xlv; cf. Boyle to [Benjamin Worsley?], n.d. (probably late 1640s), *ibid.,* Vol. VI, pp. 39–40.

[27] Boyle Papers, Royal Society (hereafter **Boyle Papers**), Vol. VIII, fol. 128, quoted in Jacob, "Boyle, Young Theodicean" (cit. n. 16), p. 158 (quotation). See also Harold Fisch, "The Scientist as Priest: A Note on Robert Boyle's Natural Theology," *Isis,* 1953, *44*:252–265; Shapin and Schaffer, *Leviathan and the Air-Pump* (cit. n. 4), p. 319; and, on Sunday experiments, Jacob, "Boyle, Young Theodicean," pp. 153–154. In one of Boyle's later notebooks he recorded how many experiments he had performed day by day for periods between 1684 and 1688. By then, the rule had clearly become "never on Sunday": Boyle Papers, Commonplace Book, 190, fols. 167–171. See also Maddison, *Life of Boyle* (cit. n. 16), p. 187; and Maddison, "Studies in the Life of Robert Boyle, F.R.S., Part IV: Robert Boyle and Some of His Foreign Visitors," *Notes Rec. Roy. Soc.,* 1954, *11*:38–53, on p. 38.

[28] Boyle to Benjamin Worsley, n.d., in Boyle, *Works,* Vol. VI, pp. 39–41; and Boyle to Lady Ranelagh, 13 Nov. ?, *ibid.,* pp. 43–44. (Both letters probably date from the late 1640s.) See also Marie Boas, *Robert Boyle and Seventeenth-Century Chemistry* (Cambridge: Cambridge Univ. Press, 1958), pp. 15–16, 19, 21.

[29] Note the monastic flavor of the Cowley, Evelyn, and Petty plans for philosophical colleges. Evelyn's referred explicitly to a Carthusian model: Evelyn, *Diary and Correspondence* (cit. n. 12), Vol. III, p. 118.

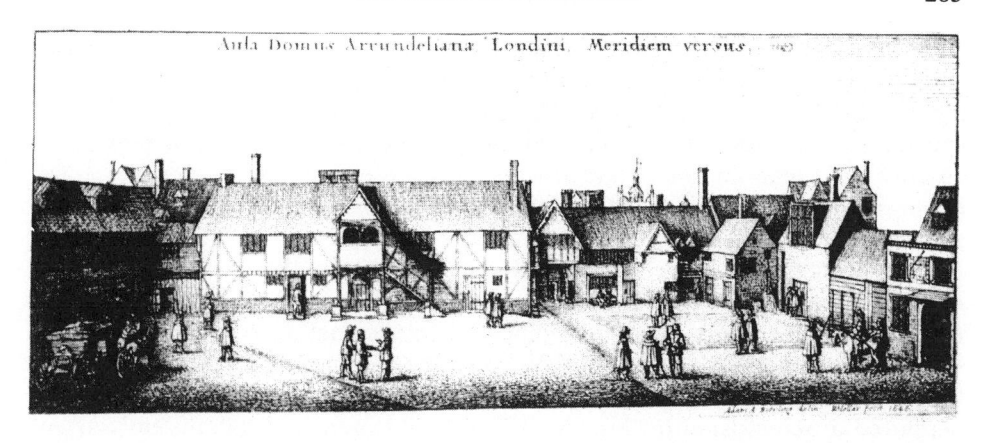

Figure 3. Arundel House in the Strand, London; engraving from about 1646 by Wenceslaus Hollar. Arundel House was the meeting place of the Royal Society during 1667–1674 and London home of the Howards, Earls of Arundel and Dukes of Norfolk. Arundel House began to be pulled down from 1678. From Arthur M. Hind, Wenceslaus Hollar and His Views of London and Windsor in the Seventeenth Century *(London, 1922).*

the world, and lived abstracted from all pleasures, designs, and interests."[30] At the same time, Boyle effectively secured the persona of a man to whom justified access was freely available. He was entitled by birth and by wealth (even as diminished by the Irish wars) to a public life, and, indeed, there were forces that acted to ensure that he did live in the public realm. He advertised the public status of experimental work and, from his first publication, condemned unwarranted secrecy and intellectual unsociability.[31] Yet he chose much solitude, was seen to do so, and was drawn only fitfully into the company of fellow Christian virtuosi, extended exposure to which drove him once more to solitude. In constructing his life and making it morally legitimate Boyle was endeavoring to

---

[30] Gilbert Burnet, *History of His Own Time,* 6 vols. (Oxford: Oxford Univ. Press, 1833), Vol. I, p. 351; cf. Burnet, *Select Sermons* (cit. n. 20), pp. 202, 210: Boyle "had neither designs nor passions."
[31] [Robert Boyle], "An Epistolical Discourse of Philaretus to Empiricus, . . . inviting All True Lovers of Vertue and Mankind, to a Free and Generous Communication of Their Secrets and Receits in Physick" (prob. written 1647), in Hartlib, comp., *Chymical, Medicinal, and Chyrurgical Addresses* (cit. n. 10), pp. 113–150, rpt. in Margaret E. Rowbottom, "The Earliest Published Writing of Robert Boyle," *Annals of Science,* 1948–1950, 6:376–389, on pp. 380–385.

define the nature of a space in which experimental work might be practically situated and in which experimental knowledge would be seen as authentic. Such a space did not then clearly exist. The conditions of access to it and the form of social relations within it had to be determined and justified. This space had necessarily to be carved out of and rearranged from existing domains of accepted public and private activity and existing stipulations about the proper uses of spaces.

Many contemporary commentators remarked upon the ease of access to Boyle's laboratory. John Aubrey wrote about Boyle's "noble laboratory" at Lady Ranelagh's house as a major object of intellectual pilgrimage: "When foreigners come to hither, 'tis one of their curiosities to make him a Visit." This was the laboratory that was said to be "constantly open to the Curious, whom he permitted to see most of his Processes." In 1668 Lorenzo Magalotti, emissary of the Florentine experimentalists, traveled especially to Oxford to see Boyle and boasted that he was rewarded with "about ten hours" of his discourse, "spread over two occasions." John Evelyn noted that Boyle "had so universal an esteeme in Foraine parts; that not any Stranger of note or quality; Learn'd or Curious coming into England, but us'd to Visite him." He "was seldome without company" in the afternoons, after his laboratory work was finished.[32]

But the strain of maintaining quarters "constantly open to the curious" told upon him and was seen to do so. As an overwrought young man he besought "deare Philosophy" to "come quickly & releive Your Distresst Client" of the "vaine Company" that forms a "perfect Tryall of my Patience." Experimental philosophy might rescue him "from some strange, hasty, Anchoritish Vow"; it could save him from his natural "Hermit's Aversenesse to Society."[33] When, during the plague, members of the Royal Society descended upon him in Oxford, he bolted for the solitude of his village retreat at Stanton St. John, complaining of "ye great Concourse of strangers," while assuring Oldenburg that "I am not here soe neere a Hermite" but that some visitors were still welcome. Even as John Evelyn praised Boyle's accessibility, he recorded that the crowding "was sometimes so incomodious that he now and then repair'd to a private Lodging in another quarter [of London], and at other times" to Leese or elsewhere in the country "among his noble relations."[34]

---

[32] On foreigners: Aubrey, *Brief Lives* (cit. n. 19), p. 198; Eustace Budgell, *Memoirs of the Lives and Characters of the Illustrious Family of the Boyles*, 3rd ed. (London, 1737), p. 144; R. E. W. Maddison, "Studies in the Life of Robert Boyle, F.R.S., Part I: Robert Boyle and Some of His Foreign Visitors," *Notes Rec. Roy. Soc.*, 1951, 9:1–35, on p. 3; Birch, "Life of Boyle" (cit. n. 19), p. cxlv. For Magalotti: W. E. Knowles Middleton, "Some Italian Visitors to the Early Royal Society," *Notes Rec. Roy. Soc.*, 1978–1979, 33:157–173, on p. 163; *Lorenzo Magalotti at the Court of Charles II: His "Relazione d'Inghilterra" of 1668*, ed. and trans. Middleton (Waterloo, Ontario: Wilfrid Laurier Univ. Press, 1980), p. 8. Similar hospitality was extended in 1669, when Magalotti escorted the Grand Duke of Tuscany to Boyle's Pall Mall laboratory: [Lorenzo Magalotti], *Travels of Cosmo the Third, Grand Duke of Tuscany* (London, 1821), pp. 291–293; and R. W. Waller, "Lorenzo Magalotti in England 1668–1669," *Italian Studies*, 1937, 1:49–66. For Evelyn's comments: John Evelyn to William Wotton, 29 Mar. 1696, quoted in Maddison, "Studies in the Life of Boyle, Part IV" (cit. n. 27), p. 38.

[33] Boyle Papers, Vol. XXXVII, fol. 166 (no date but probably mid to late 1640s). For background to this manuscript ("The Gentleman"), see J. R. Jacob, *Robert Boyle and the English Revolution: A Study in Social and Intellectual Change* (New York: Burt Franklin, 1977), pp. 48–49.

[34] Boyle to Oldenburg, 8, c. 16 and 30 Sept. and 9 Dec. 1665; Oldenburg to Spinoza, 12 Oct. 1665; Sir Robert Moray to Oldenburg, 4 Dec. 1665, in Oldenburg, *Correspondence* (cit. n. 9), Vol. II, pp. 502, 509, 537, 563, 568, 627, 639; and Evelyn to Wotton, 29 Mar. 1696, quoted in Maddison, "Studies in the Life of Boyle, Part IV," p. 38; and Maddison, *Life of Boyle* (cit. n. 16), pp. 186–188.

Toward the end of his life Boyle took drastic and highly visible steps to restrict access to his drawing room and laboratory. It is reported that when he was at work trying experiments in the Pall Mall laboratory and did not wish to be interrupted, he caused a sign to be posted on his door: "Mr. Boyle cannot be spoken with to-day." In his last years and in declining health, he issued a special public advertisement "to those of his ffriends & Acquaintance, that are wont to do him the honour & favour of visiting him," to the effect that he desired "to be excus'd from receiving visits" except at stated times, "(unless upon occasions very extraordinary)."[35] Bishop Burnet said that Boyle "felt his easiness of access" made "great wasts on his time," but "thought his obligation to strangers was more than bare civility."[36]

That obligation was a powerful constraint. The forces that acted to keep Boyle's door ajar were social forces. Boyle was a gentleman as well as an experimental philosopher. Indeed, as a young man he had reflected systematically upon the code of the gentleman and his own position in that code. The place where Boyle worked was also the residence of the son of the first Earl of Cork. It was a point of honor that the private residence of a gentleman should be open to the legitimate visits of other gentlemen. Seventeenth-century handbooks on the code of gentility stressed this openness of access: one such text noted that "Hospitalitie" was "one of the apparentest Signalls of *Gentrie*." Modern historians confirm the equation between easy access and gentlemanly standing: "generous hospitality was the hallmark of a gentleman"; "so long as the habit of open hospitality persisted, privacy was unobtainable, and indeed unheard of." And as the young Boyle himself confided in his *Commonplace Book,* a "Noble Descent" gives "the Gentleman a Free Admittance into many Companys, whence Inferior Persons (tho never so Deserving) are . . . excluded."[37] Other gentlemen knew who was a gentleman, they knew the code regulating access to his residence, and they knew that Boyle was obliged to operate under this code. But they did not know, nor could they, what an experimental scientist was, nor what might be the nature of a different code governing admittance to his laboratory. In the event, as Marie Boas wrote, they might plausibly come to the conclusion that Boyle "was only a virtuoso, amusing himself with science, [that] he could be interrupted at any time. . . . There was always a swarm of idle gentlemen and ladies who wanted to see amusing and curious experiments."[38] When, however, Boyle wished to shut his door to these distractions, he was able to draw upon widely understood moral patterns that enabled others to recognize what he was doing and why it might be

---

[35] On the sign: Weld, *History of the Royal Society* (cit. n. 21), Vol. I, p. 136n; Maddison, *Life of Boyle,* pp. 177–178; cf. *Diary of Hooke,* in Gunther, *Early Science in Oxford* (cit. n. 15), Vol. X, p. 139 (entry for 29 July 1689): "To Mr. Boyle. Not to be spoken wth." For the advertisement: Boyle Papers, Vol. XXXV, fol. 194. There are a number of other drafts of this advertisement in the Boyle Papers; one is printed in Maddison, *Life of Boyle,* p. 177.

[36] Burnet, *Select Sermons* (cit. n. 20), p. 201; Maddison, "Studies in the Life of Boyle, Part I" (cit. n. 32), pp. 2–3.

[37] Richard Brathwait, *The English Gentleman* (London, 1630), pp. 65–66, sig. Nnn2r; Lawrence Stone and Jeanne C. Fawtier Stone, *An Open Elite? England 1540–1880* (Oxford: Clarendon Press, 1984), pp. 307–310; and Boyle Papers, Vol. XXXVII, fol. 160v.

[38] Boas, *Boyle and Seventeenth-Century Chemistry* (cit. n. 28), p. 207. On the role of the virtuoso and the expectation that his collections would be accessible to the visits of others, see Walter E. Houghton, Jr., "The English Virtuoso in the Seventeenth Century," *Journal of the History of Ideas,* 1942, *3*:51–73, 190–219; and Oliver Impey and Arthur MacGregor, eds., *The Origins of Museums: The Cabinet of Curiosities in Sixteenth- and Seventeenth-Century Europe* (Oxford: Clarendon Press, 1985).

legitimate. The occasional privacy of laboratory work could be assimilated to the morally warrantable solitude characteristic of the religious isolate.[39]

RIGHTS OF PASSAGE

What were the formal conditions of entry to experimental spaces? We do have some information concerning the policy of the early Royal Society, particularly regarding access of English philosophers to foreign venues. It was evidently common for the society's council to give "intelligent persons, whether Fellows of the Society or not, what are styled 'Letters Recommendatory.' " These documents, in Latin, requested "that all persons in authority abroad would kindly receive the bearer, who was desirous of cultivating science, and show him any attention in their power."[40] Similarly, in 1663 the society drafted a statute regulating access to its own meetings. As soon as the president took the chair, "those persons that [were] not of the society, [were to] withdraw." There was, however, an exemption for certain classes of persons to remain if they chose, that is, "for any of his majesty's subjects . . . having the title and place of a baron, or any of his majesty's privy council . . . , or for any foreigner of eminent repute, with the allowance of the president." Other persons might be permitted to stay with the explicit consent of the president and fellows in attendance. Barons and higher-ranking aristocrats could become fellows on application, without the display of philosophical credentials.[41]

Too much should not be made of such fragmentary evidence of formal conditions granting or withholding rights of entry to experimental sites. It is noteworthy how sparse such evidence is, even for a legally incorporated body like the Royal Society. In the main, the management of access to experimental spaces, even those of constituted organizations, was effected more informally. For example, there was the letter of introduction to an experimentalist, a number of which survive. In 1685 one visitor carried with him a letter of introduction, from someone presumably known to Boyle, which identified the bearer as "ambitious to be known to you, whose just character of merit is above his quality . . . , being the eldest son of [a diplomat] and brother-in-law to the king of Denmark's envoy."[42] In this instance and in others like it, it was not stated that the proposed entrant to Boyle's society possessed any particular technical competences, nor even that he was "one of the curious," merely that he was a gentleman of quality and merit, as

---

[39] There was, of course, another model available in principle to identify the conditions of privacy. This was the alchemist's laboratory, but Boyle worked hard to discredit that model, even as he spent time in the relatively open laboratories of the Hartlib circle. See, e.g., Samuel Hartlib to Boyle, 28 Feb. 1654 and 14 Sept. 1658, in Boyle, *Works*, Vol. VI, pp. 78–83, 114–115.

[40] Weld, *History of the Royal Society* (cit. n. 21), Vol. I, pp. 224–225.

[41] Birch, *History of the Royal Society* (cit. n. 23), Vol. I, pp. 264–265. On the society's increasing concern for secrecy and the limitation of public access in the 1670s, see Michael Hunter and Paul B. Wood, "Towards Solomon's House: Rival Strategies for Reforming the Early Royal Society," *History of Science*, 1986, 24:49–108, on pp. 74–75. The test for admittance to fellowship of a special claim to scientific knowledge was not formalized until 1730: Maurice Crosland, "Explicit Qualifications as a Criterion for Membership of the Royal Society: A Historical Review," *Notes Rec. Roy. Soc.*, 1983, 37:167–187.

[42] Joseph Hill to Boyle, 20 Apr. 1685, in Boyle, *Works*, Vol. VI, p. 661. At about the same time, an eminent cleric wrote announcing his imminent visit but assured the weary Boyle that it would be sufficient to have "some servant of yours" delegated "to shew me your laboratory": Bishop of Cork to Boyle, 12 June 1683, *ibid.*, p. 615.

vouched for by the correspondent. In other cases, "curiosity" was explicitly stipulated as a sufficient criterion for entry.

Generally speaking, it appears that access to most experimental venues (and especially those located in private residences) was achieved in a highly informal manner, through the tacit system of recognitions, rights, and expectations that operated in the wider society of gentlemen. If we consider Boyle's laboratories and drawing rooms, it seems that entry was attained if one of three conditions could be met: if the applicant was (1) known to Boyle by sight and of a standing that would ordinarily give rights of access; (2) known to Boyle by legitimate reputation; (3) known to Boyle neither by sight nor by reputation, but arriving with (or with an introduction from) someone who satisfied condition (1) or (2). These criteria can be expressed much more concisely: access to experimental spaces was managed by calling upon the same sorts of conventions that regulated entry to gentlemen's houses, and the relevant rooms within them, in general.[43] These criteria were not codified and written down because they did not need to be. They would be known and worked with by every gentleman. Indeed, they would almost certainly be known and worked with by those who were not gentlemen, shaping their understanding of the grounds for denying entry. Standing gave access. Boyle was perhaps unusual among English gentlemen in reflecting explicitly upon this largely tacit knowledge: "A man of meane Extraction (tho never so advantag'd by Greate meritts) is seldome admitted to the Privacy & the secrets of greate ones promiscuously & scarce dares pretend to it, for feare of being censur'd saucy, or an Intruder."[44]

I have alluded to some formal criteria governing entry to the rooms of the Royal Society. For all the significance of such considerations, informal criteria operated there as well, just as they did in the case of Boyle's laboratories, to manage passage across the society's threshold. These almost certainly encompassed not only the informal criteria mentioned in connection with Boyle, for whom standing gave access, but also other sorts of tacit criteria. When Lorenzo Magalotti visited England in 1668 (after the Royal Society's removal to Arundel House), his arrival at the society's weekly meeting was apparently expected and special experiments had been made ready to be shown. But Magalotti had second thoughts about the advisability of attending and the terms on which he thought entry was offered. "I understood," he wrote to an Italian prelate, "that one is not permitted to go in simply as a curious passer-by, [and] I would not agree to take my place there as a scholar, for one thing because I am not one. . . . Thus,

---

[43] One would like to know much more about the specific sites within the house where experimental work was done and where experimental discourses were held. In addition to rooms designated for public use, the *closet*, where many virtuosi kept their curiosities (including scientific instruments) and where intimate conversations often took place, should be of particular interest. The closet was a room variously located and variously employed in the seventeenth century, but many examples were situated off the bedroom, meaning that this was a private space, access to which acknowledged or accorded intimacy. For the closet and divisions of domestic space generally, see Mark Girouard, *Life in the English Country House: A Social and Architectural History* (Harmondsworth: Penguin, 1980), esp. pp. 129–130; and Gervase Jackson-Stops and James Pipkin, *The English Country House: A Grand Tour* (London: Weidenfeld & Nicolson, 1985), Ch. 9. For an analysis of the internal layout of the house in relation to social structure, see Norbert Elias, *The Court Society,* trans. Edmund Jephcott (Oxford: Basil Blackwell, 1983), Ch. 3.

[44] Robert Boyle, "An Account of Philaretus [i.e., Mr. R. Boyle] during His Minority," in Boyle, *Works,* Vol. I, pp. xii–xxvi, on p. xiii.

therefore, I got as far as the door and then went away, and if they do not want to permit me to go and be a mere spectator without being obliged to give opinions like all the others, I shall certainly be without the desire to do so."[45] There are reasons to doubt the absolute reliability of Magalotti's testimony. Nevertheless, he pointed to a crucially important tacit criterion of entry. Magalotti was, of course, the sort of person, carrying the sort of credentials, who would have had unquestioned access to the Royal Society meeting at Arundel House, or, indeed to Arundel House itself. His claims indicate, however, that the experimental activities that went on within its interior imposed further informal criteria regulating entry. These included the uncodified expectation that, once admitted, one would act *as a participant*. The notion of participation followed from a distinction, customary but not absolute, between *spectating* and *witnessing*. The Royal Society expected those in attendance to validate experimental knowledge as participants, by giving witness to matters of fact, rather than to play the role of passive spectators to the doings of others.[46] But there was a further consideration: those granted entry were tacitly enjoined to employ the conventions of deportment and discourse deemed appropriate to the experimental enterprise, rather than those current in, say, hermetic, metaphysical, or rationalist practice. Those unwilling to observe these conventions could exclude themselves. These are the grounds on which one might rightly say that a philosopher like Thomas Hobbes was, in fact, excluded from the precincts of the Royal Society, even though there is no evidence that he sought entry and was turned away. 'Espinasse was therefore quite correct in saying that the society was "open to all classes rather in the same way as the law-courts and the Ritz," and Quentin Skinner was also right to characterize it as "like a gentlemen's club," even if he unnecessarily contrasted that status with the society's ostensible role as "the conscious centre of all genuinely scientific endeavour."[47]

RELATIONS IN PUBLIC

If we are able to recognize what kind of space we are in, we find we already possess implicit knowledge of how it is customary to behave there. But in the middle of the seventeenth century the experimental laboratory and the places of experimental discourse did not have standard designations, nor did people who found themselves within them have any tacit knowledge of the behavioral norms obtaining there. On the one hand, publicists of the experimental program offered detailed guidance on the social relations deemed appropriate to experimental spaces; on the other, there is virtual silence about some of the most basic fea-

---

[45] Lorenzo Magalotti to Cardinal Leopold, 14 Feb. 1668, in Middleton, "Some Italian Visitors to the Royal Society," p. 160; Waller, "Magalotti in England," pp. 52–53; and *Magalotti at the Court of Charles II,* ed. and trans. Middleton, p. 8 (all cit. n. 32).

[46] For divergences of opinion among the fellowship about whether or not mere spectators ought to be encouraged, see Hunter and Wood, "Towards Solomon's House," p. 71, 87–92. Hooke took a particularly strong line that *participants* only were wanted; see also *Philosophical Experiments and Observations of the Late Eminent Dr. Robert Hooke,* ed. W. Derham (London, 1726), pp. 26–27.

[47] Margaret 'Espinasse, "The Decline and Fall of Restoration Science," *Past and Present,* 1958, *14*:71–89, on p. 86; and Quentin Skinner, "Thomas Hobbes and the Nature of the Early Royal Society," *Historical Journal,* 1969, *12*:217–239, on p. 238. See also Michael Hunter, *The Royal Society and Its Fellows 1660–1700: The Morphology of an Early Scientific Institution* (Chalfont St. Giles, Bucks.: British Society for the History of Science, 1982), p. 8. On Hobbes and the Royal Society see Shapin and Schaffer, *Leviathan and the Air-Pump* (cit. n. 4), pp. 131–139.

tures of these places. The situation is about what one would expect if new patterns of behavior in one domain were being put together out of patterns current in others.

In 1663 the Royal Society was visited by two Frenchmen, Samuel Sorbière, physician and informal emissary of the Montmor Academy, and the young Lyonnaise scholar Balthasar de Monconys. Both subsequently published fairly detailed accounts of the society's procedures. Sorbière recorded that the meeting room at Gresham was some sort of "Amphitheatre," possibly the college reading hall or an adaptation of a living room of Gresham's sixteenth-century cloistered house to make it suitable for public lecturing. The president sat at the center of a head table, with the secretary at his side and chairs for distinguished visitors. The ordinary fellows sat themselves on plain wooden benches arranged in tiers, "as they think fit, and without any Ceremony."[48] An account dating from around 1707, toward the end of the society's stay at Gresham, gives a description of three rooms in which it conducted its affairs but omits any detail of the internal arrangements or of social relations within them.[49]

When Magalotti visited the Royal Society at Arundel House in February 1668, he described the assembly room off the gallery, "in the middle of which is a large round table surrounded by two rows of seats, and nearer to it by a circle of plush stools for strangers." On his second visit in April 1669 he recorded that the president sat "on a seat in the middle of the table of the assembly."[50] No visitor, or any other commentator, provides a detailed account of the physical and social arrangements attending the performance of experiments in the Royal Society. Monconys offers a recitation of experiments done, without describing the circumstances in which they were done. Sorbière mentions only that there was brief discussion of "the Experiments proposed by the Secretary." Magalotti records that he saw experiments performed, demonstrated by "a certain Mr Hooke."

---

[48] Samuel Sorbière, *A Voyage to England, containing Many Things relating to the State of Learning, Religion, and Other Curiosities of that Kingdom* (London: 1709; trans. of 1664 French original), pp. 35–38; and *Journal des voyages de Monsieur de Monconys* (Lyons, 1666), separately paginated "Seconde Partie: Voyage d'Angleterre," p. 26. See also Thomas Molyneux to William Molyneux, 26 May 1683, in K. Theodore Hoppen, "The Royal Society and Ireland: William Molyneux, F.R.S. (1656–1698)," *Notes Rec. Roy. Soc.*, 1963, *18*:125–135, on p. 126; and Maddison, "Studies in the Life of Boyle, Part I" (cit. n. 32), pp. 14–21. There is a brief "official" account of the society's rooms in Thomas Sprat, *History of the Royal Society* (London, 1667), p. 93.

[49] *Account of the Proceedings in the Council of the Royal Society, in Order to Remove from Gresham College* (London, 1707?), partly rpt. in Weld, *History of the Royal Society*, Vol. I, pp. 82–83; and Martin, "Former Homes of the Royal Society" p. 13 (both cit. n. 21). A brief account of the society's rooms in the last year of its occupancy of Gresham College is in *London in 1710: From the Travels of Zacharias Conrad von Uffenbach*, trans. and ed. W. H. Quarrell and Margaret Mare (London: Faber & Faber, 1934), pp. 97–102. Uffenbach described the "wretchedly ordered" repository and library, and the "very small and wretched" room where the society usually met. At that time it was sparsely decorated with portraits of its members (including, Uffenbach claimed, a picture of Hooke, about which he was mistaken or which was subsequently lost), two globes, a model of a contrivance for rowing, and a large pendulum clock. There is no pictorial record of an internal space occupied by the Royal Society in the seventeenth century. For an engraving (of doubtful date) recording a Crane Court meeting see T. E. Allibone, *The Royal Society and Its Dining Clubs* (Oxford: Pergamon Press, 1976), frontispiece.

[50] Magalotti to Cardinal Leopold, 21 Feb. 1668, in Middleton, "Some Italian Visitors to the Royal Society," pp. 160–161; and Magalotti, *Travels of Cosmo*, pp. 185–186 (both cit. n. 32). Too much weight should not, perhaps, be given to Magalotti's evidence. He seems to have been confused about where exactly he was: in the 1669 account he says that he went with the Grand Duke "to Arundel House, in the interior of Gresham College." Moreover, he seems to have derived portions of his version of the society from Sorbière's earlier account.

These were set up on a table in the corner of the meeting room at Arundel House. When working properly, experiments were transferred to a table in the middle of the room and displayed, "each by its inventor." Experimental discussion then ensued.[51] By the 1670s it is evident that experimental "discourse," or formal presentation setting forth and interpreting experiments tried elsewhere, was much more central to the society's affairs than experiments tried and displayed within its precincts.

Sorbière, Monconys, Magalotti, and other observers all stressed the civility of the Royal Society's proceedings. The president, "qui est toujours une personne de condition," was clearly treated with considerable deference, by virtue of his character, his office, and, most important, his function in guaranteeing good order. Patterns provided by procedure in the House of Commons are evident. Fellows addressed their speech to the president, and not to other fellows, just as members of the House of Commons conventionally addressed the Speaker. Thus, the convenient fiction was maintained that it was always the matter and not the man that was being addressed. Both Sorbière and Magalotti noted that fellows removed their hats when speaking, as a sign of respect to the president (again following Commons practice). Whoever was speaking was never interrupted, "and Differences of Opinion cause no manner of Resentment, nor as much as a disobliging Way of Speech."[52] An English observer said that the society "lay aside all set Speeches and Eloquent Haranques (as fit to be banisht out of all Civil Assemblies, as a thing found by woful experience, especially in *England*, fatal to Peace and good Manners)," just as the reading of prepared speeches was (and is) conventionally deprecated in Commons. "Opposite opinions" could be maintained without "obstinacy," but with good temper and "the language of civility and moderation."[53]

This decorum was the more remarkable in that it was freely entered into and freely sustained. Sorbière said that "it cannot be discerned that any Authority prevails here"; and Magalotti noted that at "their meetings, no precedence or distinction of place is observed, except by the president and the secretary." As in the seventeenth-century House of Commons, the practice of taking any available seat (with the exception of the president and the secretaries, who, like the Commons Speaker, his clerks, and privy councillors, sat at the head of the room) constituted a visible symbol of the equality in principle of all fellows and of the absence of sects, even if the reality, in both houses, might be otherwise.[54] All

---

[51] *Journal des voyages de Monconys* (cit. n. 48), pp. 26–28, 47, 55–57; Sorbière, *Voyage to England* (cit. n. 48), p. 37; Maddison, "Studies in the Life of Boyle, Part I" (cit. n. 32), pp. 16–19 (Monconys), 21 (Sorbière); and Magalotti to Cardinal Leopold, 21 Feb. 1668, in Middleton, "Some Italian Visitors to the Royal Society" (cit. n. 32), pp. 161–162.

[52] *Journal des voyages de Monconys*, p. 26 (quoted); Magalotti, *Travels of Cosmo*, pp. 186–187; and Sorbière, *Voyage to England*, pp. 36–37. For Commons practice, see Sir Thomas Smith, *De republica anglorum* (London, ca. 1600), pp. 51–52. (This practice differs from that of the House of Lords, where speakers address "My lords.")

[53] On set speeches: Edward Chamberlayne, *Angliae notitia: or the Present State of England* (7th ed., London, 1673), p. 345. Cf. Sir Thomas Erskine May, *A Treatise on the Law, Privileges, Proceedings and Usage of Parliament*, ed. T. Lonsdale Webster and William Edward Grey, 11th ed. (London: William Clowes & Sons, 1906), pp. 310, 314–315, 344–345; and Lord Campion, *An Introduction to the Procedure of the House of Commons* (London: Macmillan, 1958), pp. 190, 192. On "opposite opinions": Magalotti, *Travels of Cosmo*, pp. 187–188. Cf. J. E. Neale, *The Elizabethan House of Commons* (London: Jonathan Cape, 1949), pp. 404–407; and Smith, *De republica anglorum* (cit. n. 52), p. 52. (Needless to say, these were stipulations of ideal behavior: violations of the norms in Commons were frequent.)

[54] Sorbière, *Voyage to England*, p. 38; and Magalotti, *Travels of Cosmo*, p. 187. Cf. George Henry

visitors found it worth recording that the society's mace, laid on the table before the president when the meetings were convened, was an emblem of the source of order. Again, as in Commons, the mace indicated that the ultimate source was royal. The king gave the society its original mace even as he replaced the Commons mace that had disappeared in the Interregnum. The display of the mace in the Royal Society confirmed that its authority flowed from, and was of the same quality as, that of the king. Nevertheless, Thomas Sprat took violent exception to any notion that mace ceremonials constituted rituals of authority: "The *Royal Society* itself is so careful that such ceremonies should be just no more than what are necessary to avoid Confusion."[55] Sprat took the view that the space occupied and defined by the fellowship was truly novel: it was regulated by no traditional set of rituals, customs, or conventions. An anonymous fellow writing in the 1670s agreed: the society's job was "not to whiten the walls of an old house, but to build a new one; or at least, to enlarge the old, & really to mend its faults."[56]

Yet no type of building, no type of society is wholly new. And despite the protestations of early publicists, it is evident that the social relations and patterns of discourse obtaining within the rooms of the Royal Society were rearrangements and revaluations of existing models. Aspects of a parliamentary pattern have already been mentioned. The relationship between the proceedings of the early Royal Society and the Interregnum London coffeehouse merits extended discussion, most particularly in connection with the rules of good order in a mixed assembly. Other elements resonate of the monastery, the workshop, the club, the college, and the army.[57] Yet the most potent model for the society's social relations was drawn from the type of space in which they actually occurred. The code which is closest to that prescribed for the experimental discourses of the Royal Society was that which operated within the public rooms of a gentleman's private house.

THE EXPERIMENTAL PUBLIC

What was the experimental public like? How many people, and what sorts of people, composed that public? In order to answer these questions we have to

---

Jennings, comp., *An Anecdotal History of the British Parliament* . . . (London: Horace, Cox, 1880), p. 433; Vernon F. Snow, *Parliament in Elizabethan England: John Hooker's* Order and Usage (New Haven, Conn.: Yale Univ. Press, 1977), p. 164; and Neale, *Elizabethan House of Commons* (cit. n. 53), p. 364.

[55] *Journal des voyages de Monconys*, p. 26; Sorbière, *Voyage to England*, p. 36; Magalotti, *Travels of Cosmo*, p. 186; Thomas Sprat, *Observations on Mons. de Sorbière's Voyage into England* (1665; London, 1708), pp. 164–165; Sprat, *History of the Royal Society* (cit. n. 48), p. 94. For the Commons mace: Erskine May, *Usage of Parliament*, p. 155; Campion, *Procedure of Commons*, pp. 54, 73; and for the society's mace: Margery Purver, *The Royal Society: Concept and Creation* (London: Routledge & Kegan Paul, 1967), p. 140.

[56] Quoted in Hunter and Wood, "Towards Solomon's House" (cit. n. 41), p. 81. The same author described the council as "the Societys Parliament" (pp. 68, 83).

[57] The directed coordination of the society's labors prompted Robert Hooke to compare it to "a Cortesian army well Disciplined and regulated though their number be but small" (quoted in Hunter and Wood, "Towards Solomon's House," p. 87). On coffeehouses see, e.g., Aytoun Ellis, *The Penny Universities: A History of the Coffee-Houses* (London: Secker & Warburg, 1956), esp. pp. 46–47 on rules of order. The Royal Society "club" of the 1670s held much of its conversation in City coffeehouses like Garraway's and Jonathan's. On occasion, experimental performances were even staged at coffeehouses. For the connections between the late-Interregnum Harringtonian Rota club, meeting at Miles's coffeehouse, and the early Royal Society, see Anna M. Strumia, "Vita istituzionale della Royal Society seicentesca in alcuni studi recenti," *Rivista Storica Italiana*, 1986, 98:500–523, on pp. 520–523. (I owe this reference to Michael Hunter.)

distinguish rhetoric from reality. When, for example, Sprat referred to the Royal Society's experimental public as being made up of "the concurring Testimonies of threescore or an hundred" and pointed to "many sincere witnesses standing by" experimental performances, he was, it seems, referring to an ideal state. The Royal Society was, of course, the most populated experimental space of Restoration England, but its effective attendance at weekly meetings probably averaged no more than two score, and by the 1670s meetings were being canceled for lack of attendance.[58] More intimate groups assembled as "clubs" of the society, centered particularly upon Hooke and usually meeting at coffeehouses near Gresham College.

In the event, historians have rightly questioned whether the rooms of the Royal Society should properly be regarded as a major experimental site.[59] Most actual experimental research was performed elsewhere, most notably in private residences like Boyle's Oxford and Pall Mall laboratories and in Hooke's quarters. Unsurprisingly, evidence about the population in these places is scarce. Boyle frequently named his experimental witnesses, and in no case does that named number exceed three. We do also have commentators' testimony about the throngs of visitors, but these are probably best regarded as genuine spectators rather than witnesses.[60] I shall mention the circumstances of experimental work in Hooke's lodgings later, but his laboratory was certainly more thinly populated than that of his patron. Apart from Hooke himself, the population of Hooke's laboratory seems mainly to have been composed of his various assistants, technicians, and domestics.

I need in this connection to make a distinction between a real and a relevant experimental public, between the population actually present at experimental scenes and those whose attendance was deemed by authors to be germane to the making of knowledge. We have, for example, conclusive evidence of the presence in Boyle's laboratories of technicians and assistants of various sorts. As we might say, their role was vital, since Boyle himself had little if anything to do with the physical manipulation of experimental apparatus, and since at least several of these technicians were far more than mere laborers.[61] Yet their presence was scarcely acknowledged in the scenes over which Boyle presided. Two of them, Hooke and Denis Papin, were named and responsible elements in those scenes, although even here Boyle's account probably understates their contribution. Toward the end of his career, Boyle acknowledged Papin's responsibility for the writing of experimental narratives as well as for the physical conduct of air-pump trials. "I had," he wrote, "cause enough to trust his skill and dili-

---

[58] Sprat, *History of the Royal Society*, pp. 73, 100; and, on attendance, Hunter, *The Royal Society and Its Fellows* (cit. n. 47), pp. 16–19; and J. L. Heilbron, *Physics at the Royal Society during Newton's Presidency* (Los Angeles: William Andrews Clark Memorial Library, 1983), p. 4.

[59] Hunter, *Science and Society in Restoration England* (cit. n. 15), p. 46.

[60] The only record of a crowd scene in one of Boyle's laboratories is an account of the visit in 1677 of the German chemist Johann Daniel Kraft, when the display of phosphorus attracted "the confused curiosity of many spectators in a narrow compass." However, "no strangers were present" when the secret of the phosphorus was later revealed: Robert Hooke, *Lectures and Collections made by Robert Hooke* . . . (London, 1678), pp. 273–282; see also J. V. Golinski, "A Noble Spectacle: Research on Phosphorus and the Public Culture of Science in the Early Royal Society," *Isis*, 1989.

[61] I have prepared an extended study of the role and identity of technicians in seventeenth-century England; see also R. E. W. Maddison, "Studies in the Life of Robert Boyle, F.R.S., Part V: Boyle's Operator: Ambrose Godfrey Hanckwitz, F.R.S.," *Notes Rec. Roy. Soc.*, 1955, *11*:159–188 (see p. 159 for a partial list of Boyle's technicians).

gence." But Boyle still insisted on his own ultimate responsibility for the knowledge produced, and the manner in which he did so is instructive: Boyle asked Papin to "set down in writing all the experiments and the phaenomena arising therefrom, as if they had been made and observed by his own skill. . . . But I, myself, was always present at the making of the chief experiments, and also at some of those of an inferior sort, to observe whether all things were done according to my mind." Certain interpretations of experiments were indeed left to Papin: "Some few of these inferences owe themselves more to my assistant than to me."[62] Still, Boyle, not Papin, was the author of this text.

For the most part, however, Boyle's host of "laborants," "operators," "assistants," and "chemical servants" were invisible actors. They were not a part of the relevant experimental public. They made the machines work, but they could not make knowledge. Indeed, their greatest visibility (albeit still anonymous) derived from the capacity of their *lack* of skill to sabotage experimental operations. Time after time in Boyle's texts, technicians appear as sources of trouble. They are the unnamed ones responsible for pumps exploding, materials being impure, glasses not being ground correctly, machines lacking the required integrity.[63]

Technicians had skill but lacked the qualifications to make knowledge. This is why they were rarely part of the relevant experimental public, and when they were part of that public, it was because they were only ambiguously functioning in the role of technician. Ultimately, their absence from the relevant experimental public derived from their formal position in scenes presided over by others. Boyle's technicians, including those of mixed status like Hooke and Papin, were paid by him to do jobs of experimental work, just as both were paid to do similar tasks by the gentlemen of the Royal Society. As Boyle noted in connection with his disinclination to become a cleric, those that were paid to do something were open to the charge that this was *why* they did it.[64] A gentleman's word might be relied upon partly because what he said was without consideration of remuneration. Free verbal action, such as giving testimony, was credible by virtue of its freedom. Technicians, as such, lacked that circumstance of credibility. Thus, so far as their capacity to give authentic experimental testimony was concerned, they were truly not present in experimental scenes. Technicians were not *there* in roughly the same way, and for roughly the same reasons, that allowed Victorian families to speak in front of the servants. It did not matter that the servants might hear: if they told what they heard to other servants, it did not signify; and if they told it to gentlemen, it would not be credited.

THE CONDITION OF GENTLEMEN

The early Royal Society set itself the task of putting together, justifying, and maintaining a relevant public for experiment. Its publicist Thomas Sprat reflected

---

[62] Robert Boyle, "A Continuation of New Experiments Physico-Mechanical, touching the Spring and Weight of the Air . . . The Second Part" (1680), in Boyle, *Works,* Vol. IV, pp. 505–593, on pp. 506–507.

[63] Among very many examples, see esp. a note on the ineptitude of John Mayow (identified only by his initials): Robert Boyle, "A Continuation of New Experiments Physico-Mechanical touching the Spring and Weight of the Air" (1669), in Boyle, *Works,* Vol. III, pp. 175–276, on p. 187.

[64] Burnet, *Select Sermons* (cit. n. 20), p. 200; Birch, "Life of Boyle" (cit. n. 19), p. lx: "The irreligious fortified themselves against all that was said by the clergy, with this, that *it was their trade,* and that *they were paid for it.*"

at length on the social composition of this public and its bearing on the integrity of knowledge-making practices. Historians are now thoroughly familiar with the Royal Society's early insistence that its company was made up of "many eminent men of all Qualities," that it celebrated its social diversity, and that it pointed to the necessary participation in the experimental program of "vulgar hands." Nevertheless, this same society deemed it essential that "the farr greater Number are Gentlemen, free, and unconfin'd." In the view of Sprat and his associates the condition of gentlemen was the condition for the reliability and objectivity of experimental knowledge.[65]

There were two major reasons for this. First, an undue proportion of merchants in the society might translate into a search for present profit at the expense of luciferous experimentation and even into an insistence upon trade secrecy, both of which would distort the search for knowledge. This is what Glanvill meant in praising the society for its freedom from "sordid Interests."[66] More important, the form of the social relations of an assembly composed of unfree men, or, worse, a society divided between free and unfree, would corrupt the processes by which experimental knowledge ought to be made and evaluated, and by which that knowledge might be advertised as reliable. Unfree men were those who lacked discretionary control of their own actions. Technicians, for example, belonged to this class—the class of servants—because their scientific labor was paid for. Merchants might be regarded as compromised in that their actions were geared to achieving the end of present profit. One could not be sure that their word corresponded to their state of belief. Put merchants and servants in an assembly with gentlemen and you would achieve certain definite advantages. But there was also a risk in the shape of the knowledge-making social relations that might be released. Inequalities of rank could, in Sprat's view, corrode the basis of free collective judgment on which the experimental program relied.[67]

As Sprat said, the trouble with existing intellectual communities was the master-servant relationship upon which their knowledge-constituting practices were founded, the scheme by which "*Philosophers* have bin always *Masters, & Scholars;* some imposing, & all the other submitting; and not as equal observers without dependence." He judged that "very mischievous . . . consequences" had resulted because "the Seats of Knowledg, have been for the most part hereto-

---

[65] Sprat, *History of the Royal Society* (cit. n. 48), pp. 63–67, 76, 407, 427, 431, 435; also Robert Hooke, *Micrographia* (London, 1665), "Preface," sig. g1v. For visitors' accounts of social diversity in the society, see, e.g., Magalotti, *Travels of Cosmo* (cit. n. 32), pp. 186–188; Sorbière, *Voyage to England* (cit. n. 48), p. 37.

[66] Joseph Glanvill, *Scepsis scientifica* (London, 1665), "Preface," sig. c1r; also sig. b4v for "a Society *of persons that can command both* Wit *and* Fortune." Peter Dear rightly notes evidence that the testimony of "lowly folk" might be credible because their accounts were less likely than those of the educated to be colored by theoretical commitments. This view was, however, rarely expressed and, as Dear says, was more than counterbalanced by the consideration that "gentlemen were trustworthy just because they *were* gentlemen": Dear, *"Totius in verba"* (cit. n. 7), pp. 156–157, emphasis in original.

[67] Sprat, *History of the Royal Society* (cit. n. 48), pp. 65–67. On seventeenth-century English thought about the master-servant relationship and the political significance of servitude, see C. B. Macpherson, *The Political Theory of Possessive Individualism: Hobbes to Locke* (Oxford: Oxford Univ. Press, 1970), esp. Ch. 3; Christopher Hill, "Pottage for Freeborn Englishmen: Attitudes to Wage-Labour," in Hill, *Change and Continuity in Seventeenth-Century England* (Cambridge, Mass.: Harvard Univ. Press, 1975), Ch. 10.

fore, not *Laboratories,* as they ought to be; but onely *Scholes,* where some have *taught,* and all the rest subscrib'd." Thus the schoolroom was a useful resource in modeling a proper experimental space, precisely because it exemplified those conventional social relations deemed grossly inappropriate to the new practice: "The very inequality of the Titles of *Teachers,* and *Scholars,* does very much suppress, and tame mens Spirits; which though it should be proper for Discipline and Education; yet it is by no means consistent with a free Philosophical Consultation. It is undoubtedly true; that scarce any man's mind, is so capable of *thinking strongly,* in the presence of one, whom he *fears* and *reverences;* as he is, when that restraint is taken off."[68]

The solution to the practical problem thus resolved into the description and construction of a social space that was both free and disciplined. Sprat said that the "cure" for the disease afflicting current systems of knowledge "must be no other, than to form an *Assembly* at one time, whose privileges shall be the same; whose gain shall be in common; whose *Members* were not brought up at the feet of each other." Such disinterested free men, freely mobilizing themselves around experimental phenomena and creating the witnessed matter of fact, could form an intellectual polity "upon whose labours, mankind might . . . freely rely."[69] The social space that Sprat was attempting to describe was a composite of a number of existing and past spaces, real and ideal. Still, one model for such a space was, perhaps, more pertinent than any other, precisely because, as I have shown, it corresponded to the type of space within which experimental discourses typically occurred. This was the gentleman's private residence and, within it, its public rooms. The conventions regulating discourse in the drawing room were readily available for the construction of the new space and for making morally visible the social relations appropriate to it. It was the acknowledged freedom of the gentleman's action, the honor accorded to his word, the moral discipline he imposed upon himself, and the presumed moral equality of the company of gentlemen that guaranteed the reliability of experimental knowledge. In other words, gentlemen in, genuine knowledge out.

Gentlemen were bound to credit the word of their fellows or, at least, to refrain from publicly discrediting it.[70] These expectations and obligations were grounded in the face-to-face relations obtaining in concrete spaces. The obligation to tell the truth, like the consequences of questioning that one was being told the truth, were intensified when one looked the other "in the face," and particularly when it was done in the public rooms of the other's house. The disastrous effects of violating this code were visible to the Royal Society in the quarrel between Gilles

[68] Sprat, *History of the Royal Society,* pp. 67–69. Cf. John Webster, *Academiarum examen* (London, 1654), p. 106, where it is recommended that youth be educated "so they may not be sayers, but doers, not idle spectators, but painful operators; . . . which can never come to pass, unless they have Laboratories as well as Libraries, and work in the fire, better than build Castles in the air."

[69] Sprat, *History of the Royal Society,* p. 70. On fellows' freedom of judgment: Lotte Mulligan and Glenn Mulligan, "Reconstructing Restoration Science: Styles of Leadership and Social Composition of the Early Royal Society," *Social Studies of Science,* 1981, *11*:327–364, on p. 330 (quoting William Croone); on the moral economy of the experimental community generally: Shapin and Schaffer, *Leviathan and the Air-Pump* (cit. n. 4), pp. 310–319, 332–344; and for Hobbes's suggestion that the Royal Society did indeed have "masters": *ibid.,* pp. 112–115. On the presumed equality of all gentlemen: J. C. D. Clark, *English Society 1688–1832: Ideology, Social Structure and Political Practice during the Ancien Regime* (Cambridge: Cambridge Univ. Press, 1985), p. 103.

[70] See, e.g., Peacham, *The Complete Gentleman* (cit. n. 22), p. 24; and Brathwait, *The English Gentleman* (cit. n. 37), pp. 83–84.

Roberval and Henri-Louis Habert de Montmor in the latter's Parisian town house. As Ismael Boulliau told the story to Christiaan Huygens, Roberval

> has done a very stupid thing in the house of M. de Montmor who is as you know a man of honor and position; he was so uncivil as to say to him in his own house . . . , that he had more wit than he, and that he was less only in worldly goods. . . . Monsieur de Montmor, who is very circumspect, said to him that he could and should behave more civilly than to quarrel with him and treat him with contempt in his own house.

Roberval never returned to the Montmor Academy, and the group never recovered. The Parisians tried to learn a lesson: as this dispute was over doctrine, they resolved to move "towards the study of nature and inventions," in which civility could be more easily maintained since the price of dissenting publicly from a gentleman's testimony on matters of fact would dissuade others from the contest.[71]

The code relating to face-to-face interactions in the house could be, and was, extended to the social relations of experimental knowledge production generally. It was rare indeed for any gentleman's testimony on a matter of experimental fact to be gainsaid. In the early 1670s Henry More disputed Boyle's report of a hydrostatical matter of fact. The manner of Boyle's response is telling: "Though [More] was too civil to give me, *in terminus,* the lye; yet he did indeed deny the matter of fact to be true. Which I cannot easily think, the experiment having been tried both before our whole society, and very critically, by its royal founder, his majesty himself."[72] Boyle appealed to the honor of a *company* of gentlemen, and, ultimately, to the greatest gentleman of all. In 1667 Oldenburg specifically cautioned fellows not to deny within the society's rooms experimental testimony deriving from foreign philosophers. Oldenburg took an offending fellow aside afterward and asked him "how he would resent it, if he should communicate upon his own knowledge an unusual experiment to [those foreign experimenters], and they brand it in public with the mark of falsehood: that such expressions in so public a place, and in so mixed an assembly, would certainly prove very destructive to all philosophical commerce."[73]

The same relationship of trust that was enjoined to govern experimental discourse in the drawing room was constitutive of transactions between public and private rooms of the experimental house. I noted at the beginning the central problem posed in empiricist practice by the indispensable role of testimony and trust. The Royal Society was evidently quite aware that the population of direct witnesses to experimental trials in the laboratory was limited by practical considerations if by nothing else. Nevertheless, the trajectory of a successful candidate for the status of matter of fact necessarily transited the public spaces in which it

---

[71] Ismael Boulliau to Christiaan Huygens, 6 Dec. 1658, in Harcourt Brown, *Scientific Organizations in Seventeenth Century France (1620–1680)* (Baltimore: Williams & Wilkins, 1934), pp. 87–89; see also pp. 108, 119, 126–127 (on civility), and p. 96 (for Oldenburg's familiarity with proceedings at Montmor's house).

[72] Boyle, "Hydrostatical Discourse" (cit. n. 8), p. 615. For this episode, and for Boyle-More relations generally, see Shapin and Schaffer, *Leviathan and the Air-Pump,* pp. 207–224.

[73] The episode concerned reports by physicians in Danzig regarding the transfusion of animal blood into humans: see Oldenburg to Boyle, 10 Dec. 1667, in Boyle, *Works,* Vol. VI, pp. 254–255; and Oldenburg, *Correspondence* (cit. n. 9), Vol. IV, pp. 26–28.

was validated. The practical solution offered by the society was the acceptance of a division of experimental labor and the protection of a relationship of trust between those within and without the laboratory threshold. Sprat said that there was a natural division of labor among the fellowship: "Those that have the best faculty of *Experimenting,* are commonly most averse from reading Books; and so it is fit, that this *Defect* should be supply'd by others pains." Those that actually performed experimental trials, and those that accompanied them as direct witnesses, were necessarily few in number, but they acted as representatives of all the rest. One could, and ought to, trust them in the way one could trust the evidence of one's own senses: "Those, to whom the conduct of the *Experiment* is committed . . . do (as it were) carry the eyes, and the imaginations of the whole company into the *Laboratory* with them." Their testimony of what had been done and found out in the laboratory, undoubted because of their condition and quality, formed the basis of the assembly's discursive work, "which is to *judg, and resolve* upon the matter of Fact," sometimes accompanied by a showing of the experiment tried in the laboratory, sometimes on the basis of narrative alone. Only when there was clear agreement ("the concurring Testimonies") was a matter of fact established. Such procedures were advertised as morally infallible. Glanvill reckoned that "the relations of your Tryals may be received as undoubted Records of certain events, and as securely be depended on, as the Propositions of Euclide." The very transition from private to public space that marked the passage from opinion to knowledge was a remedy for endemic tendencies to "over-hasty" causal conjecturing, to "finishing the *roof,* before the *foundation* has been well laid." Sprat assured his readers that "though the *Experiment* was but the private task of one or two, or some such small number; yet the *conjecturing,* and *debating* on its *consequences,* was still the employment of their full, and solemn Assemblies."[74] An item of experimental knowledge was not finished until it had, literally, come out into society.

TRYING IT AT HOME

A house contains many types of functionally differentiated rooms, each with its conditions of access and conventions of appropriate conduct within. Social life within the house involves a circulation from one sort of room to another. The career of experimental knowledge is predicated upon the same sort of circulation. Thus far I have spoken of the making of experimental knowledge in a loose way, scarcely differentiating between its production and its evaluation. I must now deal more systematically with the stages of knowledge making and relate these to the physical and social spaces in which they take place.

In mid to late seventeenth-century England there was a linguistic distinction the force and sense of which seem to have escaped most historians of science. This was the discrimination between "trying" an experiment, "showing" it, and "discoursing" upon it. In the common usage of the main experimental actors of this setting, the distinction between these terms was both routine and rigorous. The trying of an experiment corresponds to research proper, getting the thing to work, possibly attended with uncertainty about what constitutes a working

[74] Sprat, *History of the Royal Society* (cit. n. 48), pp. 97–102; Glanvill, *Scepsis scientifica* (cit. n. 66), "Preface," sig. c1r.

experiment. Showing is the display to others of a working experiment, what is commonly called demonstration.[75] And experimental discourses are the range of expatiatory and interpretative verbal behaviors that either accompany experimental shows or refer to shows or trials done at some other time or place. I want to say that trying was an activity that in practice occurred within relatively private spaces, whereas showing and discoursing were events in relatively public space. The career of experimental knowledge is the circulation between private and public spaces.[76]

We can get a purchase upon this notion by considering a day in the experimental life of Robert Hooke. I have noted that Hooke lived where he worked, in rooms at Gresham College with an adjacent laboratory, rooms that were little visited by fellow experimentalists, English or foreign. He rose and then dined early, usually at home and frequently with his technicians, some of whom lodged with him. Before issuing forth, Hooke worked at home, trying experiments, as his diary records: "tryd experiment of fire," or "tryd experiment of gunpowder." Some of these, Hooke noted, were preparations for displays at the Royal Society, either next door or, during the Arundel House period, a mile and a half away. It was in the assembly rooms of the society that these experiments were to be shown and discoursed on: "tryd expt of penetration of Liquors . . . shewd it at Arundell house." Experimental discourses could also take place elsewhere. When Hooke left his rooms, he would invariably resort to the local coffeehouses or taverns, where he would expect to meet a small number of serious and competent philosophers for experimental discussion. In the evenings he was a fixture at the tables of distinguished fellows of the society, notably at Boyle's, Christopher Wren's, and Lord William Brouncker's houses, where further experimental discourse occurred.[77]

Kuhn has written about what he sees as a crucial difference between the role of experiment in mid seventeenth-century England and preceding practices. In the experimental program of Boyle, Hooke, and their associates, Kuhn says, experiments were seldom performed "to demonstrate what was already known. . . . Rather they wished to see how nature would behave under previously unobserved, often previously nonexistent, circumstances."[78] Broadly speaking, the point is a legitimate one. However, it applies only to one stage of

---

[75] There are judicial resonances here, but they should not be overemphasized. What was most often being "tried" in experiment was some hypothesis or other explanatory item. In law the trial is of matter of fact, and the jury's judgment is of what counts as fact. However, the best parallel is between the experimental show and the "show trial," where the matter of fact is known (or decided upon) in advance. Both bear the same relation to their respective genuine trials. For the judicial process of "bringing matters to a trial," see Martin, " 'Knowledge Is Power' " (cit. n. 7), Ch. 3.

[76] In this connection see David Gooding's excellent work on Faraday at the Royal Institution. Gooding studies the passage from the basement laboratory to the ground floor lecture theater as movement in the epistemological status of experimental phenomena: Gooding, " 'In Nature's School': Faraday as an Experimentalist," in Gooding and Frank A. L. James, eds., *Faraday Rediscovered: Essays on the Life and Work of Michael Faraday, 1791–1867* (Basingstoke: Macmillan, 1985), pp. 106–135. See also H. M. Collins, "Public Experiments and Displays of Virtuosity: The Core-Set Revisited," *Soc. Stud. Sci.,* 1988, *18* (in press).

[77] 'Espinasse, *Robert Hooke* (cit. n. 24), pp. 106–147; and Shapin, "Who Was Robert Hooke?" (cit. n. 25). The quotations, instances of which could be multiplied indefinitely, are from Hooke's 1672–1680 *Diary* (cit. n. 25), pp. 15, 37.

[78] Thomas S. Kuhn, "Mathematical versus Experimental Traditions in the Development of Physical Science," in Kuhn, *The Essential Tension: Selected Studies in Scientific Tradition and Change* (Chicago: Univ. Chicago Press, 1977), pp. 31–65, on p. 43.

experimentation and to one site at which experimental activity occurs. Hooke and Boyle might, indeed, undertake experimental trials without substantial foreknowledge of their outcome, although they could scarcely have done so without *any* foreknowledge, since they would then have been unable to distinguish between experimental success and failure. An experimental trial could fail; indeed, trials usually did fail, in the sense that an outcome was achieved out of which the desired sense could not be made. Thus, Hooke's diary records, among many other instances: "Made tryall of Speculum. not good"; "Made tryall upon Speculum it succeeded not"; "at home all day trying the fire expt but could not make it succeed." So far as trials are concerned, a failure might legitimately be attributed to one or more of a number of causes: the experimenter was inept or blundered in some way; the equipment was defective or the materials impure; relevant background circumstances, not specifiable or controllable at the time of trial, were unpropitious, and so on.[79] However, a further possibility was open and, indeed, sometimes considered by experimenters, namely that the theory, hypothesis, or perspective that informed one's sense of what counted as a successful outcome was itself incorrect. In a trial it was therefore always possible that an outcome deemed unsuccessful might come to be regarded as the successful realization of another theory of nature. In this way, the definition of what counted as a well-working experimental trial was, in principle, open-ended. In the views of the relevant actors, nature might perhaps speak unexpected words, and the experimenter would be obliged to listen.

The notion of the experimental trial therefore carried with it a sense of indiscipline: the experimenter might not be fully in control of the scene. The thing might fail. It might fail for lack of technical competence on the part of the experimenter, or it might fail for want of theoretical resources required to display the phenomena as docile.[80] Trials were undisciplined experiments, and these, like undisciplined animals, children, and strangers, might be deemed unfit to be displayed in public. This is why experimental trials were, in fact, almost invariably performed in relatively private spaces (such as Hooke's rooms and Boyle's laboratory) rather than in the public rooms of the Royal Society.

The weekly meetings of the Royal Society required not trials but shows and discourses.[81] It was Hooke's job as curator of experiments to prepare these performances for the society's deliberation, instruction, and entertainment. His notes entitled "Dr. Hook's Method of Making Experiments" stipulate that the curator was to make the trial "with Care and Exactness," then to be "diligent, accurate, and curious" in "shewing to the Assembly of Spectators, such Circumstances and Effects . . . as are material." Even a visitor like Magalotti observed that he who was in charge of the society's experiments "does not come to make

---

[79] Quotations are from Hooke's 1672–1680 *Diary*, pp. 27–29, 33. For Boyle's views on what counted as an experimental failure, see Shapin and Schaffer, *Leviathan and the Air-Pump*, pp. 185–201.

[80] For uses of Foucauldian notions of "discipline" and "docile bodies" in the sociology of scientific knowledge: Michael Lynch, "Discipline and the Material Form of Images: An Analysis of Scientific Visibility," *Soc. Stud. Sci.*, 1985, *15*:37–66; Bruno Latour, *Science in Action: How to Follow Scientists and Engineers through Society* (Milton Keynes: Open Univ. Press, 1987), Ch. 3.

[81] Historians concerned with other issues have known this for some time, e.g., Hunter, *Science and Society in Restoration England* (cit. n. 15), p. 46; Hunter and Wood, "Towards Solomon's House" (cit. n. 41), p. 76; and Penelope M. Gouk, "Acoustics in the Early Royal Society 1660–1680," *Notes Rec. Roy. Soc.*, 1981–1982, *36*:155–175, on p. 170.

them in public before having made them at home."[82] Hooke had specific directions to this effect. For instance, in connection with a set of magnetic experiments, "It was ordered, that Mr. Hooke . . . try by himself a good number of experiments . . . and draw up an account of their success, and to communicate it to the Society, so that they might call for such of them as they should think good to be shewn before them." And in the case of a transfusion trial, Hooke and others were "appointed to be curators of this experiment, first in private by themselves, and then, in case of success, in public before the society."[83] Hooke did labor assiduously "at home," disciplining the trials and, when they had been made docile, bringing them to be shown.

He was a success at his job. His first biographer said that his experiments for the Royal Society were "performed with the least Embarrassment, clearly, and evidently."[84] There was always the risk of "embarrassment" precisely because these were to be not trials but shows, performed not in private but in public. "Embarrassment" was avoided, and the society had a successful meeting, when "the experiments succeeded," that is, when they met the shared expectations attending their outcome (and, presumably, when they offered a certain amount of amusement and entertainment).

But even Hooke did not always succeed. When an experimental show failed, the reasons were more circumscribed than in the case of a trial. With any event labeled as, and intended to be, a show, failure could mean only that the experimenter or the materials under his direction were in some way wanting. Accordingly, the Royal Society was not tolerant of failed shows. Hooke's wrist was smartly slapped when he produced in public the undisciplined phenomena that abounded in private settings: "The operator was ordered to make his compressing engine very staunch; and for that end to try it often by himself, that it might be in good order against the next meeting"; "Mr. Hooke was ordered to try this by himself at home"; "He made an experiment of the force of falling bodies to raise a weight; but was ordered to try it by himself, and then to shew it again in public."[85]

The relations between trials and shows, between activities proper to private and to public spaces, were, however, inherently problematic. The status of what had been produced or witnessed was a matter for judgment. A clear example of this is the case of the so-called anomalous suspension of water. In the early 1660s there was serious dispute in the Royal Society over the factual existence and correct interpretation of this phenomenon. (Water that is well purged of air bubbles will not descend from its initial standing in the Torricellian apparatus when it is placed in an evacuated air pump. Boyle had pointed to descent as crucial

---

[82] *Philosophical Experiments of Hooke*, ed. Derham (cit. n. 46), pp. 26–28; and Magalotti to Cardinal Leopold, 20 Feb. 1668, in Maddison, "Some Italian Visitors to the Royal Society" (cit. n. 32), p. 161.

[83] Birch, *History of the Royal Society* (cit. n. 23), Vol. III, p. 124 (entry for 12 Feb. 1674), and Vol. II, p. 115 (entry for 26 Sept. 1666).

[84] Richard Waller, "The Life of Dr. Robert Hooke," in *The Posthumous Works of Robert Hooke . . .* , ed. Waller (London, 1705), pp. i–xxviii, on p. iii (also quoted in Weld, *History of the Royal Society* [cit. n. 21], Vol. I, p. 438). Weld was one of the first historians to note this characteristic of Royal Society experiments: "These experiments were generally repetitions of experiments already made in private and exhibited afterwards for the satisfaction and information of the Society": p. 136n.

[85] Birch, *History of the Royal Society* (cit. n. 23), Vol. I, pp. 177, 194, 260 (entries for 14 Jan., 11 Feb., and 17 June 1663).

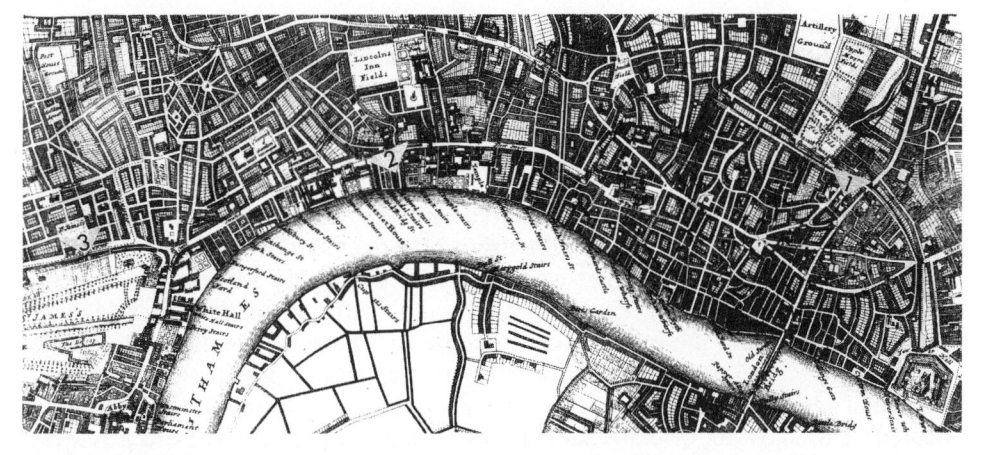

Figure 4. Detail from a map of London in the early eighteenth century, by John Strype, showing the relative locations of Gresham College (1), Arundel House (2), and Boyle's house and laboratory in Pall Mall (3). From John Stow, A Survey of the Cities of London and Westminster (London, 1720); courtesy of Special Collections, University of Edinburgh Library.

confirmation of his hypothesis of the air's spring.) Huygens had produced the alleged phenomenon in Holland, and Boyle disputed its status as an authentic fact of nature by suggesting that nondescent was due to the leakage of external air into Huygens's pump. Hooke was directed to prepare the experiment for the Royal Society. During the early phases of the career of anomalous suspension in England, the experimental leaders of the Royal Society were of the opinion that no such phenomenon legitimately existed. Any experiment that showed it was considered to have been incompetently performed—the apparatus leaked. Since members of the society had considerable experience of Hooke's bringing them experiments in pumps that were not "sufficiently tight," they readily concluded that Hooke's first productions of anomalous suspension were instances of experimental failure.[86] The experimental phenomena had not been made sufficiently docile. Hooke had indeed tried the experiment at home and had deemed it ready to be shown. The leaders of the society concluded otherwise: Hooke had produced only a trial, a failed show. What Hooke claimed to be knowledge, the society rejected as artifact. They disputed his claim by stipulating that the thing was not proper to be shown in a public place.[87]

When the Royal Society was at Arundel House, its curator Robert Hooke was continually ordered to bring the air pump to their meetings from its permanent lodgings in Hooke's rooms a mile and a half away at Gresham (see Figure 4). In the course of being trundled back and forth, the brittle seals that ensured the machine against leakage were liable to crack, so that the curator's experimental

---

[86] *Ibid.*, pp. 139, 212, 218, 220, 238, 248, 254–255, 268, 274–275, 286–287, 295, 299–301, 305, 310, 386.

[87] The story of anomalous suspension is told in Shapin and Schaffer, *Leviathan and the Air-Pump* (cit. n. 4), Ch. 6. The society ultimately came round to the view that anomalous suspension authentically existed, and, therefore, that experiments *not* revealing it were incompetent. This shift crucially involved Boyle's personal experience of the phenomenon and Huygens's visit to London to produce anomalous suspension before witnesses. For similar doubt of Hooke's experimental testimony, see "An Account of the Experiment made by Mr. *Hook,* of Preserving Animals Alive by Blowing through Their Lungs with Bellows," *Phil. Trans.*, 21 Oct. 1667, No. 28, pp. 539–540.

shows sometimes failed. Hooke made a modest proposal. He suggested that, in this one instance and for this circumscribed practical reason, the honorable fellows who wished to satisfy themselves how matters stood should come to him, instead of Hooke and the machine going to them. Hooke "moved that . . . a committee might be appointed to see some experiments made with [the air pump] at his lodgings."[88]

An *ad hoc* committee was constituted and the visit to Hooke's rooms was made. In this instance, the normal pattern of movement in seventeenth-century experimental science was reversed: those who wanted to witness experimental knowledge in the making came to where the instruments permanently lived, rather than obliging the instruments to come to where witnesses lived. This inversion of the usual hierarchical ordering of public and private spaces was exceptional in seventeenth-century practice, and, in the event, it was rarely repeated. The showing of experimental phenomena in public spaces to a relevant public of gentlemen witnesses was an obligatory move in that setting for the construction of reliable knowledge. What underwrote assent to knowledge claims was the word of a gentleman, the conventions regulating access to a gentleman's house, and the social relations within it.

The contrast with more modern patterns is evident. The disjunction between places of residence and places where scientific knowledge is made is now almost absolute. The separation between the laboratory and the house means that a new privacy surrounds the making of knowledge whose status as open and public is often insisted upon. The implications of this disjunction are both obvious and enormously consequential. Public assent to scientific claims is no longer based upon public familiarity with the phenomena or upon public acquaintance with those who make the claims. We now believe scientists not because we know them, and not because of our direct experience of their work. Instead, we believe them because of their visible display of the emblems of recognized expertise and because their claims are vouched for by other experts we do not know. Practices used in the wider society to assess the creditworthiness of individuals are no longer adequate to assess the credibility of scientific claims. We can, it is true, make the occasional trip to places where scientific knowledge is made. However, when we do so, we come as visitors, as guests in a house where nobody lives.

---

[88] Birch, *History of the Royal Society,* Vol. II, p. 189 (entry for 25 July 1667).

Maria Winkelmann at the Berlin Academy

A Turning Point for Women in Science

By Londa Schiebinger

> If one considers the reputations of Madame Kirch [Maria Winkelmann] and Mlle Cunitz, one must admit that there is no branch of science . . . in which women are not capable of achievement, and that in astronomy, in particular, Germany takes the prize above all other states in Europe.
> —ALPHONSE DES VIGNOLES, vice president of the Berlin Academy, 1721

EUROPEAN SCIENCE WAS in many ways a new enterprise in the seventeenth and eighteenth centuries, an enterprise that (at least ideologically) welcomed a broad participation. The regulations of the newly founded Berlin Academy stressed that modern science could flourish only with contributions from men of all social classes, nationalities, and religions.[1] Was this ideological largess to be extended to women as well? After centuries of proscribing women from active participation, were centers of European intellectual life now to open their doors to them?

The major European academies of science were founded in the seventeenth century—the Royal Society of London in 1662, the French Académie Royale des Sciences in 1666, and the Berlin Akademie der Wissenschaften in 1700.[2] Women were not, however, to become regular members of these academies for three

---

Support for this research and for my larger project on women and the origins of modern science was provided by the Deutscher Akademischer Austauschdienst, the Rockefeller Foundation, and the National Endowment for the Humanities. My thanks to those friends and critics who read and commented on this essay: Robert Proctor, Richard Kremer, Margaret Rossiter, Roger Hahn, Merry Wiesner, Robert Westman, and Lyndal Roper.

[1] "General-Instructions für die Societät der Wissenschaften vom 11. Juli 1700," in Adolf von Harnack, *Geschichte der Königlich Preussischen Akademie der Wissenschaften zu Berlin,* 3 vols. (1900; Hildesheim: Georg Olms, 1970), Vol. II, p. 106. Thomas Sprat made a similar point in *The History of the Royal Society of London* (London, 1667), pp. 62–63, 72. Harnack, who was the father of German feminist Agnes von Zahn-Harnack, gave the highlights of Winkelmann's story in his three-volume work. Such a sensitivity to the plight of women is rarely found in official (and essentially laudatory) histories of national institutions.

[2] The Berlin Academy first bore a Latin name, the Societas Regia Scientiarum. From its founding, it was also commonly known as the Brandenburgische or Berlin Societät der Wissenschaften. In the 1740s it took a French name, Académie Royale des Sciences et Belles-Lettres. In the 1780s it became the Königlich Preussische Akademie der Wissenschaften, which it remained until its reorganization after World War II, when it took its present name, the Akademie der Wissenschaften der Deutschen Demokratischen Republik. For simplicity I refer to the Societät der Wissenschaften as the Berlin Academy or the Academy of Sciences.

centuries. At the Royal Society in London, the first women members—Marjory Stephenson and Kathleen Lonsdale—were not elected until 1945. The prestigious Académie des Sciences in Paris did not admit a woman as a full member until 1979 (Yvonne Choquet-Bruhat). Even Marie Curie, the first person ever to win two Nobel prizes, was denied membership. No woman scientist was awarded membership at Berlin until 1949, when Lise Meitner was made a corresponding member.[3]

Why were women denied membership in the scientific academies of Europe? It would be a mistake to think there were no qualified women scientists when the academies first opened their doors. There were, in fact, a significant number of women trained in the sciences. The exclusion of women was not a foregone conclusion but resulted from a process of extended negotiation between these women and the academy officials.

The case of Maria Winkelmann at the Academy of Sciences in Berlin was a decisive and pivotal one.[4] In 1712 she lost her year-long battle to become Academy astronomer. Already a seasoned astronomer when her husband, the Academy astronomer Gottfried Kirch, died in 1710, Winkelmann asked to be appointed in her husband's stead. But although Leibniz was among her backers, her request was denied, setting an important precedent for women's participation in the scientific work of the Academy.

The story of Maria Winkelmann's rejection by the Academy is a compelling one. But more important, it illustrates patterns in women's participation in early modern science. When Winkelmann petitioned the Academy in the early years of the eighteenth century, she was caught in conflicting social trends. Craft traditions, on the one hand, fostered women's participation in science. Through apprenticeships, women gained access to the secrets and tools of a trade, whether illustrating manuscripts or using telescopes. As we will see, Winkelmann's petition to become Academy astronomer drew legitimacy from guild traditions that recognized the right of a widow to carry on the family business. Craft traditions, however, were counterbalanced by other trends, both old and new. For centuries, women had been excluded from universities, as they were to be excluded from the new scientific academies. In many ways, the new trend of professionalization served to reaffirm the traditional exclusion of women from intellectual culture. Indeed, there were those at the Berlin Academy who judged it improper

---

[3] See *Notes and Records of the Royal Society of London*, 1946, *4*:39–40; and Lise Meitner "The Status of Women in the Professions," *Physics Today*, 1960, *13*(8):16–21. See also Kathleen Lonsdale's thoughtful "Women in Science: Reminiscences and Reflections," *Impact of Science on Society*, 1970, *20*(1):45–59. The Berlin Academy, unlike its fraternal counterparts in London and Paris, awarded honorary membership to a few women of high social standing—Catherine the Great (elected 1767), Duchess Juliane Giovane (1794), and Maria Wentzel (1900) (see Section VIII below).

[4] When writing the history of women, we immediately face problems even in such small matters as what name to use for our main character. To employ the married name reveals a nineteenth- and twentieth-century bias. I have elected to use Maria Winkelmann's maiden name throughout since this is the name she used for her publications. (Her name as it appears here has been modernized from the original Winckelmannin.) The use of the maiden name is also consistent with the practice of the day. Astronomer Maria Cunitz, for example, published under her maiden name; midwives listed in eighteenth-century Prussian address lexicons also used their maiden names (their married names were given in parentheses). The use of Winkelmann's maiden name also makes it easier to distinguish her from her husband without falling into the somewhat degrading habit of using her first name. Winkelmann did also refer to herself as "Kirchin" (the feminine form of Kirch, her husband's name) in correspondence with Leibniz and Academy officials, playing (I assume) on her husband's name in her quest for employment. Academy officials referred to her as "Kirchin" or the "widow Kirch."

for a woman to practice the art of astronomy and suggested that Winkelmann return to her "distaff" and "spindle."[5]

I. WOMEN AND CRAFT TRADITIONS IN EARLY MODERN ASTRONOMY

Edgar Zilsel was among the first to point to the importance of craft skills for the development of modern science in the West. Zilsel located the origin of modern science in the fusion of three traditions: the tradition of letters exemplified by the literary humanists; the tradition of logic and mathematics exemplified by the Aristotelian scholastics; and the tradition of practical experiment and application exemplified by the empirical artist-engineers. Astronomy drew from each of these traditions. It was, however, the craft aspects of astronomy that were especially important in the sixteenth and seventeenth centuries. The astronomer was both theoretician and technician; he or she was versed not just in Copernican theory and mathematics, but also in the arts of glass grinding, copperplate engraving, and instrument making. These were skills of the artisan, not of the scholar. Thus the astronomer of the seventeenth century bore a close resemblance to the guild master or apprentice.[6]

Astronomers were never, of course, officially organized into guilds. Yet craft traditions that molded all aspects of working life in early modern Europe were very much alive in astronomical practices. In many ways the legal and political structures of the guild merely codified these wider practices and traditions. This was especially true in Germany, where stirrings of industrialization came late. Whereas in England and Holland guilds declined from the mid-seventeenth century, in Germany they remained an important economic and cultural force well into the nineteenth century.[7]

The new value attached to the traditional skills of the artisan allowed for broader participation in the sciences. Of the various institutional homes of astronomy, only the artisanal workshop welcomed women. Women were not newcomers to the workshop: it was in craft traditions that the fifteenth century writer

---

[5] Alphonse des Vignoles, "Eloge de Madame Kirch à l'occasion de laquelle on parle de quelques autres Femmes & d'un Paison Astronomes," *Bibliothèque germanique*, 1721, *3*:115–183, on p. 181.

[6] See Edgar Zilsel, "The Sociological Roots of Modern Science," *American Journal of Sociology*, 1942, *47*:544–562, on pp. 545–546; see also Arthur Clegg, "Craftsmen and the Origin of Science," *Science and Society*, 1979, *43*:186–201; Paolo Rossi, *Philosophy, Technology, and the Arts of the Early Modern Era*, trans. Salvator Attansasio (New York: Harper & Row, 1970); and Rupert Hall, "The Scholar and the Craftsman in the Scientific Revolution," *Critical Problems in the History of Science*, ed. Marshall Clagett (Madison: Univ. Wisconsin Press, 1959), pp. 3–23.

Astronomers also held chairs as mathematicians in universities and served powerful patrons at royal courts. Noble patrons themselves became avid amateurs of astronomy. The astronomer has also been likened to a feudal lord: see Robert Westman, "The Astronomer's Role in the Sixteenth Century: A Preliminary Study," *History of Science*, 1980, *18*:105–147, on pp. 124–125; and Ernst Zinner, *Die Geschichte der Sternkunde von den ersten Anfängen bis zu Gegenwart* (Berlin: Julius Springer, 1931), pp. 587–590.

[7] Jean Quataert has warned against conflating important distinctions between guilds and households; see Quataert, "The Shaping of Women's Work in Manufacturing: Guilds, Households, and the State in Central Europe, 1648–1870," *American Historical Review*, 1985, *90*(5):1122–1148, on p. 1134. For the case of astronomy, however, the larger danger has been to ignore almost entirely both these forms of production. Here I use the term *craft* to refer to household production, and *guild* to refer to regulated crafts. Guilds emerged throughout Europe between the twelfth and fifteenth centuries, when they dominated most urban economies. See Anthony Black, *Guilds and Civil Society in European Political Thought from the Twelfth Century to the Present* (Ithaca: Cornell Univ. Press, 1984), p. 123; and Quataert, "Women's Work in Manufacturing," p. 1125.

Christine de Pizan had located women's greatest innovations in the arts and sciences—the spinning of wool, silk, and linen, and "creating the general means of civilized existence."[8] In the workshop, women's (like men's) contributions depended less on book learning and more on practical innovations in illustrating, calculating, or observing. It was the intersection of these traditions—craft traditions in astronomy and the tradition of women in the crafts—that fostered women's participation in astronomy.

The craft aspects of astronomy, then, are crucial for understanding the prominence of women in the field. It is important to keep in mind that Winkelmann was not the lone woman astronomer. Between 1650 and 1720, women constituted a little over fourteen percent of German astronomers.[9] The strength of the artisan in Germany may also explain the observation by Alphonse des Vignoles, quoted in the epigraph, that for women's accomplishments in astronomy, Germany took the prize. There were more women astronomers in Germany at the turn of the century than in any other European country.

Though guild traditions gave women access to the practice of science, it is important not to see this in romantic terms. Women's position in astronomy was similar to their position in the guilds—important, but subordinate. Only a few women, such as Maria Cunitz or Maria Winkelmann, directed and published their own work. More often a woman served in various support positions, editing her husband's writings or performing astronomical calculations. While guild traditions gave women an initial toehold in the sciences, the limitations built into those traditions allowed women a creative role only in exceptional cases.

II. WINKELMANN'S EDUCATION

The apprentice system provided the key to women's training in astronomy. Maria Margaretha Winkelmann was born in 1670 at Panitzsch (near Leipzig), the daughter of a Lutheran minister. She was educated privately by her father and, after his death, by her uncle. The young Winkelmann made great progress in the arts and letters; from an early age, she took a special interest in astronomy. She received advanced training in astronomy from the self-taught Christoph Arnold, who lived in the neighboring town of Sommerfeld. Like Winkelmann, Arnold benefited from the openness of astronomy. It was characteristic of this period that the farmer and self-taught Arnold was recognized for his contributions.[10] At

---

[8] Christine de Pizan, *The Book of the City of Ladies*, trans. Earl Jeffrey Richards (1405; New York: Persea Books, 1982), pp. 70–80.

[9] In his work on artists in Nuremberg (among whom he included astronomers and the like), Joachim von Sandrart recorded the names of three women among some fifty entries: *Teutsche Academie der Edlen Bau-, Bild-und Mahlerey-Künste* (Frankfurt, 1675). In his section on mathematicians and astronomers in Silesia, Friedrich Luce recorded one woman's name along with the names of about ten men; see Luce, *Fürsten Kron oder eigentliche wahrhaffte Beschreibung ober und nieder Schlesiens* (Frankfurt am Main, 1685). In his *Historia astronomiae* (Wittenberg, 1741), Friedrich Weidler listed 3 women along with 22 men astronomers working in Germany. Weidler's count is based on published work and includes men only marginally involved in astronomy (such as Leibniz and Christian Wolff). It is significant that a number of the lexicons made a point of including women. The titles of both Georg Will's and Christian Jöcher's works announce that the achievements of "both sexes" are to be included; see Will, *Nürnbergisches Gelehrten-Lexicon oder Beschreibung aller Nürnbergischen Gelehrten beyderley Geschlechtes* . . . (Nürnberg, 1755–1758); and Jöcher, *Allgemeines Gelehrten-Lexicon, darinne die Gelehrten aller Stände sowohl männ- als weiblichen Geschlechts* (Leipzig, 1760).

[10] Arnold discovered the comets of 1683 and 1686, and observed the transit of Mercury across the

Arnold's house, Winkelmann served as an unofficial apprentice, learning the art of astronomy through hands-on experience in observation and calculation.

Winkelmann's education followed a pattern commonly found in the trades, and a common one for women. In guild families, household and economic concerns were closely associated, if not identical. The importance of the household as a social and economic unit conferred on daughters and wives important responsibilities and privileges. Women's position in the guilds was stronger than has generally been appreciated. Before 1600 in Nuremberg, for example, women were active in nearly all areas of production. Craftswomen in fifteenth-century Cologne also held strong positions. Of the nearly forty guilds that Margret Wensky described in her study of working women in Cologne, between twenty and twenty-four had women members.[11] Women entered these guilds either as apprentices or through marriage.

There were, however, also differences in male and female apprenticeships. Women trained as apprentices but did not have the journeymen years. Journeymen might travel from master to master. Young women, in contrast, took what training was available in their homes. If a young woman did not have good training available at her doorstep, it is unlikely that she could have traveled to the master of her choice. At least within the sciences, there is no example of a woman apprentice traveling from master to master. In the seventeenth and early eighteenth centuries, the most important factor determining a woman's future in science was her father. Winkelmann trained with Arnold, outside her home; but her case was extraordinary because she was an orphan.

Astronomy in late seventeenth-century Germany was not, however, organized entirely along guild lines, and women's exclusion from university education created additional differences between women's and men's preparation. Had Maria Winkelmann been male, she would probably have continued her studies at the nearby universities of Leipzig or Jena. Leading male astronomers—Johannes Hevelius, Georg Eimmart, Gottfried Kirch—held university degrees, though not degrees in astronomy. Hevelius, for example, was by profession a brewer and was educated in jurisprudence. This was not uncommon; mathematics and astronomy in this period were not autonomous disciplines. Consequently, most astronomers studied law, theology, or medicine.[12]

---

sun. For his efforts he was awarded a pension by the Senate of Leipzig: Vignoles, "Eloge de Madame Kirch" (cit. n. 5), pp. 169–171. Although many scientific papers of the Kirch family have survived, few of their personal papers have. On the whereabouts of the Kirch papers see Diedrich Wattenberg, "Zur Geschichte der Astronomie in Berlin im 16. bis 18. Jahrhundert, II," *Die Sterne*, 1972, 49(2):104–116. Most of what we know of Winkelmann's life comes from the eulogy (cit. n. 5) by Alphonse des Vignoles, vice president of the Berlin Academy and a family friend. Most biographical notes follow Vignoles. Parts of this eulogy were reprinted in the *Neue Zeitungen von gelehrten Sachen*, August 1722, pp. 642–647. See also Gottfried Kirch and Maria Margaretha Winkelmann, *Das älteste Berliner Wetter-Buch 1700–1701*, ed. G. Hellmann (Berlin, 1893); and P. Aufgebauer, "Die Astronomenfamilie Kirch," *Die Sterne*, 1971, 47(6):241–247.

[11] Margret Wensky, *Die Stellung der Frau in der stadtkölnischen Wirtschaft im Spätmittelalter: Quellen und Darstellungen zur hansischen Geschichte* (Cologne: Böhlau, 1981), pp. 318–319. For women's role in guilds in Germany see Ute Gerhard, *Verhältnisse und Verhinderungen: Frauenarbeit, Familie und Rechte der Frauen im 19. Jahrhundert* (Frankfurt am Main: Suhrkamp, 1978); Carl Bücher, *Die Frauenfrage im Mittelalter* (Tübingen, 1882); Peter Ketsch, *Frauen im Mittelalter: Quellen und Materialien*, 2 vols. (Düsseldorf: Schwann, 1983) Vol. I, Ch. 6.1; and Merry E. Wiesner, *Working Women in Renaissance Germany* (New Brunswick, N.J.: Rutgers Univ. Press, 1986), esp. Ch. 5.

[12] See Johann Westphal, *Leben, Studien und Schriften des Astronomen Johann Hevelius* (Königsberg, 1820). Similarly, Copernicus studied law and medicine; see Westman, "Astronomer's Role"

Though women's exclusion from universities set limits to their participation in astronomy, it did not exclude them entirely.[13] Debates over the nature of the universe filled university halls, yet the practice of astronomy—the actual work of observing the heavens—took place largely outside the universities. In the seventeenth century, the art of observation was learned under the watchful eye of a master. Gottfried Kirch, for example, studied at Hevelius's private observatory in Danzig; this was as important for his astronomical career as his study of mathematics with Erhard Weigel at the University of Jena. An astronomical apprenticeship was also important for Kirch's son. When Christfried Kirch applied for the position of astronomer at the Berlin Academy, his training with his father (and mother) was a more important credential than his year at university.[14]

After her apprenticeship, a scientifically minded woman often married a scientist in order to continue practicing her trade. It was at the astronomer Christoph Arnold's house, where Maria Winkelmann served her unofficial apprenticeship, that she met Gottfried Kirch, Germany's leading astronomer. Though Winkelmann's uncle wanted her to marry a young Lutheran minister, he consented to her marriage to Kirch.[15] By marrying Kirch—a man some thirty years her senior—Winkelmann secured her place in astronomy. Knowing she would have no opportunity to practice astronomy as an independent woman, she moved, in typical guild fashion, from being an assistant to Arnold to becoming an assistant to Kirch. Kirch also benefited from this marriage. In Winkelmann, he found a second wife who could care for his domestic affairs, and also a much-needed astronomical assistant who could help with calculations, observations, and the making of calendars.[16]

In 1700 Kirch and Winkelmann took up residence in Berlin, the newly expanding cultural center of Brandenburg. This move represented an advance in social standing for both husband and wife. Yet in the late seventeenth century, the route to Berlin was very different for men and for women. A university education at Jena and apprenticeship to the well-known astronomer Hevelius afforded Kirch the opportunity to move from the household of a tailor in the small town of Guben to the position of astronomer at the Royal Academy of Sciences. Maria Winkelmann's mobility, on the other hand, came not through education, but through marriage. Though coming via different routes, both served at the Berlin Academy: Gottfried as Academy astronomer, Maria as an unofficial but recognized assistant to her husband.[17]

(cit. n. 7), p. 117. See also Diedrich Wattenberg, "Zur Geschichte der Astronomie in Berlin im 16. bis 18. Jahrhundert, I," *Die Sterne*, 1972, *48*(3):161–172, on p. 161.

[13] Zinner has shown that a university education was not absolutely required for the practice of astronomy: *Die Geschichte der Sternkunde* (cit. n. 7), p. 590.

[14] F. Herbert Weiss, "Quellenbeiträge zur Geschichte der Preussischen Akademie der Wissenschaften," *Jahrbuch der Preussischen Akademie der Wissenschaften*, 1939, pp. 214–224, on pp. 221–222.

[15] Vignoles, "Eloge de Madame Kirch" (cit. n. 5), p. 173.

[16] *Ibid.*, p. 172; and "Lebens Umstände und Schicksale des ehemahles berühmten Gottfried Kirchs, "Königl. Preuß. Astronomi der Societät der Wissenschafften zu Berlin," *Dresdenische gelehrte Anzeigen*, 1761, No. 49, pp. 769–777, on p. 775.

[17] See Erik Amburger, *Die Mitglieder der deutschen Akademie der Wissenschaft zu Berlin 1700–1950* (Berlin: Akademie-Verlag, 1950), p. 173: "1700–1710 Kirch, Gottfried, Astronom, unterstützt von seiner Frau, Kirch, Maria Margaretha geb. Winkelmann, geb. Panitzsch bei Leipzig 25.2.1670, gest. Berlin 29.12.1720. 1716–1720 Kirch, Christfried, Astronom, unterstützt von seiner Mutter."

III. COMETS AND CALENDARS: WINKELMANN'S SCIENTIFIC ACHIEVEMENT

In 1710 Winkelmann petitioned the Academy of Sciences for a position as calendar maker. Was she merely a wifely assistant engaged at the periphery in what has been defined as "women's work"? Or was she a qualified astronomer capable of setting and carrying out her own researches?

Though Maria Winkelmann is little known today, she was well regarded in her time.[18] Her scientific accomplishments during her first decade at the Berlin Academy were many and varied. Every evening, as was her habit, she observed the heavens beginning at nine o'clock.[19] During the course of an evening's observations in 1702, she discovered a previously unknown comet—a discovery that should have secured her position in the astronomical community. (Her husband's position at the Academy rested partly on his discovery of the comet of 1680.) There is no question about Winkelmann's priority in the discovery. In the 1930s F. H. Weiss published her original report of the sighting of the comet (see Figure 1).[20] In his notes from that night, Kirch also recorded that his wife found the comet while he slept:

> Early in the morning (about 2:00 A.M.) the sky was clear and starry. Some nights before, I had observed a variable star, and my wife (as I slept) wanted to find and see it for herself. In so doing, she found a comet in the sky. At which time she woke me, and I found that it was indeed a comet. . . . I was surprised that I had not seen it the night before.[21]

---

[18] In early histories of astronomy, Winkelmann received at least a mention. During her lifetime, she was cited in the German editions of Christian Wolff's *Mathematisches Lexicon* (1716, in Wolff, *Gesammelte Werke,* ed. J. E. Hofmann [Hildesheim: Georg Olms, 1965]), Pt. I, Vol. XI, p. 972. Wolff reported that "of special glory to the German nation is Kirch's widow who is well studied in astronomical observation and calculation"; this tribute was dropped, however, in the 1741 Latin edition of the work. Friedrich Weidler picked up the review of her publications from the *Acta eruditorum* in his *Historia astronomiae* (cit. n. 9) of 1741, p. 556, as did Joseph Jérôme Le Français de Lalande in his *Bibliographie astronomique; avec L'histoire de l'astronomie* (Paris, 1803), p. 359. Jérôme de Lalande included Winkelmann in the short history of women astronomers that introduced his popular astronomy textbook for women, *Astronomie des dames* (Paris, 1786), "Préface historique". J. E. Bode, astronomer of the Berlin Academy, gave her an entry in his "Chronologisches Verzeichniss der berühmtesten Astronomen, seit dem dreizehnten Jahrhundert, ihrer Verdienste, Schriften und Endeckungen," *Astronomisches Jahrbuch für das Jahr 1816* (Berlin, 1813), p. 113. In the nineteenth and twentieth centuries, Winkelmann appeared most often in popular histories of astronomy or in articles about women astronomers. In his semipopular *Geschichte der Astronomie,* Rudolf Wolf gave much attention to women's achievements, beginning with those of Hypatia (Munich, 1877, p. 458). See also E. Lagrange, "Les femmes-astronomes," *Ciel et terre,* 1885, *5*:513–527, on pp. 515–516; Alphonse Rebière reprinted Lagrange's account in his *Les femmes dans la science* (2d ed., Paris, 1897), pp. 153–154; H. J. Mozans (pseud. of John Augustine Zahm) also repeated this account in his *Woman in Science,* (1913; rpt. Cambridge, Mass.: MIT Press, 1974), pp. 173–174. For more recent accounts, see R. V. Rizzo, "Early Daughters of Urania," *Sky and Telescope,* 1954, *14*(1):7–9, on p. 8; and Diedrich Wattenberg, "Frauen in der Astronomie," *Vorträge und Schriften,* 1963, *14*:1–8. Lettie S. Multhauf included highlights from Winkelmann's life in an article on the Kirch family in the *Dictionary of Scientific Biography,* ed. Charles Gillispie, 16 vols. (New York: Scribner's, 1973), Vol. VII, pp. 373–374. Winkelmann also received a note in her husband's entry in the *Allgemeine deutsche Biographie* (Berlin, 1882), Vol. XV, p. 788, and in the more recent *Neue deutsche Biographie* (Berlin, 1972), Vol. XI, pp. 634–635. Though we have official Academy portraits of her husband and son, we have no portraits of her or her daughters.

[19] Winkelmann to G. W. Leibniz, Leibniz Archive, Niedersächsische Landesbibliothek, Hannover, Kirch, No. 472, p. 11. I thank Gerda Utermöhlen and staff of the Leibniz Archive for their assistance.

[20] Weiss, "Quellenbeiträge zur Geschichte der Preussischen Akademie" (cit. n. 14), pp. 223–224, from his private collection. A copy of Winkelmann's report can be found in the Kirch papers, Paris Observatory, MS A.B. 3.7, No. 83, 41, B. I thank the staff of the Paris Observatory for their assistance.

[21] "Früh um 2 Uhr war es hell gestirnet. Meine Ehefrau hat (nach dem ich etliche Nächte zuvor

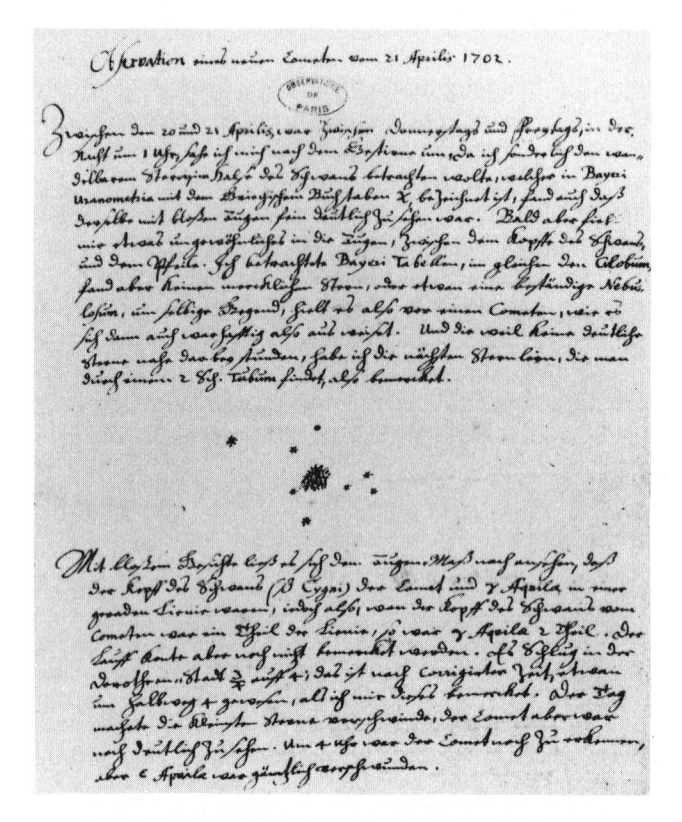

Figure 1. Winkelmann's report of her discovery of the comet of 1702. Reproduced with kind permission of the Observatoire de Paris.

As the first "scientific" achievement of the young Academy, a report of the comet was sent immediately to the king. The report, however, bore Kirch's, not Winkelmann's name.[22] Published accounts of the comet also bore Kirch's name, which unfortunately led many historians to attribute the discovery to him alone.[23]

Why did Winkelmann let this happen? Surely, she knew that recognition for her achievements could be important to her future career. Nor was she hesitant about publishing; she was to publish three tracts under her own name between 1709 and 1711. Her inability to claim recognition for her discovery hinged, in part, on her lack of training in Latin—the shared scientific language in Germany at the time—which made it difficult for her to publish her discovery in the *Acta eruditorum,* then Germany's only scientific journal. Her own publications were all in German.

More important to the problem of credit for the initial sighting of the comet, however, was the fact that Maria and Gottfried worked closely together. The labor of husband and wife did not divide along modern lines: he was not fully

---

dem wandelbaren Stern am Halse das Schwans observiert, und sie ihn, als ich noch schlieff, auch gerne sehen, und selbst finden wolte) einen Comentan am Himmel gefunden. Worauff sie mich auffwachetet, da ich denn fand, dass es warhaftig ein Comet war. . . . Es wondert mich doch, dass ich den Cometan die vorigen Nächte nicht gesehen habe." Kirch papers, Paris Observatory, MS A.B. 3.5, No. 81 B, p. 33.

[22] Adolf von Harnack, "Berichte des Secretars der brandenburgischen Societät der Wissenschaften J. Th. Jablonski an der Präsidenten G. W. Leibniz (1700–1715) nebst einigen Antworten von Leibniz," *Philos.-histor. Abhandlungen der königlichen Akademie der Wissenschaften zu Berlin,* 1897, *3,* Letter No. 22.

[23] See Wattenberg, "Zur Geschichte der Astronomie in Berlin II," (cit. n. 10), p. 107.

professional, working in an observatory outside the home; she was not fully a housewife, confined to hearth and home. Nor were they independent professionals, each holding a chair of astronomy. Instead, they worked very much as a team and on common problems. As Vignoles put it, they took turns observing so that their observations followed night after night without interruption. At other times they observed together, dividing the work (he observing to the north, she to the south) so that they could make observations that a single person could not make accurately.[24] After Winkelmann's sighting of the comet on 21 April, both Kirch and Winkelmann followed its course until 5 May.

Though Gottfried Kirch published the report under his own name and as if he alone had made the discovery, it would be too simple to fault him for "expropriating" his wife's achievement. According to Vignoles, a family friend, Kirch was timid about acknowledging his wife's contributions to their common work and so published the first report of the comet without mentioning her. Later, however, someone (we do not know who) told him "that he could feel free to acknowledge her contributions." Thus when the report of the comet was reprinted eight years later in the first volume of the journal of the Berlin Academy, *Miscellanea Berolinensia,* Kirch mentioned Winkelmann's part in the discovery. This report opened with the words: "My wife . . . beheld an unexpected comet."[25]

In addition to their scientific work, Kirch and Winkelmann took an active interest in the development of astronomical facilities at the Academy. The Academy of Sciences in Berlin was founded primarily to promote astronomy. In 1696 Sophie Charlotte, electress of Brandenburg, later queen of Prussia, had directed her minister Johann Theodor Jablonski to build an observatory, a project that took a decade to complete.[26] The Kirch family struggled long and hard, squeezing money from Academy and royal purses, to create the conditions necessary for good astronomical observations. Winkelmann took an active part in these efforts. On 4 November 1707 she wrote to Leibniz (adviser to Sophie Charlotte and president of the Academy), describing her sighting of the northern lights. In her letter, Winkelmann enticed Leibniz with reports of northern lights "the likes of which my husband has never seen," yet her real motive in writing was to secure housing for the astronomers more convenient to the observatory. She asked for Leibniz's intervention.[27]

During the years of their acquaintance at the Berlin Academy, Leibniz had expressed a high regard for Winkelmann's scientific abilities. Though none of his letters to her have been preserved, her letters to him reveal his interest in her scientific observations.[28] In 1709 Leibniz presented her to the Prussian court,

---

[24] Vignoles, "Eloge de Madame Kirch" (cit. n. 5), p. 174.

[25] See Gottfried Kirch, "Observationes cometae novi," *Acta eruditorum,* 21 Apr., 1702, pp. 256–258; Vignoles, "Eloge de Madame Kirch," pp. 175–176; and Gottfried Kirch, "De cometa anno 1702: Berolini observato," *Miscellanea Berolinensia,* 1710, *1*:213–214.

[26] The observatory could not be used until 1706, and the official opening did not take place until 1711: Harnack, *Geschichte der Akademie zu Berlin* (cit. n. 1), Vol. I, pp. 48–49.

[27] Winkelmann (as Kirchin) to Leibniz, 4 Nov. 1707, Leibniz Archive, Kirch, No. 472, pp. 11–12. The astronomers eventually received quarters in 1708: Harnack, *Geschichte der Akademie zu Berlin* (cit. n. 1), Vol. I, p. 152.

[28] Winkelmann often sent Leibniz special reports of her observations. She knew him well enough to drop by to announce that her book was finished and would arrive from the publisher in a few hours. See Winkelmann to Leibniz, n.d. Leibniz Archive, Kirch, No. 472, p. 10.

where Winkelmann was to explain her sighting of sunspots. In a letter of introduction Leibniz wrote:

> There is [in Berlin] a most learned woman who could pass as a rarity. Her achievement is not in literature or rhetoric but in the most profound doctrines of astronomy. ... I do not believe that this woman easily finds her equal in the science in which she excels. ... She favors the Copernican system (the idea that the sun is at rest) like all the learned astronomers of our time. And it is a pleasure to hear her defend that system through the Holy Scripture in which she is also very learned. She observes with the best observers, she knows how to handle marvelously the quadrant and the telescope (*grandes lunettes d'approche*).

He added that if only she had been sent to the Cape of Good Hope instead of Peter Kolb (Baron von Krosigk's apprentice astronomer), the Academy would have received more reliable observations.[29]

Maria Winkelmann apparently made a good impression at the court of Frederick I. On 17 July she reported in a letter to Leibniz that the ambassador of Denmark had visited the Royal Observatory and had praised her for the aid and assistance she offered her husband in his astronomical work. While at court, Winkelmann also distributed copies of her astrological pamphlet "Vorstellung des Himmels bey der Zusammenkunfft dreyer Grossmächtigsten Könige."[30] Leibniz, commenting on Winkelmann's tract, remarked on "an astrological note that on the second of that month that the sun, Saturn and Venus would be in a straight line. One supposes that there is significance in this."[31]

Maria Winkelmann's three pamphlets published between 1709 and 1711 were all astrological. In his 1721 eulogy, Vignoles tried to explain away her interest in astrology. "Madame Kirch," as he called her, "prepared horoscopes at the request of her friends, but always against her will and in order not to be unkind to her patrons."[32] Perhaps Winkelmann's interest in astrology was purely financial, as Vignoles suggested. Yet her correspondence with Leibniz reveals her belief that nature was something more than matter in motion. In her description of the extraordinary northern lights of 4 November 1707 she wrote to Leibniz, "I am not sure what nature was trying to tell us."[33] Another of Winkelmann's

---

[29] Leibniz to Sophie Charlotte, Jan. 1709, in Gottfried Wilhelm Leibniz, *Die Werke von Leibniz*, ed. Onno Klopp, 11 vols. (Hannover: Klindworths, 1864–1888), Vol. IX, p. 295–296. Leibniz is referring to the attempt to get an exact measurement of the lunar parallax, which failed because Kolb was irresponsible and only occasionally made observations; see Hans Ludendorff, "Zur Frühgeschichte der Astronomie in Berlin," *Vorträge und Schriften der Preussischen Akademie der Wissenschaften*, 1942, 9:3–23, on p. 15. Vignoles also reported that Leibniz often tested Winkelmann's knowledge of certain subjects; she was, he wrote, a zealous partisan of the Copernican system. See Vignoles, "Eloge de Madame Kirch" (cit. n. 5), p. 182.

[30] Winkelmann to Leibniz, 17 July 1709, Leibniz Archive, Jablonski, No. 440, pp. 111–112; and Maria Margaretha Winkelmann, "Vorstellung des Himmels bey der Zusammenkunfft dreyer Grossmächtigsten Könige" (Potsdam, 1709). This pamphlet was originally housed in the Preussische Staatsbibliothek, where one still finds a card for it in the catalogue; through the vagariés of war the only extant copy is in the Biblioteka Jagielloński in Krakow.

[31] Leibniz's note in the margin of Winkelmann's letter to him, 17 July 1709, Leibniz Archive; reprinted in Harnack, "Berichte des Secretars Jablonski an der Präsidenten Leibniz" (cit. n. 22), Letter No. 87.

[32] Vignoles, "Eloge de Madame Kirch" (cit. n. 5), p. 182.

[33] Winkelmann to Leibniz, 4 Nov. 1707, Leibniz Archive, Kirch, No. 472, pp. 11–12. Though astrology had begun losing ground in Germany in the sixteenth century, it continued to exercise considerable influence even within scientific circles; see Zinner, *Die Geschichte der Sternkunde* (cit. n. 6), pp. 558–564. Erhard Weigel, Gottfried Kirch's teacher at the University of Jena, prepared

pamphlets, "Die Vorbereitung zur grossen Opposition," predicting the appearance of a new comet, was reviewed favorably in the *Acta eruditorum*. The reviewer praised her talents, ranking her "skill in observation and astronomical calculation" as equal to that of her husband. Even though Winkelmann made "concessions" to the art of astrology, the reviewer judged her work valuable. The review closed with a lavish tribute to this woman who "understood matters . . . that are not understood without the force of intelligence and the zeal of hard work."[34]

Several months after her pamphlet appeared in 1711, Academy Secretary Jablonski reported favorably that Winkelmann was becoming famous. Nowhere is there a hint that Academy authorities objected to her astrological work. In fact, her son, Christfried Kirch, continued to publish astrological calendars some years later during his tenure as Academy astronomer.[35]

Winkelmann mixed astrology and astronomy in calendar making, a project of both scientific and monetary interest for her and the Academy. Unlike many major European courts, the Prussian court did not yet have its own calendar. In 1700 the Reichstag at Regensburg ruled that an improved calendar similar to the Gregorian calendar was to be used in German lands.[36] Thus the production of an astronomically accurate calendar became a major project for the Academy of Sciences, founded in the same year. In addition to fixing the days and months, each calendar predicted the position of the sun, moon, and planets (calculated using the Rudolphine tables); the phases of the moon; eclipses of the sun or moon to the hour; and the rising and setting of the sun within a quarter of an hour for each day.

The monopoly on the sale of calendars was one of the two monopolies granted to the Academy by the king in 1700. Throughout the eighteenth century, the Berlin Academy of Science derived a large part of its revenues from the sale of calendars. This income (some 2,500 talers per year in the early 1700s; over 19,000 talers in the 1770s, with the added income of the Silesian calendar) made the position of astronomer particularly important.[37]

---

horoscopes and held astrological beliefs. Though he later turned against calendar makers who profited from "ill-founded" astrological predictions, Weigel continued to believe that comets were omens of good or ill fortune. See Weigel, *Unterschiedliche Beschreibung-und Bedeutung sowohl der Cometen insgemein als in Sonderheit des Wunder-Cometen* (1681); and Otto Knopf, *Die Astronomie an der Universität Jena von der Gründung der Universität im Jahre 1558 bis zur Entpflichtung des Verfassers im Jahre 1927* (Jena: Gustav Fischer, 1937), pp. 59–63.

[34] Maria Margaretha Winkelmann, *Vorbereitung, zur grossen Opposition, oder merckwürdige Himmels-Gestalt im 1712* (Cölln an der Spree, 1711). To my knowledge, the only extant copy of this pamphlet is at the Paris Observatory, acquired (I assume) along with other papers of the Kirch family by Joseph-Nicolas Delisle. For the review of her work see "Praeparatio ad Oppositionem magnam, sive notabilis Coelisacies ad Annum 1712, quam sequenti 1713 excipit oppositio triplex Saturni & Jovis, delineata a Maria Margaretha Winkelmannia, Kirchii Vidua, Astronomiae & Astrologiae Cultrice," *Acta eruditorum*, 1712, pp. 77–79.

[35] See Christfried Kirch, "Bestehende in einem Prognostico Astronomico-Astrologico Vorinnen Diejenigen merckwürdigste Vergebenheiten welcher sich an Sonne, Mond, und Sternen, in diesem 1726 Jahr . . . ," *Curieuser Astronomischer und Historischer Kalender Auf das Jahr Christi 1726 Berechnet, und Auf der Stadt Danzig und umliegender Ort Horizont mit Fleiss gerichtet* (Danzig, 1726), appendix.

[36] Since the Gregorian calendar reform of 1582, Catholics and Protestants had used calendars which differed by ten days: Wattenberg, "Zur Geschichte der Astronomie in Berlin, I" (cit. n. 12), p. 165. The "improved" Protestant calendar prepared by the Academy astronomer was similar to the Gregorian calendar except that Easter was calculated differently.

[37] Aufgebauer, "Die Astronomenfamilie Kirch" (cit. n. 10), p. 246. The other monopoly was silk, which never produced the desired revenues for the Academy.

Though calendrical reform was an important scientific project for the Academy, the sale of calendars depended on their more popular aspects. Calendars—which Leibniz called "the library of the common man"—had been issued since at least the fourteenth century and drew much of their popular appeal from astrology. Until 1768 there was little distinction between calendars and farmer's almanacs; both predicted the best times for haircutting, bloodletting, conceiving children, planting seeds, and felling timber.[38]

Weather prediction was also an important part of the Academy calendars and an important part of the duties of the Academy astronomer. Between 1697 and 1774 different members of the Kirch family kept a daily record of the weather.[39] Winkelmann's weather prediction partook of the "art of astrology."[40] In her *Wetter-Buch* of 1701 she wrote: "In God's name I have recorded the weather daily with diligent attention, and in order to see from which aspects [of the sky] the changes of weather may come." Her weather predictions also had an empirical basis. Her daily observations were made with the aid of a "weather-glass," a term used at that time for both the barometer and thermometer. Daily observation, she noted, sharpens prediction and can be very useful, especially in agriculture and navigation. It was Winkelmann's hope that "weather can be more accurately forecast, if more diligence is applied."[41]

IV. THE ATTEMPT TO BECOME ACADEMY ASTRONOMER

Gottfried Kirch died in 1710. It fell to the executive council of the Academy—President Leibniz, Secretary J. Th. Jablonski, his brother and Court Pastor D. E. Jablonski, and Librarian Cuneau—to appoint a new astronomer. The council needed to make the appointment quickly, as the Academy depended on the yearly revenues from the calendar; but apart from one in-house candidate, Jablonski could think of no one qualified for the position.[42] Ten years earlier the

---

[38] Harnack, *Geschichte der Akademie zu Berlin* (cit. n. 1), Vol. I, p. 124; Wolf, *Geschichte der Astronomie* (cit. n. 18), pp. 94–105. An excellent explanation of the calendar is given in a manual for women calendar users written in 1737 by Sidonia Hedwig Zäunemannin. See her *Curieuser und immer wahrender Astronomisch Meteorologisch-Oeconomischer-Frauenzimmer-Reise-und Hand-Kalender* (Erfurt, 1737). For more information about the Prussian calendars see Ludendorff, "Frühgeschichte der Astronomie" (cit. n. 29), pp. 19–22; and Knopf, *Astronomie an der Universität Jena* (cit. n. 33), p. 49. In 1702 there were six kinds of calendars sold by the Academy: the Improved Astronomical and Economic Calendar, the Improved Calendar of Curiosities, the Improved Conversation Calendar, the Improved Writing Calendar, and the Improved Historical and Geographical Calendar. The Academy also produced a wall calendar; see *Verbesserter Haushaltungs-Kalender* (Berlin, 1702), foreword.

[39] Kirch and Winkelmann traded off observing the weather from 1697 to 1702. Winkelmann then observed daily from 1702 to 1714 and 1716 to 1720. Her daughter Christine carried on after her mother's death, observing the weather daily from 1720 to 1751, 1755 to 1759, and 1760 to 1774: Hellmann in Kirch and Winkelmann, *Das älteste Berliner Wetter-Buch* (cit. n. 10), p. 12. The offices of astronomer and meteorologer were split after Winkelmann's death in 1720; see Alphonse des Vignoles, "Eloge de M. Kirch le Fils, Astronome de Berlin," *Journal litteraire d'Allemagne de Suisse et du Nord*, 1741, *1*:300–351, on p. 328.

[40] See Societät der Wissenschaften, *Historische-und Geographischer Calender* (Berlin, 1729), appendix. The conjunction of planets was thought to influence winds and temperatures. Thus the conjunction of "cold and dry" Saturn with "cold and moist" Venus forebodes much snow and hail in winter, while "warm and moist" Jupiter with "hot and dry" Mars brings warm and thundery weather: Zäunemannin, *Curieuser und immer wahrender Kalender* (cit. n. 38), pp. 112, 118.

[41] Kirch and Winkelmann, *Das älteste Berliner Wetter-Buch* (cit. n. 10), pp. 20, 12, 20–21.

[42] Harnack, "Berichte des Secretars Jablonski an den Präsidenten Leibniz" (cit. n. 22), No. 112. From 1700 to 1711 the Berlin Society of Sciences rarely held meetings, and no proceedings of meetings were kept or preserved. Thus these letters, which passed between Jablonski, the Academy

council had settled on Gottfried Kirch, who despite his advanced age (sixty-one) was the best in the field. Though there were few candidates, Maria Winkelmann's name did not enter the deliberations in 1710.[43] This is even more surprising when one considers that her qualifications at that time were not that different from her husband's earlier. They both had long years of experience preparing calendars (before coming to the Berlin Academy of Sciences, Kirch had earned his living by selling Christian, Jewish, and Turkish calendars); they had both discovered comets—Kirch in 1680, Winkelmann in 1702; and they both prepared ephemerides and recorded numerous other observations. What Winkelmann did not have, which nearly every member of the Academy did, was a university degree.

Kirch died in July; in August, since her name had not come up in discussions about the appointment, Winkelmann submitted it herself, along with her credentials. In a letter to Secretary Jablonski, she asked that she and her son be appointed assistant astronomers in charge of preparing the Academy calendar (see Figure 2).[44] Winkelmann made it clear that she was asking only for a position as assistant calendar maker. I would not, she wrote, be so "bold as to suggest that I take over completely the office [of astronomer]." Her argument for her candidacy was twofold. First, she argued, she was well qualified, since she was instructed in astronomical calculation and observation by her husband. Second, and more important, she had been engaged in astronomical work since her marriage and had, de facto, been working for the Academy since her husband's appointment ten years earlier. Indeed, she reported, "for some time, while my dear departed husband was weak and ill, I prepared the calendar from his calculations and published it under his name." She also reminded Jablonski that he had had occasion to remark on how she "lent a helping hand to her husband's astronomical work"—work for which she was paid a wage. In addition, she asked to be allowed to stay in the astronomer's quarters. For Winkelmann, a position at the Berlin Academy was not just an honor, it was a way to support herself and her four children. Her husband, she reported, had died leaving her with no means of support.

Secretary Jablonski was aware that the Academy's handling of the Winkelmann case would set important precedents for the role of women in it. In September 1710 he cautioned Leibniz: "You should be aware that this approaching decision could serve as a precedent. We are tentatively of the opinion that this case must be judged not only on its present merits but also as it could be judged for all time, for what we concede to her could serve as an example in the future."[45] Winkelmann's repeated requests for an official appointment at the obser-

---

secretary, and Leibniz, president *in absentia,* provide a rare written record of the Academy's early activities; see *ibid.,* pp. 5–6.

[43] The director of the Astrophysical Observatory in Potsdam in the 1940s, Hans Ludendorff, considered the only men qualified to be Academy astronomer at the turn of the century to have been E. Weigel of Jena (already dead in 1699), G. C. Eimmart of Nuremberg (already too old), and J. P. von Wurtzelbau of Nuremberg. He judged Winkelmann to be among the leading astronomers of her day. Ludendorff, "Frühgeschichte der Astronomie" (cit. n. 29), p. 12.

[44] Winkelmann (as Kirchin) to the Berlin Academy, 2 Aug. 1710; original in Kirch papers, Archives of the Akademie der Wissenschaften der DDR, I–III, 1, pp. 46–48; copy in the Leibniz Archive, Jablonski, 440, pp. 154–165. I would like to thank Christa Kirsten, Director, and the staff of the Zentrales Akademie-Archiv, der Akademie der Wissenschaften der DDR, for assistance.

[45] "Wenn E. Excell. Dero Gedanken darüber zu eröfnen belieben, so können dieselben bei bevorstehender überlegung derselben zur Richtschnur dienen. Hier ist man vorläufig der Meinung, daß

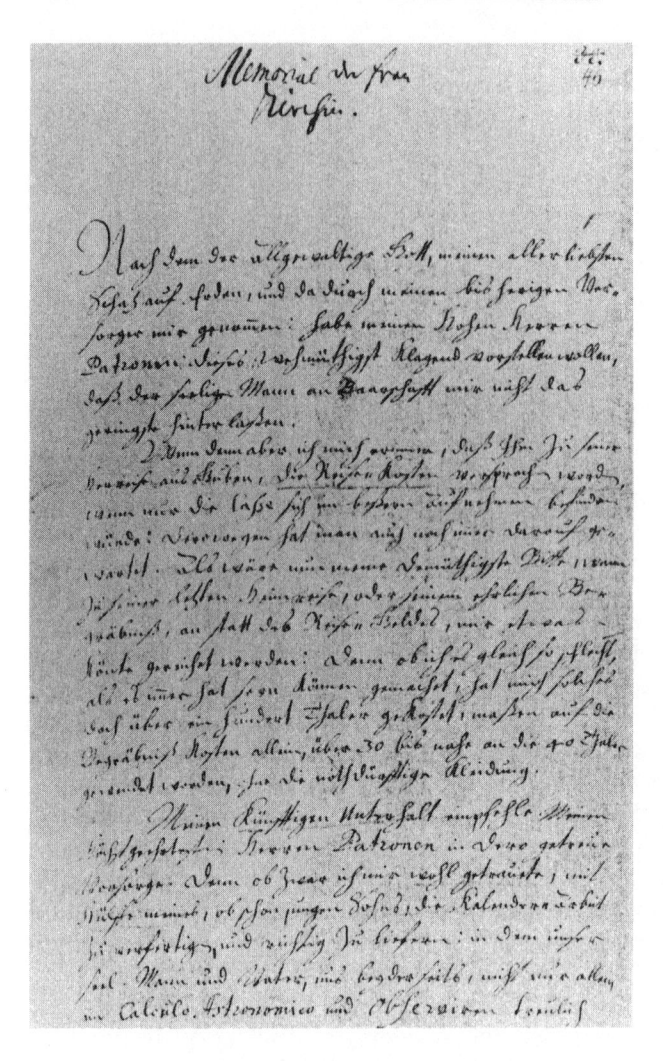

Figure 2. The first page of Winkelmann's six-page letter to the Academy asking to be appointed assistant astronomer. Reproduced with kind permission of the Zentrales Akademie-Archiv, Akademie der Wissenschaften der DDR.

vatory were not welcomed by the Berlin Academy because of concern about the effect on its reputation of hiring a woman. Jablonski wrote Leibniz: "That she be kept on in an official capacity to work on the calendar or to continue with observations simply will not do. Already during her husband's lifetime the society was burdened with ridicule because its calendar was prepared by a woman. If she were now to be kept on in such a capacity, mouths would gape even wider." By rejecting Winkelmann's candidacy, the Academy ensured that the social stigma attached to women would not further tarnish its already dull reputation.[46] In 1667

---

man die Sache so ansehen müsse, wie sie nicht nur gegenwärtig, sondern auch in Zukunft allezeit bestehen könne, immassen wass ihr eingeräumet würde, denen Künftigen zum Exempel dienen werde." Harnack, "Berichte des Secretars Jablonski an der Präsidenten Leibniz" (cit. n. 22), No. 115. Unfortunately, Leibniz's response to Jablonski has not been preserved.

[46] Jablonski to Leibniz, 1 Nov. 1710, in Harnack, "Berichte des Secretars Jablonski an der Präsidenten Leibniz" (cit. n. 22), No. 116. "Der Frau Kirchin gönnet Jedermann alles Gutes und wird ihr Niemand entgegen sein, ihr alles zuzuwenden, was möglich und anständig ist. Das sie aber bei der Calenderarbeit oder Observiren gebraucht und beibehalten werde, würde sich darum so viel weniger schicken, weil schon zu ihres Mannes Lebzeit sich spötter gefunden, so der Societät aufgebürdet,

a similar fear had prompted the members of the Royal Society of London to think long and hard before allowing Margaret Cavendish, Duchess of Newcastle, to visit a session.[47]

Leibniz was one of the few at the Academy who supported Winkelmann. In the council meeting of 18 March 1711 (one of the last meetings at which he presided before leaving Berlin), Leibniz argued that the Academy, considered as either a religious or an academic body, should provide a widow with housing and salary for six months as was customary. At Leibniz's behest, the Academy granted Winkelmann the right to stay in its housing a while longer; the proposal that she be paid a salary, however, was defeated. Instead the council paid her forty talers for her husband's observation notebooks. Later that year, the Academy showed some goodwill toward Winkelmann by presenting her with a medal.[48]

After Leibniz left Berlin, Winkelmann took her case to the king, and again her petition to be appointed Academy astronomer was placed before the council. With Leibniz gone, however, the council became more adamant in denying her requests. In 1712, after one and a half years of active petitioning, Winkelmann received a final rejection. Concerning her request for appointment as assistant astronomer, the council reported: "Frau Kirch's request is in many ways unseemly (*ungereimt*) and inadmissable (*unzulässig*). We must try and persuade her to be content and to withdraw of her own accord; otherwise we must definitely say no."[49]

The Academy never spelled out its reasons for refusing to appoint her to an official position, but Winkelmann traced her misfortunes to her sex. In a poignant passage, she recounted her husband's assurances that God would show his grace through influential patrons. This, she wrote, does not hold true for the "female sex." Her disappointment was deep: "Now I go through a severe desert, and because . . . water is scarce, . . . the taste is bitter." It was about this time that Winkelmann felt compelled to defend women's intellectual abilities in the preface to one of her scientific works. Citing Biblical authority, she argued that the "female sex as well as the male possesses talents of mind and spirit." With experience and diligent study, she wrote, a woman can become as "skilled as a man at observing and understanding the skies."[50]

Thus although Winkelmann had been involved in preparing the calendar for ten years and knew the work well, the position of Academy astronomer was awarded to Johann Heinrich Hoffmann.[51] Hoffmann had been a member of the Academy

---

dass ihre Calender durch ein Weib verfertiget werden, denen man hiermit das Maul noch weiter aufsperren würde." In 1706, the Academy was already coming under attack for its inactivity; see Harnack, *Geschichte der Akademie zu Berlin* (cit. n. 1), Vol. I, pp. 155–156.

[47] Samuel Pepys wrote in his diary that there was "much debate, *pro* and *con*, it seems many being against it, and we do believe the town will be full of ballads of it." See Samuel Mintz, "The Duchess of Newcastle's Visit to the Royal Society," *Journal of English and Germanic Philology*, 1952, *51*(2):168–176.

[48] Protokollum Concilii, Societatis Scientiarum, 15 Dec. 1710, 18 Mar. 1711, 9 Sept. 1711, DDR Academy Archives, I, IV, 6, Pt. 1, pp. 54, 65–66, 93. Unfortunately, we do not know why Winkelmann received a medal.

[49] *Ibid.*, 3 Feb. 1712, p. 106.

[50] Winkelmann to the council of the Berlin Academy, 3 Mar. 1711, DDR Academy Archives, Kirch papers, I, III, 1, p. 50; and Winkelmann, *Vorbereitung zur grossen Opposition* (cit. n. 34), pp. 3–4.

[51] Hoffmann has been almost entirely forgotten today. His name does not appear in the comprehensive, multivolumed *Allgemeine deutsche Biographie* or the *Neue deutsche Biographie*. Nor is he

since its founding in 1700 and had long hoped to be appointed Academy astronomer. Yet, his tenure was not a happy one. By December 1711 he was already behind in his work. Jablonski wrote to Leibniz complaining that Hoffmann was guilty of neglecting his work. Jablonski suggested that perhaps Hoffmann needed an assistant; ironically he suggested "Frau Kirchin, for example, who would spur him on a bit." In 1712 Jablonski again had occasion to complain to Leibniz about Hoffmann's performance. Hoffmann had not completed the yearly observations as he should have, and the calendar was still not ready. Hoffmann was officially censured by the Academy for his poor performance. While Hoffmann was being reprimanded, Winkelmann was becoming, as Jablonski reported, "rather well known" for her pamphlet on the conjunction of Saturn and Jupiter.[52]

During this period, conflict arose between Winkelmann and Hoffmann, each of whom considered the other a competitor at the observatory. Jablonski reported to Leibniz that Winkelmann had complained that "Hoffmann used her help secretly, yet denounced her publicly, and never let her use the observatory." Unemployed and unappreciated for her scientific skills, Winkelmann moved across Berlin in October 1712 to the private observatory of Baron Bernhard Friedrich von Krosigk. This did not end Hoffmann's problems with the Academy, however. In 1715 Jablonski complained once again to Leibniz that Hoffmann was neglecting his duties.[53]

V. THE CLASH BETWEEN CRAFT TRADITIONS AND PROFESSIONAL SCIENCE

Did Winkelmann have a legitimate claim to the post of assistant astronomer? How was it possible in 1700 for a woman to hold a semiofficial position (as Winkelmann did) as assistant to her husband at the Berlin Academy? How did she imagine that her requests to continue on the calendar project would be taken seriously?

Winkelmann owed her position at the Academy to the perpetuation of guild traditions. These were as alive in the Academy as they were in Germany as a whole. Wolfram Fischer has argued that the relation of apprentice-journeyman-master provided a model for many German institutions. Fischer gave the example of the masons; W. V. Farrar has developed the example of the universities. According to Farrar, the guild character of the university system survived longer in Germany than elsewhere.[54]

But while retaining vestiges of the guild system, the Berlin Academy incorporated other traditions. We should distinguish two levels of participation in the Academy. At the top was a tier of university-educated, internationally renowned

---

mentioned in histories of astronomy such as Wolf's *Geschichte der Astronomie*. Maria Winkelmann is recognized for her work in each of these sources.

[52] Harnack, "Berichte des Secretars Jablonski an der Präsidenten Leibniz, Nos. 112, 133, 143, 144. In this last Jablonski was probably referring to Winkelmann's 1711 *Vorbereitung zur grossen Opposition*.

[53] Krosigk's observatory was built in 1705. See Jablonski to Leibniz, 29 Oct. 1712, in Harnack, "Berichte des Secretars Jablonski and der Präsidenten Leibniz," No. 143; and *ibid.*, No. 167.

[54] Wolfram Fischer, *Handwerksrecht und Handwerkswirtschaft um 1800* (Berlin: Duncker & Humblot, 1955), p. 18. W. V. Farrar found the corporation of master-tradesmen of the guild analogous to the academic body of professors, the journeyman's *Wanderjahre* similar to the student's traveling from university to university, and the masterpiece of the tradesman similar to the university M.A. See Farrar, "Science and the German University System: 1790–1850," in *The Emergence of Science in Western Europe*, ed. Maurice Crosland (London: Macmillan, 1975), p. 181.

members. This aspect of the organization had nothing in common with the guilds; rather, class standing was important for membership at this level. Like members of the Royal Society in London and the Académie des Sciences in Paris, many "gentlemen" members of the Berlin Academy were of noble standing. It was its financial structure that set the Berlin Academy apart from its counterparts in Paris or London and nearer craft traditions. Roger Hahn has shown that members of the Académie des Sciences in Paris drew pensions directly from the king's purse in order to distance themselves from traditional trades and professions, considered "mere occupations."[55] The Berlin Academy, in contrast, drew much of its revenues directly from two trades—calendar making and silk making—and hired artisans, the second tier of participants in its activities, to carry out the tasks required.

The Academy astronomer was in fact caught between the two tiers of the Academy hierarchy: as a university-educated mathematician, he was a distinguished gentleman; as calendar maker, he was an artisan who worked for his employer. The "gentlemen" of the Academy (except the president and secretary) were not paid, nor did they pay for their membership. The astronomer, however, like the other artisans of the Academy, derived his living (500 talers per year) from its coffers. It should be noted that though Maria Winkelmann asked to continue as Academy calendar maker, she never asked to become a member of the Academy (nor was she granted membership).[56]

It was as the wife of an artisan-astronomer that Winkelmann enjoyed a modest measure of respect at the Academy. When she petitioned the council to continue as assistant calendar maker, she was invoking (although not explicitly) age-old principles well established in the organized crafts and free arts. In most cases, guild regulations gave a widow the right to run the family business after the death of her husband. Guild regulations were local and varied from region to region, craft to craft; yet general patterns can be identified. In her study of thirty-eight Cologne guilds in the late Middle Ages, Margret Wensky found that eighteen of those guilds allowed a widow to continue the family business after her husband's death.[57] The rights of widows followed three general patterns. In some guilds, the widow was allowed to serve as an independent master as long as she lived. In others, she was allowed to continue the family business but only with the help of journeymen or apprentices. In still others, she filled in for one or two years to provide continuity until her oldest son came of age.[58] Within lower echelons of the Academy, widows were allowed to continue in their husband's position. Pont, widow of the keeper of the Academy mulberry trees, was allowed to complete the last four years of her husband's six-year contract.[59]

This is what Maria Winkelmann also tried to do. After the death of her husband, she tried to carry on the "family" business of calendar making as an inde-

---

[55] Roger Hahn, *The Anatomy of a Scientific Institution: The Paris Academy of Science 1660–1803* (Berkeley/Los Angeles: Univ. California Press, 1971), p. 39.
[56] Harnack, *Geschichte der Akademie zu Berlin* (cit. n. 1), Vol. I, p. 370.
[57] Wensky, "Die Stellung der Frau" (cit. n. 11), pp. 58–59.
[58] See, e.g., in Ketsch, *Frauen im Mittelalter* (cit. n. 11), Vol. I, p. 210: the regulations for the Lübeck dyers guild (No. 296); p. 204: the regulations of the Cologne hatmakers' guild (No. 276); and p. 29.
[59] Protokollum Concilii, Societatis Scientiarum, 23 Sept. 1716, DDR Academy Archives, I, IV, 6, Pt. 2, pp. 230–232.

pendent master. Yet, as we have seen, she found that traditions which had once secured women a (limited) role in science were not to apply in the new institutions.

Though the Academy retained vestiges of an older order, it also contained the seeds of a new. The founding of the Academy in 1700 represented a first step in the professionalization of astronomy in Germany. Earlier observatories—those of Hevelius in Danzig and Eimmart in Nuremberg—had been private. The Academy's observatory, however, was a public ornament of the Prussian state. Astronomers were no longer owners and directors of their own observatories, but employees of the Academy, selected by a patron on the basis of personal merit rather than family tradition. This shift of the character of scientific institutions from private to public had dramatic implications for the role of women in science. As astronomy moved more and more out of the private observatories and into the public world, women lost their toehold in modern science.

VI. A BRIEF RETURN TO THE ACADEMY

Although Winkelmann could not remain at the Berlin Academy, she did continue her astronomical work. In October 1712 she moved with her family to Baron von Krosigk's private observatory, where she and Gottfried Kirch had worked while the Academy observatory was under construction. She was thus able to keep in touch with astronomical work in Berlin.

At Krosigk's observatory, Winkelmann reached the height of her career. With her husband dead and her son away at university, she enjoyed the rank of "master" astronomer. She continued her daily observations and—now the master—had two students to assist her. The published reports of their joint observations bear her name. Many of Winkelmann's observations from this period—the conjunction of Saturn and Mars, several eclipses of the moon, and several sightings of sunspots—were published in her son's *Ephemeriden* of 1714 and 1715.[60] During this period, she also supported herself and her daughters by preparing calendars for Breslau and Nuremberg.[61]

When Krosigk died in 1714, Maria Winkelmann left his observatory, taking a position in Danzig as assistant to a professor of mathematics.[62] This part of her life remains sketchy. When this position fell through, Winkelmann again found a patron. The family of Hevelius (Gottfried Kirch's teacher) invited her and her son, Christfried, now a student in Leipzig, to reorganize the deceased astronomer's observatory and to use it to continue their own observations.

In 1716 the Winkelmann-Kirch family received an invitation from Peter the Great of Russia to become astronomers in Moscow.[63] The family decided instead

---

[60] Christfried Kirch, *Teutsche Ephemeris* (Nuremberg, 1715), p. 82; *ibid.* (1714), pp. 76–77, 80; and *ibid.* (1715), pp. 78–80, 82–84.

[61] Winkelmann was given permission by the Berlin Academy to prepare these calendars: Protokollum Concilii, Societatis Scientiarum, 18 Oct. 1717, DDR Academy Archives, I, IV, 6, Pt. 2, p. 280. Though Herbert Weiss has suggested that Winkelmann did not sign her name as author of these calendars because a woman astronomer was accorded little respect ("Quellenbeiträge zur Geschichte der Preussischen Akademie" [cit. n. 14], p. 216), a look at calendars of the eighteenth and nineteenth century shows that very few were signed.

[62] The only report on this is from Vignoles, "Eloge de Madame Kirch" (cit. n. 5), p. 180.

[63] *Ibid.*

to return to Berlin when Christfried was appointed one of two observers for the Academy following the death of Hoffmann.[64] The Academy had grave reservations about the abilities of their newly appointed astronomers: Christfried Kirch was not well grounded in astronomical theory and could not express himself decently in German or Latin; J. W. Wagner was weak in astronomical calculation. Academy funds, however, were insufficient to support the appointment of a "celebrated" astronomer who would require a higher salary, better housing, and assistants. Under these circumstances, a factor weighing in Kirch's favor was that, along with him, the Academy received an extra astronomical hand—Winkelmann—with skills very similar to those of Kirch and Wagner. Thus, Winkelmann returned once again to the work of observation and calendar making for the Academy, this time as assistant to her son.[65]

But all was not well. The opinion was still prevalent that women should not do astronomy, at least not in a public capacity.[66] In 1717 Winkelmann was reprimanded by the Academy council for talking too much to visitors at the observatory. The council cautioned her to "retire to the background and leave the talking to Wagner and her son." A month later, the Academy again reported that "Frau Kirch meddles too much with Society matters and is too visible at the observatory when strangers visit." Again the council warned Winkelmann "to let herself be seen at the observatory as little as possible, especially on *public* occasions."[67] As Vignoles reported, there were those who found it wrong for a woman to practice astronomy. Maria Winkelmann was forced to make a choice. She could either continue to badger the Academy for a position of her own, or, in the interest of her son's reputation, she could retire, as the Academy requested, to the background. Vignoles reported that she chose the latter option. Academy records show, however, that the choice was not hers to make. On 21 October 1717 the Academy resolved to remove Winkelmann—who apparently had paid little heed to their warnings—from Academy grounds. She was forced to leave her house and the observatory. The Academy did not, however, want her to abandon her duties as mother; officials expressed the hope that Winkelmann "could find a house nearby so that Herr Kirch could continue to eat at her table."[68]

In 1717 Winkelmann quit the Academy's observatory and continued her observations only at home, as was thought appropriate, "behind closed doors," a move which Vignoles judged detrimental to the progress she might have made in astronomy. With few scientific instruments at her disposal, she was forced to

---

[64] Christfried was first appointed Academy "astronomer" in 1728. For his letters of application see Weiss, "Quellenbeiträge zur Geschichte der Preussischen Akademie" (cit. n. 14), pp. 219–222. In the first letter, Kirch suggested that he do the calendar work while the society found another astronomer. Among his qualifications, he mentioned his long experience preparing calendars. In his second letter, he asked for the job of astronomer, playing very much on the good reputation of his father. The position of Academy astronomer was largely hereditary: J. E. Bode, Academy astronomer at the turn of the eighteenth century, was also related to the Kirch family.

[65] Protokollum Concilii, Societatis Scientiarum, 8 Oct. 1716, 6 Apr. 1718, DDR Academy Archives, I, IV, 6, Pt. 2, pp. 236, 318.

[66] Vignoles, "Eloge de Madame Kirch" (cit. n. 5), pp. 181.

[67] Protokollum Concilii, Societatis Scientiarum, 18 Aug. 1717, DDR Academy Archives, I, IV, 6, Pt. 2, pp. 269, 272–273 (emphases added).

[68] Vignoles, "Eloge de Madame Kirch," p. 181; Protokollum Concilii, Societatis Scientiarum, 21 Oct. 1717, DDR Academy Archives, I, IV, 6, Pt. 2, pp. 275–276.

quit astronomical science. Maria Winkelmann died of fever in 1720. In Vignoles's opinion, "she merited a fate better than the one she received."[69]

VII. WOMEN ASTRONOMERS IN GERMANY

Maria Winkelmann was not the only woman astronomer in late seventeenth-century Germany. Between 1650 and 1710 a surprisingly large number of women—Maria Cunitz (1610–1664), Elisabetha Hevelius (1647–1693), Maria Klara Eimmart (1676–1707), Maria Winkelmann (1670–1720) and her daughters Christine Kirch (1696–1782) and Margaretha (active in the 1740s)—worked in German astronomy. The group comprised, as noted above, fourteen percent of German astronomers for this period.[70] All these women worked in family observatories: Hevelius built his private observatory in 1640 and again in 1687; Eimmart built his in 1678.[71] Of this group, only Maria Cunitz was not the daughter or wife of an astronomer who, in guildlike fashion, assisted a master in his trade. As in the case of Winkelmann, the perpetuation of craft traditions allowed these women access to the secrets and tools of the astronomical trade.

It is perhaps unfair to include the example of Maria Cunitz among women working within the crafts tradition, for her father was a landowner. Nonetheless, her education too depended on training given her by her father, the learned medical doctor Heinrich Cunitz, lord of the estates of Kunzendorf and Hoch Giersdorf near Schweidnitz in Silesia. Sometimes called the "second Hypatia," Cunitz learned from her father six languages—Hebrew, Greek, Latin, Italian, French, and Polish—as well as history, medicine, mathematics, painting, poetry, and music.[72] Her principal occupation, however, was astronomy. In 1630 she married Eliae von Lowen, a medical doctor and amateur astronomer. During the Thirty Years' War her family took refuge in Poland, where she prepared her astronomical tables, published in 1650 as *Urania propitia*. The main purpose of this work was to simplify Kepler's Rudolphine Tables, used for calculating the position of

---

[69] Vignoles, "Eloge de Madame Kirch," pp. 181, 182.

[70] See note 9 above.

[71] Ernst Zinner, *Deutsche und niederländische astronomische Instrumente des 11.–18. Jahrhunderts* (Munich: C. H. Beck, 1956), pp. 221–223. Hevelius's observatory spanned the roofs of three adjoining houses; Eimmart's was built on the city wall. See also Wolf, *Geschichte der Astronomie* (cit. n. 18), p. 458.

[72] Like a number of women of her time, Cunitz provided some biographical information in the preface of her book. See Maria Cunitz, *Urania Propitia, sive Tabulae Astronomica mirè faciles, vim hypothesium physicarum à Kepplero proditarum complexae; facillimo calculandi compendio, sine ullą Logarithmorum mentione, phaenomenis satisfacientes* (Oels, 1650), esp. p. 147. See also Joanne Hallervordio, *Bibliotheca curiosa* (Frankfurt, 1676), p. 260; and Vignoles, "Eloge de Madame Kirch" (cit. n. 5), pp. 163–168. In his *Mathematisches Lexicon* (cit. n. 18), Christian Wolff reported that Cunitz simplified the Rudolphine Tables (p. 1360); he also mentioned Cunitz in his *Elementa matheseos universae*, Vol. IV, in *Gesammelte Werke*, Vol. XXXIII, p. 112. See also Weidler, *Historia astronomiae* (cit. n. 18), pp. 489–490. Cunitz received a note under Winkelmann's entry in Bode's *Astronomisches Jahrbuch* (cit. n. 18), p. 113. Since Cunitz was not the daughter or wife of an astronomer, she is one of the few women to receive her own entry in many histories of astronomy. See also Wolf, *Geschichte der Astronomie*, pp. 305–306; Lagrange, "Les femmes-astronomes" (cit. n. 18), p. 517; Mozans, *Woman in Science* (cit. n. 18), pp. 170–171, who gives the reference to a "second Hypatia"; Rizzo, "Early Daughters of Urania" (cit. n. 18), p. 8; and Ingrid Guentherodt, "Maria Cunitz und Maria Sibylla Merian: Pionierinnen der deutschen Wissenschaftssprache im 17. Jahrhundert," *Zeitschrift für germanistische Linguistik*, 1986, *14*(1):23–49. For an excellent bibliographic source on women in this period see Jean Woods and Maria Fürstenwald, *Schriftstellerinnen, Künstlerinnen und gelehrte Frauen des deutschen Barock: Ein Lexikon* (Stuttgart: J. B. Metzler, 1984).

the planets; but Maria Cunitz was not merely a calculator: her book also treated the art and theory of astronomy.[73]

Maria Klara Eimmartin-Müller, another woman practicing astronomy at the end of the seventeenth century, fits squarely into craft traditions.[74] The daughter of Georg Christoph Eimmart, astronomer and director of the Nuremberg Academy of Art from 1699 to 1704, Maria Eimmart learned French, Latin, drawing, and mathematics from her father. As a young girl she also learned the art of astronomy at her father's observatory, where she worked alongside his other students. She owed her place in astronomy largely to the strong position of women in the graphic arts. Much of Eimmart's scientific achievement depended on her ability to make exact sketches of the sun and moon. Between 1693 and 1698, she prepared 250 drawings of phases of the moon in a continuous series, thus laying the groundwork for a new lunar map. She also made two drawings of the total eclipse of 1706.[75] A few sources claim that in 1701 Eimmart published a work on ancient views of the sun, *Ichnographia nova contemplationum de Sole*, under her father's name, but there is no evidence that this was her work.[76]

After training as an apprentice to her father, the scientifically minded Eimmart secured her position in astronomy by marrying the astronomer Johann Heinrich Müller in 1706. Müller was professor of physics at a Nuremberg *Gymnasium* and since 1705 director of her father's observatory. Through this marriage Maria Eimmart ensured that she could continue her astronomical work at her father's observatory, now as wife of the director. Johann Müller also benefited from this marriage. Through the principle of daughter's rights, the Eimmart observatory became part of his daughter's inheritance, passing through the daughter to her husband.[77] Maria Müller's astronomical career was cut short when she died in childbirth in 1707.

Elisabetha Koopman (later Hevelius) of Danzig also married with care to ensure her career in astronomy. In 1663 she married a leading astronomer, Johannes Hevelius, a man thirty-six years older than herself. Hevelius, a brewer by trade, took over the lucrative family beer business in 1641. His first wife, Catherina Rebeschke, had managed the household and brewery, leaving Hevelius free to serve in city government and to pursue his avocation, astronomy. When she died in 1662, Hevelius married Elisabetha Koopman, who had been interested in

---

[73] As is common in German women's books of the period, the text was given in Latin and German. The work also presented a guide to astronomy for the layperson. Astronomy, Cunitz taught, has four parts: observation, which must be carefully recorded; mechanics or the craft of making instruments; hypotheses or theory of the heavens; and calculus or tables of predictions. See also Vignoles, "Eloge de Madame Kirch" (cit. n. 5), pp. 148–149.

[74] See the mention of Maria Klara Eimmart under her father's entry in Weidler, *Historia astronomiae* (cit. n. 18), p. 543. See also Wolf, *Geschichte der Astronomie* (cit. n. 18), p. 104; and Kurt Pilz, *600 Jahre Astronomie in Nürnberg* (Nuremberg: Carl, 1977).

[75] Johann Gabriel Doppelmayr, *Historische Nachricht von den nürnbergischen Mathematicis und Künstlern* (Nuremberg, 1730; rpt. Hildesheim: Olms, 1972), pp. 259–260; and Jöcher, *Allgemeines Gelehrten-Lexicon* (cit. n. 9), Vol. III, p. 743.

[76] See, e.g., J. C. Poggendorff, *Handwörterbuch zur Geschichte der exacten Wissenschaften* (Leipzig: Barth, 1863), Vol. I, p. 651. Eighteenth-century lexicons, however, which list her works in great detail, attribute the *Ichnographia* to her father. See Doppelmayr, *Historische Nachricht* (cit. n. 75), p. 126; and Will, *Nürnbergisches Gelehrten-Lexicon* (cit. n. 9).

[77] According to Peter Ketsch, family trades passed more often to the daughter than to the son: Ketsch, Frauen im Mittelalter (cit. n. 11), Vol. I, p. 29.

astronomy for many years.[78] In appropriate guild fashion, Elisabetha Hevelius served as chief assistant to her husband, both in the family business and in the family observatory.

Margaret Rossiter has defined and described the notion of "women's work" in nineteenth- and twentieth-century science, and especially in astronomy.[79] Women's work in science—tedious computation, support positions, and the like—is a legacy of the guild wife. Elisabetha Hevelius is perhaps the best example of a wife who served as chief assistant to her astronomer husband. The role of the guild wife, however, cannot be collapsed into that of a mere assistant. The very different structure of the workplace—in the seventeenth century the observatory was in the home, not part of a university—gave the wife a more comprehensive role. For twenty-seven years Elisabetha Hevelius collaborated with her husband, observing the heavens in the cold of night by his side (see Figure 3).[80] After his death, Elisabetha Hevelius edited and published their joint work, *Prodromus astronomiae,* a catalogue of 1,888 stars and their positions.[81]

The "astronomical wife" was not an exception, but an established tradition. When Gottfried Kirch studied with Hevelius in Danzig, he learned through the example of Elisabetha Hevelius the difference an astronomical wife could make.

VIII. "INVISIBLE ASSISTANTS": WOMEN'S PARTICIPATION IN THE BERLIN ACADEMY

Maria Winkelmann was not the only woman present at the founding of the Berlin Academy of Sciences. Sophie Charlotte, queen of Prussia, was important as an ambassador of scientific ideas at the court in Berlin. Working closely with Leibniz and her ministers, Sophie Charlotte carried forth plans and negotiations for the founding of the Berlin Academy with such vigor that Leibniz claimed it "the role of women of elevated mind more properly than men to cultivate knowledge."[82] Frederick II, her grandson, credited her with establishing the Academy. He wrote that "she founded the royal Academy and brought Leibniz and many other learned men to Berlin. She wanted always to know the first principle of things." Since she died shortly after its founding, it remains unclear whether Sophie Charlotte intended to take an active part in the Academy or to serve merely as a patron.[83]

---

[78] Eugene McPike, *Hevelius, Flamsteed, and Halley* (London: Taylor & Francis, 1937), pp. 4–5. Elisabetha Hevelius receives short entries in Rebière, Mozans, and Rizzo (see n. 18).

[79] Margaret Rossiter, "Women's Work in Science, 1880–1910," *Isis,* 1980, 71(258):381–398; see also Rossiter, *Women Scientists in America: Struggles and Strategies to 1940* (Baltimore: Johns Hopkins Univ. Press, 1982), pp. 51–72.

[80] Elisabetha Hevelius is shown in three plates; two depict her working at the sextant with Johannes, and a third shows her using a telescope: Johannes Hevelius, *Machina Coelestis, pars prior; Organographiam, sive Instrumentorum Astronomicorum omnium, quibus Auctor hactenus Sidera rimatus* (Danzig, 1673), plates following pp. 222, 254, 450.

[81] Johannes Hevelius, *Prodromus astronomiae* (Danzig, 1690).

[82] Sophie Charlotte, princess of Hannover, was privately tutored by Leibniz from an early age, well read in Latin, well traveled, and a devotee of French culture. See Leibniz to Sophie Charlotte, Nov. 1697, in Harnack, *Geschichte der Akademie zu Berlin* (cit. n. 1), Vol. II, p. 44.

[83] Frederick II, "Mémoire de l'Académie," 1748, reprinted in Jean-Pierre Erman, *Mémoire pour servir à l'histoire de Sophie Charlotte, reine de Prusse* (Berlin, 1801), p. 382. In a nineteenth-century Academy calendar Sophie Charlotte was credited with giving the order for the founding of the Academy according to Leibniz's plan; see *Adress Calender* (Berlin, 1845), p. 113. Harnack argued that it

The founding statutes of the Berlin Academy of Sciences did not bar women from membership. In fact, Leibniz thought women should benefit from participation. In his sketch of Academy regulations of 1700, he wrote that a scientific academy would foster good taste, solid understanding, and an appreciation of God's handiwork, not only among German nobility, "but among other people of high standing (as well as among women)."[84] Yet despite his intentions, women were not admitted. Perhaps the decision to use the scientific societies of London and Paris as models for the Academy in Berlin reinforced the exclusion of women. Although neither the London nor the Paris society had regulations excluding women, neither society admitted them.

The fate of Winkelmann's daughters—Christine and Margaretha—reveals a process of privatization of women within the Academy. Trained (in guild fashion) in astronomy from the age of ten, both Kirch daughters worked for the Academy as assistants to their brother, Christfried. According to Vignoles, "Margaretha, the youngest, usually took a telescope; Christine, the oldest, most often took the pendulum in order to mark exactly the time of each individual observation."[85] Yet having witnessed the lost battles of their mother, Christine and Margaretha did not ask (as Winkelmann had) for official positions. Nor did they exude the fire of their mother, badgering the Academy for housing or greeting foreign visitors. Rather, they molded their behavior to fit Academy prescriptions, becoming "invisible helpers" to their brother. Again, Vignoles describes the sisters' situation: "They helped their brother carry out his professional duties; . . . nonetheless they remained very private and spoke with no one but their close friends. By the same modesty, they avoided going to the observatory when there was to be an eclipse or other observation that might attract strangers."[86]

When Christfried died in 1740, the Kirch sisters lost their male protector and were forced to observe more often at home. Although they watched the heavens daily, conditions made serious astronomical work almost impossible. When Christine sent their observations of the comets of 1742 and 1743 to Joseph Nicolas Delisle, director of the Paris Observatory, she complained: "We observed daily [the course of the comet] as well as we could . . . but our observations were done under very bad conditions and with inferior instruments, namely with a two foot (*zwei Schühe*) telescope. . . . We could not use a larger telescope because our house had no window large enough to accommodate it."[87]

---

was not Leibniz but Sophie Charlotte who initiated plans for the Academy: *Geschichte der Akademie zu Berlin* (cit. n. 1), Vol. I, pp. 48–49. Perhaps the decision in Maria Winkelmann's case would have gone in her favor had Sophie Charlotte been alive to intervene.

[84] "Leibnizens Denkschrift in Bezug auf die Einrichtung einer Societas Scientiarum et Artium in Berlin vom 26. März 1700, bestimmt für den Kurfürsten," in Harnack, *Geschichte der Akademie zu Berlin* (cit. n. 1), Vol. II, p. 80. I thank Gerda Utermöhlen of the Leibniz Archive, Hannover, for calling this passage to my attention.

[85] Vignoles, "Eloge de M. Kirch le Fils" (cit. n. 39), p. 349.

[86] *Ibid*.

[87] Christine Kirch to Delisle, 24 July 1744, Paris Observatory, Delisle papers, MS A. B. 1. IV, No. 12a; and 28 Apr. 1745, No. 42. There are eight letters from Christine Kirch to Delisle at the Paris Observatory. Delisle initiated the correspondence in an attempt to buy the observation notebooks of Gottfried and Christfried Kirch. It should be noted that he did not ask for her astronomical observations; she volunteered them. Hers are the only letters from a woman in his sixteen volumes of correspondence. On Delisle see Roger Jaquel, "L'astronome Français Joseph-Nicolas Delisle (1688–1768) et Christfried Kirch (1694–1740), Directeur de l'Observatoire de Berlin (1716–1740)," *Actes du 97e Congrès National des Sociétés Savantes*, 1972, pp. 407–432.

Figure 3. Like Gottfried Kirch and Maria Winkelmann, Elisabeth and Johannes Hevelius collaborated in astronomical work. This illustration shows them working together with the sextant. From Johannes Hevelius, *Machinae coelestis* (Danzig, 1673), facing page 222. By permission of the Houghton Library, Harvard University.

Though Christine and Margaretha Kirch had little opportunity to go to the observatory after their brother's death, Christine continued to prepare the Academy calendar—silently and behind the scenes—from at least 1720 until her death in 1782. This is not surprising. By the 1740s, calendar making was no longer on the cutting edge of astronomical science, but tedious and time-consuming work. Never married, Christine supported herself through her calendar work, for which she received a small pension of 400 talers per year.[88]

After Christine Kirch retired, no other women did scientific work for the Berlin Academy of Sciences until well into the twentieth century.[89] During the eighteenth century, the Academy did, however, grant honorary membership to some women of the noble classes. The first to be granted honorary membership at the (then) Académie Royale des Sciences et Belles-Lettres was one of the most powerful persons in Europe at the time, Catherine the Great of Russia. Rank still spoke loudly in Prussia, and the prestige of her rank outweighed the liabilities of her sex. Catherine's position in the Academy was wholly honorary.[90] After Frederick the Great's tenure as president, few women were elected. One exception was poet and writer Duchess Juliane Giovane, who was awarded honorary membership in 1794. No other woman was elected for 106 years, and even then it was for purely nonscientific reasons: in 1900 Maria Wentzel was awarded honorary membership for her gift of 1,500,000 marks.[91]

It is clear that before 1949 only women of the very highest social standing were admitted to membership in the Berlin Academy of Sciences. Though Catherine the Great and Juliane Giovane were women of intellectual stature, they were also women of social rank. Maria Winkelmann, however, was a tradeswoman who dirtied her hands in the actual work of astronomy (she was referred to by Academy officials as a "Weib," and not a "Frauenzimmer"). The election of a woman purely on scientific merit had to wait until 1949, when the physicist Lise Meitner was elected, but as only a corresponding member. Meitner was followed by the chemist Irène Joliot-Curie, daughter of Marie Curie, and then by the medical

---

[88] Harnack, *Geschichte der Akademie zu Berlin* (cit. n. 1), Vol. I, p. 491.

[89] I am unaware of other women doing scientific work like that of the Kirch sisters. This is not surprising since it was craft traditions that secured their positions, and by the 1780s, when Christine Kirch died, these traditions were growing feeble.

[90] It should be pointed out that Catherine was elected in 1767, when Frederick the Great, as president of the Academy, personally oversaw all appointments. The following year, Frederick decreed that her membership in the Academy should be elevated from honorary status to that of a regular foreign member: Harnack, *Geschichte der Akademie zu Berlin* (cit. n. 1), Vol. I, pp. 369, 473; and Werner Hartkopf, *Die Akademie der Wissenschaften der DDR: Ein Beitrag zu ihrer Geschichte* (Berlin: Akademie-Verlag, 1983), pp. 219–220. During Frederick's presidency several women had their work read before the Academy. On the evening of 26 Jan. 1769, e.g., the Countess of Skorzewska's "Considérations sur l'origine des Polonais" was read at a public session. Though she was present at the reading, the assembly would not seat her: Harnack, *Geschichte der Akademie zu Berlin* (cit. n. 1), Vol. I, p. 370. In 1770 another woman—the French anatomist and writer Marie Geneviève Charlotte Thiroux d'Arconville—held a discourse on "L'amour-propre" at an Academy session ("Essai sur l'amour-propre envisagé comme principe de morale," *Discours prononcé a l'assemblée ordinaire de l'académie royale des sciences et belles-lettres de Prusse*, 1770, pp. 1–32). This essay has also been attributed to Frederick the Great. As one might also expect, women of royalty attended some Academy sessions, not as members but as guests. In 1772, for example, a host of royalty including the queen of Sweden (sister of Frederick), the princess Amélie of Prussia, and the Abbess of Quedlimbourg were present at a public session (*Nouveaux Mémoires de l'Académie Royale des Sciences et Belles-Lettres*, 1772, p. 5).

[91] Maria Wentzel's endowment of 1894 was in honor of her architect husband and factory-owning father: Harnack, *Geschichte der Akademie zu Berlin* (cit. n. 1), Vol. I, p. 1019.

doctor Cécilie Vogt in 1950. The first woman to be awarded full membership was the historian Liselotte Welskopf, in 1964.

Since the founding of the Academy of Sciences in Berlin in 1700, only fourteen of its twenty-nine hundred members have been women. Of those fourteen, only four have enjoyed full membership.[92] As of 1983, no woman had ever served in any leadership role as president, vice president, general secretary, or head of any of the various scientific sections.

IX. THE CONSEQUENCES FOR WOMEN'S PARTICIPATION IN SCIENCE

As the case of Maria Winkelmann illustrates, the poor representation of women in the Berlin Academy cannot be traced simply to an absence of qualified women.[93] Instead, the exclusion of women resulted from policies consciously implemented at an early period in the Academy's history. These decisions, made in the early eighteenth century, held serious consequences for women's later participation.

The Academy did not, however, make its decisions in a vacuum. Larger developments in both science and society set parameters within which the Academy maneuvered. The professionalization of the sciences (a gradual process which took place over the span of two centuries) weakened craft traditions within the sciences and weakened, in turn, women's position in science. With the gradual professionalization of astronomy, astronomers ceased working in family attics doubling as observatories as they had done in the days of the Kirch family. As late as 1704, Gottfried Kirch recorded in his diary: "July 4, [The sky was] light early. But I was unable to use the floor [to make observations through windows in the ceiling] since the washing from two households was hanging there. It was a pity, because I missed the conjunction of Jupiter and Venus."[94] Kirch's complaint reveals a striking juxtaposition of science and private life that began to disappear in the course of the eighteenth century. With the increasing polarization of public and private life, the family moved into the private sphere of hearth and foyer, while science migrated to the public sphere of the university and industry.[95] This polarization held important consequences for women's participation in science.

With the privatization of the family, husbands and wives ceased to be partners in the family business, and women were increasingly confined to the domestic role of wives and mothers. A wife like Maria Winkelmann-Kirch could no longer

---

[92] For information on Academy membership see Hartkopf, *Die Akademie der Wissenschaften der DDR* (cit. n. 90).

[93] In addition to the fourteen women who became Academy members over the past three and a half centuries, fifteen other women won Academy prizes: *ibid.*

[94] Quoted in Zinner, *Die Geschichte der Sternkunde* (cit. n. 6), p. 583, and Wattenberg, "Zur Geschichte der Astronomie in Berlin, I" (cit. n. 12), p. 166.

[95] See Jean-Louis Flandrin, *Families in Former Times: Kinship, Household, and Sexuality*, trans. Richard Southern (1975; Cambridge: Cambridge Univ. Press, 1979); Edward Shorter, *Making of the Modern Family* (New York: Basic Books, 1975); Werner Conze, ed., *Sozialgeschichte der Familie in der Neuzeit Europas* (Stuttgart: Klett, 1976); Michael Mitterauer and Reinhard Sieder, *The European Family: Patriarchy to Partnership from Middle Ages to Present*, trans. Karla Oosterveen and Manfred Hoerzinger (1977; Chicago: Univ. Chicago Press, 1982); Richard Evans and W. R. Lee, *The German Family: Essays on the Social History of the Family in Nineteenth- and Twentieth-Century Germany* (London: Croom Helm, 1981); and Heidi Rosenbaum, *Formen der Familie* (Frankfurt am Main: Suhrkamp Verlag, 1982).

become assistant astronomer to a scientific academy through marriage. Such positions became reserved for those with public certification of their qualifications.

With the changes in the social structure of science, women's participation changed. On the one hand, women attempted to follow the course of public instruction and certification through the universities, like their male counterparts. These attempts, however, were not successful until nearly two centuries later, at the turn of the twentieth century.[96] A second option open to women was to continue to participate within the (now private) family sphere as increasingly "invisible" assistants to a scientific husband or brother. These invisible assistants are difficult to distinguish from the unpaid artisan wife and represent a legacy of that tradition.

Changes in both the structure of science and the structure of the family served to distance wifely assistants from the professional world of science. Whereas Gottfried Kirch acknowledged his wife's work in scientific publications in 1710, Hermann von Helmholtz in the mid-nineteenth century praised his wife for her help in his experiments only privately and never publicly acknowledged her help either in his books or in his papers.[97]

In eighteenth-century Germany, modern science was a new enterprise forging new institutions and norms. With respect to the problem of women, science (and society) at this time may be seen as standing at a fork in the road. Science could either affirm and broaden practices inherited from craft traditions and welcome women as full participants, or it could reaffirm academic traditions and continue to exclude them. As the case of Maria Winkelmann demonstrates, the Berlin Academy of Science chose to follow the latter path.

---

[96] Women were not formally admitted to European universities until the 1860s in Switzerland, 1870s in England, 1880s in France, and 1900s in Germany. See Rita McWilliams-Tullberg, "Women and Degrees at Cambridge University 1862–1897," in *A Widening Sphere: Changing Roles of Victorian Women,* ed. Martha Vincinus (Bloomington: Indiana Univ. Press, 1977), pp. 117–146; and Laetitia Böhm, "Von dem Anfängen des akademischen Frauenstudiums in Deutschland," *Historisches Jahrbuch,* 1958, 77:2298–2327.

[97] I thank Richard Kremer, Department of History, Dartmouth College, for information on Helmholtz.

APPENDIX

ARTICLES ON EARLY MODERN EUROPEAN SCIENCE IN *ISIS*, 1970–1996

The selection of articles for the present reader reflects, among other things, a concern to provide balance in topical treatment. The following list, restricted to the last quarter century (and not including brief items such as Notes), contains many excellent additional pieces. I have had to use fallible judgment in determining when the period should be taken to have started and when (especially) it ends.

1970

Fletcher, John E. "Astronomy in the Life and Correspondence of Athansius Kircher." (61:52–67)
Carozzi, Albert V. "Robert Hooke, Rudolf Erich Raspe, and the Concept of 'Earthquakes.'" (61: 85–91)
Aston, Margaret E. "The Fiery Trigon Conjunction: An Elizabethan Astrological Prediction." (61: 159–187)
Stuewer, Roger H. "A Critical Analysis of Newton's Work on Diffraction." (61:188–205)
Baird, A.W.S. "Pascal's Idea of Nature." (61: 297–320)
Shea, William R. "Galileo, Scheiner, and the Interpretation of Sunspots." (61:498–519)

1971

Iltis, Carolyn. "Leibniz and the *Vis Viva* Controversy." (62:21–35)
Holmes, Frederic L. "Analysis by Fire and Solvent Extractions: The Metamorphosis of a Tradition." (62:129–148)
Marx, Jacques. "Alchimie et Palingénésie." (62: 275–289)
Albury, William R. "Halley and the *Traité de la lumière* of Huygens: New Light on Halley's Relationship with Newton." (62:445–468)

1972

Drake, Stillman. "The Uniform Motion Equivalent to a Uniformly Accelerated Motion from Rest (Galileo Gleanings XX)." (63:28–38)
Ariotti, Piero. "Benedetto Castelli: Early Systematic Experiments and Theory of the Differential Absorption of Heat by Colors." (63:79–87)
Westfall, Richard S. "Circular Motion in Seventeenth-Century Mechanics." (63:184–189)
Palter, Robert. "Early Measurements of Magnetic Force." (63:544–558)

1973

Debus, Allen G. "Motion in the Chemical Texts of the Renaissance." (64:5–17)
Hine, William L. "Mersenne and Copernicanism." (64:18–32)
Kitcher, Philip. "Fluxions, Limits, and Infinite Littlenesse: A Study of Newton's Presentation of the Calculus." (64:33–49)
Drake, Stillman. "Galileo's Experimental Confirmation of Horizontal Inertia: Unpublished Manuscripts (Galileo Gleanings XXII)." (64:291–305)
Iltis, Carolyn. "The Decline of Cartesianism in Mechanics: The Leibnizian-Cartesian Debates." (64: 356–373)
MacLachlan, James. "A Test of an 'Imaginary' Experiment of Galileo's." (64:374–379)
Osler, Margaret J. "Galileo, Motion, and Essences." (64:504–509)

1974

Van Helden, Albert. "The Telescope in the Seventeenth Century." (65:38–58)
Finocchiaro, Maurice A. "Newton's Third Rule of Philosophizing: A Role for Logic in Historiography." (65:66–73)
Guédon, Jean-Claude. "Protestantisme et chimie: Le milieu intellectuel de Nicolas Lémery." (65: 212–228)
Rosińska, Grażyna. "Naṣīr al-Din al-Ṭūsī and Ibn al-Shāṭir in Cracow?" (65:239–243)
Carré, Marie-Rose. "A Man between Two Worlds: Pierre Borel and His *Discours nouveau prouvant la pluralité des mondes* of 1657." (65:322–335)

1975

Westman, Robert S. "The Melanchthon Circle, Rheticus, and the Wittenberg Interpretation of the Copernican Theory." (66:165–193)
Shapiro, Alan E. "Newton's Definition of a Light Ray and the Diffusion Theory of Chromatic Dispersion." (66:194–210)
Arrioti, Piero. "Bonaventura Cavalieri, Marin Mersenne, and the Reflecting Telescope." (66:303–321)
Ross, Richard P. "Oronce Fine's *De sinibus libri II*: The First Printed Trigonometric Treatise of the French Renaissance." (66:379–386)

1976

Drake, Stillman. "The Evolution of *De motu* (Galileo Gleanings XXIV)." (67:239–250)

Naylor, Ronald. "Galileo: Real Experiment and Didactic Demonstration." (67:398–419)

1977

Agassi, Joseph. "Robert Boyle's Anonymous Writings." (68:284–287)

Clarke, Desmond M. "The Impact Rules of Descartes' Physics." (68:55–66)

Buck, Peter. "Seventeenth-Century Political Arithmetic: Civil Strife and Vital Statistics." (68:67–84)

Moyer, Albert. "Robert Hooke's Ambiguous Presentation of 'Hooke's Law.'" (68:266–275)

Babb, Stanley E. "Accuracy of Planetary Theories, Particularly for Mars." (68:426–436)

Blackwell, Richard J. "Christiaan Huygens' *The Motion of Colliding Bodies*." (68:574–597)

1978

Frankel, Henry R. "The Importance of Galileo's Nontelescopic Observations concerning the Size of the Fixed Stars." (69:77–82)

Wallace, William A. "Causes and Forces in Sixteenth-Century Physics." (69:400–412)

Grant, Edward. "The Principle of the Impenetrability of Bodies in the History of Concepts of Separate Space from the Middle Ages to the Seventeenth Century." (69:551–571)

1979

Breiner, Laurence A. "The Career of Cockatrice." (70:30–47)

Tamny, Martin. "Newton, Creation, and Perception." (70:48–58)

Christianson, J.R. "Tycho Brahe's German Treatise on the Comet of 1577: A Study in Science and Politics." (70:110–140)

Hill, David K. "A Note on a Galilean Worksheet." (70:269–271)

Zetterberg, J. Peter. "Hermetic Geocentricity: John Dee's Celestial Egg." (70:385–393)

1980

Westfall, Richard S. "Newton's Marvelous Years of Discovery and Their Aftermath: Myth versus Manuscript." (71:109–121)

Shapiro, Alan E. "The Evolving Structure of Newton's Theory of White Light and Color." (71:211–235)

Jacob, James R.; Jacob, Margaret C. "The Anglican Origins of Modern Science: The Metaphysical Foundations of the Whig Constitution." (71:251–267)

Mulligan, Lotte. "Puritans and English Science: A Critique of Webster." (71:456–469)

Naylor, Ronald H. "Galileo's Theory of Projectile Motion." (71:550–570)

1981

Shapin, Steven. "Of Gods and Kings: Natural Philosophy and Politics in the Leibniz-Clarke Disputes." (72:187–215)

Jobe, Thomas Harmon. "The Devil in Restoration Science: The Glanvill-Webster Witchcraft Debate." (72:342–356)

1982

Sakellariadis, Spyros. "Descartes's Use of Empirical Data to Test Hypotheses." (73:68–76)

Steneck, Nicholas. "Greatrakes the Stroker: The Interpretation of Historians." (73:161–177)

Kaplan, Barbara Beigun. "Greatrakes the Stroker: The Interpretation of His Contemporaries." (73:178–185)

Hutchison, Keith. "What Happened to Occult Qualities in the Scientific Revolution?" (73:233–253)

Cohen, I. Bernard. "Newton's Copy of Leibniz's *Theodice*: With Some Remarks on the Turned-Down Pages of Books in Newton's Library." (73:410–414)

Dobbs, B.J.T. "Newton's Alchemy and His Theory of Matter." (73:511–528)

1983

Freudenthal, Gad. "Theory of Matter and Cosmology in William Gilbert's *De magnete*." (74:22–37)

Lennox, James G. "Robert Boyle's Defense of Teleological Inference in Experimental Science." (74:38–52)

Guerlac, Henry. "Can We Date Newton's Early Optical Experiments?" (74:74–80)

1984

Wisan, Winifred Lovell. "Galileo and the Process of Scientific Creation." (75:269–286)

Park, Katharine. "Bacon's 'Enchanted Glass.'" (75:290–302)

Daston, Lorraine J. "Galilean Analogies: Imagination at the Bounds of Sense." (75:302–310)

Galison, Peter. "Descartes's Comparisons: From the Visible to the Invisible." (75:311–326)

Eamon, William; Paheau, Françoise. "The Accademia Segreta of Girolamo Ruscelli: A Sixteenth-Century Italian Scientific Society." (75:327–342)

Bulmer-Thomas, Ivor. "Guldin's Theorem—or Pappus's?" (75:348–352)

Eastwood, Bruce Stansfield. "Descartes on Refraction: Scientific versus Rhetorical Method." (75:481–502)

1985

Westfall, Richard S. "Science and Patronage: Galileo and the Telescope." (76:11–30)

Dear, Peter. "*Totius in verba*: Rhetoric and Authority in the Early Royal Society." (76:145–161)

McGuire, J.E.; Tamny, Martin. "Newton's Astronomical Apprenticeship: Notes of 1664/5." (76:349–365)

Smith, A. Mark. "Galileo's Proof for the Earth's Motion from the Movement of Sunspots." (76: 543–551)

1986

Stewart, Larry. "Public Lectures and Private Patronage in Newtonian England." (77:47–58)
Alexandre, Josette. "La comete de Halley à travers les ouvrages et manuscrits de l'Observatoire de Paris." (77:79–84)
McMullin, Ernan. "Giordano Bruno at Oxford." (77: 85–94)
Hill, David K. "Galileo's Work on 116v: A New Analysis." (77:283–291)
Wisan, Winifred Lovell. "Galileo and God's Creation." (77:473–486)

1987

McMullin, Ernan. "Bruno and Copernicus." (78: 55–74)
Schiebinger, Londa. "Maria Winkelmann at the Berlin Academy: A Turning Point for Women in Science." (78:174–200)
Newman, William. "Newton's 'Clavis' as Starkey's 'Key.'" (78:564–574)

1988

Meinel, Christoph. "Early Seventeenth-Century Atomism: Theory, Epistemology, and the Insufficiency of Experiment." (79:68–103)
Shapin, Steven. "The House of Experiment in Seventeenth-Century England." (79:373–404)
Hill, David K. "Dissecting Trajectories: Galileo's Early Experiments on Projectile Motion and the Law of Fall." (79:646–668)

1989

Golinski, J.V. "A Noble Spectacle: Phosphorus and the Public Cultures of Science in the Early Royal Society." (80:11–39)
Segre, Michael. "Viviani's Life of Galileo." (80: 207–231)
Meyer, Eric. "Galileo's Cosmogonical Calculations." (80:456–468)

1990

Hutchison, Keith. "Sunspots, Galileo, and the Orbit of the Earth." (81:68–74)
Biagioli, Mario. "Galileo the Emblem Maker." (81: 230–258)
Dear, Peter. "Miracles, Experiments, and the Ordinary Course of Nature." (81:663–683)
Naylor, Ronald H. "Galileo's Method of Analysis and Synthesis." (81:695–707)

1991

Mancosu, Paolo; Vailati, Ezio. "Torricelli's Infinitely Long Solid and Its Philosophical Reception in the Seventeenth Century." (82:50–70)

Daston, Lorraine. "History of Science in an Elegiac Mode: E. A. Burtt's *Metaphysical Foundations of Modern Physical Science* Revisited." (82: 522–531)
Laird, W.R. "Archimedes among the Humanists." (82:629–638)
Cormack, Lesley B. "'Good Fences Make Good Neighbors': Geography as Self-Definition in Early Modern England." (82:639–661)

1992

Winkler, Mary G.; Van Helden, Albert. "Representing the Heavens: Galileo and Visual Astronomy." (83:195–217)
Swerdlow, N.M. "Annals of Scientific Publishing: Johannes Petreius's Letter to Rheticus." (83: 270–274)
Copenhaver, Brian P. "Did Science Have a Renaissance?" (83:387–407)

1993

Sarasohn, Lisa T. "Nicolas-Claude Fabri de Peiresc and the Patronage of the New Science in the Seventeenth Century." (84:70–90)
Shackelford, Jole. "Tycho Brahe, Laboratory Design, and the Aim of Science: Reading Plans in Context." (84:211–230)
Feingold, Mordechai. "Newton, Leibniz, and Barrow Too: An Attempt at a Reinterpretation." (84:310–338)

1994

Smith, Pamela H. "Alchemy as a Language of Mediation at the Habsburg Court." (85:1–25)
Westman, Robert S. "Two Cultures or One? A Second Look at Kuhn's *The Copernican Revolution*." (85:79–115)
Principe, Lawrence M. "Style and Thought of the Early Boyle: Discovery of the 1648 Manuscript of 'Seraphic Love.'" (85:247–260)
Shapiro, Alan E. "Artists' Colors and Newton's Colors." (85:600–630)
Dobbs, B.J.T. "Newton as Final Cause and First Mover." (85:633–643)

1995

Navarro Brotòns, Vìctor. "The Reception of Copernicanism in Sixteenth-Century Spain: The Case of Diego de Zúñiga." (86:52–78)

1996

Methuen, Charlotte. "Maestlin's Teaching of Copernicus: The Evidence of His University Textbook and Disputations." (87:230–247)

CONTRIBUTORS AND EDITORS

Lesley B. Cormack is associate professor in the Department of History and Classics at the University of Alberta. Her new book, *Charting an Empire: Geography at the English Universities 1580–1620*, is due to be published by the University of Chicago Press in 1997.

Peter Dear is associate professor of history and of science and technology studies at Cornell University. He is the author of *Mersenne and the Learning of the Schools* (1988) and *Discipline and Experience: The Mathematical Way in the Scientific Revolution* (1995).

B. J. T. Dobbs was, until her death in 1994, professor of history at the University of California, Davis. Her two important books on Newton's alchemy are *The Foundations of Newton's Alchemy: Or, "The Hunting of the Greene Lyon"* (1975) and *The Janus Faces of Genius: The Role of Alchemy in Newton's Thought* (1991).

Bruce S. Eastwood is professor of history at the University of Kentucky. He has written many articles on ancient, medieval, and early modern optics and astronomy, some of which are collected in his *Astronomy and Optics from Pliny to Descartes: Texts, Diagrams and Conceptual Structures* (1989).

Owen Hannaway is professor of the history of science at the Johns Hopkins University. He is the author of *The Chemists and the Word: The Didactic Origins of Chemistry* (Baltimore: Johns Hopkins University Press, 1975) and recent studies on Georgius Agricola.

Keith Hutchison, of the Department of History and Philosophy of Science at the University of Melbourne, has written numerous studies of natural philosophy in the sixteenth and seventeenth centuries, including *Supernaturalism and the Mechanical Philosophy* (1983) and *Dormitive Virutes, Scholastic Qualities, and the New Philosophies* (1991).

Thomas S. Kuhn, who died in 1996, was emeritus professor of the Massachusetts Institute of Technology, having, prior to his retirement, been the Laurence S. Rockefeller Professor of Philosophy. He was the author of *The Structure of Scientific Revolutions* (1962), *The Essential Tension: Selected Studies in Scientific Tradition and Change* (1977), and *Black-Body Theory and the Quantum Discontinuity* (1978).

Christoph Meinel is professor in the Institute of Philosophy at the University of Regensburg. Among his publications on seventeenth-century topics are *Die Bibliothek des Joachim Jungius: Ein Beitrag zur "Historia litteraria" der fruhen Neuzeit* (1992) and *Naturals moralische Anstalt: Die Meteorologia philosophico-politico des Franz Reinzer, S.J.: Ein naturwissenschaftsliches Emblembuch aus dem Jahre 1698* (1987).

Margaret J. Osler is a member of the history department at the University of Calgary. She recently published *Divine Will and the Mechanical Philosophy: Descartes and Gassendi on Contingency and Necessity in the Created World* (1994).

Londa Schiebinger, professor of history and women's studies at Pennsylvania State University, is the author of *The Mind Has No Sex? Women in the Origins of Modern Science* (1989) and *Nature's Body: Gender in the Making of Modern Science* (1993).

Steven Shapin, professor in the Department of Sociology and in the science studies program at the University of California, San Diego, is the author of *A Social History of Truth: Civility and Science in Seventeenth-Century England* (1994), *The Scientific Revolution* (1996), and coauthor (with Simon Schaffer) of *Leviathan and the Air-Pump: Hobbes, Boyle, and the Experimental Life* (1985).

Albert Van Helden, professor of history at Rice University, is the author of many works on the history of astronomy, among them *The Invention of the Telescope* (1977) and *Measuring the Universe: Cosmic Dimensions from Aristarchus to Halley* (1985). He has also published a new translation of *Galileo's Sidereal Messenger* (1989).

Richard S. Westfall, who died in 1996, was for many years professor of the history of science at Indiana University. He was the author of *Force in Newton's Physics: The Science of Dynamics in the Seventeenth Century* (1971) and *Never at Rest: A Biography of Isaac Newton* (1980), as well as many other books and articles on seventeenth-century science.

Robert S. Westman is professor in the Department of History and in the program in science studies at the University of California, San Diego. Among his many studies on early modern astronomy are "The Astronomer's Role in the Sixteenth Century: A Preliminary Study" (1980) and (with Owen Gingerich), *The Wittich Connection: Conflict and Priority in Late Sixteenth-Century Astronomy* (1988).